KEW INDEX
FOR 1986

KEW INDEX FOR 1986

NAMES OF SEED-BEARING PLANTS, FERNS, AND FERN ALLIES AT THE RANK OF FAMILY AND BELOW PUBLISHED DURING 1986 WITH SOME OMISSIONS FROM EARLIER YEARS

COMPILED BY

R. A. DAVIES

and

K. M. LLOYD

AT THE HERBARIUM OF THE ROYAL BOTANIC GARDENS

KEW

CLARENDON PRESS · OXFORD

1987

Oxford University Press, Walton Street, Oxford OX2 6DP

Oxford New York Toronto
Delhi Bombay Calcutta Madras Karachi
Petaling Jaya Singapore Hong Kong Tokyo
Nairobi Dar es Salaam Cape Town
Melbourne Auckland

and associated companies in
Beirut Berlin Ibadan Nicosia

Oxford is a trade mark of Oxford University Press

Published in the United States
by Oxford University Press, New York

British Library Cataloguing in Publication Data
Kew index for . . . : names of seed-bearing plants, ferns, and fern allies at the rank of family and below published during . . . with some omissions from earlier years.—1986–
1. Phanerogams 2. Botany—Nomenclators
I. Index Kewensis
582'.0014 QK11
ISBN 0–19–854227–5

ISSN 0951–7502

Typeset by Promenade Graphics Ltd., Cheltenham
Printed and bound in
Great Britain by Biddles Ltd.
Guildford and King's Lynn

FOREWORD

We are pleased to present the first of a new series of annual indexes, presaged in the Preface to *Index Kewensis Supplement* XVIII, also published this year.

The successful installation of Kew's PRIME computer has provided a vastly more efficient means of compiling *Index Kewensis,* and it allows rapid production of annual compilations which have been in demand for some years.

In consultation with our publishers, we have decided to call the new publication *Kew Index,* since it differs from the established publication in other ways than simply in being annual.

(1) In an appendix we are providing entries for ferns and fern allies which have not been included in *Index Kewensis* before; this is the basic material which, when expanded and further edited, will become the next *Index Filicum*.

(2) The closing date for *Index Kewensis* has for many years been the end of February (of the sixth year after the start of a Supplement); the closing date for *Kew Index* is the end of the calendar year, i.e. literature received in the *Kew Index* office after New Year 1987 will not be included in the *Index* for 1986 regardless of the publication date of that literature.

(3) The page size has been produced to make the book more manageable on the shelf.

In other respects *Kew Index* resembles *Index Kewensis*.

Entries from the new annuals will, it is intended, be brought together and produced as supplements of *Index Kewensis* (without the fern appendix) as in the past.

As in *Supplement* XVIII, Cyrillic characters have been transliterated in accordance with British Standard 2979 (1958): *Transliteration of Cyrillic and Greek Characters*.

Names at all taxonomic ranks from family downwards are included.

Names at ranks above genus are arranged alphabetically under families in a separate sequence at the beginning of the volume.

The dagger (†) indicates the corrected re-entry of a name cited in an earlier volume.

Books and journals received at Kew during 1986 or seen at the British Museum (Natural History) have been scanned. A few extra entries missed in previous years have also been included.

We are also pleased to acknowledge our indebtedness to Dr R. K. Brummitt for much nomenclatural advice and to the Kew Computer Section, and especially Mr M. Jackson, for their contribution to the success of the new systems.

Royal Botanic Gardens
Kew
January 1987

E. ARTHUR BELL
Director

CONTENTS

KEW INDEX FOR 1986
NAMES OF ALL RANKS ABOVE GENUS UP TO AND INCLUDING FAMILY

CHENOPODIACEAE

subtribus **Physandreae** (Bunge) U. Pratov, Rod Climacoptera Botsch.: 61 (1986), without basionym ref.: subdiv. Physandreae.

GUMILLEACEAE

fam. **Gumilleaceae** G. Kunkel, Pl. Human Consumption: 174 (1984), provisional name without latin descr.

NELSONIACEAE

fam. **Nelsoniaceae** G. Kunkel, Pl. Human Consumption: 245 (1984), nom. nov. but without repl. syn. date: Acanthaceae I. Nelsonioideae Engler

ORCHIDACEAE

subfam. **Neuwiedioideae** P. Burns-Balogh & V.A. Funk in Smithsonian Contrib. Bot., 61: 53 (1986).

tribus **Prasophylleae** (Schlechter) P. Burns-Balogh & V.A. Funk in Smithsonian Contrib. Bot., 61: 54 (1986), without basionym ref.: Basionym not stated.

tribus **Pterostylideae** (Schlechter) P. Burns-Balogh & V.A. Funk in Smithsonian Contrib. Bot., 61: 56 (1986), without basionym ref.: Basionym not stated.

PALMAE

subtribus **Archontophoenicinae** J. Dransfield & N.W. Uhl in Principes, 30(1): 8 (1986).

subtribus **Beccariophoenicinae** J. Dransfield & N.W. Uhl in Principes, 30(1): 10 (1986).

subtribus **Coryphinae** J. Dransfield & N.W. Uhl in Principes, 30(1): 4 (1986).

subtribus **Cyrtostachydinae** J. Dransfield & N.W. Uhl in Principes, 30(1): 8 (1986).

subtribus **Euterpeinae** J. Dransfield & N.W. Uhl in Principes, 30(1): 7 (1986).

subtribus **Leopoldiniinae** J. Dransfield & N.W. Uhl in Principes, 30(1): 7 (1986).

subtribus **Manicariinae** J. Dransfield & N.W. Uhl in Principes, 30(1): 7 (1986).

subtribus **Oncocalaminae** J. Dransfield & N.W. Uhl in Principes, 30(1): 5 (1986).

subtribus **Oraniinae** J. Dransfield & N.W. Uhl in Principes, 30(1): 7 (1986).

subtribus **Pigafettinae** J. Dransfield & N.W. Uhl in Principes, 30(1): 5 (1986).

subtribus **Plectocomiinae** J. Dransfield & N.W. Uhl in Principes, 30(1): 5 (1986).

tribus **Podococceae** J. Dransfield & N.W. Uhl in Principes, 30(1): 6 (1986).

subtribus **Roystoneinae** J. Dransfield & N.W. Uhl in Principes, 30(1): 7 (1986).

subtribus **Sclerospermatinae** J. Dransfield & N.W. Uhl in Principes, 30(1): 9 (1986).

PAPAVERACEAE

subtribus **Discocapninae** M. Lidén in Op. Bot., 88: 104 (1986).

subtribus **Sarcocapninae** M. Lidén in Op. Bot., 88: 29 (1986).

REHMANNIACEAE

fam. **Rehmanniaceae** G. Kunkel, Pl. Human Consumption: 304 (1984), provisional name without latin descr.

SMILACACEAE

subfam. **Luzuriagoideae** (Engl.) J.G. Conran & H.T. Clifford in Fl. Australia, 46: 231 (1986): Liliaceae subfam. Luzuriagoideae.

subfam. **Petermannioideae** (Hutch.) J.G. Conran & H.T. Clifford in Fl. Australia, 46: 230 (1986): fam. Petermanniaceae.

subfam. **Ripogonoideae** (Conran & Cliff.) J.G. Conran & H.T. Clifford in Fl. Australia, 46: 230 (1986): fam. Ripogonaceae.

KEW INDEX FOR 1986
NAMES AT GENERIC RANK AND BELOW

A

ABAREMA (Legumin.).
moniliformis *(Ducke) R.C. Barneby & J.W. Grimes in Acta Amazonica,* 14(1–2 Suppl.): 95 (1984 publ. 1986): Pithecellobium moniliforme.
piresii *R.C. Barneby & J.W. Grimes in Acta Amazonica,* 14(1–2 Suppl.): 96 (1984 publ. 1986)—Brazil.

ABELMOSCHUS (Malvac.).
officinalis *Endl., Cat. Horti Vindob.,* 2: 253 (1842); *cf. D.J. Mabberley in Taxon,* 33(3): 434 (1984)— Locality not stated.
pseudomanihot *(DC.) Endl., Cat. Horti Vindob.,* 2: 253 (1842); *cf. D.J. Mabberley in Taxon,* 33(3): 434 (1984): Hibiscus pseudomanihot.

ABIES (Pinac.).
fabri
subsp. minensis *(Bordères-Rey & Gaussen) K.D. Rushforth in Notes Roy. Bot. Gard. Edinburgh,* 43(2): 273 (1986): A. minensis.

ABRONIA (Nyctaginac.).
nana
var. harrisii *S.L. Welsh in Great Basin Nat.,* 46(2): 258 (1986)—U.S.A. (Utah).

ABUTILON (Malvac.).
neelgherrense
var. fischeri *T.K. Paul & M.P. Nayar in Bull. Bot. Surv. India,* 25(1–4): 183 (1983 publ. 1985)—India.

ACACIA (Legumin.).
angustifolia *Sterler, Hort. Nymph.:* 1 (1821); *cf. D.J. Mabberley in Taxon,* 33(3): 440 (1984)— Locality not stated.
callosa *Bertol., Hort. Bonon.,* 1826: 3 (1826) *et Giorn. Arcad.,* 29: 342 (1826); *cf. D.J. Mabberley in Taxon,* 32(1): 82 (1983)— West Indies.
eremophiloides *L. Pedley & P.I. Forster in Austrobaileya,* 2(3): 277 (1986)—Australia (Queensland).
maconochieana *L. Pedley in Austrobaileya,* 2(3): 235 (1986)—Australia (N. Terr., W. Australia).
metrosiderifolia *Sterler, Hort. Nymph.:* 2 (1821); *cf. D.J. Mabberley in Taxon,* 33(3): 440 (1984)— Locality not stated.

ACACIA :–
platyacantha *Bertol., Hort. Bot. Bonon.,* 1826: 3 (1826) *et Giorn. Arcad.,* 29: 342 (1826); *cf. D.J. Mabberley in Taxon,* 32(1): 82 (1983)— Brazil.
velutina *Bertol., Syll. Pl. Hort. Bot. Bonon.,* 1827: 3 (1827); *cf. D.J. Mabberley in Taxon,* 32(1): 82 (1983)— West Indies.

ACAENA (Rosac.).
berteriana *Bernh., Sel. Sem. Hort. Erfurt.,* 1835; *Linnaea 11, Lit.:* 85 (1837); *cf. D.J. Mabberley in Taxon,* 32(1): 84 (1983)— Locality not stated.

ACALYPHA (Euphorbiac.).
elizabethiae *R.A. Howard in Phytologia,* 61(1): 1 (1986), as 'elizabethae'—Windward Is.
×koraensis *A. Radcliffe-Smith in Kew Bull.,* 41(2): 451 (1986)—Kenya.

ACANTHOCARPUS (Xanthorrhoeac.).
canaliculatus *A.S. George in Fl. Australia,* 46: 221, 96 (1986)—Australia (W. Australia).
humilis *A.S. George in Fl. Australia,* 46: 221, 95 (1986)—Australia (W. Australia).
parviflorus *A.S. George in Fl. Australia,* 46: 221, 96 (1986)—Australia (W. Australia).
robustus *A.S. George in Fl. Australia,* 46: 222, 95 (1986)—Australia (W. Australia).
rupestris *A.S. George in Fl. Australia,* 46: 222, 96 (1986)—Australia (W. Australia).
verticillatus *A.S. George in Fl. Australia,* 46: 222, 94 (1986)—Australia (W. Australia)

ACANTHOGILIA A.G. Day & R. Moran in Proc. Calif. Acad. Sci., Ser. 4, 44(7): 111 (1986). *POLEMONIACEAE.*
gloriosa *(Brandegee) A.G. Day & R. Moran in Proc. Calif. Acad. Sci., Ser. 4,* 44(7): 115 (1986): Gilia gloriosa.

ACANTHOLIMON (Plumbaginac.).
echinus
subsp. creticum *(Boiss.) K. Papanicolaou & S. Kokkini in A. Strid (ed.), Mount. Fl. Greece,* 1: 748 (1986): A. androsaceum var. creticum.

ACANTHUS (Acanthac.).
ebracteatus
subsp. ebarbatus *R.M. Barker in J. Adelaide Bot. Gard.,* 9: 73 (1986)—Australia (N. Terr., Queensland, W. Australia).

ACER (Acerac.).
changii *Z.R. Xu in Acta Sci. Nat. Univ. Sunyatseni,* 1986(2): 100 (1986), without type—China.
cordatum
var. jinggangshanense *Z.X. Yu in Bull. Bot. Res. North-East. Forest. Inst.,* 6(1): 151 (1986)—China.
dissectum *C.P. Thunberg ex A. Murray in Linnaeus, Syst. Veg., ed. 14:* 911 (1784)—Japan. †
fabri
var. tongguense *Z.X. Yu in Bull. Bot. Res. North-East. Forest. Inst.,* 6(1): 152 (1986)—China.
gracilifolium *Fang & C.C. Fu in Fl. Tsinlingensis,* 1(3): 453, 228 (1981)—China.
japonicum *C.P. Thunberg ex A. Murray in Linnaeus, Syst. Veg., ed. 14:* 911 (1784)—Japan. †
lauyuense *Fang ex C.C. Fu in Fl. Tsinlingensis,* 1(3): 454, 231 (1981)—China.
neapolitanum *Ten., Cat. Pl. Hort. Reg. Neap., App.,* 1: 75 (1815); *cf. D.J. Mabberley in Taxon,* 33(3): 440 (1984)— Locality not stated.
opalus
var. reginae-amaliae *(Orph. ex Boiss.) B. Aldén in A. Strid (ed.), Mount. Fl. Greece,* 1: 582 (1986): A. reginae-amaliae.
palmatum *C.P. Thunberg ex A. Murray in Linnaeus, Syst. Veg., ed. 14:* 911 (1784)—Japan. †
pictum *C.P. Thunberg ex A. Murray in Linnaeus, Syst. Veg., ed. 14:* 912 (1784)—Japan. †
septemlobum *C.P. Thunberg ex A. Murray in Linnaeus, Syst. Veg., ed. 14:* 912 (1784)—Japan. †
shenkanense *Fang ex C.C. Fu in Fl. Tsinlingensis,* 1(3): 452, 220 (1981)—China.
tricaudatum *Fang & C.C. Fu in Fl. Tsinlingensis,* 1(3): 452, 224 (1981)—China.
trifidum *C.P. Thunberg ex A. Murray in Linnaeus, Syst. Veg., ed. 14:* 912 (1784)—Japan. †

ACERAS (Orchidac.).
anthropomorphum *(Willd.) Sm. in Rees, Cyclop.,* 39(1): Aceras n. 2 (1818); *cf. D.J. Mabberley in Taxon,* 32(1): 81 (1983), as 'anthropophorum': Basionym not stated.
anthropophorum
f. purpurata *M. Balayer in Bull. Soc. Bot. France, Lett. Bot.,* 133(3): 280 (1986)—France.

ACETOSA (Polygonac.).
thyrsiflora
subsp. intermedia *(DC.) Á. Löve in Taxon,* 35(3): 612 (1986): Rumex intermedius.

ACETOSELLA (Polygonac.).
gracilescens *(K.H. Rech.) Á. Löve in Taxon,* 35(3): 612 (1986): Rumex paucifolius var. gracilescens.
paucifolia *(Nutt.) Á. Löve in Taxon,* 35(3): 612 (1986): Rumex paucifolius.

ACHILLEA (Compos.).
aucheri
subsp. glabra *A. Huber-Morath in Fl. Iran.,* 158: 57 (1986)—Iran.
croatica *Hort. Vindob. ex Hornem., Hort. Hafn.:* 838 (1815); *cf. D.J. Mabberley in Taxon,* 33(3): 436 (1984)— Locality not stated.
×huber-morathii *K.H. Rechinger in Fl. Iran.,* 158: 71 (1986)—Iran. (A. bibersteinii × nobilis subsp. neilreichii).
millefolium
subsp. chitralensis *A. Huber-Morath in Fl. Iran.,* 158: 64 (1986)—Pakistan.
subsp. elbursensis *A. Huber-Morath in Fl. Iran.,* 158: 63 (1986)—Iran.
pindicola
subsp. integrifolia *(Halácsy) R. Franzén in Willdenowia,* 16(1): 20 (1986): A. clavennae var. integrifolia.
sambucifolia *C.C. Gmel., Hort. Mag. Duc. Bad. Carlsruh.:* 3 (1811); *cf. D.J. Mabberley in Taxon,* 33(3): 440 (1984)— Locality not stated.
sprengeliana *Huber in Revue horticole,* 48: 455 (1876); *cf. D.J. Mabberley in Taxon,* 34(3): 450 (1985)— Locality not stated.
talagonica
var. oxylepis *(Boiss. & Hausskn.) A. Huber-Morath in Fl. Iran.,* 158: 59 (1986): A. oxylepis.

ACHYRANTHES (Amaranthac.).
sessilis *(L.) Besser, Cat. Jard. Bot. Krz.:* 12 (1810); *cf. D.J. Mabberley in Taxon,* 30(1): 14 (1981): Basionym not stated.
virgata *(Jacq.) Zeyher, Verz. Saemm. Gewaech. Schwezingen:* 11 (1818); *cf. D.J. Mabberley in Taxon,* 33(3): 440 (1984): Celosia virgata.

ACHYROCLINE (Compos.).
peruviana *M.O. Dillon & A. Sagástegui-Alva in Phytologia,* 60(2): 107 (1986)—Peru.

ACIDOSASA (Gramin.).
dayongensis *T.P. Yi in Bull. Bot. Res. North-East. Forest. Inst.,* 6(4): 25 (1986)—China.

ACMISPON (Legumin.).
sect. Simpeteria *(Ottley) P. Lassen in Willdenowia,* 16(1): 107 (1986): Lotus sect. Simpeteria.
roudairei *(Bonnet) P. Lassen in Willdenowia,* 16(1): 108 (1986): Lotus roudairei.

ACONITUM (Ranunculac.).
benzilanense *T.L. Ming in Acta Bot. Yunnanica,* 7(3): 304 (1985)—China.

ACONITUM :-
elliotii
var. glabrescens *W.T. Wang & L.Q. Li in Acta Bot. Yunnanica,* 8(3): 259 (1986)—Xizang.
var. pilopetalum *W.T. Wang & L.Q. Li in Acta Bot. Yunnanica,* 8(3): 260 (1986)—Xizang.
hemsleyanum
var. puberulum *W.T. Wang & L.Q. Li in Acta Bot. Yunnanica,* 8(3): 260 (1986)—Xizang.
huizenense *T.L. Ming in Acta Bot. Yunnanica,* 7(3): 301 (1985)—China.
japonicum *C.P. Thunberg ex A. Murray in Linnaeus, Syst. Veg., ed. 14:* 504 (1784)—Japan. †
longtouense *T.L. Ming in Acta Bot. Yunnanica,* 7(3): 302 (1985)—China.
maximum
f. album *K. Sato in J. Jap. Bot.,* 61(5): 138 (1986)—Japan.
pseudokongboense *W.T. Wang & L.Q. Li in Acta Bot. Yunnanica,* 8(3): 260 (1986)—Xizang.

ACONOGONUM (Polygonac.).
pangianum *G.D. Pal & G.G. Maiti in Bull. Bot. Surv. India,* 26(1–2): 95 (1984 publ. 1985)—India.

ACORUS (Arac.).
latifolius *Z.Y. Zhu in Acta Bot. Bor.-Occid. Sin.,* 5(2): 118 (1985)—China.
xiangyeus *Z.Y. Zhu in Acta Bot. Bor.-Occid. Sin.,* 5(2): 119 (1985)—China.

ACOSTA (Compos.).
avilae *(Pau) J. Fernández Casas & A. Susanna de la Serna in Fontqueria,* 2: 21 (1982): Centaurea avilae.
×beltranii *(Pau) J. Fernández Casas & A. Susanna de la Serna in Fontqueria,* 2: 21 (1982): Centaurea tenuifolia subsp. beltranii.
boissieri
subsp. atlantica *(Font Quer) J. Fernández Casas & A. Susanna de la Serna in Fontqueria,* 2: 21 (1982): Centaurea boissieri var. atlantica.
subsp. calvescens *(Maire) J. Fernández Casas & A. Susanna de la Serna in Fontqueria,* 2: 21 (1982): Centaurea boissieri var. calvescens.
subsp. funkii *(Schultz Bip. ex Willk.) J. Fernández Casas & A. Susanna de la Serna in Fontqueria,* 2: 21 (1982): Centaurea funkii.
subsp. mariolensis *(Rouy) J. Fernández Casas & A. Susanna de la Serna in Fontqueria,* 2: 22 (1982): Centaurea mariolensis.
subsp. prostrata *(Cosson) J. Fernández Casas & A. Susanna de la Serna in Fontqueria,* 2: 22 (1982): Centaurea prostrata.

ACOSTA :-
subsp. willkommii *(Schultz Bip. ex Willk.) J. Fernández Casas & A. Susanna de la Serna in Fontqueria,* 2: 22 (1982): Centaurea willkommii.
citricolor *(Font Quer) J. Fernández Casas & A. Susanna de la Serna in Fontqueria,* 2: 22 (1982): Centaurea citricolor.
jaennensis *(Degen & Debeaux) J. Fernández Casas & A. Susanna de la Serna in Fontqueria,* 2: 22 (1982): Centaurea jaennensis.
paui *(Loscos ex Willk.) J. Fernández Casas & A. Susanna de la Serna in Fontqueria,* 2: 22 (1982): Centaurea paui.
pinae *(Pau) J. Fernández Casas & A. Susanna de la Serna in Fontqueria,* 2: 22 (1982): Centaurea pinae.
pinnata *(Pau) J. Fernández Casas & A. Susanna de la Serna in Fontqueria,* 2: 22 (1982): Centaurea pinnata.
resupinata *(Cosson) J. Fernández Casas & A. Susanna de la Serna in Fontqueria,* 2: 22 (1982): Centaurea resupinata.
subsp. degenii *(Sennen) J. Fernández Casas & A. Susanna de la Serna in Fontqueria,* 2: 22 (1982): Centaurea degenii.
subsp. lagascae *(Nyman) J. Fernández Casas & A. Susanna de la Serna in Fontqueria,* 2: 22 (1982): Centaurea lagascae.
subsp. spachii *(Schultz Bip. ex Willk.) J. Fernández Casas & A. Susanna de la Serna in Fontqueria,* 2: 23 (1982): Centaurea spachii.
rouyi *(Coincy) J. Fernández Casas & A. Susanna de la Serna in Fontqueria,* 2: 23 (1982): Centaurea rouyi.
sagredoi *(Blanca) J. Fernández Casas & A. Susanna de la Serna in Fontqueria,* 2: 23 (1982): Centaurea sagredoi.

ACOURTIA (Compos.).
acevedoi *M. González Elizondo in Phytologia,* 61(2): 117 (1986)—Mexico.

ACRIOPSIS (Orchidac.).
densiflora
var. borneensis *(Ridley) M.E. Minderhoud & E.F. de Vogel in Orchid Monogr.,* 1: 7 (1986): A. borneensis.
gracilis *M.E. Minderhoud & E.F. de Vogel in Orchid Monogr.,* 1: 7 (1986)—Borneo (Sabah).
javanica
var. auriculata *M.E. Minderhoud & E.F. de Vogel in Orchid Monogr.,* 1: 14 (1986)—Borneo (Sarawak, Sabah, Brunei, Kalimantan), Burma, Vietnam, Malaya, Sumatra, Java.
var. floribunda *(Ames) M.E. Minderhoud & E.F. de Vogel in Orchid Monogr.,* 1: 13 (1986): A. floribunda.

ACRONYCHIA (Rutac.).
oblongifolia *(Hook.) Endl., Cat. Horti Vindob.,* 2: 402 (1842); *cf. D.J. Mabberley in Taxon,* 33(3): 434 (1984): Basionym not stated.

ACTAEA (Ranunculac.).
japonica *C.P. Thunberg ex A. Murray in Linnaeus, Syst. Veg., ed. 14:* 488 (1784)—Japan. †

ACTINIDIA (Actinidiac.).
arguta
var. giraldii *(Diels) C.F. Liang ex C.W. Chang in Fl. Tsinlingensis,* 1(3): 294 (1981): A. giraldii.

ACTINOCARPUS (Alismatac.).
damasonium *Sweet, Hort. Suburb. Lond.:* 79 (1818); *cf. D.J. Mabberley in Taxon,* 31(1): 69 (1982)— Locality not stated.

ADENANTHOS (Proteac.).
×pamela *E.C. Nelson in Glasra,* 9: 2 (1986)—Australia (W. Australia). (A. detmoldii × obovata).

ADENOPHORA (Campanulac.).
amurica *C.X. Fu & M.Y. Liu in Bull. Bot. Res. North-East. Forest. Inst.,* 6(1): 159 (1986)—China.

ADENOPODIA (Legumin.).
floribunda *(Kleinhoonte) J.P.M. Brenan in Kew Bull.,* 41(1): 83 (1986): Piptadenia floribunda.
gymnantha *J.P.M. Brenan in Kew Bull.,* 41(1): 87 (1986)—Mexico.
minutiflora *(Ducke) J.P.M. Brenan in Kew Bull.,* 41(1): 84 (1986): Piptadenia minutiflora.
oaxacana *M. Sousa in Kew Bull.,* 41(1): 85 (1986)—Mexico.
patens *(Hook. & Arn.) J. Dixon ex J.P.M. Brenan in Kew Bull.,* 41(1): 80 (1986): Inga patens.
rotundifolia *(Harms) J.P.M. Brenan in Kew Bull.,* 41(1): 79 (1986): Entada rotundifolia.
scelerata *(A. Chev.) J.P.M. Brenan in Kew Bull.,* 41(1): 76 (1986): Entada scelerata.
schlechteri *(Harms) J.P.M. Brenan in Kew Bull.,* 41(1): 78 (1986): Piptadenia schlechteri.
uaupensis *(Spruce ex Benth.) J.P.M. Brenan in Kew Bull.,* 41(1): 82 (1986): Piptadenia uaupensis.

ADENOROPIUM (Euphorbiac.).
opiferum *(Mart.) Mart., Consp. Reg. Veg.:* 54 (1835); *cf. D.J. Mabberley in Taxon,* 30(1): 14 (1981): Basionym not stated.

ADINANDRA (Theac.).
grandifolia *Nguyen Huu Hien & G.P. Yakovlev in Bot. Zhurn.,* 71(8): 1119 (1986)—Vietnam.

ADINANDRA :-
lienii *Nguyen Huu Hien & G.P. Yakovlev in Bot. Zhurn.,* 71(8): 1118 (1986)—Veitnam.
milletii
var. dalatensis *Nguyen Huu Hien & G.P. Yakovlev in Bot. Zhurn.,* 71(8): 1119 (1986)—Vietnam.

AECHMEA (Bromeliac.).
alegrensis *W. Weber in Feddes Repert.,* 97(3–4): 110 (1986)—Brazil.
bocainensis *E. Pereira & E.M.C. Leme in Rev. Brasil. Biol.,* 45(4): 634 (1985 publ. 1986)—Brazil.
hellae *W. Weber in Feddes Repert.,* 97(3–4): 111 (1986)—Brazil.
marauensis *E.M.C. Leme in J. Bromeliad Soc.,* 36(6): 266 (1986)—Brazil.
seideliana *W. Weber in Feddes Repert.,* 97(3–4): 113 (1986)—Brazil.
tuitensis *P. Magaña & E.J. Lott in Phytologia,* 59(4): 221 (1986)—Mexico.

AEGONYCHON (Boraginac.).
thessalicum *(Aldén) J. Holub in Preslia,* 58(4): 301 (1986): Lithospermum goulandriorum subsp. thessalicum.

AEOLLANTHUS (Labiat.).
sect. Cucullilabium *O. Ryding in Symb. Bot. Upsal.,* 26(1): 139 (1986).
sect. Cyathobasis *O. Ryding in Symb. Bot. Upsal.,* 26(1): 60 (1986).
sect. Maculobractea *O. Ryding in Symb. Bot. Upsal.,* 26(1): 103 (1986).
sect. Ovatodiscus *O. Ryding in Symb. Bot. Upsal.,* 26(1): 108 (1986).
sect. Rotundobasis *O. Ryding in Symb. Bot. Upsal.,* 26(1): 122 (1986).
alternatus *O. Ryding in Symb. Bot. Upsal.,* 26(1): 102 (1986)—Zambia.
angolensis *O. Ryding in Symb. Bot. Upsal.,* 26(1): 132 (1986)—Angola.
densiflorus *O. Ryding in Symb. Bot. Upsal.,* 26(1): 137 (1986)—Ethiopia, Sudan, Uganda, Kenya, Tanzania.
lisowskii *O. Ryding in Symb. Bot. Upsal.,* 26(1): 101 (1986)—Zaïre.
petiolatus *O. Ryding in Symb. Bot. Upsal.,* 26(1): 129 (1986)—Zambia.
plicatus *O. Ryding in Symb. Bot. Upsal.,* 26(1): 134 (1986)—Angola.
subacaulis
var. ericoides *(De Wild.) O. Ryding in Symb. Bot. Upsal.,* 26(1): 100 (1986): A. ericoides.
var. linearis *(Burk.) O. Ryding in Symb. Bot. Upsal.,* 26(1): 98 (1986): Icomum lineare.
subintegrus *O. Ryding in Symb. Bot. Upsal.,* 26(1): 104 (1986)—Zaïre, Zambia, Angola, Tanzania.
trifidus *O. Ryding in Symb. Bot. Upsal.,* 26(1): 136 (1986)—Cameroun, Nigeria.
viscosus *O. Ryding in Symb. Bot. Upsal.,* 26(1): 132 (1986)—Zimbabwe, Mozambique.

AEONIUM (Crassulac.).
×beltranii *A. Bañares Baudet in Vieraea,* 16(1–2): 69 (1986)—Canary Is. (A. decorum × subplanum).
×castellodecorum *A. Bañares Baudet in Vieraea,* 16(1–2): 64 (1986)—Canary Is. (A. castellopaivae × decorum).
×cilifolium *A. Bañares Baudet in Vieraea,* 16(1–2): 59 (1986)—Canary Is. (A. ciliatum × sedifolium).
×holospathulatum *A. Bañares Baudet in Vieraea,* 16(1–2): 66 (1986)—Canary Is. (A. holochrysum × spathulatum var. spathulatum).
×nogalesii *A. Bañares Baudet in Vieraea,* 16(1–2): 61 (1986)—Canary Is. (A. palmensis × sedifolium).
×sanchezii *A. Bañares Baudet in Vieraea,* 16(1–2): 66 (1986)—Canary Is. (A. rubrolineatum × spathulatum var. spathulatum).
×wildpretii *A. Bañares Baudet in Vieraea,* 16(1–2): 61 (1986)—Canary Is. (A. palmensis × vestitum).

AERANGIS (Orchidac.).
subgen. Uniflorae *K. Senghas in Schlechter Orchideen,* 1(16–18): 1051 (1986).
ellisii
var. grandiflora *J. Stewart in Amer. Orchid Soc. Bull.,* 55(8): 799 (1986)—Madagascar.
punctata *J. Stewart in Amer. Orchid Soc. Bull.,* 55(11): 1120 (1986)—Madagascar.

AESCULUS (Hippocastanac.).
carnea *Zeyher, Verz. Saemm. Gewaech. Schwezingen:* 12 (1818); *cf. D.J. Mabberley in Taxon,* 33(3): 440 (1984)— Locality not stated.
chuniana *Hu & Fang in Act. Sc. Nat. Univ. Szech.,* 1960(3): 101, t. 11 (1962); *cf. W.P. Fang in Fl. Reipubl. Popul. Sin.,* 46: 284 (1981)— China. †
lantsangensis *Hu & Fang in Act. Sc. Nat. Univ. Szech.,* 1960(3): 87, t. 3 (1962); *cf. W.P. Fang in Fl. Reipubl. Popul. Sin.,* 46: 279 (1981)— China. †
megaphylla *Hu & Fang in Act. Sc. Nat. Univ. Szech.,* 1960(3): 104, t. 12 (1962); *cf. W.P. Fang in Fl. Reipubl. Popul. Sin.,* 46: 287 (1981)— China. †
polyneura *Hu & Fang in Act. Sc. Nat. Univ. Szech.,* 1960(3): 89, t. 4 (1962); *cf. W.P. Fang in Fl. Reipubl. Popul. Sin.,* 46: 277 (1981)— China. †
tsiangii *Hu & Fang in Act. Sc. Nat. Univ. Szech.,* 1960(3): 93, t. 6, 7 (1962); *cf. W.P. Fang in Fl. Reipubl. Popul. Sin.,* 46: 279 (1981)— China. †

AETHIOCARPA K. Vollesen in Kew Bull., 41(4): 959 (1986). *STERCULIACEAE.*
lepidota *K. Vollesen in Kew Bull.,* 41(4): 959 (1986)—Somali Republic.

AFGEKIA (Legumin.).
filipes *(Dunn) R. Geesink, Scala Millettiearum (Leiden Bot. Ser., 8):* 77 (1984): Adinobotrys filipes.

AFROTYSONIA (Boraginac.).
glochidiata *(R. Mill) R.R. Mill in Notes Roy. Bot. Gard. Edinburgh,* 43(3): 470 (1986): Tysonia glochidiata.
pilosicaulis *R.R. Mill in Notes Roy. Bot. Gard. Edinburgh,* 43(3): 472 (1986)—Tanzania.

AGALINIS (Scrophulariac.).
gypsophila *B.L. Turner in Phytologia,* 59(5): 319 (1986)—Mexico.

AGAPANTHUS (Alliac.).
globosus *Bull, Cat. 1899:* 3 (1899); *cf. D.J. Mabberley in Taxon,* 34(3): 456 (1985)— Locality not stated.
insignis *Bull, Cat.,* 377: ii (tab.), 1 (1904); *cf. D.J. Mabberley in Taxon,* 34(3): 456 (1985)— Locality not stated.

AGATHYRSUS (Compos.).
acuminatus *(Willd.) Sweet, Hort. Brit., ed. 2:* 277 (1830); *cf. D.J. Mabberley in Taxon,* 30(1): 14 (1981): Basionym not stated.
macrophyllus *(Willd.) Sweet, Hort. Brit., ed. 2:* 277 (1830); *cf. D.J. Mabberley in Taxon,* 30(1): 14 (1981): Basionym not stated.
sonchifolius *(Willd.) Sweet, Hort. Brit., ed. 2:* 277 (1830); *cf. D.J. Mabberley in Taxon,* 31(1): 69 (1982): Basionym not stated.

AGAVE (Agavac.).
cantala *(Haw.) Roxb. ex Salm-Dyck, Index Pl. Succ. Hort. Dyck.:* 1 (1822); *cf. D.J. Mabberley in Taxon,* 30(1): 14 (1981): Basionym not stated.
madagascariensis *(Haw.) Salm-Dyck, Index Pl. Succ. Hort. Dyck.:* 1 (1822); *cf. D.J. Mabberley in Taxon,* 30(1): 14 (1981): Basionym not stated.
papyriocarpa
subsp. macrocarpa *A. Alvarez de Zayas in Rev. Jard. Bot. Nacion. Univ. Habana,* 5(3): 7 (1984 publ. 1985)—Cuba.
tubulata
subsp. brevituba *A. Alvarez de Zayas in Rev. Jard. Bot. Nacion. Univ. Habana,* 5(3): 7 (1984 publ. 1985)—Cuba.

AGERATINA (Compos.).
asclepiadea *(L.f.) R.M. King & H. Robinson in Phytologia,* 60(1): 80 (1986): Cacalia asclepiadea.
grandifolia *(Regel) R.M. King & H. Robinson in Phytologia,* 60(1): 80 (1986): Eupatorium grandifolium.
grashoffii *B.L. Turner in Phytologia,* 61(2): 77 (1986)—Mexico.
sundbergii *B.L. Turner in Phytologia,* 61(2): 78 (1986)—Mexico.

AGERATUM (Compos.).
album *Hort. Berol. ex Hornem., Hort. Hafn.*: 792 (1815); *cf. D.J. Mabberley in Taxon,* 33(3): 436 (1984)— Locality not stated.
foetens *C. Huber in Revue horticole,* 46: 467 (1874); *cf. D.J. Mabberley in Taxon,* 34(3): 450 (1985)— Locality not stated.

AGLAIA (Meliac.).
beddomei *(Kosterm.) S.S. Jain & R.C. Gaur in J. Econ. Taxon. Bot.,* 7(2): 465 (1985 publ. 1986): Amoora beddomei.
mannii *(King ex Brandis) S.S. Jain & R.C. Gaur in J. Econ. Taxon. Bot.,* 7(2): 466 (1985 publ. 1986): Amoora mannii.

AGRAULUS (Gramin.).
flavescens *(Host) Sweet, Hort. Brit., ed. 2:* 556 (1830); *cf. D.J. Mabberley in Taxon,* 31(1): 69 (1982): Basionym not stated.

AGROSTIS (Gramin.).
ciliata *C.P. Thunberg ex A. Murray in Linnaeus, Syst. Veg., ed. 14:* 111 (1784)—Japan. †
diffusa *S.M. Phillips in Kew Bull.,* 41(1): 137 (1986)—Ethiopia.
mannii
subsp. aethiopica *S.M. Phillips in Kew Bull.,* 41(1): 134 (1986)—Ethiopia.
munroana
subsp. indica *S. Bhattacharya & S.K. Jain in Bull. Bot. Surv. India,* 25(1–4): 205 (1983 publ. 1985)—India.
novogaliciana *R. McVaugh, Fl. Novo-Galiciana,* 14: 41 (1983)—Mexico.
pilosula
f. ciliata *(Trin.) S. Bhattacharya & S.K. Jain in Bull. Bot. Surv. India,* 25(1–4): 210 (1983 publ. 1985): A. ciliata.
f. filifolia *(Bor) S. Bhattacharya & S.K. Jain in Bull. Bot. Surv. India,* 25(1–4): 210 (1983 publ. 1985): A. pilosula var. filifolia.
f. wallichiana *(Steud.) S. Bhattacharya & S.K. Jain in Bull. Bot. Surv. India,* 25(1–4): 210 (1983 publ. 1985): A. wallichiana.
tungnathii *S. Bhattacharya & S.K. Jain in Bull. Bot. Surv. India,* 25(1–4): 204 (1983 publ. 1985)—India.

AIDIA (Rubiac.).
rugulosa *(Thw.) D.D. Tirvengadum in Ceylon J. Sci., Biol. Sci.,* 14(1–2): 3 (1981), without basionym ref.: Basionym not stated.

AJUGA (Labiat.).
decumbens *C.P. Thunberg ex A. Murray in Linnaeus, Syst. Veg., ed. 14:* 525 (1784)—Japan. †
orientalis
subsp. aenesia *(Heldr.) D. Phitos & J. Damboldt in Bot. Chronika,* 5(1–2): 111 (1985): A. orientalis var. aenesia.

ALANGIUM (Alangiac.).
chinense
var. pauciflorum *Fang ex Y.C. Ho in Fl. Tsinlingensis,* 1(3): 455, 347 (1981)—China.

ALBIZIA (Legumin.).
julibrissin
f. albiflora *J. Ōhara in J. Phytogeogr. Taxon.,* 33(2): 72 (1985)—Japan.

ALCHEMILLA (Rosac.).
colura *O.M. Hilliard in Notes Roy. Bot. Gard. Edinburgh,* 43(3): 370 (1986)—South Africa (Cape Province, Natal), Lesotho.

ALCHORNEA (Euphorbiac.).
janeirensis *Casar. in Atti 3 Riun. Sci. Ital.:* 515 (1841); *cf. D.J. Mabberley in Taxon,* 30(1): 14 (1981)— Locality not stated.

ALECTRA (Scrophulariac.).
sessiliflora
f. barbata *(Hiern) O.M. Hilliard & B.L. Burtt in Notes Roy. Bot. Gard. Edinburgh,* 43(3): 375 (1986): Melasma barbatum.

ALEXA (Legumin.).
herminiana *N. Ramirez in Ernstia,* 37: 41 (1986)—Venezuela.

ALEXGEORGEA (Restionac.).
nitens *(Nees) S. Carlquist in Aliso,* 11(2): 158 (1985 publ. 1986): Restio nitens.
nitens *(Nees) L.A.S. Johnson & B.G. Briggs in Telopea,* 2(6): 781 (1986): Restio nitens.

ALIELLA M. Qaiser & H.W. Lack in Bot. Jahrb., 106(4): 488 (1986). *COMPOSITAE.*
embergeri *(Humbert & Maire) M. Qaiser & H.W. Lack in Bot. Jahrb.,* 106(4): 493 (1986): Phagnalon embergeri.
helichrysoides *(Ball) M. Qaiser & H.W. Lack in Bot. Jahrb.,* 106(4): 492 (1986): Gnaphalium helichrysoides.
subsp. nitidum *(Emb.) M. Qaiser & H.W. Lack in Bot. Jahrb.,* 106(4): 493 (1986): Phagnalon helichrysoides var. nitidum.
platyphylla *(Maire) M. Qaiser & H.W. Lack in Bot. Jahrb.,* 106(4): 490 (1986): Gnaphalium helichrysoides var. platyphyllum.

ALKANNA (Boraginac.).
lutea *Moris, Enum. Pl. Nov. Reg. Hort. Bot. Taur.,* 1845; *Ann. Sci. Nat. III,* 5: 364 (1846); *cf. D.J. Mabberley in Taxon,* 33(3): 441 (1984)— Locality not stated.

ALLIUM (Alliac.).
barszczewskii
f. niveum *L.S. Krasovskaya in L.S. Krasovskaya & I.G. Levichev, Fl. Chatkal'skogo Zapovednika:* 166 (1986)—U.S.S.R. (Middle Asia).

ALLIUM :-
f. violaceum *L.S. Krasovskaya in L.S. Krasovskaya & I.G. Levichev, Fl. Chatkal'skogo Zapovednika:* 167 (1986)—U.S.S.R. (Middle Asia).
brevidentatum *F.Z. Li in Bull. Bot. Res. North-East. Forest. Inst.*, 6(1): 170 (1986)—China.
cornutum *Clem. in Atti 3 Riun. Sci. Ital.:* 517 (1841); *cf. D.J. Mabberley in Taxon,* 31(1): 69 (1982)— Locality not stated.
costatovaginatum *R. Kam. & I.G. Levichev in L.S. Krasovskaya & I.G. Levichev, Fl. Chatkal'skogo Zapovednika:* 167 (1986)—U.S.S.R. (Middle Asia).
curtum
var. negevense *F. Kollmann in N. Feinbrun-Dothan, Fl. Palaestina,* 4: 397, 93 (1986)—Israel.
eldivanense *N. Özhatay in Notes Roy. Bot. Gard. Edinburgh,* 44(1): 147 (1986)—Turkey.
ilgazense *N. Özhatay in Notes Roy. Bot. Gard. Edinburgh,* 44(1): 147 (1986)—Turkey.
lopadusanum *G. Bartolo, S. Brullo & P. Pavone in Willdenowia,* 16(1): 89 (1986)—Sicily.
motor *R. Kam. & I.G. Levichev in L.S. Krasovskaya & I.G. Levichev, Fl. Chatkal'skogo Zapovednika:* 168 (1986)—U.S.S.R. (Middle Asia).
×tokaliense *R. Kam. & I.G. Levichev in L.S. Krasovskaya & I.G. Levichev, Fl. Chatkal'skogo Zapovednika:* 170 (1986)—U.S.S.R. (Middle Asia).
truncatum *(Feinbr.) F. Kollmann & D. Zohary in N. Feinbrun-Dothan, Fl. Palaestina,* 4: 90 (1986): A. ampeloprasum var. truncatum.

ALLOPLECTUS (Gesneriac.).
bicolor *G. Don f. in Loud., Encycl. Pl., new ed.:* 1402 (1855); *cf. D.J. Mabberley in Taxon,* 30(1): 14 (1981)— Locality not stated.

ALOE (Aloeac.).
inconspicua *D.C.H. Plowes in Aloe,* 23(2): 32 (1986)—South Africa (Natal).

ALOPECURUS (Gramin.).
baptarrhenius *S.M. Phillips in Kew Bull.,* 41(4): 1027 (1986)—Ethiopia.

ALSINE (Caryophyllac.).
cupaniana
subsp. postii *(Holmboe) J. Holub in Preslia,* 58(4): 301 (1986): Stellaria media subsp. postii.
molluginea *Hort. Vindob. ex Hornem., Hort. Hafn.:* 295 (1813); *cf. D.J. Mabberley in Taxon,* 33(3): 436 (1984)— Locality not stated.
nodosa *(Chaub. & Bory) Vis. in Atti 2 Riun. Sci. Ital.:* 180 (1841); *cf. D.J. Mabberley in Taxon,* 31(1): 69 (1982): Basionym not stated.

ALSOPHILA
insulana *(Holtt.) R. Tryon in Contrib. Gray Herb.,* 200: 34 (1970): Cyathea insulana Holtt.

ALSTROEMERIA (Alstroemeriac.).
hookeri *Sweet, Hort. Brit.:* 408 (1827); *cf. D.J. Mabberley in Taxon,* 30(1): 15 (1981)— Locality not stated.

ALTERNANTHERA (Amaranthac.).
ficoidea *Sm. in Rees, Cyclop.,* 39(1): Alternanthera no. 8 (Dec. 1818); *cf. D.J. Mabberley in Taxon,* 30(1): 15 (1981)— Locality not stated.
polygonoides *R. Br. ex Sweet, Hort. Suburb. Lond.:* 48 (June-July 1818); *cf. D.J. Mabberley in Taxon,* 30(1): 15 (1981)— Locality not stated.

ALYSICARPUS (Legumin.).
luteovexillatus *V.N. Naik & D.S. Pokle in J. Econ. Taxon. Bot.,* 7(3): 670 (1985 publ. 1986)—India.

ALYSSOIDES (Crucif.).
utriculata
subsp. bulgarica *(Sagorski) P. Hartvig in A. Strid (ed.), Mount. Fl. Greece,* 1: 277 (1986): Vesicaria utriculata subsp. bulgarica.

ALYSSUM (Crucif.).
arenarium *Kitaib. ex Spreng., Nachtr. Bot. Gart. Univ. Halle:* 10 (1801); *cf. D.J. Mabberley in Taxon,* 32(1): 84 (1983)— Locality not stated.
gustavssonii *P. Hartvig in A. Strid (ed.), Mount. Fl. Greece,* 1: 293 (1986)—Greece.
malacitanum *(Rivas Goday) T.R. Dudley in Feddes Repert.,* 97(3–4): 140 (1986): A. serpyllifolium subsp. malacitanum.
micropetalum *Fisch. ex Hornem., Hort. Hafn., Suppl.:* 70 (1819); *cf. D.J. Mabberley in Taxon,* 33(3): 436 (1984)— Locality not stated.
nebrodense
subsp. tenuicaule *P. Hartvig in A. Strid (ed.), Mount. Fl. Greece,* 1: 300 (1986)—Greece.
pintodasilvae *T.R. Dudley in Feddes Repert.,* 97(3–4): 136 (1986)—Portugal.

AMARYLLIS (Amaryllidac.).
pyrrhocroa *Bull, Retail List,* 72: 14 (1872); *cf. D.J. Mabberley in Taxon,* 34(3): 456 (1985)— Locality not stated.
spathacea *(Sims) Sweet, Hort. Brit.:* 402 (1827); *cf. D.J. Mabberley in Taxon,* 30(1): 15 (1981): Basionym not stated.
striatifolia *Sweet, Hort. Brit.:* 402 (1827); *cf. D.J. Mabberley in Taxon,* 30(1): 15 (1981)— Locality not stated.
subbarbata *(Hook.) Sweet, Hort. Brit.:* 403 (1827); *cf. D.J. Mabberley in Taxon,* 30(1): 15 (1981): Basionym not stated.

AMBERBOA (Compos.).
libyca *(Viv.) S.A. Alavi in Fl. Libya,* 107: 267 (1983): Lacellia libyca.

AMHERSTIA (Legumin.).
nobilis *Wall. in Taylor & Phillips, Phil. Mag., n.s.,* 3: 223 (1826); *cf. D.J. Mabberley in Taxon,* 31(1): 67 (1982)— Locality not stated.

AMITOSTIGMA (Orchidac.).
keiskeoides *(Gagnep.) L.A. Garay & W. Kittredge in Bot. Mus. Leafl. Harvard Univ.,* 30(3): (47) 181 (1985 publ. 1986): Habernaria keiskeoides.

AMMANNIA (Lythrac.).
incarnata *Koen. ex Rottb., Pl. Hort. Univ. Rar. Prog.:* 9 (1773); *cf. D.J. Mabberley in Taxon,* 33(3): 438 (1984)— Locality not stated.

AMOMUM (Zingiberac.).
mioga *C.P. Thunberg ex A. Murray in Linnaeus, Syst. Veg., ed. 14:* 51 (1784)— Japan. †

AMOORA (Meliac.).
tsangii *(Merr.) X.M. Chen in J. Wuhan Bot. Res.,* 4(2): 180 (1986): Aglaia tsangii.

AMORPHOPHALLUS (Arac.).
coaetaneus *S.Y. Liu & S.J. Wei in Guihaia,* 6(3): 183 (1986)—China.
coudercii *(Bogner) J. Bogner in Aroideana,* 8(3): 75 (1985 publ. 1986): Plesmonium coudercii.
nicolsonianus *M. Sivadasan in Pl. Syst. Evol.,* 153(3–4): 165 (1986)—India.

AMPELOPSIS (Vitac.).
mollifolia *W.T. Wang in Bull. Bot. Res. North-East. Forest. Inst.,* 6(4): 21 (1986)—China.

AMPHIBROMUS (Gramin.).
macrorhinus *S.W.L. Jacobs & L. Lapinpuro in Telopea,* 2(6): 723 (1986)—Australia (Victoria, New South Wales, S. Australia, W. Australia), Tasmania.
pithogastrus *S.W.L. Jacobs & L. Lapinpuro in Telopea,* 2(6): 724 (1986)—Australia (New South Wales, Victoria).
sinuatus *S.W.L. Jacobs & L. Lapinpuro in Telopea,* 2(6): 727 (1986)—Australia (New South Wales, Victoria).
vickeryae *S.W.L. Jacobs & L. Lapinpuro in Telopea,* 2(6): 725 (1986)—Australia (W. Australia).

AMYGDALUS (Rosac.).
davidiana
var. potanini *(Batal.) T.T. Yü & L.T. Lu in Fl. Reipubl. Popul. Sin.,* 38: 22 (1986): Prunus persica var. potanini.
ferganensis *(Kost. & Rjab.) T.T. Yü & L.T. Lu in Fl. Reipubl. Popul. Sin.,* 38: 20 (1986): Prunus persica subsp. ferganensis.

AMYGDALUS :–
mira *(Koehne) T.T. Yü & L.T. Lu in Fl. Reipubl. Popul. Sin.,* 38: 23 (1986): Prunus mira.
persica
var. aganonucipersica *(Schübler & Martens) T.T. Yü & L.T. Lu in Fl. Reipubl. Popul. Sin.,* 38: 18 (1986): A. persica s. aganonucipersica.
var. compressa *(Loud.) T.T. Yü & L.T. Lu in Fl. Reipubl. Popul. Sin.,* 38: 19 (1986): Persica vulgaris var. compressa.
var. scleronucipersica *(Schübler & Martens) T.T. Yü & L.T. Lu in Fl. Reipubl. Popul. Sin.,* 38: 19 (1986): A. persica η scleronucipersica.
var. scleropersica *(Reich.) T.T. Yü & L.T. Lu in Fl. Reipubl. Popul. Sin.,* 38: 19 (1986): A. persica b. scleropersica.
reticulata *Runemark ex M. Khatamsaz in Iranian J. Bot.,* 3(1): 78 (1985)—Iran.
×rhodia *K. Browicz in Ann. Mus. Goulandris,* 7: 34 (1986)—Rhodes. (A. communis × graeca).
nothovar. elongata *K. Browicz in Ann. Mus. Goulandris,* 7: 36 (1986)—Rhodes.

ANACYCLUS (Compos.).
anthemoides *(L.) Lag., Elenchus:* 2 (1816); *cf. D.J. Mabberley in Taxon,* 31(1): 69 (1982): Basionym not stated.
pyrethrum *(L.) Lag., Elenchus:* 2 (1816); *cf. D.J. Mabberley in Taxon,* 31(1): 69 (1982): Basionym not stated.

ANAXAGOREA (Annonac.).
rheophytica *P.J.M. Maas & L.Y.Th. Westra in Proc. Kon. Nederl. Akad. Wetensch., C,* 89(1): 75 (1986)—Venezuela.

ANCHUSA (Boraginac.).
milleri *Lam. ex Spreng., Nachtr. Bot. Gart. Univ. Halle:* 11 (1801); *cf. D.J. Mabberley in Taxon,* 32(1): 84 (1983)— Locality not stated.
rupestris *(Pallas) Sweet, Hort. Suburb. Lond.:* 30 (1818); *cf. D.J. Mabberley in Taxon,* 31(1): 69 (1982): Basionym not stated.

ANDROMEDA (Ericac.).
japonica *C.P. Thunberg ex A. Murray in Linnaeus, Syst. Veg., ed. 14:* 407 (1784)— Japan. †

ANDROPOGON (Gramin.).
ciliatus *C.P. Thunberg ex A. Murray in Linnaeus, Syst. Veg., ed. 14:* 903 (1784)— Japan. †
cotulifer *C.P. Thunberg ex A. Murray in Linnaeus, Syst. Veg., ed. 14:* 903 (1784)— Japan. †
crinitus *C.P. Thunberg ex A. Murray in Linnaeus, Syst. Veg., ed. 14:* 903 (1784)— Japan. †

ANDROPOGON :–
glomeratus
var. scabriglumus *C.S. Campbell in Syst. Bot.*, 11(2): 291 (1986)—U.S.A. (New Mexico).
serratus *C.P. Thunberg ex A. Murray in Linnaeus, Syst. Veg., ed. 14:* 903 (1784)—Japan. †
virginicus
var. decipiens *C.S. Campbell in Syst. Bot.*, 11(2): 290 (1986)—U.S.A. (Florida).

ANDROSACE (Primulac.).
sect. Mirabiles *(Hand.-Mazz.) Y.C. Yang & R.F. Huang in Acta Phytotax. Sin.*, 24(3): 219 (1986): A. subsect. Mirabiles.
alaschanica
var. zadoensis *Y.C. Yang & R.F. Huang in Acta Phytotax. Sin.*, 24(3): 231 (1986)—China.
bisulca
var. aurata *(Petitm.) Y.C. Yang & R.F. Huang in Acta Phytotax. Sin.*, 24(3): 228 (1986): A. aurata.
cernuiflora *Y.C. Yang & R.F. Huang in Acta Phytotax. Sin.*, 24(3): 230 (1986)—China.
henryi
subsp. simulans *C.M. Hu & Y.C. Yang in Acta Phytotax. Sin.*, 24(3): 218 (1986)—China.
hookeriana
var. mairei *(Lévl.) J.X. Yang & Huang in Fl. Tsinlingensis,* 1(4): 41 (1983): A. mairei.
laxa *C.M. Hu & Y.C. Yang in Acta Phytotax. Sin.*, 24(3): 224 (1986)—China.
medifissa *Chen & Y.C. Yang in Acta Phytotax. Sin.*, 24(3): 216 (1986)—China.
minor *(Hand.-Mazz.) C.M. Hu & Y.C. Yang in Acta Phytotax. Sin.*, 24(3): 222 (1986): A. rigida var. minor.
ovalifolia *Y.C. Yang in Acta Phytotax. Sin.*, 24(3): 221 (1986)—Xizang.
pomeiensis *C.M. Hu & Y.C. Yang in Acta Phytotax. Sin.*, 24(3): 228 (1986)—Xizang.
tanggulashanensis *Y.C. Yang & R.F. Huang in Acta Phytotax. Sin.*, 24(3): 226 (1986)—Xizang, China.

ANDROTRICHUM (Cyperac.).
montevidense *(Link) Schrad., Ind. Sem. Hort. Acad. Gott.*, 1835: 1 (1835); *Linnaea 11, Lit.:* 87 (1837); *cf. D.J. Mabberley in Taxon,* 32(1): 84 (1983): Basionym not stated.

ANEMONE (Ranunculac.).
cernua *C.P. Thunberg ex A. Murray in Linnaeus, Syst. Veg., ed. 14:* 510 (1784)—Japan. †
multilobulata *W.T. Wang & L.Q. Li in Acta Bot. Yunnanica,* 8(3): 263 (1986)—China.
nutantiflora *W.T. Wang & L.Q. Li in Acta Bot. Yunnanica,* 8(3): 265 (1986)—China.
pulsatilla
var. hispanica *(Zimmermann) O. de Bolòs & J. Vigo, Fl. Països Catalans:* 239 (1984): Pulsatilla rubra subsp. hispanica.

ANEMONE :–
var. montsicciana *O. de Bolòs & J. Vigo, Fl. Països Catalans:* 239 (1984)—Spain.
var. nana *(Aichele & Schwegler) O. de Bolòs & J. Vigo, Fl. Països Catalans:* 239 (1984): Pulsatilla rubra var. nana.

ANGELICA (Umbellif.).
chinghaiensis *Shan ex K.T. Fu in Fl. Tsinlingensis,* 1(3): 462, 421 (1981)—China.
ostruthium *(L.) Lag., Elenchus:* 2 (1816); *cf. D.J. Mabberley in Taxon,* 31(1): 69 (1982): Basionym not stated.
sachalinensis
var. glabra *(Koidz.) T. Yamazaki in J. Jap. Bot.*, 61(8): 243 (1986): A. matsumurae var. glabra.
var. pubescens *T. Yamazaki in J. Jap. Bot.*, 61(8): 243 (1986)—Japan.
tsinlingensis *K.T. Fu in Fl. Tsinlingensis,* 1(3): 461, 420 (1981)—China.

ANGOPHORA (Myrtac.).
bakeri
subsp. crassifolia *G.J. Leach in Telopea,* 2(6): 766 (1986)—Australia (New South Wales).
subsp. paludosa *G.J. Leach in Telopea,* 2(6): 764 (1986)—Australia (New South Wales).
costata
subsp. euryphylla *L. Johnson ex G.J. Leach in Telopea,* 2(6): 759 (1986)—Australia (New South Wales).
subsp. leiocarpa *L. Johnson ex G.J. Leach in Telopea,* 2(6): 760 (1986)—Australia (New South Wales, Queensland).

ANGRAECUM (Orchidac.).
longicalcar *(Bosser) K. Senghas in Schlechter Orchideen,* 1(16–18): 1004 (1986): A. eburneum var. longicalcar.

ANISACANTHUS (Acanthac.).
quadrifidus
var. brevilobus *(Hagen) J. Henrickson in Sida,* 11(3): 298 (1986): A. wrightii var. brevilobus.
var. potosinus *J. Henrickson in Sida,* 11(3): 296 (1986)—Mexico.
var. wrightii *(Torr.) J. Henrickson in Sida,* 11(3): 296 (1986): Drejera wrightii.

ANISEIA (Convolvulac.).
stenantha
var. macrostephana *Y.H. Zhang in Acta Phytotax. Sin.*, 24(2): 155 (1986)—China.

ANISOPAPPUS (Compos.).
bampsianus *S. Lisowski in Bull. Jard. Bot. Nation. Belg.*, 56(1–2): 199 (1986)—Zaïre.
lawalreanus *S. Lisowski in Bull. Jard. Bot. Nation. Belg.*, 56(1–2): 200 (1986)—Zaïre.
petitianus *S. Lisowski in Bull. Jard. Bot. Nation. Belg.*, 56(1–2): 202 (1986)—Zaïre.

ANNESLEA (Theac.).
paradoxa *Nguyen Huu Hien & G.P. Yakovlev in Bot. Zhurn.*, 71(8): 1120 (1986)—Vietnam.

ANNONA (Annonac.).
cherimola *Miller, Gard. Dict.*, ed.8, without page no. (1768)—Peru. †

ANOECTOCHILUS (Orchidac.).
luzonensis *(Ames) J.J. Wood in Kew Bull.*, 41(4): 815 (1986): Adenostylis luzonensis.
petola *(Lindl.) Hereman, Paxt. Bot. Dict.*: 600 (1868); *cf. D.J. Mabberley in Taxon,* 32(1): 87 (1983): Macodes petola.

ANOMATHECA (Iridac.).
capitata *(Ker) De Vriese, Hort. Spaarn-Berg.*: 5 (1839); *cf. D.J. Mabberley in Taxon,* 33(3): 441 (1984): Aristea capitata.

ANREDERA (Basellac.).
scandens *Sm. in Rees, Cyclop.*, 39(1): Anredera n. 1 (1818); *cf. D.J. Mabberley in Taxon,* 32(1): 81 (1983)— Locality not stated.

ANTENNARIA (Compos.).
margaritacea *(L.) Sweet, Hort. Brit.*: 221 (1826); *cf. D.J. Mabberley in Taxon,* 30(1): 15 (1981): Basionym not stated.
plantaginea *Sweet, Hort. Brit.*: 221 (1826); *cf. D.J. Mabberley in Taxon,* 30(1): 15 (1981)—Locality not stated.
roezlii *Huber, Cat. Gén. 1871 & 1872:* 3 (1871); *cf. D.J. Mabberley in Taxon,* 34(3): 453 (1985)— Locality not stated.

ANTHEMIS (Compos.).
altissima
var. discoidea *M. Iranshahr in Fl. Iran.*, 158: 24 (1986), without latin descr.—Iran.
cretica
subsp. columnae *(Ten.) R. Franzén in Willdenowia,* 16(1): 40 (1986): A. montana var. columnae.
laconica *R. Franzén in Willdenowia,* 16(1): 41 (1986)—Greece.
triumfettii
subsp. decumbens *M. Iranshahr in Fl. Iran.*, 158: 21 (1986)—Iran.
subsp. khorasanica *(Rech. f.) M. Iranshahr in Fl. Iran.*, 158: 20 (1986): A. khorasanica.

ANTHERICUM (Anthericac.).
japonicum *C.P. Thunberg ex A. Murray in Linnaeus, Syst. Veg., ed. 14:* 331 (1784)—Japan. †

ANTHOCHORTUS (Restionac.).
capensis *E.E. Esterhuysen in Bothalia,* 15(3–4): 484 (1985)—South Africa (Cape Province).
crinalis *(Mast.) H.P. Linder in Bothalia,* 15(3–4): 486 (1985): Restio crinalis.
graminifolius *(Kunth) H.P. Linder in Bothalia,* 15(3–4): 486 (1985): Restio graminifolius.

ANTHOCHORTUS :-
insignis *(Mast.) H.P. Linder in Bothalia,* 15(3–4): 487 (1985): Phyllocomos insignis.
laxiflorus *(Nees) H.P. Linder in Bothalia,* 15(3–4): 487 (1985): Hypolaena laxiflora.

ANTHOLYZA (Iridac.).
bicolor *Gasp. ex Vis., Ort. Bot. Padova:* 56, 134 (1842); *cf. D.J. Mabberley in Taxon,* 33(3): 441 (1984)— Locality not stated.

ANTHOSPERMUM (Rubiac.).
basuticum *C. Puff in Fl. S. Afr.*, 31(1:2): 20 (1986)—Lesotho, South Africa (Cape Province, Orange Free State, Natal).
bicorne *C. Puff in Fl. S. Afr.*, 31(1:2): 36 (1986)—South Africa (Cape Province).
comptonii *C. Puff in Fl. S. Afr.*, 31(1:2): 32 (1986)—South Africa (Cape Province).
dregei
subsp. ecklonis *(Sond.) C. Puff in Fl. S. Afr.*, 31(1:2): 32 (1986): A. ecklonis.
esterhuysenianum *C. Puff in Fl. S. Afr.*, 31(1:2): 33 (1986)—South Africa (Cape Province).
var. hirsutum *C. Puff in Fl. S. Afr.*, 31(1:2): 33 (1986)—South Africa (Cape Province).
galioides
subsp. reflexifolium *(Kuntze) C. Puff in Fl. S. Afr.*, 31(1:2): 30 (1986): A. aethiopicum var. reflexifolium.
ibityense *C. Puff, Biosyst. Study African & Madagascan Rubiac.-Anthosperm. (Pl. Syst. Evol., Supp. 3):* 298 (1986)—Madagascar.
isaloense *Homolle ex C. Puff, Biosyst. Study African & Madagascan Rubiac.-Anthosperm. (Pl. Syst. Evol., Supp. 3):* 218 (1986)—Madagascar.
longisepalum *Homolle ex C. Puff, Biosyst. Study African & Madagascan Rubiac.-Anthosperm. (Pl. Syst. Evol., Supp. 3):* 299 (1986)—Madagascar.
madagascariense *Homolle ex C. Puff, Biosyst. Study African & Madagascan Rubiac.-Anthosperm. (Pl. Syst. Evol., Supp. 3):* 215 (1986)—Madagascar.
monticola *C. Puff in Fl. S. Afr.*, 31(1:2): 19 (1986)—South Africa (Cape Province, Orange Free State, Natal), Lesotho.
palustre *Homolle ex C. Puff, Biosyst. Study African & Madagascan Rubiac.-Anthosperm. (Pl. Syst. Evol., Supp. 3):* 324 (1986)—Madagascar.
perrieri *Homolle ex C. Puff, Biosyst. Study African & Madagascan Rubiac.-Anthosperm. (Pl. Syst. Evol., Supp. 3):* 280 (1986)—Madagascar.
pumilum
subsp. rigidum *(Eckl. & Zeyh.) C. Puff in Fl. S. Afr.*, 31(1:2): 26 (1986): A. rigidum.
spathulatum
subsp. ecklonianum *(Cruse) C. Puff in Fl. S. Afr.*, 31(1:2): 18 (1986): A. aethiopicum var. ecklonianum.

ANTHOSPERMUM :–

subsp. saxatile *C. Puff in Fl. S. Afr.*, 31(1:2): 18 (1986)—South Africa (Cape Province).

subsp. tulbaghense *C. Puff in Fl. S. Afr.*, 31(1:2): 19 (1986)—South Africa (Cape Province).

subsp. uitenhagense *C. Puff in Fl. S. Afr.*, 31(1:2): 17 (1986)—South Africa (Cape Province).

streyi *C. Puff in Fl. S. Afr.*, 31(1:2): 28 (1986)—South Africa (Natal).

ternatum

subsp. randii *(S. Moore) C. Puff, Biosyst. Study African & Madagascan Rubiac.-Anthosperm. (Pl. Syst. Evol., Supp. 3):* 285 (1986): A. randii.

thymoides

subsp. antsirabense *C. Puff, Biosyst. Study African & Madagascan Rubiac.-Anthosperm. (Pl. Syst. Evol., Supp. 3):* 328 (1986)—Madagascar.

villosicarpum *(Verdc.) C. Puff, Biosyst. Study African & Madagascan Rubiac.-Anthosperm. (Pl. Syst. Evol., Supp. 3):* 292 (1986): A. herbaceum var. villosicarpum.

zimbabwense *C. Puff, Biosyst. Study African & Madagascan Rubiac.-Anthosperm. (Pl. Syst. Evol., Supp. 3):* 201 (1986)—Zimbabwe, Mozambique.

ANTHURIUM (Arac.).

alticola *T.B. Croat, Rev. Gen. Anthurium (Arac.) ... Pt. 2: Panama (Monogr. Syst. Bot. Missouri Bot. Gard. 14):*29 (1986)—Panama, Colombia.

angustilobum *T.B. Croat, Rev. Gen. Anthurium (Arac.) ... Pt. 2: Panama (Monogr. Syst. Bot. Missouri Bot. Gard. 14):*33 (1986)—Panama.

antonioanum *T.B. Croat, Rev. Gen. Anthurium (Arac.) ... Pt. 2: Panama (Monogr. Syst. Bot. Missouri Bot. Gard. 14):*34 (1986)—Panama.

aroense *G.S. Bunting in Phytologia,* 60(5): 293 (1986)—Venezuela.

barryi *T.B. Croat, Rev. Gen. Anthurium (Arac.) ... Pt. 2: Panama (Monogr. Syst. Bot. Missouri Bot. Gard. 14):*36 (1986)—Panama.

bernardii *T.B. Croat in Aroideana,* 8(4): 120 (1985 publ. 1986)—Venezuela.

berryi *G.S. Bunting in Phytologia,* 60(5): 293 (1986)—Venezuela.

bicollectivum *T.B. Croat, Rev. Gen. Anthurium (Arac.) ... Pt. 2: Panama (Monogr. Syst. Bot. Missouri Bot. Gard. 14):*37 (1986)—Panama.

brevispadix *T.B. Croat, Rev. Gen. Anthurium (Arac.) ... Pt. 2: Panama (Monogr. Syst. Bot. Missouri Bot. Gard. 14):*40 (1986)—Panama.

caloveboranum *T.B. Croat, Rev. Gen. Anthurium (Arac.) ... Pt. 2: Panama (Monogr. Syst. Bot. Missouri Bot. Gard. 14):*42 (1986)—Panama.

cartiense *T.B. Croat, Rev. Gen. Anthurium (Arac.) ... Pt. 2: Panama (Monogr. Syst. Bot. Missouri Bot. Gard. 14):*44 (1986)—Panama.

ANTHURIUM :–

cerropirrense *T.B. Croat, Rev. Gen. Anthurium (Arac.) ... Pt. 2: Panama (Monogr. Syst. Bot. Missouri Bot. Gard. 14):*47 (1986)—Panama.

chorranum *T.B. Croat, Rev. Gen. Anthurium (Arac.) ... Pt. 2: Panama (Monogr. Syst. Bot. Missouri Bot. Gard. 14):*49 (1986)—Panama.

chromostachyum *T.B. Croat, Rev. Gen. Anthurium (Arac.) ... Pt. 2: Panama (Monogr. Syst. Bot. Missouri Bot. Gard. 14):*50 (1986)—Panama.

cineraceum *T.B. Croat, Rev. Gen. Anthurium (Arac.) ... Pt. 2: Panama (Monogr. Syst. Bot. Missouri Bot. Gard. 14):*52 (1986)—Colombia, Panama.

cinereopetiolatum *T.B. Croat, Rev. Gen. Anthurium (Arac.) ... Pt. 2: Panama (Monogr. Syst. Bot. Missouri Bot. Gard. 14):*53 (1986)—Panama, ?Colombia.

collinsii *T.B. Croat, Rev. Gen. Anthurium (Arac.) ... Pt. 2: Panama (Monogr. Syst. Bot. Missouri Bot. Gard. 14):*59 (1986)—Panama.

colonense *T.B. Croat, Rev. Gen. Anthurium (Arac.) ... Pt. 2: Panama (Monogr. Syst. Bot. Missouri Bot. Gard. 14):*60 (1986)—Panama.

coloradense *T.B. Croat, Rev. Gen. Anthurium (Arac.) ... Pt. 2: Panama (Monogr. Syst. Bot. Missouri Bot. Gard. 14):*64 (1986)—Panama.

correae *T.B. Croat, Rev. Gen. Anthurium (Arac.) ... Pt. 2: Panama (Monogr. Syst. Bot. Missouri Bot. Gard. 14):*68 (1986)—Panama.

crassilaminum *T.B. Croat, Rev. Gen. Anthurium (Arac.) ... Pt. 2: Panama (Monogr. Syst. Bot. Missouri Bot. Gard. 14):*69 (1986)—Panama.

crassiradix *T.B. Croat, Rev. Gen. Anthurium (Arac.) ... Pt. 2: Panama (Monogr. Syst. Bot. Missouri Bot. Gard. 14):*70 (1986)—Panama.

var. purpureospadix *T.B. Croat, Rev. Gen. Anthurium (Arac.) ... Pt. 2: Panama (Monogr. Syst. Bot. Missouri Bot. Gard. 14):*71 (1986)—Panama.

crassitepalum *T.B. Croat, Rev. Gen. Anthurium (Arac.) ... Pt. 2: Panama (Monogr. Syst. Bot. Missouri Bot. Gard. 14):*72 (1986)—Panama.

cuasicanum *T.B. Croat, Rev. Gen. Anthurium (Arac.) ... Pt. 2: Panama (Monogr. Syst. Bot. Missouri Bot. Gard. 14):*76 (1986)—Panama.

cucullispathum *T.B. Croat, Rev. Gen. Anthurium (Arac.) ... Pt. 2: Panama (Monogr. Syst. Bot. Missouri Bot. Gard. 14):*77 (1986)—Panama.

curvilaminum *T.B. Croat, Rev. Gen. Anthurium (Arac.) ... Pt. 2: Panama (Monogr. Syst. Bot. Missouri Bot. Gard. 14):*80 (1986)—Panama.

curvispadix *T.B. Croat, Rev. Gen. Anthurium (Arac.) ... Pt. 2: Panama (Monogr. Syst. Bot. Missouri Bot. Gard. 14):*83 (1986)—Panama.

dichrophyllum *T.B. Croat, Rev. Gen. Anthurium (Arac.) ... Pt. 2: Panama (Monogr. Syst. Bot. Missouri Bot. Gard. 14):*85 (1986)—Panama.

dukei *T.B. Croat, Rev. Gen. Anthurium (Arac.) ... Pt. 2: Panama (Monogr. Syst. Bot. Missouri Bot. Gard. 14):*87 (1986)—Panama.

ANTHURIUM :-

erythrostachyum *T.B. Croat, Rev. Gen. Anthurium (Arac.) ... Pt. 2: Panama (Monogr. Syst. Bot. Missouri Bot. Gard. 14):*89 (1986)—Panama, Colombia.

fernandezii *T.B. Croat in Aroideana,* 8(4): 124 (1985 publ. 1986)—Venezuela.

folsomianum *T.B. Croat, Rev. Gen. Anthurium (Arac.) ... Pt. 2: Panama (Monogr. Syst. Bot. Missouri Bot. Gard. 14):*95 (1986)—Panama.

foreroanum *T.B. Croat, Rev. Gen. Anthurium (Arac.) ... Pt. 2: Panama (Monogr. Syst. Bot. Missouri Bot. Gard. 14):*96 (1986)—Panama.

fragrantissimum *T.B. Croat, Rev. Gen. Anthurium (Arac.) ... Pt. 2: Panama (Monogr. Syst. Bot. Missouri Bot. Gard. 14):*98 (1986)—Panama, Colombia.

fusiforme *T.B. Croat, Rev. Gen. Anthurium (Arac.) ... Pt. 2: Panama (Monogr. Syst. Bot. Missouri Bot. Gard. 14):*102 (1986)—Panama.

gehrigeri *T.B. Croat in Aroideana,* 8(4): 129 (1985 publ. 1986)—Venezuela.

gentryi *T.B. Croat, Rev. Gen. Anthurium (Arac.) ... Pt. 2: Panama (Monogr. Syst. Bot. Missouri Bot. Gard. 14):*104 (1986)—Panama.

globosum *T.B. Croat, Rev. Gen. Anthurium (Arac.) ... Pt. 2: Panama (Monogr. Syst. Bot. Missouri Bot. Gard. 14):*105 (1986)—Panama.

gonzalezii *T.B. Croat in Aroideana,* 8(4): 131 (1985 publ. 1986)—Venezuela.

gracililaminum *T.B. Croat, Rev. Gen. Anthurium (Arac.) ... Pt. 2: Panama (Monogr. Syst. Bot. Missouri Bot. Gard. 14):*106 (1986)—Panama.

gracilispadix *T.B. Croat, Rev. Gen. Anthurium (Arac.) ... Pt. 2: Panama (Monogr. Syst. Bot. Missouri Bot. Gard. 14):*107 (1986)—Panama.

guanchezii *G.S. Bunting in Phytologia,* 60(5): 295 (1986)—Venezuela.

hammelii *T.B. Croat, Rev. Gen. Anthurium (Arac.) ... Pt. 2: Panama (Monogr. Syst. Bot. Missouri Bot. Gard. 14):*109 (1986)—Panama.

hebetatum *T.B. Croat, Rev. Gen. Anthurium (Arac.) ... Pt. 2: Panama (Monogr. Syst. Bot. Missouri Bot. Gard. 14):*109 (1986)—Panama.

hornitense *T.B. Croat, Rev. Gen. Anthurium (Arac.) ... Pt. 2: Panama (Monogr. Syst. Bot. Missouri Bot. Gard. 14):*114 (1986)—Panama.

huequeense *G.S. Bunting in Phytologia,* 60(5): 295 (1986)—Venezuela.

hutchisonii *T.B. Croat, Rev. Gen. Anthurium (Arac.) ... Pt. 2: Panama (Monogr. Syst. Bot. Missouri Bot. Gard. 14):*115 (1986)—Panama.

impolitum *T.B. Croat, Rev. Gen. Anthurium (Arac.) ... Pt. 2: Panama (Monogr. Syst. Bot. Missouri Bot. Gard. 14):*116 (1986)—Panama.

ANTHURIUM :-

jefense *T.B. Croat, Rev. Gen. Anthurium (Arac.) ... Pt. 2: Panama (Monogr. Syst. Bot. Missouri Bot. Gard. 14):*118 (1986)—Panama.

kallunkiae *T.B. Croat, Rev. Gen. Anthurium (Arac.) ... Pt. 2: Panama (Monogr. Syst. Bot. Missouri Bot. Gard. 14):*119 (1986)—Panama.

kamemotoanum *T.B. Croat, Rev. Gen. Anthurium (Arac.) ... Pt. 2: Panama (Monogr. Syst. Bot. Missouri Bot. Gard. 14):*120 (1986)—Panama.

lactifructum *T.B. Croat, Rev. Gen. Anthurium (Arac.) ... Pt. 2: Panama (Monogr. Syst. Bot. Missouri Bot. Gard. 14):*122 (1986)—Panama.

lancifolium

var. albifructum *T.B. Croat, Rev. Gen. Anthurium (Arac.) ... Pt. 2: Panama (Monogr. Syst. Bot. Missouri Bot. Gard. 14):*125 (1986)—Panama.

leptocaule *T.B. Croat, Rev. Gen. Anthurium (Arac.) ... Pt. 2: Panama (Monogr. Syst. Bot. Missouri Bot. Gard. 14):*127 (1986)—Panama.

llanense *T.B. Croat, Rev. Gen. Anthurium (Arac.) ... Pt. 2: Panama (Monogr. Syst. Bot. Missouri Bot. Gard. 14):*128 (1986)—Panama.

longissimum

subsp. nirguense *G.S. Bunting in Phytologia,* 60(5): 295 (1986)—Venezuela.

longistipitatum *T.B. Croat, Rev. Gen. Anthurium (Arac.) ... Pt. 2: Panama (Monogr. Syst. Bot. Missouri Bot. Gard. 14):*129 (1986)—Panama.

madisonianum *T.B. Croat, Rev. Gen. Anthurium (Arac.) ... Pt. 2: Panama (Monogr. Syst. Bot. Missouri Bot. Gard. 14):*132 (1986)—Panama.

malianum *T.B. Croat, Rev. Gen. Anthurium (Arac.) ... Pt. 2: Panama (Monogr. Syst. Bot. Missouri Bot. Gard. 14):*134 (1986)—Colombia/Panama border.

melastomatis *T.B. Croat, Rev. Gen. Anthurium (Arac.) ... Pt. 2: Panama (Monogr. Syst. Bot. Missouri Bot. Gard. 14):*137 (1986)—Panama.

nervatum *T.B. Croat, Rev. Gen. Anthurium (Arac.) ... Pt. 2: Panama (Monogr. Syst. Bot. Missouri Bot. Gard. 14):*141 (1986)—Panama.

niqueanum *T.B. Croat, Rev. Gen. Anthurium (Arac.) ... Pt. 2: Panama (Monogr. Syst. Bot. Missouri Bot. Gard. 14):*143 (1986)—Panama.

oxystachyum *T.B. Croat, Rev. Gen. Anthurium (Arac.) ... Pt. 2: Panama (Monogr. Syst. Bot. Missouri Bot. Gard. 14):*145 (1986)—Panama.

pageanum *T.B. Croat, Rev. Gen. Anthurium (Arac.) ... Pt. 2: Panama (Monogr. Syst. Bot. Missouri Bot. Gard. 14):*146 (1986)—Panama.

ANTHURIUM :-
panamense *T.B. Croat, Rev. Gen. Anthurium (Arac.) ... Pt. 2: Panama (Monogr. Syst. Bot. Missouri Bot. Gard. 14):*148 (1986)—Panama.
papillilaminum *T.B. Croat, Rev. Gen. Anthurium (Arac.) ... Pt. 2: Panama (Monogr. Syst. Bot. Missouri Bot. Gard. 14):*152 (1986)—Panama.
pauciflorum *T.B. Croat, Rev. Gen. Anthurium (Arac.) ... Pt. 2: Panama (Monogr. Syst. Bot. Missouri Bot. Gard. 14):*153 (1986)—Panama.
pendens *T.B. Croat, Rev. Gen. Anthurium (Arac.) ... Pt. 2: Panama (Monogr. Syst. Bot. Missouri Bot. Gard. 14):*153 (1986)—Panama, Colombia.
perijanum *G.S. Bunting in Phytologia,* 60(5): 296 (1986)—Venezuela.
pirrense *T.B. Croat, Rev. Gen. Anthurium (Arac.) ... Pt. 2: Panama (Monogr. Syst. Bot. Missouri Bot. Gard. 14):*156 (1986)—Panama.
pittieri
var. morii *T.B. Croat, Rev. Gen. Anthurium (Arac.) ... Pt. 2: Panama (Monogr. Syst. Bot. Missouri Bot. Gard. 14):*159 (1986)—Panama.
platyrhizum *T.B. Croat, Rev. Gen. Anthurium (Arac.) ... Pt. 2: Panama (Monogr. Syst. Bot. Missouri Bot. Gard. 14):*160 (1986)—Panama.
protensum
subsp. arcuatum *T.B. Croat, Rev. Gen. Anthurium (Arac.) ... Pt. 2: Panama (Monogr. Syst. Bot. Missouri Bot. Gard. 14):*164 (1986)—Panama.
pseudospectabile *T.B. Croat, Rev. Gen. Anthurium (Arac.) ... Pt. 2: Panama (Monogr. Syst. Bot. Missouri Bot. Gard. 14):*165 (1986)—Panama.
redolens *T.B. Croat, Rev. Gen. Anthurium (Arac.) ... Pt. 2: Panama (Monogr. Syst. Bot. Missouri Bot. Gard. 14):*169 (1986)—Panama.
roseospadix *T.B. Croat, Rev. Gen. Anthurium (Arac.) ... Pt. 2: Panama (Monogr. Syst. Bot. Missouri Bot. Gard. 14):*170 (1986)—Panama.
rubrifructum *T.B. Croat, Rev. Gen. Anthurium (Arac.) ... Pt. 2: Panama (Monogr. Syst. Bot. Missouri Bot. Gard. 14):*173 (1986)—Panama.
rupicola *T.B. Croat, Rev. Gen. Anthurium (Arac.) ... Pt. 2: Panama (Monogr. Syst. Bot. Missouri Bot. Gard. 14):*173 (1986)—Panama.
sagawae *T.B. Croat, Rev. Gen. Anthurium (Arac.) ... Pt. 2: Panama (Monogr. Syst. Bot. Missouri Bot. Gard. 14):*175 (1986)—Panama.
sapense *T.B. Croat, Rev. Gen. Anthurium (Arac.) ... Pt. 2: Panama (Monogr. Syst. Bot. Missouri Bot. Gard. 14):*178 (1986)—Panama.

ANTHURIUM :-
smithii *T.B. Croat in Aroideana,* 8(4): 134 (1985 publ. 1986)—Venezuela, Colombia.
subrotundum *T.B. Croat, Rev. Gen. Anthurium (Arac.) ... Pt. 2: Panama (Monogr. Syst. Bot. Missouri Bot. Gard. 14):*183 (1986)—Panama.
subscriptum *G.S. Bunting in Phytologia,* 60(5): 296 (1986)—Venezuela.
supraglandulum *T.B. Croat, Rev. Gen. Anthurium (Arac.) ... Pt. 2: Panama (Monogr. Syst. Bot. Missouri Bot. Gard. 14):*184 (1986)—Panama.
sytsmae *T.B. Croat, Rev. Gen. Anthurium (Arac.) ... Pt. 2: Panama (Monogr. Syst. Bot. Missouri Bot. Gard. 14):*185 (1986)—Panama.
tacarcunense *T.B. Croat, Rev. Gen. Anthurium (Arac.) ... Pt. 2: Panama (Monogr. Syst. Bot. Missouri Bot. Gard. 14):*186 (1986)—Panama.
teribense *T.B. Croat, Rev. Gen. Anthurium (Arac.) ... Pt. 2: Panama (Monogr. Syst. Bot. Missouri Bot. Gard. 14):*187 (1986)—Panama.
tutense *T.B. Croat, Rev. Gen. Anthurium (Arac.) ... Pt. 2: Panama (Monogr. Syst. Bot. Missouri Bot. Gard. 14):*192 (1986)—Panama.
tysonii *T.B. Croat, Rev. Gen. Anthurium (Arac.) ... Pt. 2: Panama (Monogr. Syst. Bot. Missouri Bot. Gard. 14):*193 (1986)—Panama.
wedelianum *T.B. Croat, Rev. Gen. Anthurium (Arac.) ... Pt. 2: Panama (Monogr. Syst. Bot. Missouri Bot. Gard. 14):*196 (1986)—Panama.
subsp. viridispadix *T.B. Croat, Rev. Gen. Anthurium (Arac.) ... Pt. 2: Panama (Monogr. Syst. Bot. Missouri Bot. Gard. 14):*199 (1986)—Panama.
xanthoneurum *G.S. Bunting in Phytologia,* 60(5): 298 (1986)—Venezuela.

ANTHYLLIS (Legumin.).
henoniana
subsp. valentina *(Esteve) O. de Bolòs & J. Vigo, Fl. Països Catalans:* 626 (1984), without exact basionym page: A. sericea subsp. valentina.

ANTIRRHINUM (Scrophulariac.).
hispanicum
subsp. australe *(Rothm.) B.E. Smythies in Englera,* 3(3): 500 (1986): A. australe.
subsp. graniticum *(Rothm.) B.E. Smythies in Englera,* 3(3): 500 (1986): A. graniticum.
meonanthum
subsp. onubensis *J. Fernández Casas in Fontqueria,* 2: 27 (1982)—Spain.
ternatum *J. Fernández Casas in Fontqueria,* 2: 27 (1982)—Morocco.

ANURUS (Legumin.).
nissolia
subsp. futakii *(Chrtková) J. Soják in Čas. Nár. Muz. (Prague),* 152(3): 160 (1983): Lathyrus nissolia subsp. futakii.
subsp. pubescens *(Beck) J. Soják in Čas. Nár. Muz. (Prague),* 152(3): 160 (1983): Lathyrus nissolia α var. pubescens.

APARGIA (Compos.).
pratensis *Hornem., Hort. Hafn.:* 758 (1815); *cf. D.J. Mabberley in Taxon,* 33(3): 436 (1984)— Locality not stated.

APHANAMIXIS (Meliac.).
chittagonga *(Miq.) K. Haridasan & R.R. Rao, Forest Fl. Meghalaya:* 204 (1985): Aglaia chittagonga.
wallichii *(King) K. Haridasan & R.R. Rao, Forest Fl. Meghalaya:* 206 (1985): Amoora wallichii.

APHELANDRA (Acanthac.).
scabra *(Vahl) Sm. in Rees, Cyclop.,* 39(1): Aphelandra n. 3 (1818); *cf. D.J. Mabberley in Taxon,* 32(1): 81 (1983): Justicia scabra.

APIOS (Legumin.).
perennis *Vahl ex Hornem., Hort. Hafn.:* 682 (1815); *cf. D.J. Mabberley in Taxon,* 33(3): 436 (1984)— Locality not stated.

APODOLIRION (Amaryllidac.).
amyanum *D. Müller-Doblies in Willdenowia,* 15(2): 466 (1986)—South Africa (Cape Province).
cedarbergense *D. Müller-Doblies in Willdenowia,* 15(2): 466 (1986)—South Africa (Cape Province).

APONOGETON (Aponogetonac.).
abyssinicus
var. albiflorus *K.A. Lye in Lidia,* 1(2): 73 (1986)—Tanzania, Kenya.
var. cordatus *K.A. Lye in Lidia,* 1(2): 75 (1986)—Kenya, Somali Republic.
var. glandulifera *K.A. Lye in Lidia,* 1(2): 77 (1986)—Tanzania.
var. graminifolius *K.A. Lye in Lidia,* 1(2): 79 (1986)—Tanzania.

×**APOROHELIOCHIA** P.V. Heath in Epiphytes, 8(30): 40 (1984). *CACTACEAE.* (Aporocactus × Heliocereus × Nopalxochia).

×**APOROPHYLLUM** (Cactac.).
frieburgensis *(Weingart) P.V. Heath in Epiphytes,* 8(30): 39 (1984): Cereus frieburgensis.

APOSTASIA (Orchidac.).
ramifera *S.C. Chen & K.Y. Lang in Acta Phytotax. Sin.,* 24(5): 349 (1986)—Hainan.

AQUILARIA (Thymelaeac.).
agallochum *(Lour.) Roxb. ex Finlayson, Mission Siam, Hué:* 94 (1826); *cf. D.J. Mabberley in Taxon,* 32(1): 87 (1983): Basionym not stated.
yunnanensis *S.C. Huang in Acta Bot. Yunnanica,* 7(3): 277 (1985)—China.

AQUILEGIA (Ranunculac.).
flavescens
var. rubicunda *(Tidestr.) S.L. Welsh in Great Basin Nat.,* 46(2): 259 (1986): A. rubicunda.
formosa
var. fosteri *S.L. Welsh in Great Basin Nat.,* 46(2): 259 (1986)—U.S.A. (Utah).
ottonis
subsp. amaliae *(Orph. ex Boiss.) A. Strid in A. Strid (ed.), Mount. Fl. Greece,* 1: 227 (1986): A. amaliae.
subsp. taygetea *(Orph.) A. Strid in A. Strid (ed.), Mount. Fl. Greece,* 1: 228 (1986): A. taygetea.

ARABIDOPSIS (Crucif.).
drassiana *A.R. Naqshi & G.N. Javeid in J. Econ. Taxon. Bot.,* 7(3): 624 (1985 publ. 1986)—India.
mollissima
var. glaberrima *(Hook. f. & T. And.) A.R. Naqshi & G.N. Javeid in J. Econ. Taxon. Bot.,* 7(3): 619 (1985 publ. 1986): Sisymbrium mollissimum var. glaberrima.
pumila
var. foliosa *(Hook. f. & Th.) A.R. Naqshi & G.N. Javeid in J. Econ. Taxon. Bot.,* 7(3): 624 (1985 publ. 1986): Sisymbrium foliosum.
sarbalica *A.R. Naqshi & G.N. Javeid in J. Econ. Taxon. Bot.,* 7(3): 621 (1985 publ. 1986)—India.
tibetica *A.R. Naqshi & G.N. Javeid in J. Econ. Taxon. Bot.,* 7(3): 620 (1985 publ. 1986)—India.
tibetica *(Hook. f. & Thoms.) Lan & An ex K.C. Kuan in Fl. Xizangica,* 2: 372 (1985): Arabis tibetica.
yadungensis *K.C. Kuan & An in Fl. Xizangica,* 2: 375 (1985)—Xizang.

ARABIS (Crucif.).
vivariensis *S.L. Welsh in Great Basin Nat.,* 46(2): 263 (1986)—U.S.A. (Utah).

ARALIA (Araliac.).
japonica *C.P. Thunberg ex A. Murray in Linnaeus, Syst. Veg., ed. 14:* 300 (1784)—Japan. †
pentaphylla *C.P. Thunberg ex A. Murray in Linnaeus, Syst. Veg., ed. 14:* 300 (1784)—Japan. †

ARBELAEZASTER J. Cuatrecasas in Caldasia, 15(71–75): 1 (1986). *COMPOSITAE.*

ARBELAEZASTER :–
ellsworthii *(Cuatr.) J. Cuatrecasas in Caldasia,* 15(71–75): 2 (1986): Senecio ellsworthii.

ARBUTUS (Ericac.).
nitida *(Benth.) Jacob-Makoy, Extr. Cat. Prix Cour. 1842:* 8 (1842); *cf. D.J. Mabberley in Taxon,* 34(3): 456 (1985): Basionym not stated.

ARCUATOPTERUS M.I. Sheh & R.H. Shan in Bull. Bot. Res. North-East. Forest. Inst., 6(4): 11 (1986). *UMBELLIFERAE.*
filipedicellus *M.I. Sheh & R.H. Shan in Bull. Bot. Res. North-East. Forest. Inst.,* 6(4): 12 (1986)—China.
linearifolius *M.I. Sheh & R.H. Shan in Bull. Bot. Res. North-East. Forest. Inst.,* 6(4): 14 (1986)—China.
thalictrioideus *M.I. Sheh & R.H. Shan in Bull. Bot. Res. North-East. Forest. Inst.,* 6(4): 15 (1986)—China, Xizang.

ARDISIA (Myrsinac.).
alajuelae *(Lundell) C.L. Lundell in Phytologia,* 61(1): 62 (1986): Icacorea alajuelae.
albipedicellata *C.L. Lundell in Phytologia,* 61(1): 62 (1986), nom. nov.: Icacorea parvifolia *Lundell.*
albisepala *(Lundell) C.L. Lundell in Phytologia,* 61(1): 62 (1986): Auriculardisia albisepala.
azaharensis *C.L. Lundell in Phytologia,* 61(1): 62 (1986), nom. nov.: Auriculardisia microcalyx *Lundell.*
bekomiensis *(Lundell) C.L. Lundell in Phytologia,* 61(1): 62 (1986): Icacorea bekomiensis.
bristanii *C.L. Lundell in Phytologia,* 61(1): 62 (1986), nom. nov.: Auriculardisia parviflora *Lundell.*
canariensis *Hort. Cels. ex Fée, Cat. Méth. Pl Jard. Bot. Strasb.:* 43 (1836); *cf. D.J. Mabberley in Taxon,* 32(1): 84 (1983)—Locality not stated.
chiriquiana *(Lundell) C.L. Lundell in Phytologia,* 61(1): 62 (1986): Auriculardisia chiriquiana.
coibana *(Lundell) C.L. Lundell in Phytologia,* 61(1): 63 (1986): Graphardisia coibana.
contrerasii *C.L. Lundell in Phytologia,* 61(1): 63 (1986), nom. nov.: Gentlea crenulata *Lundell.*
cuneifolia *(Lundell) C.L. Lundell in Phytologia,* 61(1): 63 (1986): Gentlea cuneifolia.
dressleri *C.L. Lundell in Phytologia,* 61(1): 63 (1986), nom. nov.: Auriculardisia roseiflora *Lundell.*
dryeri *C.L. Lundell in Phytologia,* 61(1): 63 (1986), nom. nov.: Auriculardisia micrantha *Lundell.*
ebingeri *C.L. Lundell in Phytologia,* 61(1): 63 (1986), nom. nov.: Graphardisia purpurea *Lundell.*

ARDISIA :–
eciliata *(Lundell) C.L. Lundell in Phytologia,* 61(1): 63 (1986): Zunilia eciliata.
elliptifolia *C.L. Lundell in Phytologia,* 61(1): 63 (1986), nom. nov.: Amatlania elliptica *Lundell.*
esquipulasana *C.L. Lundell in Phytologia,* 61(1): 63 (1986), nom. nov.: Graphardisia nicaraguensis *Lundell.*
eucuneata *(Lundell) C.L. Lundell in Phytologia,* 61(1): 63 (1986): Auriculardisia eucuneata.
eurubiginosa *(Lundell) C.L. Lundell in Phytologia,* 61(1): 63 (1986): Auriculardisia eurubiginosa.
feniana *C.L. Lundell in Phytologia,* 61(1): 64 (1986), nom. nov.: Zunilia purpusii *Lundell.*
guanacastensis *(Lundell) C.L. Lundell in Phytologia,* 61(1): 64 (1986): Icacorea guanacastensis.
guinealensis *(Lundell) C.L. Lundell in Phytologia,* 61(1): 64 (1986): Icacorea guinealensis.
heterotricha *(Lundell) C.L. Lundell in Phytologia,* 61(1): 64 (1986): Auriculardisia heterotricha.
hirsutissima *C.L. Lundell in Phytologia,* 61(1): 64 (1986)—Panama.
hornitoana *C.L. Lundell in Phytologia,* 61(1): 64 (1986), nom. nov.: Icacorea reflexa *Lundell.*
hugonensis *(Lundell) C.L. Lundell in Phytologia,* 61(1): 64 (1986): Auriculardisia hugonensis.
intibucana *C.L. Lundell in Phytologia,* 61(1): 65 (1986), nom. nov.: Gentlea lancifolia *Lundell.*
ixcanensis *(Lundell) C.L. Lundell in Phytologia,* 61(1): 65 (1986): Icacorea ixcanensis.
jaliscensis *(Lundell) C.L. Lundell in Phytologia,* 61(1): 65 (1986): Icacorea jaliscensis.
jitotolana *C.L. Lundell in Phytologia,* 61(1): 65 (1986), nom. nov.: Gentlea tenuis *Lundell.*
knappii *(Lundell) C.L. Lundell in Phytologia,* 61(1): 65 (1986): Auriculardisia knappii.
latisepala *(Lundell) C.L. Lundell in Phytologia,* 61(1): 65 (1986): Auriculardisia latisepala.
leptopoda *(Lundell) C.L. Lundell in Phytologia,* 61(1): 65 (1986): Auriculardisia leptopoda.
molinae *(Lundell) C.L. Lundell in Phytologia,* 61(1): 65 (1986): Gentlea molinae.
monteverdeana *(Lundell) C.L. Lundell in Phytologia,* 61(1): 65 (1986): Icacorea monteverdeana.
morazanensis *C.L. Lundell in Phytologia,* 61(1): 65 (1986), nom. nov.: Gentlea maculata *Lundell.*
nebulosa *(Lundell) C.L. Lundell in Phytologia,* 61(1): 65 (1986): Auriculardisia nebulosa.
neohyalina *C.L. Lundell in Phytologia,* 61(1): 65 (1986), nom. nov.: Graphardisia hyalina *Lundell.*
neomirandae *C.L. Lundell in Phytologia,* 61(1): 66 (1986), nom. nov.: Zunilia mirandae *Lundell.*

ARDISIA :–
oaxacana *(Lundell) C.L. Lundell in Phytologia,* 61(1): 66 (1986): Icacorea oaxacana.
obtusata *(Lundell) C.L. Lundell in Phytologia,* 61(1): 66 (1986): Graphardisia obtusata.
parviauriculata *C.L. Lundell in Phytologia,* 61(1): 66 (1986), nom. nov.: Gentlea auriculata *Lundell.*
parvidenticulata *C.L. Lundell in Phytologia,* 61(1): 66 (1986), nom. nov.: Icacorea denticulata *Lundell.*
parvipunctata *(Lundell) C.L. Lundell in Phytologia,* 61(1): 66 (1986): Icacorea parvipunctata.
parvissima *C.L. Lundell in Phytologia,* 61(1): 66 (1986), nom. nov.: Gentlea parviflora *Lundell.*
quadrata *(Lundell) C.L. Lundell in Phytologia,* 61(1): 66 (1986): Auriculardisia quadrata.
riomonteana *C.L. Lundell in Phytologia,* 61(1): 66 (1986), nom. nov.: Graphardisia oxyphylla *Lundell.*
samalana *(Lundell) C.L. Lundell in Phytologia,* 61(1): 66 (1986): Icacorea samalana.
seranoana *(Lundell) C.L. Lundell in Phytologia,* 61(1): 66 (1986): Graphardisia seranoana.
sordida *(Lundell) C.L. Lundell in Phytologia,* 61(1): 66 (1986): Auriculardisia sordida.
squamata *(Lundell) C.L. Lundell in Phytologia,* 61(1): 67 (1986): Auriculardisia squamata.
standleyi *(Lundell) C.L. Lundell in Phytologia,* 61(1): 67 (1986): Gentlea standleyi.
steinii *C.L. Lundell in Phytologia,* 61(1): 67 (1986), nom. nov.: Auriculardisia baruana *Lundell.*
stevensii *(Lundell) C.L. Lundell in Phytologia,* 61(1): 67 (1986): Gentlea stevensii.
tarariae *C.L. Lundell in Phytologia,* 61(1): 67 (1986)—Costa Rica.
toroana *(Lundell) C.L. Lundell in Phytologia,* 61(1): 67 (1986): Auriculardisia toroana.
triangula *(Lundell) C.L. Lundell in Phytologia,* 61(1): 67 (1986): Icacorea triangula.
ustupoana *(Lundell) C.L. Lundell in Phytologia,* 61(1): 68 (1986): Graphardisia ustupoana.
utleyi *(Lundell) C.L. Lundell in Phytologia,* 61(1): 68 (1986): Icacorea utleyi.
warneri *C.L. Lundell in Phytologia,* 61(1): 68 (1986), nom. nov.: Icacorea brevipes *Lundell.*
wendtii *(Lundell) C.L. Lundell in Phytologia,* 61(1): 68 (1986): Ibarraea wendtii.
zarceroana *C.L. Lundell in Phytologia,* 61(1): 68 (1986), nom. nov.: Auriculardisia sessilifolia *Lundell.*

ARECA (Palmae).
passalacquae *C.S. Kunth in Ann. Sci. Nat. Paris,* 8: 420 (1826)—Egypt (in ancient tombs).
pumila *C.L. Blume, Rumphia,* 2: 71, t. 99, 102 (1839)—Java.

ARENARIA (Caryophyllac.).
aggregata
subsp. pseudoarmeriastrum *(Rouy) G. López González & G. Nieto Feliner in An. Jard. Bot. Madrid,* 42(2): 356 (1985 publ. 1986): A. pseudoarmeriastrum.
arcuatociliata *G. López González & G. Nieto Feliner in An. Jard. Bot. Madrid,* 42(2): 353 (1985 publ. 1986)—Spain.
cariensis *A. Carlström in Willdenowia,* 15(2): 371 (1986)—Turkey.
delaguardiae *G. López González & G. Nieto Feliner in An. Jard. Bot. Madrid,* 42(2): 351 (1985 publ. 1986)—Spain.
favargeri *(Nieto Feliner) G. López González & G. Nieto Feliner in An. Jard. Bot. Madrid,* 42(2): 358 (1985 publ. 1986): A. aggregata subsp. favargeri.
fendleri
var. aculeata *(Wats.) S.L. Welsh in Great Basin Nat.,* 46(2): 255 (1986): A. aculeata.
var. eastwoodiae *(Rydb.) S.L. Welsh in Great Basin Nat.,* 46(2): 255 (1986): A. eastwoodiae.
filicaulis
subsp. teddii *(Turrill) A. Strid in A. Strid (ed.), Mount. Fl. Greece,* 1: 86 (1986): A. teddii.
grandiflora
subsp. glabrescens *(Willk.) G. López González & G. Nieto Feliner in An. Jard. Bot. Madrid,* 42(2): 351 (1985 publ. 1986): A. incrassata var. glabrescens.
modesta
subsp. tenuis *(Gay) G. López González & G. Nieto Feliner in An. Jard. Bot. Madrid,* 42(2): 359 (1985 publ. 1986): A. tenuis.
oscensis *(Pau) G. López González & G. Nieto Feliner in An. Jard. Bot. Madrid,* 42(2): 357 (1985 publ. 1986): A. aggregata var. oscensis.

ARGOCOFFEOPSIS (Rubiac.).
eketensis *(Wernh.) E. Robbrecht in Bull. Jard. Bot. Nation. Belg.,* 56(1–2): 158 (1986): Coffea eketensis.

ARGOSTEMMA (Rubiac.).
hainanicum *H.S. Lo in Bull. Bot. Res. North-East. Forest. Inst.,* 6(4): 46 (1986)—China.
parvifolium
f. hirsutum *(Ridl.) B.C. Stone in Fed. Mus. J. (Kuala Lumpur),* 26(1): 135 (1981): A. parvifolium var. hirsutum.
yunnanense *How ex H.S. Lo in Bull. Bot. Res. North-East. Forest. Inst.,* 6(4): 45 (1986)—China.

ARGYREIA (Convolvulac.).
baoshanensis *S.H. Huang in Acta Phytotax. Sin.,* 24(1): 19 (1986)—China.

ARGYROLOBIUM (Legumin.).
zanonii
subsp. fallax *(Ball) W. Greuter in Willdenowia,* 15(2): 424 (1986): A. linnaeanum subsp. fallax.
subsp. grandiflorum *(Boiss. & Reuter) W. Greuter in Willdenowia,* 15(2): 424 (1986): A. grandiflorum.
subsp. stipulaceum *(Ball) W. Greuter in Willdenowia,* 15(2): 424 (1986): A. linnaeanum subsp. stipulaceum.

ARISAEMA (Arac.).
heterocephalum
subsp. majus *(Serizawa) J. Murata in Acta Phytotax. Geobot.,* 36(4–6): 137 (1985): A. heterocephalum var. majus.
ishizuchiense
var. alpicola *S. Serizawa in J. Jap. Bot.,* 61(1): 25 (1986)—Japan.
subsp. brevicollum *(Ohashi & J. Murata) S. Serizawa in J. Jap. Bot.,* 61(1): 25 (1986): A. ishizuchiense var. brevicollum.
monophyllum
f. akitense *(Nakai) H. Ohashi in J. Jap. Bot.,* 61(6): 172 (1986): A. akitense.
mooneyanum *M.G. Gilbert & S.J. Mayo in Kew Bull.,* 41(2): 270 (1986)—Ethiopia.
nikoense
subsp. australe *(M. Hotta) S. Serizawa in J. Jap. Bot.,* 61(1): 28 (1986): A. nikoense var. australe.
var. kaimontanum *S. Serizawa in J. Jap. Bot.,* 61(1): 28 (1986)—Japan.
ovale
var. inaense *(Serizawa) J. Murata in J. Fac. Sci. Univ. Tokyo, Bot.,* 14(1): 66 (1986): A. robustum var. inaense.
var. sadoense *(Nakai) J. Murata in J. Fac. Sci. Univ. Tokyo, Bot.,* 14(1): 64 (1986): A. sadoense.
sachalinense *(Miyabe & Kudo) J. Murata in J. Fac. Sci. Univ. Tokyo, Bot.,* 14(1): 66 (1986): A. amurense var. sachalinense.
somalense *M.G. Gilbert & S.J. Mayo in Kew Bull.,* 41(2): 274 (1986)—Somali Republic.
taiwanense *J. Murata in J. Jap. Bot.,* 60(12): 353 (1985)—Taiwan.
var. brevipedunculatum *J. Murata in J. Jap. Bot.,* 60(12): 356 (1985)—Taiwan.
ulugurense *M.G. Gilbert & S.J. Mayo in Fl. Trop. E. Afr., Arac.:* 64 (1985)—Tanzania.
undulatifolium
var. limbatum *(F. Maekawa) H. Ohashi in J. Jap. Bot.,* 61(6): 172 (1986): A. limbatum.
f. limbatum *(F. Maekawa) H. Ohashi in J. Jap. Bot.,* 61(6): 172 (1986): A. limbatum.

ARISTIDA (Gramin.).
bissei *L. Catasús Guerra in Acta Bot. Cubana,* 24: 4 (1985)—Cuba.
dominii *B.K. Simon in Austrobaileya,* 2(3): 281 (1986)—Australia (Queensland, W. Australia, N. Terr.).

ARISTIDA :–
ferrilateris *S.M. Phillips in Kew Bull.,* 41(4): 1029 (1986)—Yemen, Saudi Arabia, Ethiopia.
holathera
var. latifolia *(B. Simon) B.K. Simon in Austrobaileya,* 2(3): 283 (1986): A. browniana var. latifolia.
jaucensis *L. Catasús Guerra in Acta Bot. Cubana,* 24: 2 (1985)—Cuba.
pinifolia *L. Catasús Guerra in Acta Bot. Cubana,* 24: 3 (1985)—Cuba.
purpurescens
var. tenuispica *(Hitchc.) K.W. Allred in Rhodora,* 88(855): 383 (1986): A. tenuispica.
var. virgata *(Trin.) K.W. Allred in Rhodora,* 88(855): 383 (1986): A. virgata.
sandinensis *L. Catasús Guerra in Acta Bot. Cubana,* 24: 2 (1985)—Cuba.

ARISTOLOCHIA (Aristolochiac.).
barbourii *K. Barringer in Brittonia,* 38(2): 128 (1986), as 'barbouri'—Peru.
fosteri *K. Barringer in Brittonia,* 38(2): 129 (1986)—Peru.
hockii
subsp. tuberculata *B. Verdcourt in Fl. Trop. E. Afr., Aristolochiac.:* 7 (1986)—Tanzania.
hutchisonii *K. Barringer in Brittonia,* 38(2): 131 (1986)—Peru.

ARMENIACA (Rosac.).
communis *Besser, Cat. Jard. Bot. Krz.:* 12 (1810); *cf. D.J. Mabberley in Taxon,* 31(1): 69 (1982)— Locality not stated.
mandshurica
var. glabra *(Nakai) T.T. Yü & L.T. Lu in Fl. Reipubl. Popul. Sin.,* 38: 31 (1986): Prunus mandshurica var. glabra.
mume
var. cernua *(Franch.) T.T. Yü & L.T. Lu in Fl. Reipubl. Popul. Sin.,* 38: 32 (1986): Prunus mume var. cernua.
var. pallescens *(Franch.) T.T. Yü & L.T. Lu in Fl. Reipubl. Popul. Sin.,* 38: 32 (1986): Prunus mume var. pallescens.
vulgaris
var. ansu *(Maxim.) T.T. Yü & L.T. Lu in Fl. Reipubl. Popul. Sin.,* 38: 26 (1986): Prunus armeniaca var. ansu.

ARNEBIA (Boraginac.).
purpurea *S. Erik & H. Sümbül in Notes Roy. Bot. Gard. Edinburgh,* 44(1): 151 (1986)—Turkey.

ARNICA (Compos.).
ciliata *C.P. Thunberg ex A. Murray in Linnaeus, Syst. Veg., ed. 14:* 768 (1784)—Japan. †
frigida
subsp. griscomii *(Fern.) S.R. Downie in Canad. J. Bot.,* 64(7): 1369 (1986): A. griscomii.

ARNICA :–
japonica *C.P. Thunberg ex A. Murray in Linnaeus, Syst. Veg., ed. 14:* 768 (1784)—Japan. †
palmata *C.P. Thunberg ex A. Murray in Linnaeus, Syst. Veg., ed. 14:* 768 (1784)—Japan. †

ARNICASTRUM (Compos.).
guerrerense *J.L. Villaseñor in Syst. Bot.*, 11(2): 277 (1986)—Mexico.

ARTEMISIA (Compos.).
andersiana *D. Podlech in Fl. Iran.*, 158: 173 (1986)—Afghanistan.
capillaris *C.P. Thunberg ex A. Murray in Linnaeus, Syst. Veg., ed. 14:* 743 (1784)—Japan. †
chitralensis *D. Podlech in Fl. Iran.*, 158: 198 (1986)—Pakistan, Afghanistan.
eriopoda
var. rotundifolia *(Deb.) W. Wang ex X.D. Cui in Fl. Tsinlingensis,* 1(5): 276 (1985): A. japonica var. rotundifolia.
freitagii *D. Podlech in Fl. Iran.*, 158: 193 (1986)—Afghanistan.
ghazniensis *D. Podlech in Fl. Iran.*, 158: 213 (1986)—Afghanistan.
ghoratensis *D. Podlech in Fl. Iran.*, 158: 197 (1986)—Afghanistan.
glanduligera
subsp. kotgaiensis *D. Podlech in Fl. Iran.*, 158: 217 (1986)—Afghanistan, Pakistan.
japonica *C.P. Thunberg ex A. Murray in Linnaeus, Syst. Veg., ed. 14:* 744 (1784)—Japan. †
kandaharensis *D. Podlech in Fl. Iran.*, 158: 217 (1986)—Afghanistan.
kermanensis *D. Podlech in Fl. Iran.*, 158: 206 (1986)—Iran.
khorassanica *D. Podlech in Fl. Iran.*, 158: 210 (1986)—Iran, Afghanistan.
lactiflora
var. taibaishanensis *X.D. Cui in Fl. Tsinlingensis,* 1(5): 421, 274 (1985)—China.
multiflora *Trev., Ind. Sem. Wratislav. App. III:* 1 (1821); *cf. D.J. Mabberley in Taxon,* 32(1): 84 (1983)— China.
quettensis *D. Podlech in Fl. Iran.*, 158: 212 (1986)—Pakistan, Iran.
selengensis
var. integerrima *(Komar.) Ling & Y.R. Ling ex X.D. Cui in Fl. Tsinlingensis,* 1(5): 271 (1985): A. vulgaris var. integerrima.
sieberi
subsp. deserticola *D. Podlech in Fl. Iran.*, 158: 205 (1986)—Afghanistan.
swatensis *D. Podlech in Fl. Iran.*, 158: 189 (1986)—Pakistan.
tecti-mundi *D. Podlech in Fl. Iran.*, 158: 206 (1986)—Afghanistan, Pakistan.
tilesii
subsp. hultenii *(Maximova) V.G. Sergienko, Fl. poluostrova Kanin:* 97 (1986): A. hultenii.

ARTHRAGROSTIS (Gramin.).
aristispicula *B.K. Simon in Austrobaileya,* 2(3): 238 (1986)—Australia (Queensland).

ARTHROCEREUS (Cactac.).
itabiriticola *P.J. Braun in Kakt. And. Sukk.*, 37(11): 237 (1986)—Brazil.

ARTHROPHYLLUM (Araliac.).
kaltenbachii *H. Riedl & C. Riedl-Dorn in Linz. Biol. Beitr.*, 18(2): 374 (1986)—Vanuatu.

ARTHROPOGON (Gramin.).
bolivianus *T.S. Filgueiras in Brittonia,* 38(1): 71 (1986)—Bolivia.

ARTHROSTEMA (Melastomatac.).
subtriplinervium *(Link & Otto) Mart. in Schrank & Mart., Hort. Reg. Monac.:* 171 (1829); *cf. D.J. Mabberley in Taxon,* 33(3): 441 (1984): Basionym not stated.

ARTOCARPUS (Morac.).
macrocarpon *Dancer, Cat. Bot. Gdn. Jamaica:* 1 (1792); *cf. D.J. Mabberley in Taxon,* 30(1): 12 (1981)— Locality not stated.

ARUM (Arac.).
ternatum *C.P. Thunberg ex A. Murray in Linnaeus, Syst. Veg., ed. 14:* 827 (1784)—Japan. †

ARUNDINARIA (Gramin.).
lanshanensis *(Wen) T.H. Wen in J. Bamboo Res.,* 5(2): 19 (1986): Yushania lanshanensis.
pallidiflora *(McClure) T.H. Wen in J. Bamboo Res.,* 5(2): 19 (1986): Indocalamus pallidiflorus.
schomburgkii *Benn. in Proc. Linn. Soc.,* 1: 51 (3 Mar. 1840); *Jardine et al., Ann. Mag. Nat. Hist.,* 30(1): 15 (1981)— *cf. D.J. Mabberley in Taxon,* Locality not stated.

ARYTERA (Sapindac.).
bifoliolata *S.T. Reynolds in Fl. Australia,* 25: 198, 91 (1985)—Australia (Queensland, N. Terr.).
dictyoneura *S.T. Reynolds in Fl. Australia,* 25: 198, 93 (1985)—Australia (Queensland).
pauciflora *S.T. Reynolds in Fl. Australia,* 25: 198, 91 (1985)—Australia (Queensland).

ASARUM (Aristolochiac.).
bashanense *Z.L. Yang in Acta Bot. Bor.-Occid. Sin.,* 5(1): 50 (1985)—China.
caulescens
f. geroensis *J. Ōhara in J. Phytogeogr. Taxon.,* 33(2): 72 (1985)—Japan.
chengkouense *Z.L. Yang in Acta Bot. Bor.-Occid. Sin.,* 5(1): 54 (1985)—China.
nobilissimum *Z.L. Yang in Acta Bot. Bor.-Occid. Sin.,* 5(1): 56 (1985), as 'nobilssimum'—China.
tongjiangense *Z.L. Yang in Acta Bot. Bor.-Occid. Sin.,* 5(1): 52 (1985)—China.

ASARUM :–
wulongense *Z.L. Yang in Acta Bot. Yunnanica,* 8(3): 277 (1986)—China.

ASCLEPIAS (Asclepiadac.).
oreophila *A. Nicholas ex O.M. Hilliard & B.L. Burtt in Notes Roy. Bot. Gard. Edinburgh,* 43(2): 192 (1986)—South Africa (Natal, Cape Province).
parasitica *Wall. ex Hornem., Hort. Hafn., Suppl.:* 126 (1819); *cf. D.J. Mabberley in Taxon,* 33(3): 436 (1984)— Locality not stated.
xysmalobioides *O.M. Hilliard & B.L. Burtt in Notes Roy. Bot. Gard. Edinburgh,* 43(2): 193 (1986)—South Africa (Natal), Lesotho.

ASKIDIOSPERMA (Restionac.).
alboaristatum *(Pillans) H.P. Linder in Bothalia,* 15(3–4): 431 (1985): Chondropetalum albo-aristatum.
andreaeanum *(Pillans) H.P. Linder in Bothalia,* 15(3–4): 431 (1985): Chondropetalum andreaeanum.
chartaceum *(Pillans) H.P. Linder in Bothalia,* 15(3–4): 431 (1985): Dovea chartacea.
subsp. alticolum *E.E. Esterhuysen in Bothalia,* 15(3–4): 432 (1985)—South Africa (Cape Province).
esterhuyseniae *(Pillans) H.P. Linder in Bothalia,* 15(3–4): 432 (1985): Chondropetalum esterhuyseniae.
insigne *(Pillans) H.P. Linder in Bothalia,* 15(3–4): 432 (1985): Chondropetalum insigne.
longiflorum *(Pillans) H.P. Linder in Bothalia,* 15(3–4): 432 (1985): Chondropetalum longiflorum.
nitidum *(Mast.) H.P. Linder in Bothalia,* 15(3–4): 432 (1985): Dovea nitida.
paniculatum *(Mast.) H.P. Linder in Bothalia,* 15(3–4): 432 (1985): Dovea paniculata.
rugosum *E.E. Esterhuysen in Bothalia,* 15(3–4): 432 (1985)—South Africa (Cape Province).

ASPHODELINE (Asphodelac.).
capillaris *(DC.) Endl., Cat. Horti Vindob.,* 1: 142 (1842); *cf. D.J. Mabberley in Taxon,* 33(3): 434 (1984): Basionym not stated.
cretica *(Lam.) Endl., Cat. Horti Vindob.,* 1: 142 (1842); *cf. D.J. Mabberley in Taxon,* 33(3): 434 (1984): Basionym not stated.
taurica *(Pall.) Endl., Cat. Horti Vindob.,* 1: 142 (1842); *cf. D.J. Mabberley in Taxon,* 33(3): 434 (1984): Basionym not stated.

ASPIDOPTERYS (Malpighiac.).
cordata
var. vermae *R.C. Srivastava & N.P. Balakrishnan in Bull. Bot. Surv. India,* 25(1–4): 228 (1983 publ. 1985)—India.

ASPRELLA (Gramin.).
major *Fres., Sem. Hort. Bot. Franc.,* 1836; *Linnaea 12, Lit.:* 77 (1838-9); *cf. D.J. Mabberley in Taxon,* 32(1): 84 (1983)—Locality not stated.

ASTELMA (Compos.).
modestum *(Hook.) Forbes, Hort. Woburn.:* 184 (1833); *cf. D.J. Mabberley in Taxon,* 33(3): 441 (1984): Basionym not stated.

ASTER (Compos.).
subgen. Ascendentes *(Rydb.) J.C. Semple in Phytologia,* 58(7): 430 (1985), contrary to Art. 34.1(a) ICBN 1983; *see W.F. Lamboy in Phytologia,* 59(7): 454 (1986): A. sp.-group Ascendentes.
ageratoides
var. oophyllus *Ling ex J.Q. Fu in Fl. Tsinlingensis,* 1(5): 168 (1985), as 'nom. nov.' but contrary to Art.33.3, Note 1 (ICBN,1983)—China.
alpinus
var. serpentimontanus *(Tamam.) Ling ex J.Q. Fu in Fl. Tsinlingensis,* 1(5): 172 (1985): A. serpentimontanus.
hispidus *C.P. Thunberg ex A. Murray in Linnaeus, Syst. Veg., ed. 14:* 763 (1784)—Japan. †
moranensis
var. turneri *S. Sundberg & A.G. Jones in Bull. Torrey Bot. Club,* 113(2): 176 (1986)—Mexico.
scaber *C.P. Thunberg ex A. Murray in Linnaeus, Syst. Veg., ed. 14:* 763 (1784)—Japan. †
schaffneri *Schultz-Bip. ex S. Sundberg & A.G. Jones in Bull. Torrey Bot. Club,* 113(2): 173 (1986)—Mexico.

ASTRACANTHA (Legumin.).
adanica *(Ivanišvili) W. Greuter in Willdenowia,* 15(2): 425 (1986): Astragalus adanicus.
akscheherensis *(Freyn & Bornm.) W. Greuter in Willdenowia,* 15(2): 425 (1986): Astragalus akscheherensis.
argyrothamnos *(Boiss.) W. Greuter in Willdenowia,* 15(2): 425 (1986): Astragalus argyrothamnos.
coarctata *(Trautv.) W. Greuter in Willdenowia,* 15(2): 425 (1986): Astragalus coarctatus.
delanensis *(Sirj. & Rech. f.) W. Greuter in Willdenowia,* 15(2): 425 (1986): Astragalus delanensis.
eriocephala
subsp. elongata *(Chamberlain & V. Matthews) W. Greuter in Willdenowia,* 15(2): 425 (1986): Astragalus eriocephalus subsp. elongatus.
griseosericea *(Eig) W. Greuter in Willdenowia,* 15(2): 425 (1986): Astragalus griseosericeus.
hareftae *(Náb.) W. Greuter in Willdenowia,* 15(2): 425 (1986): Astragalus krugeanus var. hareftae.

ASTRACANTHA :–

hasbeyana *(Boiss.) W. Greuter in Willdenowia,* 15(2): 425 (1986): Astragalus hasbeyanus.
kristii *(Sirj.) W. Greuter in Willdenowia,* 15(2): 425 (1986): Astragalus kristii.
leiophylla *(Freyn & Bornm.) W. Greuter in Willdenowia,* 15(2): 425 (1986): Astragalus leiophyllus.
nebrodensis *(Guss.) W. Greuter in Willdenowia,* 15(2): 425 (1986): Astragalus siculus var. nebrodensis.
ochrochlora *(Boiss. & Hohen.) W. Greuter in Willdenowia,* 15(2): 425 (1986): Astragalus ochrochlorus.
roseocalycina *(V. Matthews) W. Greuter in Willdenowia,* 15(2): 425 (1986): Astragalus roseocalycinus.
sicula *(Raf.) W. Greuter in Willdenowia,* 15(2): 425 (1986): Astragalus siculus.
spiciformis *(Eig) W. Greuter in Willdenowia,* 15(2): 426 (1986): Astragalus spiciformis.
strictifolia *(Boiss.) W. Greuter in Willdenowia,* 15(2): 426 (1986): Astragalus strictifolius.
thracica
subsp. calabrica *(Fischer) W. Greuter in Willdenowia,* 15(2): 426 (1986): Astragalus calabricus.
subsp. cyllenea *(Fischer) W. Greuter in Willdenowia,* 15(2): 426 (1986): Astragalus cylleneus.
subsp. jankae *(Degen & Bornm.) W. Greuter in Willdenowia,* 15(2): 426 (1986): Astragalus jankae.
subsp. monachorum *(Širj.) W. Greuter in Willdenowia,* 15(2): 426 (1986): Astragalus monachorum.
subsp. parnassi *(Boiss.) W. Greuter in Willdenowia,* 15(2): 426 (1986): Astragalus parnassi.
yueksekovae *(V. Matthews) W. Greuter in Willdenowia,* 15(2): 427 (1986): Astragalus yueksekovae.
zachlensis
subsp. zebedaniensis *(Freyn & Bornm.) W. Greuter in Willdenowia,* 15(2): 427 (1986): Astragalus zebedaniensis.

ASTRAGALUS (Legumin.).

sect. Monadelphia *K.T. Fu in Acta Bot. Bor.-Occid. Sin.,* 1(2): 16 (1981).
sect. Parvistipula *K.T. Fu in Acta Bot. Bor.-Occid. Sin.,* 1(2): 15 (1981).
anserinus *N.D. Atwood, S. Goodrich & S.L. Welsh in Great Basin Nat.,* 44(2): 263 (1984)—U.S.A. (Utah, Nevada).
artemisiformis *M.R. Rasulova in Izv. Akad. Nauk Tadzh. SSR, Biol. Nauk,* 2(75): 96 (1979)—U.S.S.R. (Middle Asia).
asterias
subsp. aristidis *(Batt.) W. Greuter in Willdenowia,* 15(2): 427 (1986): A. cruciatus subsp. aristidis.
subsp. astraboides *(Pomel) W. Greuter in Willdenowia,* 15(2): 427 (1986): A. astraboides.

ASTRAGALUS :–

subsp. polyactinus *(Boiss.) W. Greuter in Willdenowia,* 15(2): 427 (1986): A. polyactinus.
subsp. radiatus *(Batt.) W. Greuter in Willdenowia,* 15(2): 427 (1986): A. cruciatus subsp. radiatus.
atraphaxifolius *M.R. Rasulova in Izv. Akad. Nauk Tadzh. SSR, Biol. Nauk,* 3(76): 108 (1979)—U.S.S.R. (Middle Asia).
atrifructus *W. Greuter & Burdet in Willdenowia,* 16(1): 108 (1986), nom. nov.: A. melanocarpus *Bunge.*
austrodarvasicus *M.R. Rasulova in Izv. Akad. Nauk Tadzh. SSR, Biol. Nauk,* 4(81): 26 (1980)—U.S.S.R. (Middle Asia).
bogensis *M.R. Rasulova in Izv. Akad. Nauk Tadzh. SSR, Biol. Nauk,* 1(78): 17 (1980)—U.S.S.R. (Middle Asia).
bor-bulakensis *M.R. Rasulova in Dokl. Akad. Nauk Tadzh. SSR,* 23(7): 402 (1980)—U.S.S.R. (Middle Asia).
camptodontus
var. lichiangensis *(Simps.) K.T. Fu in Acta Bot. Bor.-Occid. Sin.,* 6(1): 57 (1986): A. lichiangensis.
degilmonus *M.R. Rasulova in Izv. Akad. Nauk Tadzh. SSR, Biol. Nauk,* 4(81): 28 (1980)—U.S.S.R. (Middle Asia).
eremiticus
var. ampullarioides *S.L. Welsh in Great Basin Nat.,* 46(2): 262 (1986)—U.S.A. (Utah).
frigidus
var. grigorjewii *(B. Fedtsch.) B.A. Yurtsev in Arktich. Fl. SSSR,* 9(2): 32 (1986): A. grigorjewii.
gardanikaphtharicus *M.R. Rasulova in Izv. Akad. Nauk Tadzh. SSR, Biol. Nauk,* 1(78): 16 (1980)—U.S.S.R. (Middle Asia).
junussovii *M.R. Rasulova in Izv. Akad. Nauk Tadzh. SSR, Biol. Nauk,* 1(74): 86 (1979)—U.S.S.R. (Middle Asia).
karaculensis *P.N. Ovchinnikov & M. Rasulova in Izv. Akad. Nauk Tadzh. SSR, Biol. Nauk,* 4(65): 5 (1976)—U.S.S.R. (Middle Asia).
koikitaensis *M.R. Rasulova in Izv. Akad. Nauk Tadzh. SSR, Biol. Nauk,* 1(78): 93 (1980)—U.S.S.R. (Middle Asia).
kschtutensis *M.R. Rasulova in Izv. Akad. Nauk Tadzh. SSR, Biol. Nauk,* 4(77): 91 (1979)—U.S.S.R. (Middle Asia).
limnocharis
var. tabulaeus *S.L. Welsh in Great Basin Nat.,* 46(2): 261 (1986)—U.S.A. (Utah).
linearifolius *P.N. Ovchinnikov & M. Rasulova in Izv. Akad. Nauk Tadzh. SSR, Biol. Nauk,* 4(65): 4 (1976)—U.S.S.R. (Middle Asia).
longisepalus *M.R. Rasulova in Izv. Akad. Nauk Tadzh. SSR, Biol. Nauk,* 4(77): 90 (1979)—U.S.S.R. (Middle Asia).
marianus *Huber, Cat. Gén. 1871 & 1872:* 3 (1871); *cf. D.J. Mabberley in Taxon,* 34(3): 453 (1985)— U.S.A. (Colorado).

ASTRAGALUS :–
milingensis
var. heydeiodes *K.T. Fu in Acta Bot. Bor.-Occid. Sin.*, 5(1): 48 (1985)—China.
var. paucijugus *(K.T. Fu) K.T. Fu in Acta Bot. Bor.-Occid. Sin.*, 6(1): 58 (1986): A. prodigiosus var. paucijugus.
var. prodigiosus *(K.T. Fu) K.T. Fu in Acta Bot. Bor.-Occid. Sin.*, 6(1): 58 (1986): A. prodigiosus.
monodelphus
subsp. xitaibaicus *K.T. Fu in Acta Bot. Bor.-Occid. Sin.*, 1(2): 17 (1981)—China.
nitidissimus *W. Greuter & Burdet in Willdenowia*, 16(1): 108 (1986), nom. nov.: A. nitens *Boiss. & Heldr.*
pamirensis *P.N. Ovchinnikov & M. Rasulova in Izv. Akad. Nauk Tadzh. SSR, Biol. Nauk*, 4(65): 7 (1976)—U.S.S.R. (Middle Asia).
preussii
var. cutleri *R.C. Barneby in Great Basin Nat.*, 46(2): 256 (1986)—U.S.A. (Utah).
prodigiosus
var. paucijugus *K.T. Fu in Acta Bot. Bor.-Occid. Sin.*, 5(1): 48 (1985)—China.
pseudogombo *J. Fernández Casas in Fontqueria*, 2: 25 (1982)—Morocco.
ramitensis *M.R. Rasulova in Izv. Akad. Nauk Tadzh. SSR, Biol. Nauk*, 4(81): 27 (1980)—U.S.S.R. (Middle Asia).
sarygorensis *M.R. Rasulova in Izv. Akad. Nauk Tadzh. SSR, Biol. Nauk*, 2(75): 95 (1979)—U.S.S.R. (Middle Asia).
setiferus *P.N. Ovchinnikov & M. Rasulova in Izv. Akad. Nauk Tadzh. SSR, Biol. Nauk*, 4(65): 3 (1976)—U.S.S.R. (Middle Asia).
seydishehiricus *Kit Tan & H. Ocakverdi in Notes Roy. Bot. Gard. Edinburgh*, 44(1): 157 (1986)—Turkey.
subspongocarpus *P.N. Ovchinnikov & M.R. Rasulova in Dokl. Akad. Nauk Tadzh. SSR*, 23(6): 338 (1980)—U.S.S.R. (Middle Asia).
taipaishanensis *Y.C. Ho & S.B. Ho ex C.W. Chang in Fl. Tsinlingensis*, 1(3): 445, 60 (1981)—China.
tanguticus
var. albiflorus *S.W. Liu ex K.T. Fu in Acta Bot. Bor.-Occid. Sin.*, 6(1): 57 (1986)—China.
thracicus
subsp. cylleneus *(Boiss. & Heldr.) A. Strid in A. Strid (ed.), Mount. Fl. Greece*, 1: 467 (1986): A. cylleneus.
subsp. monachorum *(Širj.) A. Strid in A. Strid (ed.), Mount. Fl. Greece*, 1: 467 (1986): A. monachorum.
subsp. parnassi *(Boiss.) A. Strid in A. Strid (ed.), Mount. Fl. Greece*, 1: 466 (1986): A. parnassi.
tokachiensis *T. Yamazaki & Y. Kadota in J. Jap. Bot.*, 61(7): 199 (1986)—Japan.
tolmaczevii *B.A. Yurtsev in Arktich. Fl. SSSR*, 9(2): 178, 48 (1986)—U.S.S.R. (Siberia, Soviet Far East), Canada (N.W. Terr.).

ASTRAGALUS :–
zarokoensis *M.R. Rasulova in Izv. Akad. Nauk Tadzh. SSR, Biol. Nauk*, 4(81): 25 (1980)—U.S.S.R. (Middle Asia).

ATHANASIA (Compos.).
inopinata *(Hutch.) M. Källersjö in Nordic J. Bot.*, 5(6): 533 (1985 publ. 1986): Stilpnophytum inopinatum.
oocephala *(DC.) M. Källersjö in Nordic J. Bot.*, 5(6): 533 (1985 publ. 1986): Stilpnophyton oocephalum.

ATRACTYLIS (Compos.).
lancea *C.P. Thunberg ex A. Murray in Linnaeus, Syst. Veg., ed. 14*: 729 (1784)—Japan. †
ovata *C.P. Thunberg ex A. Murray in Linnaeus, Syst. Veg., ed. 14*: 730 (1784)—Japan. †

ATRAGENE (Ranunculac.).
japonica *C.P. Thunberg ex A. Murray in Linnaeus, Syst. Veg., ed. 14*: 511 (1784)—Japan. †
moisseenkoi *V.P. Serov in Bot. Zhurn.*, 71(8): 1128 (1986)—U.S.S.R. (Middle Asia).

ATRIPLEX (Chenopodiac.).
canescens
var. gigantea *S.L. Welsh & Stutz in Great Basin Nat.*, 44(2): 189 (1984)—U.S.A. (Utah).
var. occidentalis *(Torr. & Frem.) S.L. Welsh & Stutz in Great Basin Nat.*, 44(2): 188 (1984): Pterchiton occidentale.
gardneri
var. bonnevillensis *(C.A. Hanson) S.L. Welsh in Great Basin Nat.*, 44(2): 190 (1984): A. bonnevillensis.
var. cuneata *(A. Nels.) S.L. Welsh in Great Basin Nat.*, 44(2): 190 (1984): A. cuneata.
var. falcata *(Jones) S.L. Welsh in Great Basin Nat.*, 44(2): 191 (1984): A. nuttallii var. falcata.
var. welshii *(C.A. Hanson) S.L. Welsh in Great Basin Nat.*, 44(2): 191 (1984): A. welshii.
monoica *Weigel, Hort. Gryph.*: 36 (1782); *cf. D.J. Mabberley in Taxon*, 33(3): 441 (1984)— Locality not stated.
patula
var. triangularis *(Willd.) Thorne & S.L. Welsh in Great Basin Nat.*, 44(2): 193 (1984): A. triangularis.
saccaria
var. caput-medusae *(Eastw.) S.L. Welsh in Great Basin Nat.*, 44(2): 194 (1984): A. caput-medusae.

AUBRIETA (Crucif.).
scardica *(Wettst.) L.-Å. Gustavsson in A. Strid (ed.), Mount. Fl. Greece*, 1: 271 (1986): A. croatica var. scardica.

AULACOSPERMUM (Umbellif.).
coloratum *E.P. Korovin in Izv. Akad. Nauk Tadzh. SSR, Biol. Nauk,* 3(60): 3 (1975)—U.S.S.R. (Middle Asia).

AUSTRALOPYRUM (Gramin.).
velutinum *(Nees) B.K. Simon in Austrobaileya,* 2(3): 241 (1986): Agropyron velutinum.

AUSTROFESTUCA (Gramin.).
pubinervis *(Vickery) B.K. Simon in Austrobaileya,* 2(3): 241 (1986): Festuca pubinervis.

AUSTROSTEENISIA R. Geesink, Scala Millettiearum (Leiden Bot. Ser., 8): 78 (1984). *LEGUMINOSAE.*
blackii *(F. Muell.) R. Geesink, Scala Millettiearum (Leiden Bot. Ser., 8):* 78 (1984): Millettia blackii.
glabristyla *L.W. Jessup in Austrobaileya,* 2(3): 243 (1986)—Australia (Queensland, New South Wales).
stipularis *(C. White) L.W. Jessup in Austrobaileya,* 2(3): 243 (1986): Lonchocarpus stipularis.

AUSTROSYNOTIS C. Jeffrey in Kew Bull., 41(4): 878 (1986). *COMPOSITAE.*
rectirama *(Bak.) C. Jeffrey in Kew Bull.,* 41(4): 878 (1986): Senecio rectiramus.

AVENA (Gramin.).
agadiriana *B.R. Baum & G. Fedak in Canad. J. Bot.,* 63(8): 1379 (1985)—Morocco.
atlantica *B.R. Baum & G. Fedak in Canad. J. Bot.,* 63(6): 1057 (1985)—Morocco.
ovata *(Cav.) C.C. Gmel., Hort. Mag. Duc. Bad. Carlsruh.:* 34 (1811); *cf. D.J. Mabberley in Taxon,* 33(3): 440 (1984): Basionym not stated.

AZADIRACHTA A. Jussieu in Mirb. & Cass. apud Guill. in Bull. Sci. Nat. Geol., 23: 236 (1830); cf. D.J. Mabberley in Taxon, 31(1): 66 (1982). *MELIACEAE.*

B

BACOPA (Scrophulariac.).
pedersenii *R.A. Rossow in Parodiana,* 4(1): 176 (1986)—Argentina, Paraguay, Brazil.

BACTYRILOBIUM (Legumin.).
baccilare *(L.) Hornem., Hort. Hafn.:* 392 (1813); *cf. D.J. Mabberley in Taxon,* 33(3): 436 (1984): Basionym not stated.
javanica *(L.) Hornem., Hort. Hafn.:* 392 (1813); *cf. D.J. Mabberley in Taxon,* 33(3): 436 (1984): Cassia javanica.

BAECKEA (Myrtac.).
tuberculata *M.E. Trudgen in Nuytsia,* 5(3): 441 (1986)—Australia (S. Australia).

BALIOSPERMUM (Euphorbiac.).
calycinum
var. bracteatum *T. Chakrabarty & N.P. Balakrinshnan in J. Econ. Taxon. Bot.,* 7(2): 359 (1985 publ. 1986)—India.
densiflorum *D.G. Long in Notes Roy. Bot. Gard. Edinburgh,* 44(1): 171 (1986)—Bhutan, Burma.

BALOGHIA (Euphorbiac.).
inophylla *(Forst. f.) P.S. Green in Kew Bull.,* 41(4): 1026 (1986): Croton inophyllus.

BAMBUSA (Gramin.).
atrovirens *T.H. Wen in J. Bamboo Res.,* 5(2): 15 (1986)—China.
cacharensis *R.B. Majumder in Bull. Bot. Surv. India,* 25(1–4): 237 (1983 publ. 1985)—Assam.
pervariabilis
var. viridistriata *Q.H. Dai & X.C. Liu in Acta Phytotax. Sin.,* 24(5): 395 (1986), as 'viridi-striata'—Cultivation.

BANCROFTIA Porter in Gaceta de Venez., 370: 3 (25 Feb. 1838); cf. D.J. Mabberley in Taxon, 30(1): 11 (1981). *LABIATAE.*
decipiens *R.K. Porter in Gaceta de Venez.,* 30(1): 11 (1838); *cf. D.J. Mabberley in Taxon,* 30(1): 11 (1981)— ?Venezuela.

BANISTERIA (Malpighiac.).
volubilis *(Sims) Endl., Cat. Horti Vindob.,* 2: 375 (1842); *cf. D.J. Mabberley in Taxon,* 33(3): 434 (1984): Basionym not stated.

BANKSIA (Proteac.).
acicularis *(Vent.) Parmentier, Cat. Pl. Enghien:* 11 (1818); *cf. D.J. Mabberley in Taxon,* 33(3): 441 (1984): Basionym not stated.
pugioniformis *(Cav.) Parmentier, Cat. Pl. Enghien:* 12 (1818); *cf. D.J. Mabberley in Taxon,* 33(3): 441 (1984): Basionym not stated.
saligna *(Salisb.) Parmentier, Cat. Pl. Enghien:* 12 (1818); *cf. D.J. Mabberley in Taxon,* 33(3): 441 (1984): Basionym not stated.

BAPTISIA (Legumin.).
uniflora *(Michaux) Sm. in Rees, Cyclop.,* 39(2): Baptisia n. 2 (1819); *cf. D.J. Mabberley in Taxon,* 32(1): 81 (1983): Basionym not stated.

BARLERIA (Acanthac.).
hillcoatiae *J.R.I. Wood in Kew Bull.,* 41(1): 222 (1986)—Yemen.

BARNARDIA (Liliac.).
numidica *(Poiret) F. Speta in Linz. Biol. Beitr.,* 18(2): 402 (1986): Scilla numidica.

BASANANTHE (Passiflorac.).
merolae *F.M. Raimondo & G. Moggi in Delpinoa,* 23–24: 326 (1981–1982 publ. 1985)—Somali Republic.

BASHANIA (Gramin.).
auctiaurita *T.P. Yi in Bull. Bot. Res. North-East. Forest. Inst.,* 6(4): 27 (1986)—China.

BATOPEDINA (Rubiac.).
pulvinellata
subsp. glabrifolia *E. Robbrecht in Bull. Jard. Bot. Nation. Belg.,* 56(1–2): 148 (1986)—Zaïre.

BEADLEA (Orchidac.).
pringlei *(S. Wats.) L.A. Garay & W. Kittredge in Bot. Mus. Leafl. Harvard Univ.,* 30(3): (47) 181 (1985 publ. 1986): Spiranthes pringlei.

BEAUCARNEA (Dracaenac.).
glauca *(Lem.) Hereman, Paxt. Bot. Dict.:* 602 (1868); *cf. D.J. Mabberley in Taxon,* 32(1): 87 (1983): Basionym not stated.

BEGONIA (Begoniac.).
masoniana
var. maculata *S.K. Chen, R.X. Zheng & D.Y. Xia in Acta Bot. Yunnanica,* 8(2): 222 (1986)—China.

BELLEVALIA (Hyacinthac.).
zoharyi
var. pyricarpa *(Feinbr.) N. Feinbrun-Dothan, Fl. Palaestina,* 4: 67 (1986): B. zoharyi subsp. pyricarpa.

BENNETTIODENDRON (Flacourtiac.).
shangsiense *X.X. Chen & J.Y. Luo in Bull. Bot. Res. North-East. Forest. Inst.,* 6(1): 163 (1986)—China.

BENTLEYA E.M. Bennett in Nuytsia, 5(3): 401 (1986). *PITTOSPORACEAE.*
spinescens *E.M. Bennett in Nuytsia,* 5(3): 402 (1986)—Australia (W. Australia).

BERBERIS (Berberidac.).
ahrendtii *R.R. Rao & B.P. Uniyal in Indian J. Forest.,* 8(4): 334 (1985 publ. 1986), nom. nov.: B. lycioides *Stapf.*
brachystachys *T.S. Ying in Fl. Xizangica,* 2: 137 (1985)—Xizang.
campylotropa *T.S. Ying in Fl. Xizangica,* 2: 152 (1985)—Xizang.
chingii
subsp. subedentata *C.M. Hu in Bull. Bot. Res. North-East. Forest. Inst.,* 6(2): 9 (1986)—China.
subsp. wulingensis *C.M. Hu in Bull. Bot. Res. North-East. Forest. Inst.,* 6(2): 9 (1986)—China.
fujianensis *C.M. Hu in Bull. Bot. Res. North-East. Forest. Inst.,* 6(2): 5 (1986)—China.

BERBERIS :–
gilungensis *T.S. Ying in Fl. Xizangica,* 2: 134 (1985)—Xizang.
hypericifolia *T.S. Ying in Fl. Xizangica,* 2: 140 (1985)—Xizang.
jiangxiensis *C.M. Hu in Bull. Bot. Res. North-East. Forest. Inst.,* 6(2): 9 (1986)—China.
var. pulchella *C.M. Hu in Bull. Bot. Res. North-East. Forest. Inst.,* 6(2): 10 (1986)—China.
longispina *T.S. Ying in Fl. Xizangica,* 2: 148 (1985)—Xizang.
multicaulis *T.S. Ying in Fl. Xizangica,* 2: 147 (1985)—Xizang.
multiserrata *T.S. Ying in Fl. Xizangica,* 2: 139 (1985)—Xizang.
nullinervis *T.S. Ying in Fl. Xizangica,* 2: 141 (1985)—Xizang.
oblanceifolia *C.M. Hu in Bull. Bot. Res. North-East. Forest. Inst.,* 6(2): 12 (1986)—China.
obovatifolia *T.S. Ying in Fl. Xizangica,* 2: 146 (1985)—Xizang.
parapruinosa *T.S. Ying in Fl. Xizangica,* 2: 145 (1985)—Xizang.
photiniaefolia *C.M. Hu in Bull. Bot. Res. North-East. Forest. Inst.,* 6(2): 4 (1986)—China.
purangensis *T.S. Ying in Fl. Xizangica,* 2: 133 (1985), as 'pulangensis'—Xizang.
racemulosa *T.S. Ying in Fl. Xizangica,* 2: 129 (1985)—Xizang.
sabulicola *T.S. Ying in Fl. Xizangica,* 2: 133 (1985)—Xizang.
trichiata *T.S. Ying in Fl. Xizangica,* 2: 125 (1985)—Xizang.
umbratica *T.S. Ying in Fl. Xizangica,* 2: 135 (1985)—Xizang.
vinifera *T.S. Ying in Fl. Xizangica,* 2: 142 (1985)—Xizang.
wuyiensis *C.M. Hu in Bull. Bot. Res. North-East. Forest. Inst.,* 6(2): 7 (1986)—China.

BERINGERIA (Labiat.).
cinerea *(Desr.) Benth. ex G. Don f. in Loud., Hort. Brit.:* 483 (1830); *cf. D.J. Mabberley in Taxon,* 31(1): 69 (1982): Basionym not stated.

BERNARDIA (Euphorbiac.).
laurentii *R.A. Howard in Phytologia,* 61(1): 2 (1986)—Windward Is.

BESCHORNERIA (Agavac.).
calcicola *A. García-Mendoza in Herbertia,* 42: 28 (1986)—Mexico.

BETONICA (Labiat.).
rosea *Prien. ex Hornem., Hort. Hafn., Suppl.:* 64 (1819); *cf. D.J. Mabberley in Taxon,* 33(3): 436 (1984)— Locality not stated.

BETULA (Betulac.).
platyphylla
var. brunnea *J.X. Huang in Bull. Bot. Res. North-East. Forest. Inst.,* 6(2): 125 (1986)—China.

BIARUM (Arac.).
pyrami
var. serotinum *J. Koach & N. Feinbrun-Dothan in N. Feinbrun-Dothan, Fl. Palaestina,* 4: 398, 337 (1986)—Syria, Israel, Jordan.

BIDENS (Compos.).
alba
var. radiata *(C.H. Schultz-Bip.) R. Ballard in Amer. J. Bot.,* 73(10): 1463 (1986): B. pilosa var. radiata.
diversifolia *Schweigg., Enum. Pl. Hort. Bot. Regiom.:* 16 (1812); *cf. D.J. Mabberley in Taxon,* 33(3): 441 (1984)— Locality not stated.
ferulifolia *(Jacq.) Sweet, Hort. Brit., ed. 2:* 311 (1830); *cf. D.J. Mabberley in Taxon,* 30(1): 15 (1981): Basionym not stated.
odorata
var. calcicola *(Greenman) R. Ballard in Amer. J. Bot.,* 73(10): 1462 (1986): B. rosea var. calcicola.
var. chilpancingensis *R. Ballard in Amer. J. Bot.,* 73(10): 1463 (1986)—Mexico.
var. oaxacensis *R. Ballard in Amer. J. Bot.,* 73(10): 1462 (1986)—Mexico.

BIERMANNIA (Orchidac.).
jainiana *S.N. Hegde & A. Nageswara Rao in Bull. Bot. Surv. India,* 26(1–2): 97 (1984 publ. 1985)—India.

BIGNONIA (Bignoniac.).
grandiflora *C.P. Thunberg ex A. Murray in Linnaeus, Syst. Veg., ed. 14:* 564 (1784)—Japan. †
tomentosa *C.P. Thunberg ex A. Murray in Linnaeus, Syst. Veg., ed. 14:* 563 (1784)—Japan. †

BILLBERGIA (Bromeliac.).
amandae *W. Weber in Feddes Repert.,* 97(3–4): 115 (1986)—Brazil.
nobilis *Bull, Cat.,* 358: 1 (1901); *cf. D.J. Mabberley in Taxon,* 34(3): 456 (1985)—Locality not stated.

BISCUTELLA (Crucif.).
sempervirens
var. elvira *J.D. Olowokudejo in Willdenowia,* 16(1): 101 (1986)—Spain.

BITUMINARIA (Legumin.).
flaccida *(Náb.) W. Greuter in Willdenowia,* 16(1): 108 (1986): Psoralea flaccida.
morisiana *(Pignatti & Metlesics) W. Greuter in Willdenowia,* 16(1): 108 (1986): Psoralea morisiana.

BLADHIA (Myrsinac.).
crispa *C.P. Thunberg ex A. Murray in Linnaeus, Syst. Veg., ed. 14:* 237 (1784)—Japan. †

BLADHIA :–
villosa *C.P. Thunberg ex A. Murray in Linnaeus, Syst. Veg., ed. 14:* 237 (1784)—Japan. †

BLECHUM (Acanthac.).
angustifolium *(Sw.) Sm. in Rees, Cyclop.,* 39(2): Blechum n. 3 (1819); *cf. D.J. Mabberley in Taxon,* 32(1): 82 (1983): Basionym not stated.
erectum *Sm. in Rees, Cyclop.,* 39(2): Blechum n. 2 (1819); *cf. D.J. Mabberley in Taxon,* 32(1): 82 (1983)— Locality not stated.

BLEPHAROCALYX (Myrtac.).
eggersii *(Kiaerskou) L.R. Landrum in Syst. Bot.,* 11(1): 159 (1986): Marlieriopsis eggersii.

BOCAGEOPSIS (Annonac.).
pleiosperma *P.J.M. Maas in Proc. Kon. Nederl. Akad. Wetensch., C,* 89(3): 249 (1986)—Brazil.

BOERHAVIA (Nyctaginac.).
chinensis *(L.) Rottb., Pl. Hort. Univ. Rar. Prog.:* 4 (1773); *cf. D.J. Mabberley in Taxon,* 33(3): 439 (1984): Basionym not stated.

BOESENBERGIA (Zingiberac.).
albomaculata *S.Q. Tong in Acta Phytotax. Sin.,* 24(4): 323 (1986), as 'albo-maculata'—China.

BOISDUVALIA (Onagrac.).
densiflora *(Lindl.) Bartl., Ind. Sem. Hort. Acad. Gott.,* 1837: 2 (1837); *cf. D.J. Mabberley in Taxon,* 32(1): 84 (1983): Basionym not stated.

BOLBOSCHOENUS (Cyperac.).
maritimus
subsp. macrostachys *(Willd.) J. Soják in Čas. Nár. Muz. (Prague),* 152(1): 19 (1983): Scirpus macrostachys.
subsp. tuberosus *(Desf.) J. Soják in Čas. Nár. Muz. (Prague),* 152(1): 19 (1983): Scirpus tuberosus.

BOLDOA (Monimiac.).
fragrans *(Pers.) Endl., Cat. Horti Vindob.,* 1: 280 (1842); *cf. D.J. Mabberley in Taxon,* 33(3): 434 (1984): Basionym not stated.

BONAPARTEA (Bromeliac.).
elegans *Fée, Cat. Méth. Pl. Jard. Bot. Strasb.:* 15 (1836); *cf. D.J. Mabberley in Taxon,* 32(1): 84 (1983)— Locality not stated.

BONAVERIA (Legumin.).
securigera *(L.) Endl., Cat. Horti Vindob.,* 2: 506 (1842); *cf. D.J. Mabberley in Taxon,* 33(3): 434 (1984): Coronilla securigera.

BORRERIA (Rubiac.).
loretiana *E.L. Cabral in Parodiana,* 4(1): 134 (1986)—Argentina.

BORRERIA :–
quadrifaria *E.L. Cabral in Parodiana,* 4(1): 138 (1986)—Argentina, Paraguay, Brazil.
shandongensis *F.Z. Li & X.D. Chen in Acta Bot. Yunnanica,* 7(4): 419 (1985)—China.
sulcata *(N.M. Bacigal.) E.L. Cabral in Parodiana,* 4(1): 140 (1986): B. verticillata var. sulcata.

BOTHRIOCHLOA (Gramin.).
gracilis *W.Z. Fang in Bull. Bot. Res. North-East. Forest. Inst.,* 6(1): 100 (1986)—China.
nana *W.Z. Fang in Bull. Bot. Res. North-East. Forest. Inst.,* 6(1): 102 (1986)—China.
tuberculata *W.Z. Fang in Bull. Bot. Res. North-East. Forest. Inst.,* 6(1): 97 (1986)—China.
yunnanensis *W.Z. Fang in Bull. Bot. Res. North-East. Forest. Inst.,* 6(1): 99 (1986)—China.

BOTRYODENDRUM (Araliac.).
capitatum *(Jacq.) Endl., Cat. Horti Vindob.,* 2: 177 (1842); *cf. D.J. Mabberley in Taxon,* 33(3): 434 (1984): Basionym not stated.

BOURRERIA (Boraginac.).
rubra *E.J. Lott & J.S. Miller in Ann. Missouri Bot. Gard.,* 73(1): 216 (1986)—Mexico.

BOUTELOUA (Gramin.).
quiriegoensis *A.A. Beetle in Phytologia,* 59(4): 287 (1986)—Mexico.

BOUVARDIA (Rubiac.).
hintoniorum *B.L. Turner in Brittonia,* 38(2): 111 (1986)—Mexico.

BRACHIONIDIUM (Orchidac.).
brachycladum *C.A. Luer & R. Escobar R. in Orquideologia,* 16(3): 33 (1986)—Colombia.
dalstroemii *C.A. Luer in Lindleyana,* 1(3): 170 (1986)—Ecuador.
elegans *C.A. Luer & A.C. Hirtz in Orchidee,* 37(1). 23 (1986)—Ecuador.
escobarii *C.A. Luer in Lindleyana,* 1(3): 172 (1986)—Colombia.
hirtzii *C.A. Luer in Lindleyana,* 1(3): 174 (1986)—Ecuador.
juliani *G. Carnevali F.C. & I. Ramírez in Ernstia,* 39: 6 (1986)—Venezuela.
neblinense *G. Carnevali F.C. & I. Ramírez in Ernstia,* 39: 9 (1986)—Venezuela.
portillae *C.A. Luer in Lindleyana,* 1(3): 176 (1986)—Ecuador.
rugosum *C.A. Luer & Hirtz in Lindleyana,* 1(3): 178 (1986)—Ecuador.

BRACHYPODIUM (Gramin.).
platystachyum *Huber, Cat. Graines 1868:* 8 (1868); *cf. D.J. Mabberley in Taxon,* 34(3): 450 (1985)— Locality not stated.

BRACHYSTELE (Orchidac.).
chiangii *(Johnston) P. Burns-Balogh in Orquídea (México),* 10(1): 92 (1986): Spiranthes chiangii.

BRACHYSTELE :–
minutiflora *(A. Rich. & Gal.) P. Burns-Balogh in Orquídea (México),* 10(1): 92 (1986): Spiranthes minutiflora.
sarcoglossa *(A. Rich. & Gal.) P. Burns-Balogh in Orquídea (México),* 10(1): 92 (1986): Spiranthes sarcoglossa.
tenuissima *(L.O. Wms.) P. Burns-Balogh in Orquídea (México),* 10(1): 92 (1986): Spiranthes tenuissima.

BRACHYSTELMA (Asclepiadac.).
ciliatum *G.D. Arekal & T.M. Ramakrishna in Curr. Sci.,* 50(3): 145 (1981)—India.

BRANDELLA R.R. Mill in Notes Roy. Bot. Gard. Edinburgh, 43(3): 478 (1986). *BORAGINACEAE.*
erythraea *(Brand) R.R. Mill in Notes Roy. Bot. Gard. Edinburgh,* 43(3): 478 (1986): Adelocaryum erythraeum.

BRASILIOPUNTIA (Cactac.).
subacarpa *C.T. Rizzini & A. de Mattos-Filho in Rev. Brasil. Biol.,* 46(2): 324 (1986)—Brazil.

BRASSICA (Crucif.).
erucoides *Hornem., Hort. Hafn.:* 621 (1815); *cf. D.J. Mabberley in Taxon,* 33(3): 436 (1984)— Locality not stated.
melanosinapis *Endl., Cat. Horti Vindob.,* 2: 259 (1842); *cf. D.J. Mabberley in Taxon,* 33(3): 434 (1984)— Locality not stated.
mesopotamica *(Spreng.) Bernh., Sel. Sem. Hort. Erfurt.,* 1835; *Linnaea 11, Lit.:* 85 (1837); *cf. D.J. Mabberley in Taxon,* 32(1): 84 (1983): Basionym not stated.
praecox *Kit. ex Hornem., Hort. Hafn.:* 621 (1815); *cf. D.J. Mabberley in Taxon,* 33(3): 436 (1984)— Locality not stated.
rapa
var. sarson *(Prain) B.N. Mehrotra & B.S. Aswal in J. Econ. Taxon. Bot.,* 6(3): 587 (1985), without basionym ref.: B. campestris var. sarson.

BRASSICELLA (Crucif.).
setigera *(J. Gay ex Lange) Font Quer & Rothmaler, Sched. Fl. Iber. Select., Cent. II-III,* n. 129 (1935); *cf. C. Navarro, M. Gutiérrez-Bustillo & A.G. Bueno in Fontqueria,* 7: 11 (1985): Sinapis setigera.

BRIDELIA (Euphorbiac.).
nayarii *P. Basu in J. Econ. Taxon. Bot.,* 7(3): 634 (1985 publ. 1986)—Nicobar Is.

BROCCHINIA (Bromeliac.).
gilmartiniae *G.S. Varadarajan in J. Bromeliad Soc.,* 36(6): 251 (1986), as 'gilmartinii'—Venezuela.

BROMELIA (Bromeliac.).
gurkeniana
var. funchiana *E. Pereira & E.M.C. Leme in Rev. Brasil. Biol.,* 45(4): 635 (1985 publ. 1986)—Brazil.

BROMUS (Gramin.).
bifidus *C.P. Thunberg ex A. Murray in Linnaeus, Syst. Veg., ed. 14:* 119 (1784)—Japan. †
catharticus
var. striatus *(Hitchcock) P. Pinto-Escobar in Caldasia,* 15(71–75): 26 (1986): B. striatus.
japonicus *C.P. Thunberg ex A. Murray in Linnaeus, Syst. Veg., ed. 14:* 119 (1784)—Japan. †

BROWNETERA L.C. Rich. ex Tratt., Gen. Nov. Pl.: 19 (1825); cf. D.J. Mabberley in Taxon, 31(1): 69 (1982). *PODOCARPACEAE.*
aspleniifolia *(Lab.) Tratt., Gen. Nov. Pl.:* 20 & t. 14 (1825); *cf. D.J. Mabberley in Taxon,* 31(1): 69 (1982): Basionym not stated.

BRŬGMANSIA (Solanac.).
ceratocaula *(Ort.) Court ex Gaede, Ind. Pl. Hort. Bot. Leod.:* 29 (1828); *cf. D.J. Mabberley in Taxon,* 32(1): 84 (1983): Datura ceratocaula.
knightii *Merkus-Doornik, Hort. Spaarn-Berg.:* 81 (1849); *cf. D.J. Mabberley in Taxon,* 33(3): 441 (1984)— Locality not stated.

BRUNELLIA (Brunelliac.).
amayensis *C.I. Orozco in Caldasia,* 15(71–75): 177 (1986)—Colombia.

BRUNONIELLA (Acanthac.).
acaulis
subsp. ciliata *R.M. Barker in J. Adelaide Bot. Gard.,* 9: 103 (1986)—Australia (Queensland).
linearifolia *R.M. Barker in J. Adelaide Bot. Gard.,* 9: 106 (1986)—Australia (N. Terr., ?Queensland).

BRYONIA (Cucurbitac.).
japonica *C.P. Thunberg ex A. Murray in Linnaeus, Syst. Veg., ed. 14:* 870 (1784)—Japan. †

BRYOPHYLLUM (Crassulac.).
fedtschenkoi *(Ham. & Perr.) C.Y. Cheng in Fl. Beijing,* 1: 321 (1984), without basionym ref.: Kalanchoe fedtschenkoi.

BUBON (Umbellif.).
buctornense *Hort. Gorenk. ex Hornem., Hort. Hafn.:* 283 (1813); *cf. D.J. Mabberley in Taxon,* 33(3): 436 (1984)— Locality not stated.

BUBONIUM (Compos.).
daltonii *(Webb) T. Halvorsen in Sommerfeltia,* 3: 80 (1986): Odontospermum daltoni.
subsp. vogelii *(Webb) T. Halvorsen in Sommerfeltia,* 3: · 82 (1986): Odontospermum vogelii.

BUBONIUM :-
graveolens
subsp. odorum *(Schousb.) Wikl. ex T. Halvorsen & L. Borgen in Sommerfeltia,* 3: 72 (1986): Buphthalmum odorum.
subsp. stenophyllum *(Link) T. Halvorsen in Sommerfeltia,* 3: 74 (1986): Buphthalmum stenophyllum.
intermedium *(DC.) T. Halvorsen & Wikl. in Sommerfeltia,* 3: 66 (1986): Asteriscus sericeus var. intermedius.
sericeum *(L.f.) T. Halvorsen & Wikl. in Sommerfeltia,* 3: 68 (1986): Buphthalmum sericeum.
smithii *(Webb) T. Halvorsen in Sommerfeltia,* 3: 77 (1986): Odontospermum smithii.

BUCHNERA (Scrophulariac.).
jacoborum *J.L. Fernández-Alonso in Caldasia,* 15(71–75): 150 (1986)—Colombia.

BUDDLEJA (Buddlejac.).
latiflora *S.Y. Pao in Fl. Yunnanica,* 3: 473 (1983)—China.
sessilifolia *B.S. Sun in Fl. Yunnanica,* 3: 465 (1983)—China.

BUGLOSSOIDES (Boraginac.).
permixta *(Jord.) J. Holub in Preslia,* 58(4): 301 (1986): Lithospermum permixtum.

BULBINE (Asphodelac.).
hispida *(L.) Hornem., Hort. Hafn.:* 334 (1813); *cf. D.J. Mabberley in Taxon,* 33(3): 436 (1984): Basionym not stated.

BULBOCODIUM (Colchicac.).
caucasicum *(Bieb.) Endl., Cat. Horti Vindob.,* 1: 102 (1842); *cf. D.J. Mabberley in Taxon,* 33(3): 434 (1984): Basionym not stated.

BULBOPHYLLUM (Orchidac.).
ambrosia
subsp. nepalensis *J.J. Wood in Kew Bull.,* 41(4): 820 (1986)—Nepal.
masonii *(Senghas) J.J. Wood in Kew Bull.,* 41(4): 821 (1986): Cirrhopetalum masonii.
panigrahianum *S. Misra in Nordic J. Bot.,* 6(1): 25 (1986)—India.
reptans
var. acuta *C.L. Malhotra & B. Balodi in Bull. Bot. Surv. India,* 26(1–2): 110 (1984 publ. 1985)—India.

BULBOSTYLIS (Cyperac.).
oritrephes
subsp. australis *B.L. Burtt in Notes Roy. Bot. Gard. Edinburgh,* 43(3): 353 (1986)—South Africa (Natal, Transvaal).

BUMELIA (Sapotac.).
ekmaniana *(Urban) J. Bisse & J. Gutiérrez Amaro in Rev. Jard. Bot. Nacion. Univ. Habana,* 6(1): 22 (1985): Dipholis ekmaniana.
gymnanthifolia *J. Bisse & J. Gutiérrez Amaro in Rev. Jard. Bot. Nacion. Univ. Habana,* 6(1): 21 (1985)—Cuba.

BUMELIA :–
moaensis *J. Bisse & J. Gutiérrez Amaro in Rev. Jard. Bot. Nacion. Univ. Habana,* 6(1): 20 (1985)—Cuba.
neglecta *J. Bisse & J. Gutiérrez Amaro in Rev. Jard. Bot. Nacion. Univ. Habana,* 6(1): 21 (1985)—Cuba.

BUNIUM (Umbellif.).
tenuisectum *E.P. Korovin in Izv. Akad. Nauk Tadzh. SSR, Biol. Nauk,* 3(60): 5 (1975)—U.S.S.R. (Middle Asia).

BUPLEURUM (Umbellif.).
contractum *E.P. Korovin in Izv. Akad. Nauk Tadzh. SSR, Biol. Nauk,* 3(60): 4 (1975)—U.S.S.R. (Middle Asia).
luxiense *Y. Li & S.L. Pan in Acta Phytotax. Sin.,* 24(2): 150 (1986)—China.

BURMANNIA (Burmanniac.).
compacta *P.J.M. Maas & H. Maas-van de Kamer in Fl. Neotrop.,* 42: 53 (1986)—Venezuela.

BURMEISTERA (Campanulac.).
multipinnatisecta *G. Lozano C. & G. Galeano G. in Caldasia,* 15(71–75): 53 (1986)—Colombia.
pinnatisecta *J.L. Luteyn in Syst. Bot.,* 11(3): 474 (1986)—Colombia.

BYTTNERIA (Sterculiac.).
schunkei *C.L. Cristobal in Bol. Soc. Argent. Bot.,* 24(1–2): 125 (1985)—Peru.

C

CABRALEA A. Jussieu in Mirb. & Cass. apud Guill. in Bull. Sci. Nat. Geol., 23: 237 (1830); cf. D.J. Mabberley in Taxon, 31(1): 66 (1982). *MELIACEAE.*
canjerana
subsp. selloi *(C.DC.) H. de Souza Barreiros in Rev. Brasil. Biol.,* 46(1): 24 (1986): C. selloi.

CACALIA (Compos.).
longispica *Z.Y. Zhang & Y.H. Guo in Fl. Tsinlingensis,* 1(5): 422, 298 (1985)—China.
pinnata *Adams ex G.F. Hoffm., Hort. Mosq.:* 7 (1808); *cf. D.J. Mabberley in Taxon,* 31(1): 69 (1982)— Locality not stated.

CACTUS (Cactac.).
kagenickii *C.C. Gmel., Hort. Mag. Duc. Bad. Carlsruh.:* 48 (1811); *cf. D.J. Mabberley in Taxon,* 33(3): 440 (1984)— Locality not stated.

CAESALPINIA (Legumin.).
sepiaria
var. pubescens *Tang & Wang ex C.W. Chang in Fl. Tsinlingensis,* 1(3): 444, 8 (1981)—China.

CAJANUS (Legumin.).
sect. Fruticosa *L.J.G. van der Maesen in Agric. Univ. Wageningen Pap.,* 85(4): 47 (1985 publ. 1986).
sect. Volubilis *L.J.G. van der Maesen in Agric. Univ. Wageningen Pap.,* 85(4): 47 (1985 publ. 1986).
acutifolius *(F. v. Muell.) L.J.G. van der Maesen in Agric. Univ. Wageningen Pap.,* 85(4): 52 (1985 publ. 1986): Atylosia acutifolia.
albicans *(W. & A.) L.J.G. van der Maesen in Agric. Univ. Wageningen Pap.,* 85(4): 55 (1985 publ. 1986): Cantharospermum albicans.
aromaticus *L.J.G. van der Maesen in Agric. Univ. Wageningen Pap.,* 85(4): 61 (1985 publ. 1986)—Australia (N. Terr.).
cajanifolius *(Haines) L.J.G. van der Maesen in Agric. Univ. Wageningen Pap.,* 85(4): 91 (1985 publ. 1986): Atylosia cajanifolia.
crassicaulis *L.J.G. van der Maesen in Agric. Univ. Wageningen Pap.,* 85(4): 103 (1985 publ. 1986)—Australia (N. Terr., W. Australia).
crassus *(Prain ex King) L.J.G. van der Maesen in Agric. Univ. Wageningen Pap.,* 85(4): 105, 110 (1985 publ. 1986): Atylosia crassa.
var. burmanicus *(Collett & Hemsley) L.J.G. van der Maesen in Agric. Univ. Wageningen Pap.,* 85(4): 109 (1985 publ. 1986): Atylosia burmanica.
elongatus *(Benth.) L.J.G. van der Maesen in Agric. Univ. Wageningen Pap.,* 85(4): 115 (1985 publ. 1986): Atylosia elongata.
grandiflorus *(Benth. ex Bak.) L.J.G. van der Maesen in Agric. Univ. Wageningen Pap.,* 85(4): 125 (1985 publ. 1986): Atylosia grandiflora.
heynei *(W. & A.) L.J.G. van der Maesen in Agric. Univ. Wageningen Pap.,* 85(4): 129 (1985 publ. 1986): Dunbaria heynei.
lanceolatus *(W.V. Fitzg.) L.J.G. van der Maesen in Agric. Univ. Wageningen Pap.,* 85(4): 135 (1985 publ. 1986): Atylosia lanceolata.
lanuginosus *L.J.G. van der Maesen in Agric. Univ. Wageningen Pap.,* 85(4): 137 (1985 publ. 1986)—Australia (Queensland).
latisepalus *(Reynolds & Pedley) L.J.G. van der Maesen in Agric. Univ. Wageningen Pap.,* 85(4): 139 (1985 publ. 1986): Atylosia latisepala.
lineatus *(W. & A.) L.J.G. van der Maesen in Agric. Univ. Wageningen Pap.,* 85(4): 143 (1985 publ. 1986): Atylosia lineata.

CAJANUS :–
mareebensis *(Reynolds & Pedley) L.J.G. van der Maesen in Agric. Univ. Wageningen Pap.*, 85(4): 149 (1985 publ. 1986): Atylosia mareebensis.
mollis *(Benth.) L.J.G. van der Maesen in Agric. Univ. Wageningen Pap.*, 85(4): 154 (1985 publ. 1986): Atylosia mollis.
niveus *(Benth.) L.J.G. van der Maesen in Agric. Univ. Wageningen Pap.*, 85(4): 157 (1985 publ. 1986): Atylosia nivea.
platycarpus *(Benth.) L.J.G. van der Maesen in Agric. Univ. Wageningen Pap.*, 85(4): 160 (1985 publ. 1986): Atylosia platycarpa.
pubescens *(Ewart & Morrison) L.J.G. van der Maesen in Agric. Univ. Wageningen Pap.*, 85(4): 164 (1985 publ. 1986): Tephrosia pubescens.
reticulatus
var. grandifolius *(F. v. Muell.) L.J.G. van der Maesen in Agric. Univ. Wageningen Pap.*, 85(4): 171 (1985 publ. 1986): C. grandifolius.
var. maritimus *(Reynolds & Pedley) L.J.G. van der Maesen in Agric. Univ. Wageningen Pap.*, 85(4): 175 (1985 publ. 1986): Atylosia reticulata subsp. maritima.
rugosus *(W. & A.) L.J.G. van der Maesen in Agric. Univ. Wageningen Pap.*, 85(4): 179 (1985 publ. 1986): Atylosia rugosa.
scarabaeoides
var. pedunculatus *(Reynolds & Pedley) L.J.G. van der Maesen in Agric. Univ. Wageningen Pap.*, 85(4): 188 (1985 publ. 1986): Atylosia scarabaeoides var. pedunculata.
sericeus *(Benth. ex Bak.) L.J.G. van der Maesen in Agric. Univ. Wageningen Pap.*, 85(4): 195 (1985 publ. 1986): Atylosia sericea.
trinervius *(DC.) L.J.G. van der Maesen in Agric. Univ. Wageningen Pap.*, 85(4): 199 (1985 publ. 1986): Collaea trinervia.
villosus *(Benth. ex Baker) L.J.G. van der Maesen in Agric. Univ. Wageningen Pap.*, 85(4): 205 (1985 publ. 1986): Atylosia villosa.
viscidus *L.J.G. van der Maesen in Agric. Univ. Wageningen Pap.*, 85(4): 207 (1985 publ. 1986)—Australia (W. Australia).

CAKILE (Crucif.).
maritima
subsp. integrifolia *(Hornem.) Hyl. ex W. Greuter & H.M. Burdet in W. Greuter, H.M. Burdet & G. Long (eds.), Med-checklist*, 3: 74 (1986): C. maritima var. integrifolia.

CALADENIA (Orchidac.).
calcicola *G.W. Carr in Muelleria*, 6(3–4): 185 (1986)—Australia (Victoria).

CALADIUM (Arac.).
aturense *G.S. Bunting in Phytologia*, 60(5): 298 (1986)—Venezuela.

CALADIUM :–
lindenii
var. sylvestre *M.H. Grayum in Ann. Missouri Bot. Gard.*, 73(2): 463 (1986)—Panama, Colombia.
steyermarkii *G.S. Bunting in Phytologia*, 60(5): 300 (1986)—Venezuela.

CALAMUS (Palmae).
sect. Aulacospermus *C.F. Wei in Guihaia*, 6(1–2): 34 (1986).
sect. Brevicirrhus *C.F. Wei in Guihaia*, 6(1–2): 35 (1986), without type.
sect. Heterostachyus *C.F. Wei in Guihaia*, 6(1–2): 32 (1986).
subgen. Protocalamus *C.F. Wei in Guihaia*, 6(1–2): 23 (1986).
subgen. Rhachicirrus *C.F. Wei in Guihaia*, 6(1–2): 35 (1986).
dianbaiensis *C.F. Wei in Guihaia*, 6(1–2): 24 (1986)—China.
guangxiensis *C.F. Wei in Guihaia*, 6(1–2): 28 (1986)—China.
henryanus
var. castaneolepis *C.F. Wei in Guihaia*, 6(1–2): 32 (1986)—China.
maximus *Reinw. ex De Vriese, Hort. Spaarn-Berg.*: 35 (1839); *cf. D.J. Mabberley in Taxon*, 33(3): 441 (1984)— Locality not stated.
simplicifolius *C.F. Wei in Guihaia*, 6(1–2): 36 (1986)—China.
thysanolepis
var. polylepis *C.F. Wei in Guihaia*, 6(1–2): 24 (1986)—China.
tonkinensis
var. brevispicatus *C.F. Wei in Guihaia*, 6(1–2): 31 (1986)—China.
yuangchunensis *C.F. Wei in Guihaia*, 6(1–2): 26 (1986)—China.

CALANDRINIA (Portulacac.).
sect. Pachypodae *W.A. Kelley & J.R. Swanson in Phytologia*, 60(3): 172 (1986).
pachypoda
subsp. eyerdamii *W.A. Kelley & J.R. Swanson in Phytologia*, 60(3): 173 (1986)—Bolivia.

CALANTHE (Orchidac.).
baliensis *J.J. Wood & J.B. Comber in Kew Bull.*, 41(3): 698 (1986)—Lesser Sunda Is.

CALATHEA (Marantac.).
anderssonii *H. Kennedy in Canad. J. Bot.*, 63(6): 1145 (1985)—Ecuador.
attenuata *H. Kennedy in Nordic J. Bot.*, 6(2): 146 (1986)—Colombia, Ecuador, Brazil, Peru.
bantae *H. Kennedy in Canad. J. Bot.*, 64(7): 1321 (1986)—Ecuador, Colombia.
clivorum *H. Kennedy in Canad. J. Bot.*, 63(6): 1147 (1985)—Ecuador.
dilabens *L. Andersson & H. Kennedy in Nordic J. Bot.*, 6(4): 450 (1986)—French Guiana.

CALATHEA :-
erecta *L. Andersson & H. Kennedy in Nordic J. Bot.*, 6(4): 448 (1986)—French Guiana, Brazil.
granvillei *L. Andersson & H. Kennedy in Nordic J. Bot.*, 6(4): 451 (1986)—French Guiana, Brazil.
lagoagriana *H. Kennedy in Nordic J. Bot.*, 6(2): 148 (1986)—Ecuador, Colombia.
lanicaulis *H. Kennedy in Canad. J. Bot.*, 63(6): 1143 (1985)—Ecuador.
majestica *(Linden) H. Kennedy in Canad. J. Bot.*, 64(7): 1325 (1986): Maranta majestica.
multicincta *H. Kennedy in Canad. J. Bot.*, 64(7): 1323 (1986)—Ecuador.
pallidicosta *H. Kennedy in Nordic J. Bot.*, 6(2): 143 (1986)—Ecuador.
roseobracteata *H. Kennedy in Nordic J. Bot.*, 6(4): 459 (1986)—Ecuador.
squarrosa *L. Andersson & H. Kennedy in Nordic J. Bot.*, 6(4): 454 (1986)—French Guiana.
tinalandia *H. Kennedy in Canad. J. Bot.*, 63(6): 1141 (1985)—Ecuador.
utilis *H. Kennedy in Nordic J. Bot.*, 6(4): 457 (1986)—Ecuador.

CALCEOLARIA (Scrophulariac.).
biflora
f. falklandica *(Sp. Moore) W. Bowden in J. Ontario Rock Gard. Soc.*, June: 3 (1986): Fagelia falklandica.
lanceolata
f. acutifolia *(Witasek) W. Bowden in J. Ontario Rock Gard. Soc.*, June: 3 (1986): C. acutifolia.
f. polyrhiza *(Cav.) W. Bowden in J. Ontario Rock Gard. Soc.*, June: 3 (1986): C. polyrhiza.

CALICOTOME (Legumin.).
infesta
subsp. intermedia *(C. Presl) W. Greuter in Willdenowia*, 15(2): 428 (1986): C. intermedia.

CALLAEUM (Malpighiac.).
antifebrile *(Griseb.) D.M. Johnson in Syst. Bot.*, 11(2): 349 (1986): Banisteria antifebrilis.
chiapense *(Lundell) D.M. Johnson in Syst. Bot.*, 11(2): 346 (1986): Stigmaphyllon chiapense.
clavipetalum *D.M. Johnson in Syst. Bot.*, 11(2): 347 (1986)—Mexico.
coactum *D.M. Johnson in Syst. Bot.*, 11(2): 342 (1986)—Mexico.
macropterum *(DC.) D.M. Johnson in Syst. Bot.*, 11(2): 340 (1986): Hiraea macroptera.
malpighioides *(Turcz.) D.M. Johnson in Syst. Bot.*, 11(2): 344 (1986): Stigmaphyllon malpighioides.
psilophyllum *(Adr. Juss.) D.M. Johnson in Syst. Bot.*, 11(2): 351 (1986): Hiraea psilophylla.
reticulatum *D.M. Johnson in Syst. Bot.*, 11(2): 348 (1986)—Peru, Ecuador.

CALLAEUM :-
septentrionale *(Adr. Juss.) D.M. Johnson in Syst. Bot.*, 11(2): 343 (1986): Hiraea septentrionalis.

CALLERYA (Legumin.).
nitida *(Benth.) R. Geesink, Scala Millettiearum (Leiden Bot. Ser., 8):* 83 (1984): Millettia nitida.

CALLIANDRA (Legumin.).
slaneae *R.A. Howard in Phytologia*, 61(1): 3 (1986)—Windward Is.

CALLIANTHEMUM (Ranunculac.).
anemonoides *(Zahlbr.) Endl., Cat. Horti Vindob.*, 2: 211 (1842); *cf. D.J. Mabberley in Taxon*, 33(3): 434 (1984): Basionym not stated.

CALLICARPA (Verbenac.).
guizhouensis *H.T. Chang & Z.R. Xu in Acta Sci. Nat. Univ. Sunyatseni*, 1986(2): 99 (1986), without type—China.
japonica *C.P. Thunberg ex A. Murray in Linnaeus, Syst. Veg., ed. 14:* 153 (1784)—Japan. †
lanceolaria *Roxb. ex Hornem., Hort. Hafn., Suppl.:* 121 (1819); *cf. D.J. Mabberley in Taxon*, 33(3): 436 (1984)— Locality not stated.

CALLIOPSIS (Compos.).
diversifolia *Huber, Cat. 1866:* 4 (1866); *cf. D.J. Mabberley in Taxon*, 34(3): 450 (1985)—Locality not stated.

CALLISIA (Commelinac.).
sect. Brachyphylla *D.R. Hunt in Kew Bull.*, 41(2): 409 (1986).
sect. Cuthbertia *(Small) D.R. Hunt in Kew Bull.*, 41(2): 409 (1986): gen. Cuthbertia.
sect. Hadrodemas *(H.E. Moore) D.R. Hunt in Kew Bull.*, 41(2): 409 (1986): gen. Hadrodemas.
sect. Lauia *D.R. Hunt in Kew Bull.*, 41(2): 409 (1986).
filiformis *(Martens & Galeotti) D.R. Hunt in Kew Bull.*, 41(2): 410 (1986): Tradescantia filiformis.
gentlei
var. elegans *(Alexander ex H.E. Moore) D.R. Hunt in Kew Bull.*, 41(2): 411 (1986): C. elegans.
var. macdougallii *(Miranda) D.R. Hunt in Kew Bull.*, 41(2): 411 (1986): C. macdougallii.
rosea *(Vent.) D.R. Hunt in Kew Bull.*, 41(2): 409 (1986): Tradescantia rosea.

CALLISTEMON (Myrtac.).
pauciflorus *R.D. Spencer & P.F. Lumley in Muelleria*, 6(3–4): 295 (1986)—Australia (N. Terr.).
pearsonii *R.D. Spencer & P.F. Lumley in Muelleria*, 6(3–4): 293 (1986)—Australia (Queensland).

CALLITRIS (Cupressac.).
glaucophylla *J. Thompson & L.A.S. Johnson in Telopea,* 2(6): 731 (1986)—Australia (New South Wales, N. Terr., Queensland, S. Australia, Queensland, W. Australia).

CALOPHANOIDES (Acanthac.).
hygrophiloides *(F. Muell.) R.M. Barker in J. Adelaide Bot. Gard.,* 9: 237 (1986): Justicia hygrophiloides.

CALOPSIS (Restionac.).
adpressa *E.E. Esterhuysen in Bothalia,* 15(3–4): 465 (1985)—South Africa (Cape Province).
andreaeana *(Pillans) H.P. Linder in Bothalia,* 15(3–4): 465 (1985): Leptocarpus andreaeanus.
aspera *(Mast.) H.P. Linder in Bothalia,* 15(3–4): 465 (1985): Hypolaena aspera.
burchellii *(Mast.) H.P. Linder in Bothalia,* 15(3–4): 465 (1985): Leptocarpus burchellii.
clandestina *E.E. Esterhuysen in Bothalia,* 15(3–4): 465 (1985)—South Africa (Cape Province).
dura *E.E. Esterhuysen in Bothalia,* 15(3–4): 466 (1985)—South Africa (Cape Province).
esterhuyseniae *(Pillans) H.P. Linder in Bothalia,* 15(3–4): 466 (1985): Leptocarpus esterhuyseniae.
filiformis *(Mast.) H.P. Linder in Bothalia,* 15(3–4): 467 (1985): Hypolaena filiformis.
fruticosa *(Mast.) H.P. Linder in Bothalia,* 15(3–4): 467 (1985): Leptocarpus fruticosus.
gracilis *(Mast.) H.P. Linder in Bothalia,* 15(3–4): 467 (1985): Hypolaena gracilis.
hyalina *(Mast.) H.P. Linder in Bothalia,* 15(3–4): 467 (1985): Hypolaena hyalina.
impolita *(Kunth) H.P. Linder in Bothalia,* 15(3–4): 467 (1985): Restio impolitus.
levynsiae *(Pillans) H.P. Linder in Bothalia,* 15(3–4): 467 (1985): Leptocarpus levynsiae.
marlothii *(Pillans) H.P. Linder in Bothalia,* 15(3–4): 467 (1985): Leptocarpus marlothii.
membranacea *(Pillans) H.P. Linder in Bothalia,* 15(3–4): 469 (1985): Leptocarpus membranaceus.
monostylis *(Pillans) H.P. Linder in Bothalia,* 15(3–4): 469 (1985): Leptocarpus monostylis.
muirii *(Pillans) H.P. Linder in Bothalia,* 15(3–4): 469 (1985): Leptocarpus muirii.
nudiflora *(Pillans) H.P. Linder in Bothalia,* 15(3–4): 469 (1985): Leptocarpus nudiflorus.
pulchra *E.E. Esterhuysen in Bothalia,* 15(3–4): 469 (1985)—South Africa (Cape Province).
rigida *(Mast.) H.P. Linder in Bothalia,* 15(3–4): 470 (1985): Leptocarpus rigidus.
rigorata *(Mast.) H.P. Linder in Bothalia,* 15(3–4): 470 (1985): Leptocarpus rigoratus.
viminea *(Rottb.) H.P. Linder in Bothalia,* 15(3–4): 470 (1985): Restio vimineus.

CALPICARPUM (Apocynac.).
ornatum *Bull, Retail List,* 199: 12 (1884); *cf. D.J. Mabberley in Taxon,* 34(3): 456 (1985)— Moluccas.

CALYPSO (Orchidac.).
bulbosa
var. speciosa *(Schltr.) J.J. Wood in Kew Mag.,* 3(4): 151 (1986): C. speciosa.

CALYSTEGIA (Convolvulac.).
heterantha *(Nees & Martius) B.D. Jackson, Index Kewensis,* 1: 603 (1893): Dufourea heterantha.
sepium
var. hirsuta *K.T. Fu in Fl. Tsinlingensis,* 1(4): 398, 164 (1983)—China.

CAMBESSEDESIA (Melastomatac.).
glaziovii *Cogn. ex A.B. Martins in Rev. Brasil. Bot.,* 8(2): 177 (1985 publ. 1986)—Brazil.
hermogenesii *A.B. Martins in Rev. Brasil. Bot.,* 8(2): 178 (1985 publ. 1986)—Brazil.
membranacea
subsp. bahiana *A.B. Martins in Rev. Brasil. Bot.,* 8(2): 180 (1985 publ. 1986)—Brazil.

CAMELLIA (Theac.).
azalea *C.F. Wei in Bull. Bot. Res. North-East. Forest. Inst.,* 6(4): 141 (1986)—China.
lipoensis *H.T. Chang & Z.R. Xu in Acta Sci. Nat. Univ. Sunyatseni,* 1986(2): 98 (1986), without type—China.
microcarpa *(S.L. Mo & S.Z. Huang) S.L. Mo in Guihaia,* 6(1–2): 62 (1986): C. chrysantha var. microcarpa.
sasanqua *C.P. Thunberg ex A. Murray in Linnaeus, Syst. Veg., ed. 14:* 632 (1784)—Japan. †

CAMISSONIA (Onagrac.).
atwoodii *A. Cronquist in Great Basin Nat.,* 46(2): 258 (1986)—U.S.A. (Utah).
boothii
var. condensata *(Munz) A. Cronquist in Great Basin Nat.,* 46(2): 258 (1986): Oenothera decorticans var. condensata.
var. villosa *(Wats.) A. Cronquist in Great Basin Nat.,* 46(2): 258 (1986): Oenothera alyssoides var. villosa.
clavaeformis
var. purpurascens *(Wats.) A. Cronquist in Great Basin Nat.,* 46(2): 258 (1986): Oenothera scapoidea var. purpurascens.
scapoidea
var. utahensis *(Raven) S.L. Welsh in Great Basin Nat.,* 46(2): 258 (1986): Oenothera scapoidea subsp. utahensis.

CAMPANULA (Campanulac.).
glauca *C.P. Thunberg ex A. Murray in Linnaeus, Syst. Veg., ed. 14:* 211 (1784)—Japan. †
kastellorizana *A. Carlström in Willdenowia,* 15(2): 384 (1986)—Greece.

CAMPANULA :–
marginata *C.P. Thunberg ex A. Murray in Linnaeus, Syst. Veg., ed. 14:* 211 (1784)—Japan. †
simulans *A. Carlström in Willdenowia,* 15(2): 381 (1986)—Turkey, East Aegean Is., Greece.
tetraphylla *C.P. Thunberg ex A. Murray in Linnaeus, Syst. Veg., ed. 14:* 211 (1784)—Japan. †
triphylla *C.P. Thunberg ex A. Murray in Linnaeus, Syst. Veg., ed. 14:* 211 (1784)—Japan. †
veneris *A. Carlström in Willdenowia,* 15(2): 384 (1986)—Cyprus.

CAMPSIDIUM (Bignoniac.).
valdivianum *(Phil.) Bull, Retail List,* 59: 4 (1871); *cf. D.J. Mabberley in Taxon,* 34(3): 456 (1985): Basionym not stated.

CAMPULOCLINIUM (Compos.).
viridiflorum *Bartl., Ind. Sem. Hort. Acad. Gott.,* 1843: 2 (1843); *cf. D.J. Mabberley in Taxon,* 32:(1): 84 (1983)— Locality not stated.

CAMPYNEMANTHE (Melanthiac.).
neocaledonica *(Rendle) P. Goldblatt in Bull. Mus. Nation. Hist. Nat., B, Adansonia,* Ser. 4, 8(2): 122 (1986): Campynema neocaledonicum.
parva *P. Goldblatt in Bull. Mus. Nation. Hist. Nat., B, Adansonia,* Ser. 4, 8(2): 127 (1986)—New Caledonia.

CANARIUM (Burserac.).
australianum
var. velutinum *H.J. Hewson in Fl. Australia,* 25: 202, 169 (1985)—Australia (W. Australia, N. Terr., Queensland).

CANISTRUM (Bromeliac.).
seidelianum *W. Weber in Feddes Repert.,* 97(3–4): 117 (1986)—Brazil.

CANNA (Cannac.).
variegata *Besser, Cat. Hort. Bot. Krz.:* 25 (1810); *cf. D.J. Mabberley in Taxon,* 32(1): 84 (1983)— Locality not stated.

CANNABIS (Cannabac.).
kafiristanica *(Vav.) J. Chrtek in Čas. Nár. Muz. (Prague),* 150(1–2): 22 (1981): C. indica var. kafiristanica.

CANSCORA (Gentianac.).
diffusa
var. tetraptera *V.N. Naik & D.S. Pokle in J. Econ. Taxon. Bot.,* 7(3): 673 (1985 publ. 1986)—India.
rubiflora *X.X. Chen in Guihaia,* 6(3): 177 (1986)—China.

CAPPARIS (Capparac.).
iltisiana *T. Ruíz Z. in Ernstia,* 36: 1 (1986)—Venezuela.

CARAGANA (Legumin.).
beefensis *S. Biswas in Indian J. Forest.,* 9(1): 70 (1986)—India.
var. auriculata *S. Biswas in Indian J. Forest.,* 9(1): 71 (1986)—India.
brevispina
var. gamblei *S. Biswas in Indian J. Forest.,* 9(1): 72 (1986)—India.
spinosa *(L.) Vahl ex Hornem., Hort. Hafn.:* 694 (1815); *cf. D.J. Mabberley in Taxon,* 33(3): 436 (1984): Basionym not stated.
zahlbruckneri
subsp. litwinowii *(Kom.) G.P. Yakovlev in Bot. Zhurn.,* 71(4): 482 (1986): C. litwinowii.
var. pekinensis *(Kom.) G.P. Yakovlev in Bot. Zhurn.,* 71(4): 482 (1986): C. pekinensis.
var. pilosa *G.P. Yakovlev in Bot. Zhurn.,* 71(4): 482 (1986)—China.

CARALLIA (Rhizophorac.).
eugenioidea
var. sumatrensis *M. Ito & M. Hotta in M. Hotta (ed.), Divers. & Dynam. Pl. Life Sumatra,* 1: 73 (1986), without latin descr. or type—Sumatra.

CARALLUMA (Asclepiadac.).
adscendens
var. carinata *J.S. Sarkaria in J. Nation. Cact. Succ. Soc. India,* 4(1): 4 (1984); *cf. Asklepios,* 33: 41 (1985), without type—India.

CARDAMINE (Crucif.).
anhuiensis *D.C. Zhang & C.Z. Shao in Bull. Bot. Res. North-East. Forest. Inst.,* 6(2): 127 (1986)—China.
majovskii *K. Marhold & J. Záborský in Preslia,* 58(3): 194 (1986)—Czechoslovakia.
monteluccii *A.J.B. Brilli-Cattarini & L. Gubellini in Webbia,* 39(2): 398 (1986)—Italy, Sicily.
raphanifolia
subsp. barbareoides *(Halácsy) A. Strid in A. Strid (ed.), Mount. Fl. Greece,* 1: 257 (1986): C. barbareoides.

CARDIOSPERMUM (Sapindac.).
oliveirae *M.S. Ferrucci in Bol. Soc. Argent. Bot.,* 24(1–2): 116 (1985)—Brazil.

CARDUUS (Compos.).
×arvaticus *Font Quer & Rothmaler, Sched. Fl. Iber. Select., Cent. II-III,* n. 284 (1935); *cf. C. Navarro, M. Gutiérrez-Bustillo & A.G. Bueno in Fontqueria,* 7: 12 (1985)— Spain. (C. medius × phyllolepis).
atriplicifolius *Fisch. ex Hornem., Hort. Hafn., Suppl.:* 92 (1819); *cf. D.J. Mabberley in Taxon,* 33(3): 437 (1984)— Locality not stated.
cinerescens *Huber, Cat. Gén. 1870 & 1871:* 3 (1870); *cf. D.J. Mabberley in Taxon,* 34(3): 453 (1985)— Locality not stated.

CARDUUS :–
linearis *C.P. Thunberg ex A. Murray in Linnaeus, Syst. Veg., ed. 14:* 726 (1784)—Japan. †
picris *(Willd.) Sweet, Hort. Brit., ed. 2:* 281 (1830); *cf. D.J. Mabberley in Taxon,* 31(1): 69 (1982): Basionym not stated.
stoechadifolius *(Bieb.) Sweet, Hort. Brit., ed. 2:* 281 (1830); *cf. D.J. Mabberley in Taxon,* 31(1): 69 (1982): Basionym not stated.
subcarlinoides
subsp. gracilis *(Rouy ex Willkomm) J. Fernández Casas in Fontqueria,* 1: 10 (1982): C. granatensis var. gracilis.

CAREX (Cyperac.).
asraoi *D.M. Verma in J. Econ. Taxon. Bot.,* 7(3): 606 (1985 publ. 1986)—India.
balbii *Schkuhr ex Balbis, Cat. Stirp. Hort. Acad. Taur.:* 21 (1813); *cf. D.J. Mabberley in Taxon,* 31(1): 69 (1982)— Locality not stated.
brunnea *C.P. Thunberg ex A. Murray in Linnaeus, Syst. Veg., ed. 14:* 844 (1784)—Japan. †
brunnipes *A.A. Reznicek in Syst. Bot.,* 11(1): 82 (1986)—Mexico, Guatemala.
calcicola *T. Tang & F.T. Wang in Acta Phytotax. Sin.,* 24(3): 244 (1986)—China.
caricina *(Don) N. Ghildyal & U.C. Bhattacharyya in J. Econ. Taxon. Bot.,* 7(3): 602 (1985 publ. 1986): Cyperus caricinus.
var. glaucina *(Boeck.) N. Ghildyal & U.C. Bhattacharyya in J. Econ. Taxon. Bot.,* 7(3): 602 (1985 publ. 1986): C. glaucina.
var. leptocarpus *(Clarke) N. Ghildyal & U.C. Bhattacharyya in J. Econ. Taxon. Bot.,* 7(3): 602 (1985 publ. 1986): C. leptocarpus.
var. minor *N. Ghildyal & U.C. Bhattacharyya in J. Econ. Taxon. Bot.,* 7(3): 602 (1985 publ. 1986), nom. nov.: C. filicina var. minor *Boott.*
densifimbriata *T. Tang & F.T. Wang in Acta Phytotax. Sin.,* 24(3): 240 (1986)—China.
flexirostris *A.A. Reznicek in Syst. Bot.,* 11(1): 75 (1986)—Mexico.
harlandii
var. liuguensis *S.Y. Liang & D.Y. Liu in Bull. Bot. Res. North-East. Forest. Inst.,* 6(1): 181 (1986)—China.
incomitata *K.R. Thiele in Muelleria,* 6(3–4): 201 (1986)—Australia (Victoria, New South Wales).
ixtapalucensis *A.A. Reznicek in Syst. Bot.,* 11(1): 84 (1986)—Mexico.
japonica *C.P. Thunberg ex A. Murray in Linnaeus, Syst. Veg., ed. 14:* 845 (1784)—Japan. †
longebracteata *Hornem., Hort. Hafn.:* 881 (1815); *cf. D.J. Mabberley in Taxon,* 33(3): 437 (1984)— Locality not stated.
membranacea *Hook. in Lyon, Repulse Bay:* 195 (1825); *cf. D.J. Mabberley in Taxon,* 30(1): 15 (1981)— Locality not stated.

CAREX :–
nandadeviensis *N. Ghildyal, U.C. Bhattacharyya & P.K. Hajra in Indian J. Forest.,* 9(1): 90 (1986)—India.
pallens *(Fristedt) H. Harmaja in Ann. Bot. Fenn.,* 23(2): 148 (1986): C. digitata var. pallens.
pumila *C.P. Thunberg ex A. Murray in Linnaeus, Syst. Veg., ed. 14:* 846 (1784)—Japan. †
sahnii *N. Ghildyal & U.C. Bhattacharyya in J. Econ. Taxon. Bot.,* 7(2): 447 (1985 publ. 1986), nom. nov.: C. kingiana *Clarke.*
shangchengensis *S.Y. Liang in Acta Phytotax. Sin.,* 24(3): 242 (1986)—China.
tristachya *C.P. Thunberg ex A. Murray in Linnaeus, Syst. Veg., ed. 14:* 845 (1784)—Japan. †
zunyiensis *T. Tang & F.T. Wang in Acta Phytotax. Sin.,* 24(3): 241 (1986)—China.

CARLINA (Compos.).
acanthifolia
subsp. baetica *J. Fernández Casas & Leal in J. Fernández Casas, Exsicc. nobis distrib.,* 3: 10 (1980)—Spain.
baetica *(Fernández Casas & Leal) J. Fernández Casas in Fontqueria,* 9: 19 (1985): C. acanthifolia subsp. baetica.

CARLUDOVICA (Cyclanthac.).
sulcata *B.E. Hammel in Phytologia,* 60(1): 8 (1986)—Costa Rica, Nicaragua.

CARPACOCE (Rubiac.).
burchellii *C. Puff in Fl. S. Afr.,* 31(1:2): 59 (1986)—South Africa (Cape Province).
curvifolia *C. Puff in Fl. S. Afr.,* 31(1:2): 57 (1986)—South Africa (Cape Province).
gigantea *C. Puff in Fl. S. Afr.,* 31(1:2): 56 (1986)—South Africa (Cape Province).
scabra
subsp. rupestris *C. Puff in Fl. S. Afr.,* 31(1:2): 59 (1986)—South Africa (Cape Province).
spermacocea
subsp. orientalis *C. Puff in Fl. S. Afr.,* 31(1:2): 56 (1986)—South Africa (Cape Province).

CARPINUS (Corylac.).
dayongina *K.W. Liu & Q.Z. Lin in Bull. Bot. Res. North-East. Forest. Inst.,* 6(2): 143 (1986)—China.

CARUM (Umbellif.).
flexuosum *(With.) Sweet, Hort. Brit., ed. 2:* 246 (1830); *cf. D.J. Mabberley in Taxon,* 31(1): 70 (1982): Basionym not stated.
graecum
subsp. serpentinicum *P. Hartvig in A. Strid (ed.), Mount. Fl. Greece,* 1: 700 (1986)—Greece, Jugoslavia, ?Albania.
rigidulum
subsp. bulgaricum *P. Hartvig in A. Strid (ed.), Mount. Fl. Greece,* 1: 698 (1986)—Greece, Bulgaria, Jugoslavia.

CARUM :-
subsp. palmatum *P. Hartvig in A. Strid (ed.), Mount. Fl. Greece,* 1: 699 (1986)—Greece, ?Bulgaria.

CARYODAPHNOPSIS (Laurac.).
fosteri *H. van der Werff in Syst. Bot.,* 11(3): 415 (1986)—Peru, Ecuador.

CASSIA (Legumin.).
italica *(Mill.) Sprengel, Bot. Gart. Univ. Halle:* 21 (1800); *cf. D.J. Mabberley in Taxon,* 30(1): 9 (1981): Senna italica.
porturegalis *E.N. Bancroft in Jamaica Courant:* 13-17 (Nov. 1827) *et in Roy. Gaz.,* 49(46): 22 (1827); *cf. D.J. Mabberley in Taxon,* 30(1): 9 (1981)— ?Jamaica.
sumatrana *Roxb. ex Hornem., Hort. Hafn., Suppl.:* 135 (1819); *cf. D.J. Mabberley in Taxon,* 33(3): 437 (1984)— Locality not stated.
telfairiana *(Hook. f.) R.M. Polhill in Kew Bull.,* 41(4): 810 (1986): C. mimosoides var. telfairiana.
umbellata *Bertol., Syll. Pl. Hort. Bot. Bonon.,* 1827: 4 (1827); *cf. D.J. Mabberley in Taxon,* 32(1): 82 (1983)— Locality not stated.

CASSIPOUREA (Rhizophorac.).
acuminata *L. Liben in Bull. Jard. Bot. Nation. Belg.,* 56(1–2): 139 (1986)—Zaïre, Gabon.
louisii *L. Liben in Bull. Jard. Bot. Nation. Belg.,* 56(1–2): 140 (1986)—Zaïre, Gabon, Rwanda, Burundi.

CASTANOPSIS (Fagac.).
amabilis *W.C. Cheng & C.S. Chao in Sci. Silvae Sin.,* 8(1): 5 (1963)—China.
chunii *W.C. Cheng in Sci. Silvae Sin.,* 8(1): 5 (1963)—China.
hamata *S. Duanmu in Sci. Silvae Sin.,* 8(2): 188 (1963)—Hainan.
hupehensis *C.S. Chao in Sci. Silvae Sin.,* 8(2): 187 (1963)—China.
jianfenglingensis *S. Duanmu in Sci. Silvae Sin.,* 8(2): 187 (1963)—Hainan.
oblongifolia *W.C. Cheng & C.S. Chao in Sci. Silvae Sin.,* 8(2): 186 (1963)—China.
siamensis *S. Duanmu in Sci. Silvae Sin.,* 8(2): 189 (1963)—Thailand.

CASTILLEJA (Scrophulariac.).
parvula
var. revealii *(N. Holmgren) N.D. Atwood in Great Basin Nat.,* 46(2): 260 (1986): C. revealii.
rhexifolia
var. sulphurea *(Rydb.) N.D. Atwood in Great Basin Nat.,* 46(2): 260 (1986): C. sulphurea.

CASUARINA (Casuarinac.).
muricata *Roxb. ex Hornem., Hort. Hafn., Suppl.:* 161 (1819); *cf. D.J. Mabberley in Taxon,* 33(3): 437 (1984)— Locality not stated.

CATASETUM (Orchidac.).
denticulatum *F.E. Miranda in Lindleyana,* 1(3): 152 (1986)—Brazil.
lanceatum *F.E. Miranda in Lindleyana,* 1(3): 148 (1986)—Brazil.
multifidum *F.E. Miranda in Lindleyana,* 1(3): 154 (1986)—Brazil.
semicirculatum *F.E. Miranda in Lindleyana,* 1(3): 156 (1986)—Brazil.

CATHARTOCARPUS (Legumin.).
erubescens *W. Hamilton in Pharm. J.,* 5: 119 (1845); *cf. D.J. Mabberley in Taxon,* 30(1): 10 (1981)— ?Jamaica.

CATOBLASTUS (Palmae).
distichus *R. Bernal-Gonzalez in Principes,* 30(1): 38 (1986)—Colombia.
inconstans *(Dugand) Glassman ex R. Bernal-Gonzalez in Principes,* 30(1): 41 (1986), combination validly publ. in 1972: Catostigma inconstans.

CATOFERIA (Labiat.).
martinezii *T.P. Ramamoorthy in Kew Bull.,* 41(2): 303 (1986)—Mexico.

CAUCALIS (Umbellif.).
gummifera *(Lam.) Lag., Elenchus:* 3 (1816); *cf. D.J. Mabberley in Taxon,* 31(1): 70 (1982): Basionym not stated.
lucida *(L.) Lag., Elenchus:* 3 (1816); *cf. D.J. Mabberley in Taxon,* 31(1): 70 (1982): Basionym not stated.

CAULARTHRON (Orchidac.).
kraenzlinianum *H.G. Jones in Carib. J. Sci.,* 16(1–4): 119 (1980 publ. 1981)—Windward Is.

CAULINIA (Najadac.).
alagnensis *Poll., Cat. Pl. Hort. Bot. Veron.,* 1814: 13 (1814); *cf. D.J. Mabberley in Taxon,* 33(3): 441 (1984)— Locality not stated.

CAVENDISHIA (Ericac.).
spectabilis *Bull, Cat. 1889:* 7 (1889); *cf. D.J. Mabberley in Taxon,* 34(3): 456 (1985)—Locality not stated.

CAYLUSEA (Resedac.).
jaberi *S. Abedin in Willdenowia,* 15(2): 433 (1986)—Saudi Arabia.

CEANOTHUS (Rhamnac.).
greggii
subsp. perplexans *(Trelease) R.M. Beauchamp in Phytologia,* 59(7): 440 (1986): C. perplexans.

CECROPIA (Cecropiac.).
granvilleana *C.C. Berg in Bull. Mus. Nation. Hist. Nat., B, Adansonia,* Ser. 4, 7(3): 255 (1985 publ. 1986)—French Guiana.

CELASTRUS (Celastrac.).
alatus *C.P. Thunberg ex A. Murray in Linnaeus, Syst. Veg., ed. 14:* 237 (1784)—Japan. †
glaucophyllus
var. rugosus *(Rehd. & Wils.) C.Y. Wu ex Y.C. Ho in Fl. Tsinlingensis,* 1(3): 212 (1981): C. rugosus.
orbiculatus *C.P. Thunberg ex A. Murray in Linnaeus, Syst. Veg., ed. 14:* 237 (1784)—Japan.
punctatus *C.P. Thunberg ex A. Murray in Linnaeus, Syst. Veg., ed. 14:* 237 (1784)—Japan. †
rosthornianus
var. loeseneri *(Rehd. & Wils.) C.Y. Wu ex Y.C. Ho in Fl. Tsinlingensis,* 1(3): 213 (1981): C. loeseneri.
striatus *C.P. Thunberg ex A. Murray in Linnaeus, Syst. Veg., ed. 14:* 237 (1784)—Japan. †

CELSIA (Scrophulariac.).
coccinea *(Willd.) Parmentier, Cat. Pl. Enghien:* 17 (1818); *cf. D.J. Mabberley in Taxon,* 33(3): 441 (1984): Basionym not stated.
coromandelina *Koen. ex Rottb., Pl. Hort. Univ. Rar. Prog.:* 29 (1773); *cf. D.J. Mabberley in Taxon,* 33(3): 438 (1984)— Locality not stated.

CELTIS (Ulmac.).
conferta
subsp. amblyphylla *(F. Mueller) P.S. Green in J. Arnold Arbor.,* 67(1): 120 (1986): C. amblyphylla.

CENTAUREA (Compos.).
nothosect. Borcea *J. Fernández Casas & A. Susanna de la Serna in Fontqueria,* 7: 7 (1985), contrary to Art. H9.1 ICBN (1983).
nothosect. Chamaecentron *J. Fernández Casas & A. Susanna de la Serna in Fontqueria,* 1: 2 (1982). (C. sect. Acrocentron × sect. Chamaecyanus).
subsect. Chamaecyani *J. Fernández Casas & A. Susanna de la Serna in Treb. Inst. Bot. Barcelona,* 10: 60 (1985 publ. 1986).
subsect. Lagascanae *J. Fernández Casas & A. Susanna de la Serna in Treb. Inst. Bot. Barcelona,* 10: 93 (1985 publ. 1986).
nothosect. Lophocyanus *J. Fernández Casas & A. Susanna de la Serna in Fontqueria,* 1: 1 (1982). (C. sect. Chamaecyanus × sect. Lopholoma).
amblensis
var. tentudaica *(Rivas Goday) J. Fernández Casas & A. Susanna de la Serna in Fontqueria,* 1: 1 (1982): C. toletana subsp. tentudaica.
×andresiana *J. Fernández Casas & A. Susanna de la Serna in Fontqueria,* 1: 1 (1982)—Spain. (C. scabiosa × toletana var. argecillensis).

CENTAUREA :–
carolipauana *J. Fernández Casas & A. Susanna de la Serna in Fontqueria,* 1: 2 (1982), nom. nov.: C. prolongi var. macrocephala *Pau & Font Quer.*
×ceballosii
nothovar. andresiana *(Fernández Casas & Susanna) J. Fernández Casas & A. Susanna de la Serna in Fontqueria,* 7: 10 (1985): C. × andresiana.
×cephalariseptimae *J. Fernández Casas & A. Susanna de la Serna in Fontqueria,* 9: 16 (1985)—Spain. (C. cephalariifolia × legionis-septimae).
hanryi
subsp. isernii *(Willk.) B.E. Smythies in Englera,* 3(3): 880 (1986): C. isernii.
subsp. spinabadia *(Bubani ex Timb.-Lagr.) B.E. Smythies in Englera,* 3(3): 880 (1986): C. spinabadia.
kamciensis *H. Kočev & S. Gančev in Dokl. Bolg. Akad. Nauk,* 21(2): 151 (1968)—Bulgaria.
lagascana
f. livida *J. Fernández Casas & A. Susanna de la Serna in Treb. Inst. Bot. Barcelona,* 10: 99 (1985 publ. 1986)—Spain.
lainzii *J. Fernández Casas, Exsicc. me distrib.,* 1: 9 (1975)—Spain.
legionis-septimae *J. Fernández Casas & A. Susanna de la Serna in Fontqueria,* 9: 13 (1985)—Spain.
livonica *Weinm., Bot. Gart. Kais. Univ. Dorpat,* 1810: 38 (1810); *cf. D.J. Mabberley in Taxon,* 32(1): 84 (1983)— Locality not stated.
×piifontiana *J. Fernández Casas & A. Susanna de la Serna in Fontqueria,* 6: 5 (1984)—Spain. (C. granatensis × mariana).
resupinata
subsp. degenii *(Sennen) J. Fernández Casas & A. Susanna de la Serna in Fontqueria,* 1: 4 (1982): C. degenii.
subsp. lagascae *(Nyman) J. Fernández Casas & A. Susanna de la Serna in Fontqueria,* 1: 4 (1982): C. lagascae.
subsp. spachii *(Schultz Bip. ex Willk.) J. Fernández Casas & A. Susanna de la Serna in Fontqueria,* 1: 4 (1982): C. spachii.
×zubiae
nothovar. somedana *J. Fernández Casas & A. Susanna de la Serna in Treb. Inst. Bot. Barcelona,* 10: 115 (1985 publ. 1986)—Spain. (C. lagascana × zubiae).

CENTAURIUM (Gentianac.).
barrelieri *(Léon-Dufour) Font Quer & Rothmaler, Sched. Fl. Iber. Select., Cent. II-III,* n. 179 (1935); *cf. C. Navarro, M. Gutiérrez-Bustillo & A.G. Bueno in Fontqueria,* 7: 12 (1985): Erythraea barrelieri.
serpentinicola *A. Carlström in Willdenowia,* 16(1): 75 (1986)—Turkey.

CENTRATHERUM (Compos.).
sengaltherianum *B.M. Narayana in Curr. Sci.*, 50(6): 279 (1981)—India.

CEPA (Liliac.).
carinata *(L.) Bernh., Syst. Verz.*: 202 (1800); *cf. D.J. Mabberley in Taxon*, 33(3): 443 (1984): Allium carinatum.

CEPHALANTHERA (Orchidac.).
calcarata *S.C. Chen & K.Y. Lang in Acta Bot. Yunnanica*, 8(3): 271 (1986)—China.
rubra
f. alba *C. Raynaud, Orchid. Maroc:* 93 (1985), without type—Morocco.

CEPHALARIA (Dipsacac.).
galpiniana
subsp. simplicior *B.L. Burtt in Notes Roy. Bot. Gard. Edinburgh*, 43(3): 366 (1986)—South Africa (Natal, Cape Province), Lesotho.

CERASUS (Rosac.).
sect. Ceraseidos *(Sieb. & Zucc.) T.T. Yü & C.L. Li in Fl. Reipubl. Popul. Sin.*, 38: 67 (1986), without basionym ref.: Basionym not stated.
sect. Lobopetalum *(Koehne) T.T. Yü & C.L. Li in Fl. Reipubl. Popul. Sin.*, 38: 58 (1986): Prunus subsect. Lobopetalum.
sect. Mahaleb *(Koehne) T.T. Yü & C.L. Li in Fl. Reipubl. Popul. Sin.*, 38: 67 (1986): Prunus subsect. Mahaleb.
sect. Phyllocerasus *(Koehne) T.T. Yü & C.L. Li in Fl. Reipubl. Popul. Sin.*, 38: 52 (1986): Prunus subsect. Phyllocerasus.
sect. Phyllomahaleb *(Koehne) T.T. Yü & C.L. Li in Fl. Reipubl. Popul. Sin.*, 38: 46 (1986): Prunus subsect. Phyllomahaleb.
sect. Pseudomahaleb *(Koehne) T.T. Yü & C.L. Li in Fl. Reipubl. Popul. Sin.*, 38: 64 (1986): Prunus subsect. Pseudomahaleb.
sect. Sargentiella *(Koehne) T.T. Yü & C.L. Li in Fl. Reipubl. Popul. Sin.*, 38: 73 (1986): Prunus subsect. Sargentiella.
sect. Serrula *(Koehne) T.T. Yü & C.L. Li in Fl. Reipubl. Popul. Sin.*, 38: 76 (1986): Prunus subsect. Serrula.
campanulata *(Maxim.) T.T. Yü & C.L. Li in Fl. Reipubl. Popul. Sin.*, 38: 78 (1986): Prunus campanulata.
caudata *(Franch.) T.T. Yü & C.L. Li in Fl. Reipubl. Popul. Sin.*, 38: 68 (1986): Prunus caudata.
cerasoides
var. rubea *(C. Ingram) T.T. Yü & C.L. Li in Fl. Reipubl. Popul. Sin.*, 38: 79 (1986): Prunus cerasoides var. rubea.
clarofolia *(Schneid.) T.T. Yü & C.L. Li in Fl. Reipubl. Popul. Sin.*, 38: 54 (1986): Prunus clarofolia.
conadenia *(Koehne) T.T. Yü & C.L. Li in Fl. Reipubl. Popul. Sin.*, 38: 50 (1986): Prunus conadenia.

CERASUS :-
conradinae *(Koehne) T.T. Yü & C.L. Li in Fl. Reipubl. Popul. Sin.*, 38: 76 (1986): Prunus conradinae.
crataegifolius *(Hand.-Mazz.) T.T. Yü & C.L. Li in Fl. Reipubl. Popul. Sin.*, 38: 71 (1986): Prunus crataegifolius.
cyclamina *(Koehne) T.T. Yü & C.L. Li in Fl. Reipubl. Popul. Sin.*, 38: 58 (1986): Prunus cyclamina.
var. biflora *(Koehne) T.T. Yü & C.L. Li in Fl. Reipubl. Popul. Sin.*, 38: 59 (1986): Prunus cyclamina var. biflora.
dielsiana *(Schneid.) T.T. Yü & C.L. Li in Fl. Reipubl. Popul. Sin.*, 38: 59 (1986): Prunus dielsiana.
var. abbreviata *(Card.) T.T. Yü & C.L. Li in Fl. Reipubl. Popul. Sin.*, 38: 60 (1986): Prunus dielsiana var. abbreviata.
duclouxii *(Koehne) T.T. Yü & C.L. Li in Fl. Reipubl. Popul. Sin.*, 38: 63 (1986): Prunus duclouxii.
duerinckii *Mertens, Hort. Leeuv.*, 1840; *Linnaea 15, Lit.*: 98 (?1841); *cf. D.J. Mabberley in Taxon*, 33(3): 441 (1984)—Locality not stated.
glabra *(Pamp.) T.T. Yü & C.L. Li in Fl. Reipubl. Popul. Sin.*, 38: 60 (1986): Prunus hirtipes var. glabra.
henryi *(Schneid.) T.T. Yü & C.L. Li in Fl. Reipubl. Popul. Sin.*, 38: 64 (1986): Prunus yunnanensis var. henryi.
japonica
var. nakaii *(Lévl.) T.T. Yü & C.L. Li in Fl. Reipubl. Popul. Sin.*, 38: 86 (1986): Prunus nakaii.
mugus *(Hand.-Mazz.) T.T. Yü & C.L. Li in Fl. Reipubl. Popul. Sin.*, 38: 71 (1986): Prunus mugus.
patentipila *(Hand.-Mazz.) T.T. Yü & C.L. Li in Fl. Reipubl. Popul. Sin.*, 38: 46 (1986): Prunus patentipila.
pleiocerasus *(Koehne) T.T. Yü & C.L. Li in Fl. Reipubl. Popul. Sin.*, 38: 51 (1986): Prunus pleiocerasus.
pogonostyla *(Maxim.) T.T. Yü & C.L. Li in Fl. Reipubl. Popul. Sin.*, 38: 81 (1986): Prunus pogonostyla.
var. obovata *(Koehne) T.T. Yü & C.L. Li in Fl. Reipubl. Popul. Sin.*, 38: 82 (1986): Prunus pogonostyla var. obovata.
polytricha *(Koehne) T.T. Yü & C.L. Li in Fl. Reipubl. Popul. Sin.*, 38: 56 (1986): Prunus polytricha.
pseudocerasus *(Lindl.) hort. in Hort. Soc. Lond., Cat. Fr.*: 28 (1826); *cf. D.J. Mabberley in Taxon*, 33(3): 441 (1984): Prunus pseudocerasus.
pusilliflora *(Card.) T.T. Yü & C.L. Li in Fl. Reipubl. Popul. Sin.*, 38: 66 (1986): Prunus pusilliflora.
rufa *(Hook. f.) T.T. Yü & C.L. Li in Fl. Reipubl. Popul. Sin.*, 38: 81 (1986): Prunus rufa.
var. trichantha *(Koehne) T.T. Yü & C.L. Li in Fl. Reipubl. Popul. Sin.*, 38: 81 (1986): Prunus trichantha.

CERASUS :-
schneideriana *(Koehne) T.T. Yü & C.L. Li in Fl. Reipubl. Popul. Sin.*, 38: 60 (1986): Prunus schneideriana.
scopulorum *(Koehne) T.T. Yü & C.L. Li in Fl. Reipubl. Popul. Sin.*, 38: 61 (1986): Prunus scopulorum.
serrula *(Franch.) T.T. Yü & C.L. Li in Fl. Reipubl. Popul. Sin.*, 38: 80 (1986): Prunus serrula.
setulosa *(Batal.) T.T. Yü & C.L. Li in Fl. Reipubl. Popul. Sin.*, 38: 67 (1986): Prunus setulosa.
stipulacea *(Maxim.) T.T. Yü & C.L. Li in Fl. Reipubl. Popul. Sin.*, 38: 68 (1986): Prunus stipulacea.
subhirtella
var. pendula *(Tanaka) T.T. Yü & C.L. Li in Fl. Reipubl. Popul. Sin.*, 38: 74 (1986): Prunus subhirtella var. pendula.
susquehana *(Willd.) hort. in Hort. Soc. Lond., Cat. Fr.*: 29 (1826); *cf. D.J. Mabberley in Taxon,* 33(3): 441 (1984): Basionym not stated.
szechuanica *(Batal.) T.T. Yü & C.L. Li in Fl. Reipubl. Popul. Sin.*, 38: 49 (1986): Prunus szechuanica.
tatsienensis *(Batal.) T.T. Yü & C.L. Li in Fl. Reipubl. Popul. Sin.*, 38: 52 (1986): Prunus tatsienensis.
trichostoma *(Koehne) T.T. Yü & C.L. Li in Fl. Reipubl. Popul. Sin.*, 38: 69 (1986): Prunus trichostoma.
virginica *(Michx.) hort. in Hort. Soc. Lond., Cat. Fr.*: 30 (1826); *cf. D.J. Mabberley in Taxon,* 33(3): 441 (1984): Basionym not stated.
yedoensis *(Matsum.) T.T. Yü & C.L. Li in Fl. Reipubl. Popul. Sin.*, 38: 74 (1986): Prunus yedoensis.
yunnanensis *(Franch.) T.T. Yü & C.L. Li in Fl. Reipubl. Popul. Sin.*, 38: 64 (1986): Prunus yunnanensis.
var. polybotrys *(Koehne) T.T. Yü & C.L. Li in Fl. Reipubl. Popul. Sin.*, 38: 64 (1986): Prunus yunnanensis var. polybotrys.

CERATOCAPNOS (Papaverac.).
turbinata *(DC.) M. Lidén in Op. Bot.*, 88: 38 (1986): Fumaria turbinata.

CERATOCARYUM (Restionac.).
decipiens *(N.E. Br.) H.P. Linder in Bothalia,* 15(3–4): 479 (1985): Willdenowia decipiens.
fimbriatum *(Kunth) H.P. Linder in Bothalia,* 15(3–4): 479 (1985): Willdenowia fimbriata.
xerophilum *(Pillans) H.P. Linder in Bothalia,* 15(3–4): 480 (1985): Willdenowia xerophila.

CERATOCHILUS (Orchidac.).
insignis *(Frost) Lindl. ex G. Don f. in Loud., Hort. Brit.*: 489 (1830) *et in Sweet, Hort. Brit., ed. 2:* 490 (1830); *cf. D.J. Mabberley in Taxon,* 30(1): 15 (1981): Stanhopea insignis.

CERATOIDES (Chenopodiac.).
lanata
var. ruinina *S.L. Welsh in Great Basin Nat.*, 44(2): 196 (1984)—U.S.A. (Utah).

CERATOSTEMA (Ericac.).
fasciculatum *J.L. Luteyn in J. Arnold Arbor.*, 67(4): 491 (1986)—Ecuador.
ferreyrae *J.L. Luteyn in J. Arnold Arbor.*, 67(4): 487 (1986)—Peru.
nodosum *J.L. Luteyn in J. Arnold Arbor.*, 67(4): 489 (1986)—Ecuador.

CERATOSTYLIS (Orchidac.).
anjasmorensis *J.J. Wood & J.B. Comber in Kew Bull.*, 41(3): 697 (1986)—Java.

CERATOZAMIA (Zamiac.).
euryphyllidia *M. Vázquez Torres, S. Sabato & D.W. Stevenson in Brittonia,* 38(1): 17 (1986)—Mexico.

CERCESTIS (Arac.).
camerunensis *(Ntépé) J. Bogner in Aroideana,* 8(3): 73 (1985 publ. 1986): Rhektophyllum camerunense.
mirabilis *(N.E. Br.) J. Bogner in Aroideana,* 8(3): 73 (1985 publ. 1986): Rhektophyllum mirabile.

CEREUS (Cactac.).
speciosissimus *(Desf.) Sweet, Hort. Brit.*: 171 (1826); *cf. D.J. Mabberley in Taxon,* 30(1): 15 (1981): Basionym not stated.

CEROPEGIA (Asclepiadac.).
dichotoma
subsp. krainzii *(Sventenius) P. Bruyns in Beitr. Biol. Pfl.*, 60(3): 454 (1985 publ. 1986): C. krainzii.

CHAEROPHYLLUM (Umbellif.).
aristatum *C.P. Thunberg ex A. Murray in Linnaeus, Syst. Veg., ed. 14:* 288 (1784)—Japan. †
cyminum *Trev., Ind. Sem. Wratislav. App. III:* 1 (1821); *cf. D.J. Mabberley in Taxon,* 32(1): 84 (1983)— Locality not stated.
karsianum *Kit Tan & H. Ocakverdi in Notes Roy. Bot. Gard. Edinburgh,* 44(1): 160 (1986)—Turkey.
scabrum *C.P. Thunberg ex A. Murray in Linnaeus, Syst. Veg., ed. 14:* 289 (1784)—Japan. †

CHAETANTHERA (Compos.).
ramosissima *D. Don ex Taylor & Philips, Phil. Mag. n.s.,* 11: 391 (May 1832); *cf. D.J. Mabberley in Taxon,* 30(1): 15 (1981)—Locality not stated.

CHAMAECYTISUS (Legumin.).
hirsutus
subsp. ciliatus *(Wahlenb.) A. Strid in A. Strid (ed.), Mount. Fl. Greece,* 1: 449 (1986): Cytisus ciliatus.

CHAMAECYTISUS :–
triflorus *(Lam.) A. Skalická in Preslia,* 58(1): 23 (1986): Cytisus triflorus.

CHAMAEDOREA (Palmae).
latisecta *(H. Moore) A.H. Gentry in Ann. Missouri Bot. Gard.,* 73(1): 163 (1986): Morenia latisecta.
macroloba *G. Galeano-Garcés in Brittonia,* 38(1): 60 (1986)—Colombia.
megaphylla *A.H. Gentry in Ann. Missouri Bot. Gard.,* 73(1): 163 (1986)—Peru.
poeppigiana *(Mart.) A.H. Gentry in Ann. Missouri Bot. Gard.,* 73(1): 162 (1986): Morenia poeppigiana.
smithii *A.H. Gentry in Ann. Missouri Bot. Gard.,* 73(1): 164 (1986)—Peru.

CHAMAEROPS (Palmae).
excelsa *C.P. Thunberg ex A. Murray in Linnaeus, Syst. Veg., ed. 14:* 984 (1784)—Japan. †

CHAMAESPARTIUM (Legumin.).
sagittale
subsp. delphinense *(Verlot) O. de Bolòs & J. Vigo, Fl. Països Catalans:* 461 (1984): Genista delphinensis.

CHAMISSOA (Amaranthac.).
margaritacea *(L.) Schouw, Forteg. Kjöb. Bot. Hav. Pl.:* 123 (1847); *cf. D.J. Mabberley in Taxon,* 33(3): 439 (1984): Basionym not stated.
stricta *(Hornem.) Schouw, Forteg. Kjöb. Bot. Hav. Pl.:* 123 (1847); *cf. D.J. Mabberley in Taxon,* 33(3): 439 (1984): Celosia stricta.

CHAMOMILLA (Compos.).
pubescens *(Desf.) S.A. Alavi in Fl. Libya,* 107: 150 (1983): Cotula pubescens.

CHAPTALIA (Compos.).
callacallensis *J. Cuatrecasas in Fontqueria,* 8: 14 (1985); *et in Fontqueria,* 9: 5 (1985)—Peru.

CHASCOLYTRUM (Gramin.).
nutans *Lindl. & Nees in Nees, Del. Sem. Hort. Bot. Vratislav.,* 1840; *Linnaea 15, Lit.:* 125 (?1841); *cf. D.J. Mabberley in Taxon,* 32(1): 84 (1983)— Locality not stated.

CHEIRANTHUS (Crucif.).
pallens *(Pers.) Hornem., Hort. Hafn., Suppl.:* 73 (1819); *cf. D.J. Mabberley in Taxon,* 33(3): 437 (1984): Basionym not stated.

CHELIDONIUM (Papaverac.).
japonicum *C.P. Thunberg ex A. Murray in Linnaeus, Syst. Veg., ed. 14:* 489 (1784)—Japan. †

CHELONE (Scrophulariac.).
formosa *Thompson, Bot. Displ.,* t. 4 (1798); *cf. D.J. Mabberley in Taxon,* 30(1): 15 (1981)— Locality not stated.

CHELONE :–
laevigata *Thompson, Bot. Displ.,* sub t. 4 (1798); *cf. D.J. Mabberley in Taxon,* 30(1): 15 (1981)— Locality not stated.
pubescens *(Soland.) Parmentier, Cat. Pl. Enghien:* 18 (1818); *cf. D.J. Mabberley in Taxon,* 33(3): 441 (1984): Basionym not stated.

CHENOPODIUM (Chenopodiac.).
capitatum
var. parvicapitatum *S.L. Welsh in Great Basin Nat.,* 44(2): 199 (1984)—U.S.A. (Utah, Colorado, California, Nevada), Canada (British Columbia, Saskatchewan).

CHESNEYA (Legumin.).
antoninae *M.R. Rasulova & B.A. Sharipova in Dokl. Akad. Nauk Tadzh. SSR,* 23(8): 461 (1980)—U.S.S.R. (Middle Asia).

CHILOCARPUS (Apocynac.).
sunainaianus *S.N. Yoganarasimhan in Curr. Sci.,* 51(18): 902 (1982)—Nicobar Is.

CHILOPSIS (Bignoniac.).
linearis
subsp. arcuata *(Fosberg) J. Henrickson in Aliso,* 11(2): 194 (1985 publ. 1986): C. linearis var. arcuata.
var. tomenticaulis *J. Henrickson in Aliso,* 11(2): 191 (1985 publ. 1986)—Mexico.

CHIMONOBAMBUSA (Gramin.).
pubescens *T.H. Wen in J. Bamboo Res.,* 5(2): 20 (1986)—China.

CHIONANTHUS (Oleac.).
mala-elengi *(Dennst.) P.S. Green in Bull. Bot. Surv. India,* 26(1–2): 124 (1984 publ. 1985): Forsythia mala-elengi.

CHIRITA (Gesneriac.).
leiophylla *W.T. Wang in Guihaia,* 6(3): 159 (1986)—China.
villosissima *W.T. Wang in Guihaia,* 6(3): 161 (1986)—China.

CHIRITOPSIS (Gesneriac.).
mollifolia *D. Fang & W.T. Wang in Guihaia,* 6(1–2): 6 (1986)—China.
subulata *W.T. Wang in Guihaia,* 6(1–2): 8 (1986)—China.

CHIRONIA (Gentianac.).
ramosissima *Bernh., Syst. Verz.:* 123 (1800); *cf. D.J. Mabberley in Taxon,* 33(3): 443 (1985)— Locality not stated.

CHITONIA (Melastomatac.).
albicans *(Sw.) D. Don ex G. Don f. in Loud., Hort. Brit.:* 174 (1830); *cf. D.J. Mabberley in Taxon,* 30(1): 15 (1981): Basionym not stated.

CHLORIS (Gramin.).
berazainiae *L. Catasús Guerra in Acta Bot. Cubana,* 25: 5 (1985), as 'berazainae'—Cuba.
sagraeana
subsp. cubensis *(Hitchcock & Ekman) L. Catasús Guerra in Acta Bot. Cubana,* 25: 6 (1985), without basionym page: C. cubensis.

CHLOROSPATHA (Arac.).
croatiana *M.H. Grayum in Ann. Missouri Bot. Gard.,* 73(2): 464 (1986)—Panama, Costa Rica, Colombia.
subsp. enneaphylla *M.H. Grayum in Ann. Missouri Bot. Gard.,* 73(2): 466 (1986)—Colombia.
gentryi *M.H. Grayum in Ann. Missouri Bot. Gard.,* 73(2): 468 (1986)—Colombia.
hammeliana *M.H. Grayum & T.B. Croat in Ann. Missouri Bot. Gard.,* 73(2): 466 (1986)—Panama.

CHONDROPETALUM (Restionac.).
decipiens *E.E. Esterhuysen in Bothalia,* 15(3–4): 428 (1985)—South Africa (Cape Province).

CHRISTOLEA (Crucif.).
baiogoinensis *K.C. Kuan & An in Fl. Xizangica,* 2: 388 (1985)—Xizang.
rosularis *K.C. Kuan & An in Fl. Xizangica,* 2: 386 (1985)—Xizang.

CHRYSANTHEMUM (Compos.).
japonicum *C.P. Thunberg ex A. Murray in Linnaeus, Syst. Veg., ed. 14:* 773 (1784)—Japan. †
sinense *Sabine in Tilloch & Taylor, Phil. Mag.,* 61: 235 (1823); *cf. D.J. Mabberley in Taxon,* 31(1): 70 (1982)— Locality not stated.

CHRYSOSPLENIUM (Saxifragac.).
sect. Grayana *J.T. Pan in Acta Phytotax. Sin.,* 24(3): 204 (1986).
ser. Ramosa *J.T. Pan in Acta Phytotax. Sin.,* 24(3): 206 (1986).
sect. Trichosperma *J.T. Pan in Acta Phytotax. Sin.,* 24(3): 203 (1986).
absconditicapsulum *J.T. Pan in Fl. Xizangica,* 2: 450 (1985)—Xizang.
chinense *(Hara) J.T. Pan in Acta Phytotax. Sin.,* 24(2): 93 (1986): C. alternifolium var. chinense.
griffithii
var. intermedium *(Hara) J.T. Pan in Acta Phytotax. Sin.,* 24(2): 92 (1986): C. nudicaule var. intermedium.
hydrocotylifolium
var. emeiense *J.T. Pan in Acta Phytotax. Sin.,* 24(2): 94 (1986)—China.
lanuginosum
var. ciliatum *(Franch.) J.T. Pan in Acta Phytotax. Sin.,* 24(2): 96 (1986): C. ciliatum.

CHRYSOSPLENIUM :-
var. pilosomarginatum *(Hara) J.T. Pan in Acta Phytotax. Sin.,* 24(2): 96 (1986): C. pilosomarginatum.
lixianense *Jien ex J.T. Pan in Acta Phytotax. Sin.,* 24(3): 207 (1986)—China.
pilosum
var. pilosopetiolatum *(Jien) J.T. Pan in Acta Phytotax. Sin.,* 24(3): 206 (1986): C. pilosopetiolatum.
qinlingense *Jien ex J.T. Pan in Acta Phytotax. Sin.,* 24(3): 208 (1986)—China.
taibaishanense *J.T. Pan in Acta Phytotax. Sin.,* 24(2): 97 (1986)—China.

CHRYSOTHAMNUS (Compos.).
nauseosus
subsp. uintahensis *L.C. Anderson in Great Basin Nat.,* 44(3): 416 (1984)—U.S.A. (Utah).

CHRYSURUS (Gramin.).
aureus *(L.) Besser, Cat. Jard. Bot. Krz.:* 30 (1810); *cf. D.J. Mabberley in Taxon,* 31(1): 70 (1982): Basionym not stated.

CHUKRASIA A. Jussieu in Mirb. & Cass. apud Guill. in Bull. Sci. Nat. Geol., 23: 239 (1830); cf. D.J. Mabberley in Taxon, 31(1): 66 (1982). *MELIACEAE.*

CHYMOCARPUS D. Don ex Brewster, Taylor & Philips, Phil. Mag. & J. Sci., 2: 68 (Jan. 1833); cf. D.J. Mabberley in Taxon, 30(1): 15 (1981). *TROPAEOLACEAE.*

CHYSIS (Orchidac.).
thorvaldsenii *Liebm. in Schouw, Forteg. Kjöb. Bot. Hav. Pl.:* 26 (1847); *cf. D.J. Mabberley in Taxon,* 33(3): 439 (1984)— Locality not stated.

CICER (Legumin.).
subgen. Stenophylloma *A. Santos Guerra & G.P. Lewis in Kew Bull.,* 41(2): 459 (1986).
canariense *A. Santos Guerra & G.P. Lewis in Kew Bull.,* 41(2): 461 (1986)—Canary Is.
nigrum *Hort. Vindob. ex Zeyher, Verz. Saemm. Gewaech. Schwezingen:* 53 (1818); *cf. D.J. Mabberley in Taxon,* 33(3): 441 (1984)—Locality not stated.

CIMINALIS (Gentianac.).
variegata *V.V. Zuev in Bot. Zhurn.,* 71(10): 1406 (1986)—U.S.S.R. (Siberia, Middle Asia), Mongolia.

CINCHONA (Rubiac.).
orixensis *Roxb., Bot. Descr. Swietenia:* 21 (1793) *et in Med. Facts Obs.,* 6: 152 (1795); *cf. D.J. Mabberley in Taxon,* 31(1): 66 (1982)— India.
sanctae-luciae *Kentish, Exp. Obs. New Sp. Bark:* 51 (1784); *cf. D.J. Mabberley in Taxon,* 31(1): 66 (1982)— Locality not stated.

CINCHONA :–
stadtmannii *J.V. Thompson, Cat. Exot. Fl. Maurit., ed. 2:* 8 (1822); *cf. D.J. Mabberley in Taxon,* 31(1): 68 (1982)— Locality not stated.

CINERARIA (Compos.).
dubia *Weigel, Hort. Gryph.:* 32 (1782); *cf. D.J. Mabberley in Taxon,* 33(3): 441 (1984)— Locality not stated.
exsquamata *Sweet, Hort. Brit., ed. 2:* 305 (1830); *cf. D.J. Mabberley in Taxon,* 31(1): 70 (1982)— Locality not stated.
japonica *C.P. Thunberg ex A. Murray in Linnaeus, Syst. Veg., ed. 14:* 766 (1784)— Japan. †
petasites *(L.) Bernh., Syst. Verz.:* 146 (1800); *cf. D.J. Mabberley in Taxon,* 33(3): 443 (1984): Basionym not stated.

CINNA (Gramin.).
suaveolens *(Fries) Fries in Schouw, Forteg. Kjöb. Bot. Hav. Pl.:* 7 (1847); *cf. D.J. Mabberley in Taxon,* 33(3): 439 (1984): Basionym not stated.

CINNAMOMUM (Laurac.).
filipedicellatum *A.J.G.H. Kostermans in Bull. Bot. Surv. India,* 25(1–4): 92 (1983 publ. 1985)—India.
goaense *A.J.G.H. Kostermans in Bull. Bot. Surv. India,* 25(1–4): 94 (1983 publ. 1985)—India.
japonicum
var. chenii *(Nakai) G.F. Tao in J. Wuhan Bot. Res.,* 4(2): 164 (1986), without basionym ref.: C. chenii.
keralaense *A.J.G.H. Kostermans in Bull. Bot. Surv. India,* 25(1–4): 98 (1983 publ. 1985)—India.
walaiwarense *A.J.G.H. Kostermans in Bull. Bot. Surv. India,* 25(1–4): 119 (1983 publ. 1985)—India.

CIPURA (Iridac.).
northiana *(Schneev.) Endl., Cat. Horti Vindob.,* 1: 156 (1842); *cf. D.J. Mabberley in Taxon,* 33(3): 435 (1984): Basionym not stated.

CIRSIUM (Compos.).
hispanicum *(Lam.) Lag., Elenchus:* 5 (1816); *cf. D.J. Mabberley in Taxon,* 31(1): 70 (1982): Basionym not stated.
hispanicum *(Lam.) Font Quer & Rothmaler, Sched. Fl. Iber. Select., Cent. II-III,* n. 286 (1935); *cf. C. Navarro, M. Gutiérrez-Bustillo & A.G. Bueno in Fontqueria,* 7: 12 (1985): Carduus hispanicus.
×winkleri *Font Quer, Sched. Fl. Iber. Select., Cent. II-III,* n. 291 (1935); *cf. C. Navarro, M. Gutiérrez-Bustillo & A.G. Bueno in Fontqueria,* 7: 12 (1985)— Locality not stated. (C. flavispina × gregarium).

CISSAMPELOPSIS (Compos.).
buimalia *(Buch.-Ham. ex D. Don) C. Jeffrey & Y.L. Chen in Kew Bull.,* 41(4): 937 (1986): Senecio buimalia. †

CISSUS (Vitac.).
aubertiana *(Gage) P. Singh & B.V. Shetty in Taxon,* 35(3): 596 (1986): Vitis aubertiana.

CISTANTHE (Portulacac.).
grandiflora *(Lindl.) Schlecht., Hort. Hal.:* 10 (?1841); *cf. D.J. Mabberley in Taxon,* 33(3): 441 (1984): Calandrinia grandiflora.

CISTUS (Cistac.).
×cebennensis *P. Aubin & J. Prudhomme in Bull. Mens. Soc. Linn. Lyon,* 55(4): 136 (1986)— France. (C. pouzolzii × salvifolius).
dunalianus *Sweet, Hort. Brit.:* 34 (1826); *cf. D.J. Mabberley in Taxon,* 31(1): 70 (1982)— Locality not stated.

CITRUS (Rutac.).
japonica *C.P. Thunberg ex A. Murray in Linnaeus, Syst. Veg., ed. 14:* 697 (1784)— Japan. †
×lapinii *I.S. Kapanadze in Soobshch. Akad. Nauk Gruz. SSR,* 120(2): 390 (1985), without latin descr. or type—Cultivation.
limon *(L.) Burm.f., Fl. Ind.:* 173 (1768): Citrus medica β limon.
microcarpa *A. Bunge, Enum. Pl. China Bor. Coll.:* 10 (1833)—China. †

CLARIONEA (Compos.).
pungens *D. Don ex Taylor & Philips, Phil. Mag. & J. Sci.,* 2: 388 (May 1832); *cf. D.J. Mabberley in Taxon,* 30(1): 15 (1981)— Locality not stated.

CLARKIA (Onagrac.).
sect. Sympherica *K.E. Holsinger & H. Lewis in Ann. Missouri Bot. Gard.,* 73(2): 492 (1986).
subsect. Sympherica *K.E. Holsinger & H. Lewis in Ann. Missouri Bot. Gard.,* 73(2): 493 (1986).

CLEIDION (Euphorbiac.).
alongense *S.S.R. Bennet & S. Chandra in Indian Forester,* 111(10): 846 (1985)— India.

CLEISOSTOMA (Orchidac.).
comberi *J.J. Wood in Kew Bull.,* 41(3): 700 (1986)—Java.

CLEISTOGENES (Gramin.).
ramiflora *Keng & C.P. Wang in Bull. Bot. Res. North-East. Forest. Inst.,* 6(1): 175 (1986)— China.

CLEMATIS (Ranunculac.).
finetii
var. pedata *W.T. Wang in Acta Bot. Yunnanica,* 8(3): 269 (1986)—China.

CLEMATIS :-
florida *C.P. Thunberg ex A. Murray in Linnaeus, Syst. Veg., ed. 14:* 512 (1784)—Japan. †
iliensis *Y.S. Hou & W.H. Hou in Bull. Bot. Res. North-East. Forest. Inst.,* 6(2): 131 (1986)—Xinjiang.
japonica *C.P. Thunberg ex A. Murray in Linnaeus, Syst. Veg., ed. 14:* 512 (1784)—Japan. †
jialasaensis *W.T. Wang in Acta Bot. Yunnanica,* 8(3): 268 (1986)—Xizang.
jinzhaiensis *Z.W. Xue & X.W. Wang in Acta Phytotax. Sin.,* 24(5): 406 (1986)—China.
lingyunensis *W.T. Wang in Acta Bot. Yunnanica,* 8(3): 266 (1986)—China.
pseudootophora
var. integra *W.T. Wang in Acta Bot. Yunnanica,* 8(3): 266 (1986)—China.

CLEOME (Capparac.).
indecora *Liebm. in Schouw, Ind. Sem. Hort. Acad. Haun.,* 1845; *Linnaea,* 19: 403 (1847); *cf. D.J. Mabberley in Taxon,* 32(1): 84 (1983)— Locality not stated.
sesquiorygalis *Naud. ex C. Huber, Revue horticole,* 46: 458, 467 (1874); *cf. D.J. Mabberley in Taxon,* 34(3): 451 (1985)—Locality not stated.

CLEOMELLA (Capparac.).
palmerana
var. goodrichii *S.L. Welsh in Great Basin Nat.,* 46(2): 263 (1986)—U.S.A. (Utah).

CLERODENDRUM (Verbenac.).
cuneiforme *H.N. Moldenke in Phytologia,* 59(2): 119 (1986), nom. nov.: C. cuneatum *Gürke.*
trichotomum *C.P. Thunberg ex A. Murray in Linnaeus, Syst. Veg., ed. 14:* 578 (1784)—Japan. †

CLIDEMIA (Melastomatac.).
rubra *(Aubl.) G. Don f in Loud., Hort. Brit.:* 174 (1830); *cf. D.J. Mabberley in Taxon,* 30(1): 15 (1981): Basionym not stated.

CLIMACOPTERA (Chenopodiac.).
sect. Amblyostegia *U. Pratov, Rod Climacoptera Botsch.:* 25 (1986).
sect. Brachyphylla *Iljin ex U. Pratov, Rod Climacoptera Botsch.:* 23 (1986).
subsect. Crassae *U. Pratov, Rod Climacoptera Botsch.:* 38 (1986).
subsect. Glabrae *U. Pratov, Rod Climacoptera Botsch.:* 46 (1986).
sect. Heterotricha *Iljin ex U. Pratov, in Rod Climacoptera Botsch.:* 24 (1986).
subsect. Lanatae *U. Pratov, Rod Climacoptera Botsch.:* 43 (1986).
subsect. Oppositifoliae *U. Pratov, Rod Climacoptera Botsch.:* 26 (1986).
sect. Oxyphyllae *U. Pratov, Rod Climacoptera Botsch.:* 47 (1986).
subsect. Transoxanae *U. Pratov, Rod Climacoptera Botsch.:* 29 (1986).

CLIMACOPTERA :-
sect. Ulotricha *U. Pratov, Rod Climacoptera Botsch.:* 19 (1986).
czelekenica *U. Pratov, Rod Climacoptera Botsch.:* 21 (1986)—U.S.S.R. (Middle Asia).
narynensis *U. Pratov, Rod Climacoptera Botsch.:* 36 (1986)—U.S.S.R. (Middle Asia).
oxyphylla *U. Pratov, Rod Climacoptera Botsch.:* 47 (1986)—U.S.S.R. (Middle Asia).
pjataevae *U. Pratov, Rod Climacoptera Botsch.:* 26 (1986)—U.S.S.R. (Middle Asia).
ptiloptera *U. Pratov, Rod Climacoptera Botsch.:* 44 (1986)—U.S.S.R. (Middle Asia).
susamyrica *U. Pratov, Rod Climacoptera Botsch.:* 34 (1986)—U.S.S.R. (Middle Asia).
tyshchenkoi *U. Pratov, Rod Climacoptera Botsch.:* 32 (1986)—U.S.S.R. (Middle Asia).
ustjurtensis *U. Pratov, Rod Climacoptera Botsch.:* 27 (1986)—U.S.S.R. (Middle Asia).
vachschi *Kinz. & U. Pratov in U. Pratov, Rod Climacoptera Botsch.:* 30 (1986)—U.S.S.R. (Middle Asia).

CLOWESIA (Orchidac.).
dodsoniana *E. Aguirre León in Orquídea (México),* 10(1): 192 (1986)—Mexico.

CLYMENIA (Rutac.).
platypoda *B.C. Stone in Proc. Acad. Nat. Sci. Philadelphia,* 137(2): 223 (1985)—New Guinea (W. Irian).

CNICUS (Compos.).
altaicus *(Spreng.) Trev., Ind. Sem. Wratislav.:* 2 (1818); *cf. D.J. Mabberley in Taxon,* 31(1): 70 (1982): Basionym not stated.
diacantha *(Lab.) C.C. Gmel., Hort. Mag. Duc. Bad. Carlsruh.:* 73 (1811); *cf. D.J. Mabberley in Taxon,* 33(3): 440 (1984): Basionym not stated.
semipectinatus *Hort. Gott. ex Hornem., Hort. Hafn., Suppl.:* 93 (1819); *cf. D.J. Mabberley in Taxon,* 33(3): 437 (1984)— Locality not stated.

CNIDIUM (Umbellif.).
nullivittatum *K.T. Fu in Fl. Tsinlingensis,* 1(3): 460, 415 (1981)—China.
sinchianum *K.T. Fu in Fl. Tsinlingensis,* 1(3): 459, 415 (1981)—China.

COCCINIA (Cucurbitac.).
diversifolia *Naud. ex Huber, Cat. Print. 1864:* 6 (1864); *cf. D.J. Mabberley in Taxon,* 34(3): 452 (1985)— Locality not stated.
mackenii *Naud. ex Huber, Cat. Print. 1865:* 5 (1865); *Wochenschrift,* 8: 52 (1865); *cf. D.J. Mabberley in Taxon,* 34(3): 452 (1985)—Locality not stated.

COCCOTHRINAX (Palmae).
fagildei *A. Borhidi & O. Muñiz in Acta Bot. Hung.*, 31(1–4): 227 (1985)—Cuba.
trinitensis *A. Borhidi & O. Muñiz in Acta Bot. Hung.*, 31(1–4): 228 (1985)—Cuba.

COCHLEARIA (Crucif.).
rupicola *D.C. Zhang & J.Z. Shao in Acta Phytotax. Sin.*, 24(5): 404 (1986)—China.

COCHLEARIOPSIS Y.H. Zhang in Acta Bot. Yunnanica, 7(2): 143 (1985). *CRUCIFERAE.*
zhejiangensis *Y.H. Zhang in Acta Bot. Yunnanica,* 7(2): 144 (1985)—China.

CODIAEUM (Euphorbiac.).
peltatum *(Labill.) P.S. Green in Kew Bull.*, 41(4): 1026 (1986): Chrozophora peltata.

CODONOPSIS (Campanulac.).
ussuriensis
f. viridiflora *J. Ōhara in J. Phytogeogr. Taxon.*, 33(2): 72 (1985)—Japan.

COELESTINA (Compos.).
superbiens *Huber, Trade Cat.*: 3 (1880); *cf. D.J. Mabberley in Taxon,* 34(3): 450 (1985)— Locality not stated.

COFFEA (Rubiac.).
mufindiensis
subsp. pawekiana *(Bridson) D.M. Bridson in Kew Bull.*, 41(2): 309 (1986): C. pawekiana.
sessiliflora *D.M. Bridson in Kew Bull.*, 41(2): 307 (1986)—Kenya.
subsp. mwasumbii *D.M. Bridson in Kew Bull.*, 41(2): 308 (1986)—Tanzania.

COLA (Sterculiac.).
letouzeyana *B.-A. Nkongmeneck in Bull. Mus. Nation. Hist Nat, B, Adansonia*, Ser. 4, 7(3): 337 (1985 publ. 1986)—Cameroun.

COLEA (Bignoniac.).
celsiana *Ten., Cat. Ort. Bot. Nap.*: 87 (1845); *cf. D.J. Mabberley in Taxon,* 31(1): 70 (1982)— Locality not stated.

COLEUS (Labiat.).
osmirrhizon *W. Elliot, Fl. Andhirica:* 105 (1859), *based on Hort. malab.* 9, t. 74; *cf. D.J. Mabberley in Taxon,* 32(1): 87 (1983)—Locality not stated.

COLICODENDRON (Capparac.).
breynia *(L.) Endl., Cat. Horti Vindob.*, 2: 264 (1842); *cf. D.J. Mabberley in Taxon,* 33(3): 435 (1984): Capparis breynia.

COLLINSIA (Scrophulariac.).
parviflora
var. grandiflora *(Dougl. ex Lindl.) F.R. Ganders & G.R. Krause in Madroño,* 33(1): 69 (1986): C. grandiflora.

COLLOMIA (Polemoniac.).
renacta *E. Joyal in Brittonia,* 38(3): 243 (1986)—U.S.A. (Oregon, Nevada).

COLONA (Tiliac.).
hamannii *H. Riedl & C. Riedl-Dorn in Linz. Biol. Beitr.*, 17(2): 333 (1985)—Thailand.

COLUBRINA (Rhamnac.).
nicholsonii *A.E. van Wyk & B.D. Schrire in S. Afr. J. Bot.*, 52(4): 379 (1986)—South Africa (Cape Province).

×**COLYCEA** J. Fernández Casas & A. Susanna de la Serna in Fontqueria, 2: 21 (1982). *COMPOSITAE.* (Colymbada × Jacea).
valdesii-bermejoi *(Fdez. Casas & Susanna) J. Fernández Casas & A. Susanna de la Serna in Fontqueria,* 2: 21 (1982), without basionym page: Centaurea × valdesii-bermejoi.

×**COLYMBACOSTA** (Compos.).
genesii-lopezii *(Fdez. Casas & Susanna) J. Fernández Casas & A. Susanna de la Serna in Fontqueria,* 2: 21 (1982), without basionym page: Centaurea × genesii-lopezii.

COLYMBADA (Compos.).
amblensis *(Graells) J. Fernández Casas & A. Susanna de la Serna in Fontqueria,* 2: 19 (1982): Centaurea amblensis.
var. tentudaica *(Rivas Goday) J. Fernández Casas & A. Susanna de la Serna in Fontqueria,* 2: 19 (1982): Centaurea toletana subsp. tentudaica.
×andresiana *(Fdez. Casas & Susanna) J. Fernández Casas & A. Susanna de la Serna in Fontqueria,* 2: 20 (1982): Centaurea × andresiana.
borjae *(Valdés Bermejo & Rivas Goday) J. Fernández Casas & A. Susanna de la Serna in Fontqueria,* 2: 20 (1982): Centaurea borjae.
carolipauana *(Fdez. Casas & Susanna) J. Fernández Casas & A. Susanna de la Serna in Fontqueria,* 2: 21 (1982): Centaurea carolipauana.
×ceballosii *(Fdez. Casas) J. Fernández Casas & A. Susanna de la Serna in Fontqueria,* 2: 20 (1982): Centaurea × ceballosii.
lagascana *(Graells) J. Fernández Casas & A. Susanna de la Serna in Fontqueria,* 2: 20 (1982): Centaurea lagascana.
lainzii *(J. Fernández Casas) J. Fernández Casas & A. Susanna de la Serna in Fontqueria,* 2: 21 (1982): Centaurea lainzii.
×losana *(Pau) J. Fernández Casas & A. Susanna de la Serna in Fontqueria,* 2: 20 (1982): Centaurea × losana.
mariana *(Nyman) J. Fernández Casas & A. Susanna de la Serna in Fontqueria,* 2: 19 (1982): Centaurea mariana.

COLYMBADA :–
×tatayana *(Fdez. Casas & Susanna) J. Fernández Casas & A. Susanna de la Serna in Fontqueria,* 2: 20 (1982): Centaurea × tatayana.
toletana
var. argecillensis *(Gredilla) J. Fernández Casas & A. Susanna de la Serna in Fontqueria,* 2: 19 (1982): Centaurea argecillensis.
×zubiae *(Pau) J. Fernández Casas & A. Susanna de la Serna in Fontqueria,* 2: 20 (1982): Centaurea × zubiae.

COMBRETUM (Combretac.).
petrophilum *E. Retief in Bothalia,* 16(1): 44 (1986)—South Africa (Transvaal).

COMMELINA (Commelinac.).
clandestina *Mart., Hort. Reg. Monac. Sem.,* 1839; *Linnaea 14, Lit.:* 133 (?1841); *cf. D.J. Mabberley in Taxon,* 32(1): 84 (1983)—Locality not stated.
debilis *Ledeb., Enum. Pl. Hort. Bot. Dorp., Suppl. I:* 2 (1811); *cf. D.J. Mabberley in Taxon,* 33(3): 441 (1984)— Locality not stated.
pubescens *Hornem., Hort. Hafn.:* 950 (1815); *cf. D.J. Mabberley in Taxon,* 33(3): 437 (1984)— Locality not stated.

COMPARETTIA (Orchidac.).
coccinea
var. longicalcarata *T. Hashimoto in Ann. Tsukuba Bot. Gard.,* 4: 4 (1986)—Peru, Bolivia.
falcata
var. paulensis *(Cogn.) I. Bock in Orchidee,* 37(5): 203 (1986): C. paulensis.

CONOSTEPHIUM (Epacridac.).
marchantiorum *A. Strid in Willdenowia,* 16(1): 178 (1986)—Australia (W. Australia).

CONVALLARIA (Convallariac.).
sibirica *(Delar.) Ker in J. Sci. Arts,* 1: 182 (1816); *cf. D.J. Mabberley in Taxon,* 31(1): 70 (1982): Polygonatum sibiricum.
spicata *C.P. Thunberg ex A. Murray in Linnaeus, Syst. Veg., ed. 14:* 334 (1784)—Japan. †

CONVOLVULUS (Convolvulac.).
affinis *(Endl.) J.H. Maiden in Proc. Linn. Soc. New S. Wales,* 28(4): 711 (1903 publ. 1904), nom. illegit. non Scheele 1848: Calystegia affinis.
americanus *(Sims) J.W. Loud., Lad. Fl. Gard. Orn. Perenn.,* 2: 98 (1844); *cf. D.J. Mabberley in Taxon,* 32(1): 87 (1983): Basionym not stated.
dissectus *(Willd.) Endl., Cat. Horti Vindob.,* 2: 65 (1842); *cf. D.J. Mabberley in Taxon,* 33(3): 435 (1984): Basionym not stated.
edulis *C.P. Thunberg ex A. Murray in Linnaeus, Syst. Veg., ed. 14:* 203 (1784)—Japan. †

CONVOLVULUS :–
japonicus *C.P. Thunberg ex A. Murray in Linnaeus, Syst. Veg., ed. 14:* 200 (1784)—Japan. †
ocularis *(Bartl.) Endl., Cat. Horti Vindob.,* 2: 64 (1842); *cf. D.J. Mabberley in Taxon,* 33(3): 435 (1984): Basionym not stated.
trinervius *C.P. Thunberg ex A. Murray in Linnaeus, Syst. Veg., ed. 14:* 201 (1784)—Japan. †
willdenowii *Dehnh., Cat. Pl. Hort. Camald.:* 10 (1829); *cf. D.J. Mabberley in Taxon,* 30(1): 15 (1981)— Locality not stated.

COPROSMA (Rubiac.).
nephelephila *J. Florence in Bull. Mus. Nation. Hist. Nat., B, Adansonia, Ser. 4,* 8(1): 3 (1986)—Marquesas Is.
reticulata *J. Florence in Bull. Mus. Nation. Hist. Nat., B, Adansonia, Ser. 4,* 8(1): 6 (1986)—Marquesas Is.

CORALLORHIZA (Orchidac.).
anandae *C.L. Malhotra & B. Balodi in Bull. Bot. Surv. India,* 26(1–2): 108 (1984 publ. 1985)—India.

CORCHOROPSIS (Sterculiac.).
tomentosa
var. micropetala *Y.T. Chang in Acta Phytotax. Sin.,* 24(1): 23 (1986)—China.

CORCHORUS (Tiliac.).
brevicornutus *K. Vollesen in Kew Bull.,* 41(4): 957 (1986)—Ethiopia, Tanzania.
japonicus *C.P. Thunberg ex A. Murray in Linnaeus, Syst. Veg., ed. 14:* 501 (1784)—Japan. †

CORDIA (Boraginac.).
dardanoi *N. Taroda in Notes Roy. Bot. Gard. Edinburgh,* 44(1): 111 (1986), as 'dardani'—Brazil.
harleyi *N. Taroda in Notes Roy. Bot. Gard. Edinburgh,* 44(1): 128 (1986)—Brazil.
leucomalloides *N. Taroda in Notes Roy. Bot. Gard. Edinburgh,* 44(1): 125 (1986)—Brazil.
mayoi *N. Taroda in Notes Roy. Bot. Gard. Edinburgh,* 44(1): 129 (1986)—Brazil.

CORDYLANTHUS (Scrophulariac.).
eremicus
subsp. kernensis *T.I. Chuang & L.R. Heckard, Syst. Evol. Cordylanthus (Syst. Bot. Monog., 10):* 91 (1986)—U.S.A. (California).
kingii
subsp. densiflorus *T.I. Chuang & L.R. Heckard, Syst. Evol. Cordylanthus (Syst. Bot. Monog., 10):* 81 (1986)—U.S.A. (Utah).
subsp. helleri *(Ferris) T.I. Chuang & L.R. Heckard, Syst. Evol. Cordylanthus (Syst. Bot. Monog., 10):* 81 (1986): Adenostegia helleri.

CORDYLANTHUS :-
pilosus
subsp. hansenii *(Ferris) T.I. Chuang & L.R. Heckard, Syst. Evol. Cordylanthus (Syst. Bot. Monog., 10):* 66 (1986): Adenostegia hansenii.
subsp. trifidus *(Robinson & Greenman) T.I. Chuang & L.R. Heckard, Syst. Evol. Cordylanthus (Syst. Bot. Monog., 10):* 68 (1986): C. pilosus var. trifidus.
rigidus
subsp. involutus *(Wiggins) T.I. Chuang & L.R. Heckard, Syst. Evol. Cordylanthus (Syst. Bot. Monog., 10):* 47 (1986): C. involutus.
subsp. littoralis *(Ferris) T.I. Chuang & L.R. Heckard, Syst. Evol. Cordylanthus (Syst. Bot. Monog., 10):* 42 (1986): Adenostegia littoralis.
subsp. setigerus *T.I. Chuang & L.R. Heckard, Syst. Evol. Cordylanthus (Syst. Bot. Monog., 10):* 43 (1986)—U.S.A. (California), Mexico.
tenuis
subsp. barbatus *T.I. Chuang & L.R. Heckard, Syst. Evol. Cordylanthus (Syst. Bot. Monog., 10):* 58 (1986)—U.S.A. (California).
subsp. capillaris *(Pennell) T.I. Chuang & L.R. Heckard, Syst. Evol. Cordylanthus (Syst. Bot. Monog., 10):* 60 (1986): C. capillaris.
subsp. pallescens *(Pennell) T.I. Chuang & L.R. Heckard, Syst. Evol. Cordylanthus (Syst. Bot. Monog., 10):* 55 (1986): C. pallescens.
subsp. viscidus *(T. Howell) T.I. Chuang & L.R. Heckard, Syst. Evol. Cordylanthus (Syst. Bot. Monog., 10):* 56 (1986): Adenostegia viscida.
wrightii
subsp. kaibabensis *T.I. Chuang & L.R. Heckard, Syst. Evol. Cordylanthus (Syst. Bot. Monog., 10):* 87 (1986)—U.S.A. (Arizona).
subsp. tenuifolius *(Pennell) T.I. Chuang & L.R. Heckard, Syst. Evol. Cordylanthus (Syst. Bot. Monog., 10):* 86 (1986): C. tenuifolius.

CORDYLINE (Agavac.).
cernua *(Jacq.) Endl., Cat. Horti Vindob.,* 1: 149 (1842); *cf. D.J. Mabberley in Taxon,* 33(3): 435 (1984): Basionym not stated.
ferrea *(L.) Endl., Cat. Horti Vindob.,* 1: 149 (1842); *cf. D.J. Mabberley in Taxon,* 33(3): 435 (1984): Basionym not stated.
marginata *(Lam.) Endl., Cat. Horti Vindob.,* 1: 148 (1842); *cf. D.J. Mabberley in Taxon,* 33(3): 435 (1984): Dracaena marginata.
petiolaris *(Domin) L. Pedley in Fl. Australia,* 46: 220 (1986): C. terminalis var. petiolaris.
reflexa *(Lam.) Endl., Cat. Horti Vindob.,* 1: 148 (1842); *cf. D.J. Mabberley in Taxon,* 33(3): 435 (1984): Dracaena reflexa.

COREOPSIS (Compos.).
queretarensis *B.L. Turner in Brittonia,* 38(2): 168 (1986)—Mexico.

CORISPERMUM (Chenopodiac.).
spicatum *Schweigg., Enum. Pl. Hort. Bot. Regiom.:* 27 (1812); *cf. D.J. Mabberley in Taxon,* 33(3): 442 (1984)— Locality not stated.

CORNUS (Cornac.).
japonica *C.P. Thunberg ex A. Murray in Linnaeus, Syst. Veg., ed. 14:* 159 (1784)—Japan. †

CORONILLA (Legumin.).
moravica *A. Chrtková & E. Stavělová in Folia Geobot. Phytotax.,* 21(3): 325 (1986)—Czechoslovakia.

CORYANTHES (Orchidac.).
horichiana *R. Jenny in Orchidee,* 37(3): 126 (1986)—Costa Rica.

CORYBAS (Orchidac.).
acutus *J. Dransfield & J.B. Comber in Kew Bull.,* 41(3): 595 (1986)—Java.
calcicolus *J. Dransfield & G. Smith in Kew Bull.,* 41(3): 608 (1986)—Malaya.
calopeplos *J. Dransfield & G. Smith in Kew Bull.,* 41(3): 584 (1986)—Malaya.
comptus *J. Dransfield & G. Smith in Kew Bull.,* 41(3): 606 (1986)—Malaya.
holttumii *J. Dransfield & G. Smith in Kew Bull.,* 41(3): 590 (1986)—Malaya.
muluensis *J. Dransfield in Kew Bull.,* 41(3): 588 (1986)—Borneo (Sarawak).
piliferus *J. Dransfield in Kew Bull.,* 41(3): 588 (1986)—Borneo (Sabah).
praetermissus *J. Dransfield & J.B. Comber in Kew Bull.,* 41(3): 592 (1986)—Java.
ramosianus *J. Dransfield in Kew Bull.,* 41(3): 611 (1986)—Philippine Is.
selangorensis *J. Dransfield & G. Smith in Kew Bull.,* 41(3): 606 (1986)—Malaya.
serpentinus *J. Dransfield in Kew Bull.,* 41(3): 604 (1986)—Borneo (Sabah).
umbrosus *J. Dransfield & J.B. Comber in Kew Bull.,* 41(3): 580 (1986)—Java.
villosus *J. Dransfield & G. Smith in Kew Bull.,* 41(3): 599 (1986)—Malaya.

CORYCIUM (Orchidac.).
ingeanum *E.G.H. Oliver in S. Afr. J. Bot.,* 52(3): 256 (1986)—South Africa (Cape Province).

CORYDALIS (Papaverac.).
sect. Aulacostigma *M. Lidén in Op. Bot.,* 88: 28 (1986).
sect. Bipapillata *M. Lidén in Op. Bot.,* 88: 27 (1986).
sect. Cheilanthifoliae *M. Lidén in Op. Bot.,* 88: 28 (1986).
sect. Thalictrifoliae *(Fedde) M. Lidén in Op. Bot.,* 88: 28 (1986): C. § Thalictrifoliae.

CORYDALIS :–
ambigua
f. lineariloba *(Maxim.) C.Y. Wu & Z.Y. Su in Acta Bot. Yunnanica,* 7(3): 266 (1985): C. ambigua lusus lineariloba.
anethifolia *C.Y. Wu & Z.Y. Su in Acta Bot. Yunnanica,* 8(3): 279 (1986)—China.
bimaculata *C.Y. Wu & T.Y. Shu in Fl. Xizangica,* 2: 305 (1985)—Xizang.
corymbosa *C.Y. Wu & T.Y. Shu in Fl. Xizangica,* 2: 304 (1985)—Xizang.
crispa
var. laeviangula *C.Y. Wu & H. Chuang in Fl. Xizangica,* 2: 315 (1985)—Xizang.
var. setulosa *C.Y. Wu & H. Chuang in Fl. Xizangica,* 2: 317 (1985)—Xizang.
drakeana
var. tibetica *C.Y. Wu & H. Chuang in Fl. Xizangica,* 2: 302 (1985)—Xizang.
fangshanensis *W.T. Wang in Fl. Beijing,* 1: 670, 282 (1984)—China.
gracillima *C.Y. Wu in Fl. Xizangica,* 2: 319 (1985), nom. nov., without replaced synonym ref.: C. gracilis *Franch.*
kingii
var. minuticalcarata *C.Y. Wu in Fl. Xizangica,* 2: 284 (1985)—Xizang.
meifolia
var. ecristata *C.Y. Wu & T.Y. Shu in Fl. Xizangica,* 2: 312 (1985)—Xizang.
oligosperma *C.Y. Wu & T.Y. Shu in Fl. Xizangica,* 2: 303 (1985)—Xizang.
pallida
var. chanetii *(Lévl.) S.Y. He in Fl. Beijing,* 1: 281 (1984), without basionym ref.: C. chanetii.
paniculata *C.Y. Wu & H. Chuang in Fl. Xizangica,* 2: 298 (1985)—Xizang.
papillipes *C.Y. Wu in Fl. Xizangica,* 2: 317 (1985)—Xizang.
pseudoadoxa *C.Y. Wu & H. Chuang in Fl. Xizangica,* 2: 276 (1985), as 'pseudo-adoxa'—Xizang.
pseudothyrsiflora *C.Y. Wu & T.Y. Shu in Fl. Xizangica,* 2: 310 (1985)—Xizang.
pubescens *C.Y. Wu & H. Chuang in Fl. Xizangica,* 2: 311 (1985)—Xizang.
pubicaula *C.Y. Wu & H. Chuang in Fl. Xizangica,* 2: 266 (1985)—Xizang.
punicea *C.Y. Wu in Fl. Xizangica,* 2: 284 (1985), nom. nov., without replaced synonym ref.: C. rosea *Maxim.*
purpureocalcarata *C.Y. Wu & T.Y. Shu in Fl. Xizangica,* 2: 313 (1985)—Xizang.
remota
f. heteroclita *(K.T. Fu) C.Y. Wu & Z.Y. Su in Acta Bot. Yunnanica,* 7(3): 269 (1985): C. remota var. heteroclita.
f. lineariloba *(Maxim.) C.Y. Wu & Z.Y. Su in Acta Bot. Yunnanica,* 7(3): 268 (1985): C. remota var. lineariloba.
f. non-apiculata *(Ohwi) C.Y. Wu & Z.Y. Su in Acta Bot. Yunnanica,* 7(3): 269 (1985): C. turtschaninovii var. non-apiculata.

CORYDALIS :–
f. papillosa *(Kitag.) C.Y. Wu & Z.Y. Su in Acta Bot. Yunnanica,* 7(3): 268 (1985): C. turtschaninovii var. papillosa.
f. rotundiloba *(Maxim.) C.Y. Wu & Z.Y. Su in Acta Bot. Yunnanica,* 7(3): 268 (1985): C. remota var. rotundiloba.
semiaquilegifolia *C.Y. Wu & H. Chuang in Fl. Xizangica,* 2: 279 (1985)—China, Xizang.
tenuicalcarata *C.Y. Wu & T.Y. Shu in Fl. Xizangica,* 2: 307 (1985)—Xizang.
tibetoalpina *C.Y. Wu & T.Y. Shu in Fl. Xizangica,* 2: 294 (1985), as 'tibeto-alpina'—Xizang.
tibetooppositifolia *C.Y. Wu & T.Y. Shu in Fl. Xizangica,* 2: 299 (1985), as 'tibeto-oppositifolia'—Xizang.
yanhusuo *W.T. Wang in Acta Bot. Yunnanica,* 7(3): 269 (1985)—China.

CORYNOSTYLIS (Violac.).
albiflora *Bull, Retail List,* 54: 3 (1870); *cf. D.J. Mabberley in Taxon,* 34(3): 456 (1985)—Locality not stated.

CORYPHANTHA (Cactac.).
salm-dyckiana
var. brunnea *(Salm-Dyck) G. Unger in Kakt. And. Sukk.,* 37(5): 86 (1986): Mammillaria salm-dyckiana var. brunnea.

COSMEA (Compos.).
tenuifolia *(Lindl.) J.W. Loud., Lad. Fl. Gard. Orn. Ann.:* 185 (1840); *cf. D.J. Mabberley in Taxon,* 32(1): 87 (1983): Cosmos tenuifolia.

COSMIDIUM (Compos.).
engelmannii *Huber, Cat. Print. 1865:* 3 (1865); *Wochenschrift,* 8: 54 (1865); *cf. D.J. Mabberley in Taxon,* 34(3): 450 (1985)—Locality not stated.

COTYLEDON (Crassulac.).
acutiflora *(Andr.) Ait. f., Epit. Hort. Kew.:* 366 (1814); *cf. D.J. Mabberley in Taxon,* 31(1): 70 (1982): Basionym not stated.

COURSETIA (Legumin.).
insomniifolia *M. Lavin in Madroño,* 33(3): 182 (1986)—Mexico.

COUTAPORTLA (Rubiac.).
guatemalensis *(Standley) D.H. Lorence in Syst. Bot.,* 11(1): 210 (1986): Portlandia guatemalensis.

COWANIA D. Don ex Tilloch & Taylor, Phil. Mag., 64: 374 (Nov. 1824); cf. D.J. Mabberley in Taxon, 30(1): 15 (1981). *ROSACEAE.*

CRACCA (Legumin.).
pacifica *M. Sousa S. & M. Lavin in Brittonia,* 38(3): 302 (1986), nom. nov.: Balboa diversifolia *Liebm.*

CRAIBIODENDRON (Ericac.).
scleranthum *(Dop) W.S. Judd in J. Arnold Arbor.,* 67(4): 456 (1986): Nuihonia sclerantha.
var. kwangtungense *(S.Y. Hu) W.S. Judd in J. Arnold Arbor.,* 67(4): 457 (1986): C. kwangtungense.
vietnamense *W.S. Judd in J. Arnold Arbor.,* 67(4): 462 (1986)—Vietnam.

CRAMBE (Crucif.).
glaberrima *(Bornm.) Mouterde ex W. Greuter & Burdet in Willdenowia,* 15(2): 417 (1986): C. persica var. glaberrima.

CRASSOCEPHALUM (Compos.).
effusum *(Mattf.) C. Jeffrey in Kew Bull.,* 41(4): 907 (1986): Senecio effusus.
heteromorphum *(Hutch. & Burtt) C. Jeffrey in Kew Bull.,* 41(4): 908 (1986): Senecio heteromorphus.
paludum *C. Jeffrey in Kew Bull.,* 41(4): 906 (1986)—Uganda, Kenya, Tanzania, Sudan, Rwanda, Burundi, Zaïre, Zambia.
splendens *C. Jeffrey in Kew Bull.,* 41(4): 905 (1986)—Tanzania.

CRATAEGUS (Rosac.).
glabra *C.P. Thunberg ex A. Murray in Linnaeus, Syst. Veg., ed. 14:* 465 (1784)—Japan. †
laevis *C.P. Thunberg ex A. Murray in Linnaeus, Syst. Veg., ed. 14:* 465 (1784)—Japan. †
nivea *Host ex Ten., Cat. Ort. Bot. Nap.:* 23 (1845); *cf. D.J. Mabberley in Taxon,* 32(1): 84 (1983)— Locality not stated.
shandongensis *F.Z. Li & W.D. Peng in Bull. Bot. Res. North-East. Forest. Inst.,* 6(4): 149 (1986)—China.
villosa *C.P. Thunberg ex A. Murray in Linnaeus, Syst. Veg., ed. 14:* 465 (1784)—Japan. †

CREMASTOSPERMA (Annonac.).
macrocarpum *P.J.M. Maas in Proc. Kon. Nederl. Akad. Wetensch., C,* 89(3): 253 (1986)—Venezuela.
panamense *P.J.M. Maas in Proc. Kon. Nederl. Akad. Wetensch., C,* 89(3): 254 (1986)—Panama.

CREPIS (Compos.).
lachenalii *C.C. Gmel., Hort. Mag. Duc. Bad. Carlsruh.:* 82 (1811); *cf. D.J. Mabberley in Taxon,* 33(3): 440 (1984)— Locality not stated.
senecioides
subsp. filiformis *(Viv.) S.A. Alavi in Fl. Libya,* 107: 419 (1983): C. filiformis.
subsp. nudiflora *(Viv.) S.A. Alavi in Fl. Libya,* 107: 419 (1983): C. nudiflora.

CRINUM (Amaryllidac.).
biflorum *Rottb., Pl. Hort. Univ. Rar. Prog.:* 12 (1773); *cf. D.J. Mabberley in Taxon,* 33(3): 439 (1984)— Locality not stated.

CRINUM :–
capense *Herb., Bot. Mag.,* 47: t. 2121 5 (1820); *cf. D.J. Mabberley in Taxon,* 32(1): 87 (1983)— Locality not stated.
graciliflorum *Kunth & Bouché, Ind. Sem. Hort. Bot. Berol.,* 1844: 9 (1844); *Linnaea,* 18: 499 (1845); *cf. D.J. Mabberley in Taxon,* 32(1): 84 (1983)— Locality not stated.
toxicarium *Roxb. ex Hornem., Hort. Hafn., Suppl.:* 130 (1819); *cf. D.J. Mabberley in Taxon,* 33(3): 437 (1984)— Locality not stated.

CRITESION (Gramin.).
murinum
subsp. glaucum *(Steudel) B.K. Simon in Austrobaileya,* 2(3): 241 (1986): Hordeum glaucum.

CROTALARIA (Legumin.).
sessiliflora
var. anthylloides *(Lam.) A.A. Ansari & K. Thothathri in J. Econ. Taxon. Bot.,* 6(3): 745 (1985): C. anthylloides.
f. garhwalensis *A.A. Ansari & K. Thothathri in J. Econ. Taxon. Bot.,* 6(3): 744 (1985)—India.
f. leariformis *A.A. Ansari & K. Thothathri in J. Econ. Taxon. Bot.,* 6(3): 745 (1985)—India.
tenuifolia *Roxb. ex Hornem., Hort. Hafn., Suppl.:* 151 (1819); *cf. D.J. Mabberley in Taxon,* 33(3): 437 (1984)— Locality not stated.

CROTON (Euphorbiac.).
acutum *C.P. Thunberg ex A. Murray in Linnaeus, Syst. Veg., ed. 14:* 863 (1784)—Japan. †
casarettianus *Vis., Ort. Bot. Padova:* 73, 137 (1842); *cf. D.J. Mabberley in Taxon,* 33(3): 442 (1984)— Locality not stated.
caudatus
var. obovoideus *N.P. Balakrishnan & T. Chakrabarty in Bull. Bot. Surv. India,* 25(1–4): 190 (1983 publ. 1985)—India.
damayeshu *Y.T. Chang in Acta Phytotax. Sin.,* 24(2): 143 (1986)—China.
himalaicus *D.G. Long in Notes Roy. Bot. Gard. Edinburgh,* 44(1): 170 (1986)—India, Nepal, Assam.
japonicus *Linn.f., Suppl. Plant. Syst. Veg.:* 422 (1782), as 'Iaponicum'—Japan.
purpurascens *Y.T. Chang in Acta Phytotax. Sin.,* 24(2): 144 (1986)—China.
yanhuii *Y.T. Chang in Acta Phytotax. Sin.,* 24(2): 146 (1986)—China.

CRUCIANELLA (Rubiac.).
arabica *E. Schönbeck-Temesy & F. Ehrendorfer in Notes Roy. Bot. Gard. Edinburgh,* 44(1): 143 (1986)—Saudi Arabia.

CRYPTANTHA (Boraginac.).
cinerea
var. arenicola *L.C. Higgins & S.L. Welsh in Great Basin Nat.*, 46(2): 255 (1986)—U.S.A. (Utah).
var. laxa *(Macbr.) L.C. Higgins in Great Basin Nat.*, 46(2): 255 (1986): Oreocarya multicaulis var. laxa.
schoolcraftii *A. Tiehm in Brittonia*, 38(2): 104 (1986)—U.S.A. (Nevada).

CRYPTANTHUS (Bromeliac.).
seidelianus *W. Weber in Feddes Repert.*, 97(3–4): 119 (1986)—Brazil.

CRYPTOPUS (Orchidac.).
dissectus *(Bosser) K. Senghas in Schlechter Orchideen*, 1(16–18): 1019 (1986): C. elatus subsp. dissectus.

CTENARDISIA (Myrsinac.).
stenobotrys *(Standley) R. Fonnegra-G. & S.L. Jung-Mendaçolli in Hoehnea*, 12: 31 (1985): Ardisia stenobotrys.

CUATRESIA (Solanac.).
foreroi *A.T. Hunziker in Caldasia*, 15(71–75): 146 (1986)—Colombia.
trianae *A.T. Hunziker in Caldasia*, 15(71–75): 143 (1986)—Colombia.

CUCUMELLA (Cucurbitac.).
silentvalleyi *K.S. Manilal, T. Sabu & P. Mathew in Acta Bot. Indica*, 13(2): 283 (1985), as 'silentvalleyii'—India.

CUCUMIS (Cucurbitac.).
conomon *C.P. Thunberg ex A. Murray in Linnaeus, Syst. Veg., ed. 14:* 869 (1784)—Japan. †
erinaceus *Naud. ex Huber, Cat. Print. 1864:* 6 (1864); *cf. D.J. Mabberley in Taxon*, 34(3): 452 (1985)— Locality not stated.

CUCURBITA (Cucurbitac.).
argyrosperma *Huber, Cat. Graines 1867:* 8 (1867); *Wochenschrift*, 10: 95 (1867); *cf. D.J. Mabberley in Taxon*, 34(3): 452 (1985)— Locality not stated.
hispida *C.P. Thunberg ex A. Murray in Linnaeus, Syst. Veg., ed. 14:* 868 (1784)—Japan. †

CULCASIA (Arac.).
orientalis *S.J. Mayo in Fl. Trop. E. Afr., Arac.:* 20 (1985)—Tanzania, Kenya.

CUMINIA (Labiat.).
eriantha
var. fernandezia *(Colla) R. Harley in Kew Mag.*, 3(4): 155 (1986): C. fernandezia.

CUPANIOPSIS (Sapindac.).
dallachyi *S.T. Reynolds in Fl. Australia*, 25: 199, 58 (1985)—Australia (Queensland).

CUPHEA (Lythrac.).
bolivariensis *A. Lourteig in Phytologia*, 60(1): 30 (1986)—Venezuela.
curiosa *A. Lourteig in Phytologia*, 60(1): 38 (1986)—Venezuela.
killipii *A. Lourteig in Phytologia*, 60(1): 24 (1986)—Colombia.
cardonae *A. Lourteig in Phytologia*, 60(1): 42 (1986), nom. nov.: C. rhodocalyx var. setosa *Lourteig*.
scolnikiae *A. Lourteig in Phytologia*, 60(1): 52 (1986)—Bolivia, Brazil.

CUPRESSUS (Cupressac.).
arizonica
subsp. stephensonii *(C.B. Wolf) R.M. Beauchamp in Phytologia*, 59(7): 438 (1986): C. stephensonii.
pendula *C.P. Thunberg ex A. Murray in Linnaeus, Syst. Veg., ed. 14:* 861 (1784)—Japan. †

CURCAS (Euphorbiac.).
adansonii *Endl., Cat. Horti Vindob.*, 2: 394 (1842); *cf. D.J. Mabberley in Taxon*, 33(3): 435 (1984)— Locality not stated.
drastica *Mart. in Schrank & Mart., Hort. Reg. Monac.:* 50 (1829); *cf. D.J. Mabberley in Taxon*, 33(3): 442 (1984)— Locality not stated.

CURCUMA (Zingiberac.).
albicoma *S.Q. Tong in Acta Bot. Yunnanica*, 8(1): 38 (1986)—China.
cannanorensis *R. Ansari, V.J. Nair & N.C. Nair in Curr. Sci.*, 51(6): 293 (1982)—India.
var. lutea *R. Ansari, V.J. Nair & N.C. Nair in Curr. Sci.*, 51(6): 293 (1982)—India.
flaviflora *S.Q. Tong in Acta Bot. Yunnanica*, 8(1): 37 (1986)—China.

CYANOTIS (Commelinac.).
axillaris *D. Don ex Sweet, Hort. Brit.:* 430 (1827); *cf. D.J. Mabberley in Taxon*, 31(1): 70 (1982)— Locality not stated.

CYATHOPHYLLA Bocquet & A. Strid in A. Strid (ed.), Mount. Fl. Greece, 1: 175 (1986). *CARYOPHYLLACEAE.*
chlorifolia *(Poiret) Bocquet & A. Strid in A. Strid (ed.), Mount. Fl. Greece*, 1: 175 (1986): Cucubalus chloraefolius.

CYCAS (Cycadac.).
revoluta *C.P. Thunberg ex A. Murray in Linnaeus, Syst. Veg., ed. 14:* 926 (1784)—Japan. †

CYCLANTHERA (Cucurbitac.).
edulis *Naud. ex Huber, Cat. Graines 1872:* 3 (1872); *Belgique horticole*, 22: 83 (1872); *cf. D.J. Mabberley in Taxon*, 34(3): 452 (1985)— Locality not stated.

CYCLEA (Menispermac.).
insularis
subsp. guangxiensis *H.S. Lo in Guihaia,* 6(1–2): 57 (1986)—China.
racemosa
f. emeiensis *H.S. Lo & S.Y. Zhao in Guihaia,* 6(1–2): 58 (1986)—China.

CYCLOBALANOPSIS (Fagac.).
gracilis *(Rehd. & Wils.) W.C. Cheng & T. Hong in Sci. Silvae Sin.,* 8(1): 11 (1963): Quercus glauca f. gracilis.
hunanensis *(Hand.-Mzt.) W.C. Cheng & T. Hong in Sci. Silvae Sin.,* 8(1): 11 (1963): Quercus hunanensis.
multinervis *W.C. Cheng & T. Hong in Sci. Silvae Sin.,* 8(1): 10 (1963)—China.
pseudoglauca *Y.K. Li & X.M. Wang in Acta Bot. Yunnanica,* 7(4): 417 (1985)—China.

CYCLOPOGON (Orchidac.).
comosus *(Rchb. f.) P. Burns-Balogh & Greenwood in Orquídea (México),* 10(1): 92 (1986): Spiranthes comosa.
deminkiorum *P. Burns-Balogh & M.S. Foster in Canad. Orchid J.,* 3(3): 5 (1985)—Paraguay.

CYMBALARIA (Scrophulariac.).
acutiloba *(Boiss. & Heldr.) F. Speta in Phyton (Austria),* 26(1): 50 (1986): Linaria microcalyx var. acutiloba.
subsp. dodekanesi *(Greuter) F. Speta in Phyton (Austria),* 26(1): 51 (1986): C. microcalyx subsp. dodekanesi.
ebelii *(Cufod.) F. Speta in Phyton (Austria),* 26(1): 50 (1986): Linaria microcalyx subsp. ebelii.
microcalyx
subsp. heterosepala *(Cufod.) F. Speta in Phyton (Austria),* 26(1): 50 (1986): Linaria microcalyx var. heterosepala.
minor *(Cufod.) F. Speta in Phyton (Austria),* 26(1): 51 (1986): Linaria microcalyx var. minor.

CYMOPTERUS (Umbellif.).
acaulis
var. fendleri *(Gray) S. Goodrich in Great Basin Nat.,* 46(1): 79 (1986): C. fendleri.
var. higginsii *(Welsh) S. Goodrich in Great Basin Nat.,* 46(1): 79 (1986): C. higginsii.
var. parvus *S. Goodrich in Great Basin Nat.,* 46(1): 79 (1986)—U.S.A. (Utah).
purpureus
var. jonesii *(Coult. & Rose) S. Goodrich in Great Basin Nat.,* 46(1): 86 (1986): C. jonesii.
var. rosei *(Jones) S. Goodrich in Great Basin Nat.,* 46(1): 86 (1986): Aulospermum rosei.
terebinthinus
var. petraeus *(Jones) S. Goodrich in Great Basin Nat.,* 46(1): 87 (1986): C. petraeus.

CYNANCHUM (Asclepiadac.).
angustifolium *(Dcne.) G. Morillo in Ernstia,* 37: 3 (1986): Ditassa angustifolia.
colellae *G. Morillo in Ernstia,* 37: 3 (1986)—Venezuela.
guanchezii *G. Morillo in Ernstia,* 37: 5 (1986)—Venezuela.
julianii *G. Morillo in Ernstia,* 37: 7 (1986)—Venezuela.
multinervium *G. Morillo in Ernstia,* 37: 9 (1986)—Venezuela.
sibiricum
var. triangularilobatum *M.R. Rasulova & B.A. Sharipova in Izv. Akad. Nauk Tadzh. SSR, Biol. Nauk,* 2(83): 32 (1981), without type and as 'triangulare-lobatum'—U.S.S.R. (Middle Asia).
stellatum
var. lanceolatolobatum *M.R. Rasulova & B.A. Sharipova in Izv. Akad. Nauk Tadzh. SSR, Biol. Nauk,* 2(83): 33 (1981), without type and as 'lanceolato-lobatum'—U.S.S.R. (Middle Asia).
var. oblongolobatum *M.R. Rasulova & B.A. Sharipova in Izv. Akad. Nauk Tadzh. SSR, Biol. Nauk,* 2(83): 34 (1981), without type and as 'oblongo-lobatum'—U.S.S.R. (Middle Asia).
var. obtusilobatum *M.R. Rasulova & B.A. Sharipova in Izv. Akad. Nauk Tadzh. SSR, Biol. Nauk,* 2(83): 32 (1981), without type and as 'obtuse-lobatum'—U.S.S.R. (Middle Asia).
var. ovalilobatum *M.R. Rasulova & B.A. Sharipova in Izv. Akad. Nauk Tadzh. SSR, Biol. Nauk,* 2(83): 34 (1981), without type and as 'ovali-lobatum'—U.S.S.R. (Middle Asia).
var. ovoideolobatum *M.R. Rasulova & B.A. Sharipova in Izv. Akad. Nauk Tadzh. SSR, Biol. Nauk,* 2(83): 34 (1981), without type and as 'ovoideo-lobatum'—U.S.S.R. (Middle Asia).

CYNOGLOSSUM (Boraginac.).
afrocaeruleum *(R. Mill) H. Riedl in Linz. Biol. Beitr.,* 17(2): 320 (1985): Paracynoglossum afrocaeruleum.
alpinum *(Brand) H. Riedl in Linz. Biol. Beitr.,* 17(2): 317 (1985), as 'sp. nov.': C. montanum var. alpinum.
alpinum *(Brand) B.L. Burtt in Notes Roy. Bot. Gard. Edinburgh,* 43(3): 345 (1986): C. montanum var. alpinum.
alticola *O.M. Hilliard & B.L. Burtt in Notes Roy. Bot. Gard. Edinburgh,* 43(3): 346 (1986)—South Africa (Cape Province), Lesotho.
austroafricanum *O.M. Hilliard & B.L. Burtt in Notes Roy. Bot. Gard. Edinburgh,* 43(3): 347 (1986)—South Africa (Natal, Orange Free State).
hedbergiorum *H. Riedl in Linz. Biol. Beitr.,* 17(2): 319 (1985)—Ethiopia.

CYNOGLOSSUM :–
japonicum *C.P. Thunberg ex A. Murray in Linnaeus, Syst. Veg., ed. 14:* 187 (1784)—Japan. †
krasniqii *T. Wraber in Candollea,* 41(1): 145 (1986)—Jugoslavia.

CYNOGLOTTIS (Boraginac.).
phocidica *(L.-Å. Gustavsson) J. Holub in Preslia,* 58(4): 301 (1986): Anchusa phocidica.
serpentinicola *(Rech. f.) J. Holub in Preslia,* 58(4): 301 (1986): Anchusa serpentinicola.

CYPERUS (Cyperac.).
breedlovei *G.C. Tucker in Rhodora,* 88(856): 505 (1986)—Mexico.
matudae *G.C. Tucker in Rhodora,* 88(856): 503 (1986)—Mexico.
panamensis *(C.B. Clarke) Britton ex P.C. Standley in J. Wash. Acad. Sci.,* 15(20): 457 (1925): Mariscus panamensis.
svensonii *G.C. Tucker in Rhodora,* 88(856): 510 (1986)—Mexico, Guatemala, Honduras, Nicaragua.
wilburii *G.C. Tucker in Rhodora,* 88(856): 508 (1986)—Mexico.

CYPHOMANDRA (Solanac.).
casana *A. Child in Feddes Repert.,* 97(3–4): 143 (1986)—Ecuador.

CYPHOSTEMMA (Vitac.).
auriculata *(Roxb.) P. Singh & B.V. Shetty in Taxon,* 35(3): 596 (1986): Cissus auriculata.
jiguu *B. Verdcourt in Kew Bull.,* 41(4): 1016 (1986), nom. nov.: C. pachypus *Verdc.*

CYPRIPEDIUM (Orchidac.).
sect. Subtropica *S.C. Chen & K.Y. Lang in Acta Phytotax. Sin.,* 24(4): 317 (1986).
japonicum *C.P. Thunberg ex A. Murray in Linnaeus, Syst. Veg., ed. 14:* 817 (1784)—Japan. †
subtropicum *S.C. Chen & K.Y. Lang in Acta Phytotax. Sin.,* 24(4): 317 (1986)—Xizang.

CYRTANTHUS (Amaryllidac.).
brachysiphon *O.M. Hilliard & B.L. Burtt in Notes Roy. Bot. Gard. Edinburgh,* 43(2): 189 (1986)—South Africa (Natal).

CYSTICAPNOS (Papaverac.).
cracca *(Cham. & Schlecht.) M. Lidén in Op. Bot.,* 88: 108 (1986): Corydalis cracca.
parviflora *M. Lidén in Op. Bot.,* 88: 106 (1986)—South Africa (Cape Province).
pruinosa *(Bernh.) M. Lidén in Op. Bot.,* 88: 106 (1986): Phacocapnos pruinosa.

CYTISOPSIS (Legumin.).
ahmedii *(Battand. & Pitard) P. Lassen in Willdenowia,* 16(1): 109 (1986): Cytisus ahmedii.

CYTISUS (Legumin.).
moleroi *J. Fernández Casas, Exsicc. nobis distrib.,* 3: 4 (1980)—Spain.

CYTISUS :–
rhodopnoa *(Webb & Berth.) Schouw, Forteg. Kjöb. Bot. Hav. Pl.:* 143 (1847); *cf. D.J. Mabberley in Taxon,* 33(3): 439 (1984): Basionym not stated.

D

×DACTYLEUCORCHIS (Orchidac.).
albucina *(Ciferri & Giacomini) L.V. Aver'yanov in Bot. Zhurn.,* 71(1): 93 (1986): ×Leucororchis albucina.

DACTYLICAPNOS (Papaverac.).
sect. Minicalcara *(Khahn) M. Lidén in Op. Bot.,* 88: 20 (1986): Dicentra sect. Minicalcara.

DACTYLIS (Gramin.).
glomerata
var. marina *(Borrill) M. Speranza & G. Cristofolini in Webbia,* 39(2): 394 (1986): D. marina.
subsp. oceanica *G. Guignard in Bull. Soc. Bot. France, Lett. Bot.,* 132(4–5): 346 (1985 publ. 1986)—France.

×DACTYLODENIA (Orchidac.).
comigera *(Reichenb.) L.V. Aver'yanov in Bot. Zhurn.,* 71(1): 93 (1986): Gymnadenia comigera.
evansii *(Druce ex T. Stephenson & T.A. Stephenson) L.V. Aver'yanov in Bot. Zhurn.,* 71(1): 93 (1986): ×Orchigymnadenia evansii.
fuchsii *(G. Keller & Soó) L.V. Aver'yanov in Bot. Zhurn.,* 71(1): 93 (1986): ×Orchigymnadenia fuchsii.
gracilis *(A. Camus) L.V. Aver'yanov in Bot. Zhurn.,* 71(1): 93 (1986): ×Orchigymnadenia heinzeliana var. gracilis.
heinzeliana *(Reichardt) L.V. Aver'yanov in Bot. Zhurn.,* 71(1): 93 (1986): Orchis heinzeliana.
jeanjeanii *(G. Keller) L.V. Aver'yanov in Bot. Zhurn.,* 71(1): 93 (1986): ×Orchigymnadenia jeanjeanii.
klingeana *(Aschers. & Graebn.) L.V. Aver'yanov in Bot. Zhurn.,* 71(1): 93 (1986): ×Orchigymnadenia klingeana.
quintinii *(Godfery) L.V. Aver'yanov in Bot. Zhurn.,* 71(1): 93 (1986), as 'st. quintinii': ×Orchigymnadenia st.-quintinii.
souppensis *(E.G. Camus) L.V. Aver'yanov in Bot. Zhurn.,* 71(1): 93 (1986): Gymnadenia souppensis.
varia *(T. Stephenson & T.A. Stephenson) L.V. Aver'yanov in Bot. Zhurn.,* 71(1): 93 (1986): ×Orchigymnadenia varia.
vollmannii *(Schulze) L.V. Aver'yanov in Bot. Zhurn.,* 71(1): 93 (1986): ×Orchigymnadenia vollmannii.

×DACTYLODENIA :–
wintonii *(Druce) L.V. Aver'yanov in Bot. Zhurn.,* 71(1): 93 (1986): Habenaria wintonii.

DACTYLORHIZA (Orchidac.).
×baicalica *L.V. Aver'yanov in Bot. Zhurn.,* 71(1): 92 (1986)—U.S.S.R. (Siberia). (D. cruenta × salina).
battandieri *C. Raynaud, Orchid. Maroc:* 62 (1985)—Algeria.
elata
f. alba *(Soó) C. Raynaud, Orchid. Maroc:* 103 (1985): Orchis elata f. alba.
var. elongata *(Maire) C. Raynaud, Orchid. Maroc:* 103 (1985): Orchis elata var. elongata.
f. leucantha *(Maire & Weiller) C. Raynaud, Orchid. Maroc:* 104 (1985): Orchis elata f. leucantha.
subvar. maroccanica *(Soó) C. Raynaud, Orchid. Maroc:* 104 (1985): Orchis elata f. maroccanica.
f. pallida *(Maire & Weiller) C. Raynaud, Orchid. Maroc:* 104 (1985): Orchis elata f. pallida.
f. peltieri *(Soó) C. Raynaud, Orchid. Maroc:* 103 (1985): Orchis elata f. peltieri.
subvar. speciosissima *(Soó) C. Raynaud, Orchid. Maroc:* 103 (1985): Orchis elata f. speciosissima.
subvar. stephensonii *(Maire & Weiller) C. Raynaud, Orchid. Maroc:* 103 (1985): Orchis elata subvar. stephensonii.
elodes *(Griseb.) P. Englmaier in Abh. Zool.-Bot. Ges. Österr.,* 22: 107 (1984), without basionym page: Orchis elodes.
hebridensis *(Wilmott) L.V. Aver'yanov in Bot. Zhurn.,* 71(1): 92 (1986); *et in Spis. Rast. Gerb. Fl. SSSR,* 25(127–130): 26 (1986): Orchis hebridensis.
×hochreutineriana *(Hellmayr) L.V. Aver'yanov in Bot. Zhurn.,* 71(1): 92 (1986): Orchis hochreutineriana.
incarnata
subsp. pyrenaica *M. Balayer in Bull. Soc. Bot. France, Lett. Bot.,* 133(3): 280 (1986)—France.
insularis *(Moris) P. Englmaier in Abh. Zool. Bot. Ges. Österr.,* 22: 105 (1984): Orchis sambucina var. insularis.
×ishorica *L.V. Aver'yanov in Bot. Zhurn.,* 71(1): 93 (1986)—U.S.S.R. (European Russia). (D. baltica × incarnata).
maculata
var. pseudoelata *M. Balayer in Bull. Soc. Bot. France, Lett. Bot.,* 133(3): 280 (1986), as 'pseudo-elata'—France.
majalis
var. druceoides *M. Balayer in Bull. Soc. Bot. France, Lett. Bot.,* 133(3): 280 (1986)—France.
maurusia *(Emb. & Maire) C. Raynaud, Orchid. Maroc:* 59 (1985): Orchis maurusia.
×pardalina *(Pugsley) L.V. Aver'yanov in Bot. Zhurn.,* 71(1): 93 (1986): Orchis pardalina.

DACTYLORHIZA :–
×robertsii *L.V. Aver'yanov in Bot. Zhurn.,* 71(1): 93 (1986), without latin descr.—Great Britain.
vestita *(Lagasca & Rodriguez) L.V. Aver'yanov in Bot. Zhurn.,* 71(1): 92 (1986): Orchis vestita.

DAHLIA (Compos.).
arborea *Huber, Dahlia arborea:* 1 (1870); *Revue horticole,* 42: 203 (1870); *cf. D.J. Mabberley in Taxon,* 34(3): 453 (1985)—Locality not stated.

DALBERGIA (Legumin.).
sect. Ecastaphylla *(P. Br.) K. Thothathri in Bull. Bot. Surv. India,* 25(1–4): 169 (1983 publ. 1985): gen. Ecastaphyllum.
ser. Pinnatae *K. Thothathri in Bull. Bot. Surv. India,* 25(1–4): 170 (1983 publ. 1985).
lanceolaria
var. assamica *(Benth.) K. Thothathri in Bull. Bot. Surv. India,* 25(1–4): 171 (1983 publ. 1985): D. assamica.
var. hemsleyi *(Prain) K. Thothathri in Bull. Bot. Surv. India,* 25(1–4): 172 (1983 publ. 1985): D. hemsleyi.
var. maymyensis *(Craib) K. Thothathri in Bull. Bot. Surv. India,* 25(1–4): 172 (1983 publ. 1985): D. maymyensis.
subsp. paniculata *(Roxb.) K. Thothathri in Bull. Bot. Surv. India,* 25(1–4): 171 (1983 publ. 1985): D. paniculata.
millettii
var. mimosoides *(Franch.) K. Thothathri in Bull. Bot. Surv. India,* 25(1–4): 170 (1983 publ. 1985): D. mimosoides.
ovata
var. glomeriflora *(Kurz) K. Thothathri in Bull. Bot. Surv. India,* 25(1–4): 170 (1983 publ. 1985): D. glomeriflora.
pinnata
var. acaciaefolia *(Dalz.) K. Thothathri in Bull. Bot. Surv. India,* 25(1–4): 170 (1983 publ. 1985): D. acaciaefolia.
rimosa
var. foliacea *(Wall. ex Benth.) K. Thothathri in Bull. Bot. Surv. India,* 25(1–4): 170 (1983 publ. 1985): D. foliacea.
yunnanensis
var. collettii *(Prain) K. Thothathri in Bull. Bot. Surv. India,* 25(1–4): 171 (1983 publ. 1985): D. collettii.

DAPHNE (Thymelaeac.).
odora *C.P. Thunberg ex A. Murray in Linnaeus, Syst. Veg., ed. 14:* 372 (1784)—Japan. †
purpurascens *S.C. Huang in Acta Bot. Yunnanica,* 7(3): 280 (1985)—Xizang.

DASIPHORA (Rosac.).
arbuscula *(D. Don) J. Soják in Čas. Nár. Muz. (Prague),* 152(1): 19 (1983): Potentilla arbuscula.

DASIPHORA :–
glabra *(Lodd.) J. Soják in Čas. Nár. Muz. (Prague),* 152(1): 19 (1983): Potentilla glabra.
×glabrata *(Willd. ex Schlecht.) J. Soják in Čas. Nár. Muz. (Prague),* 152(1): 19 (1983): Potentilla glabrata.
×rehderiana *(Hand.-Mazz.) J. Soják in Čas. Nár. Muz. (Prague),* 152(1): 19 (1983): Potentilla rehderiana.
reticulata *(Bert.) J. Soják in Čas. Nár. Muz. (Prague),* 152(1): 19 (1983): Potentilla reticulata.
×veitchii *(Wils.) J. Soják in Čas. Nár. Muz. (Prague),* 152(1): 19 (1983): Potentilla veitchii.

DASYCONDYLUS (Compos.).
hirsutissimus *(Baker) R.M. King & H. Robinson in Phytologia,* 60(1): 80 (1986): Eupatorium hirsutissimum.

DATURA (Solanac.).
gigantea *Huber, Cat. Graines 1867:* 4 (1867); *Wochenschrift,* 10: 101 (1867); *cf. D.J. Mabberley in Taxon,* 34(3): 450 (1985)— Locality not stated.

DAUBENTONIA (Legumin.).
versicolor *Huber, Cat. Print. 1865:* 4 (1865); *cf. D.J. Mabberley in Taxon,* 34(3): 450 (1985)— Locality not stated.

DAUCUS (Umbellif.).
aegyptiacus *Hort. Wirc. ex Hornem., Hort. Hafn.:* 271 (1813); *cf. D.J. Mabberley in Taxon,* 33(3): 437 (1984)— Locality not stated.

DECACHAETA (Compos.).
pyramidalis *(B.L. Robins.) S. Sundberg, C.P. Cowan & B.L. Turner in Amer. J. Bot.,* 73(1): 37 (1986): Piqueria pyramidalis.

DEINBOLLIA (Sapindac.).
angustifolia *D.W. Thomas in Ann. Missouri Bot. Gard.,* 73(1): 219 (1986)—Cameroun.
unijuga *D.W. Thomas in Ann. Missouri Bot. Gard.,* 73(1): 219 (1986)—Cameroun.

DEINOCHEILOS W.T. Wang in Guihaia, 6(1–2): 1 (1986). *GESNERIACEAE.*
jiangxiense *W.T. Wang in Guihaia,* 6(1–2): 4 (1986)—China.
sichuanense *W.T. Wang in Guihaia,* 6(1–2): 2 (1986)—China.

DEIREGYNE (Orchidac.).
pyramidalis *(Lindl.) P. Burns-Balogh in Orquídea (México),* 10(1): 92 (1986): Spiranthes pyramidalis.
tenuiflora *(Greenm.) P. Burns-Balogh in Orquídea (México),* 10(1): 92 (1986): Spiranthes tenuiflora.

DELARBREA (Araliac.).
montana
subsp. arborea *(Vieill. ex R. Viguier) P.P. Lowry in Allertonia,* 4(3): 196 (1986): D. montana var. arborea.
paradoxa
subsp. depauperata *P.P. Lowry in Allertonia,* 4(3): 181 (1986)—New Caledonia.

DELISSEA (Campanulac.).
konaensis *H. St. John in Phytologia,* 59(5): 315 (1986)—Hawaiian Is.

DELPHINIUM (Ranunculac.).
andersonii
var. scaposum *(Greene) S.L. Welsh in Great Basin Nat.,* 46(2): 260 (1986): D. scaposum.
angustirhombicum *W.T. Wang in Bull. Bot. Res. North-East. Forest. Inst.,* 6(1): 9 (1986)—China.
baoshanense *W.T. Wang in Bull. Bot. Res. North-East. Forest. Inst.,* 6(1): 12 (1986)—China.
brevisepalum *W.T. Wang in Bull. Bot. Res. North-East. Forest. Inst.,* 6(1): 22 (1986)—China.
chayuense *W.T. Wang in Bull. Bot. Res. North-East. Forest. Inst.,* 6(1): 16 (1986)—Xizang.
cilicicum *P.H. Davis & Kit Tan in Notes Roy. Bot. Gard. Edinburgh,* 43(2): 267 (1986)—Turkey.
delavayi
var. lasiandrum *W.T. Wang in Bull. Bot. Res. North-East. Forest. Inst.,* 6(1): 14 (1986)—China.
dolichocentroides
var. leiogynum *W.T. Wang in Acta Bot. Yunnanica,* 8(3): 263 (1986)—China.
hueizeense *W.T. Wang in Bull. Bot. Res. North-East. Forest. Inst.,* 6(1): 18 (1986)—China.
latirhombicum *W.T. Wang in Bull. Bot. Res. North-East. Forest. Inst.,* 6(1): 7 (1986)—China.
medogense *W.T. Wang in Bull. Bot. Res. North-East. Forest. Inst.,* 6(1): 20 (1986)—Xizang.
munzianum *P.H. Davis & Kit Tan in Notes Roy. Bot. Gard. Edinburgh,* 43(2): 267 (1986)—Turkey.
ninglangshanicum *W.T. Wang in Bull. Bot. Res. North-East. Forest. Inst.,* 6(1): 10 (1986)—China.
occidentalis
var. barbeyi *(Huth) S.L. Welsh in Great Basin Nat.,* 46(2): 260 (1986): D. exaltatum var. barbeyi.
pentagynum
f. gautieri *(Rouy) C. Blanché & Molero in Fontqueria,* 9: 19 (1985): D. pentagynum subsp. gautieri.
rangtangense *W.T. Wang in Acta Bot. Yunnanica,* 8(3): 262 (1986)—China.

DELPHINIUM :–
spurium *Hort. Berol. ex Hornem., Hort. Hafn., Suppl.*: 59 (1819); *cf. D.J. Mabberley in Taxon,* 33(3): 437 (1984)— Locality not stated.
xichangense *W.T. Wang in Bull. Bot. Res. North-East. Forest. Inst.,* 6(1): 1 (1986)—China.
yajiangense *W.T. Wang in Bull. Bot. Res. North-East. Forest. Inst.,* 6(1): 5 (1986)—China.
yangii *W.T. Wang in Bull. Bot. Res. North-East. Forest. Inst.,* 6(1): 3 (1986)—China.
yulungshanicum *W.T. Wang in Bull. Bot. Res. North-East. Forest. Inst.,* 6(1): 15 (1986)—China.

DENDRANTHEMA (Compos.).
nankingense *(Hand.-Mazz.) X.D. Cui in Fl. Tsinlingensis,* 1(5): 252 (1985): Chrysanthemum nankingense.

DENDROBIUM (Orchidac.).
brevimentum *P.J. Cribb in Kew Bull.,* 41(3): 638 (1986)—Moluccas.
flexicaule *Z.H. Tsi, S.C. Sun & L.G. Xu in Bull. Bot. Res. North-East. Forest. Inst.,* 6(2): 113 (1986)—China.
guangxiense *S.J. Cheng & C.Z. Tang in Orchid Dig.,* 50(3): 95 (1986)—China.
macropus
subsp. gracilicaule *(F. Mueller) P.S. Green in J. Arnold Arbor.,* 67(1): 115 (1986): D. gracilicaule.
subsp. howeanum *(Maiden) P.S. Green in J. Arnold Arbor.,* 67(1): 115 (1986): D. gracilicaule var. howeanum.
nareshbahadurii *H.B. Naithani in Indian Forester,* 112(1): 66 (1986)—India.
subclausum
var. pandanicola *J.J. Wood in Kew Bull.,* 41(4): 817 (1986)—New Guinea (Papua New Guinea)
var. phlox *(Schltr.) J.J. Wood in Kew Bull.,* 41(4): 819 (1986): D. phlox.
var. speciosum *J.J. Wood in Kew Bull.,* 41(4): 818 (1986)—New Guinea (Papua New Guinea).

DENDROCALAMOPSIS (Gramin.).
validus *Q.H. Dai in Acta Phytotax. Sin.,* 24(5): 393 (1986)—Cultivation.

DENDROCALAMUS (Gramin.).
farinosus
f. flavostriatus *T.P. Yi in Bull. Bot. Res. North-East. Forest. Inst.,* 6(4): 28 (1986), as 'flavo-striatus'—China.

DENDROPEMON (Loranthac.).
claraensis *A.T. Leiva Sánchez in Rev. Jard. Bot. Nacion. Univ. Habana,* 5(3): 18 (1984 publ. 1985)—Cuba.

DESMANTHUS (Legumin.).
balsensis *J.L. Contreras Jiménez in Phytologia,* 60(2): 89 (1986)—Mexico.

DESMANTHUS :–
glomeratus *(Forssk.) Lag., Elenchus:* 6 (1816); *cf. D.J. Mabberley in Taxon,* 31(1): 70 (1982): Basionym not stated.

DESMODIUM (Legumin.).
elegans
var. albiflorum *P.C. Li in Fl. Xizangica,* 2: 892 (1985)—Xizang.
var. callianthum *(Franch.) P.C. Li in Fl. Xizangica,* 2: 891 (1985), without basionym ref.: D. callianthum.

DESMOS (Annonac.).
goezeanus *(F. Muell.) L.W. Jessup in Austrobaileya,* 2(3): 227 (1986): Uvaria goezeana.
wardianus *(Bailey) L.W. Jessup in Austrobaileya,* 2(3): 227 (1986): Unona wardiana.

DEUTZIA (Hydrangeac.).
compacta
var. multiradiata *J.T. Pan in Fl. Xizangica,* 2: 532 (1985)—Xizang.
corymbosa
var. dinggyensis *J.T. Pan in Fl. Xizangica,* 2: 531 (1985)—Xizang.

DEYEUXIA (Gramin.).
sect. Pungentes *Z.E. Rúgolo de Agrasar in Parodiana,* 4(1): 100 (1986).
borii *S. Bhattacharya & S.K. Jain in Bull. Bot. Surv. India,* 25(1–4): 208 (1983 publ. 1985), nom. nov.: Agrostis nagensis *Bor.*
cabrerae
var. maxima *Z.E. Rúgolo de Agrasar in Parodiana,* 4(1): 106 (1986)—Argentina.
var. trichopoda *Parodi ex Z.E. Rúgolo de Agrasar in Parodiana,* 4(1): 107 (1986)—Argentina.

DIANTHERA (Acanthac.).
japonica *C.P. Thunberg ex A. Murray in Linnaeus, Syst. Veg., ed. 14:* 64 (1784)—Japan. †

DIANTHUS (Caryophyllac.).
caucasicus *Spreng. in Henckel, Adumb. Pl. Hort. Hal.:* 23 (1806); *cf. D.J. Mabberley in Taxon,* 32(1): 84 (1983)— Locality not stated.
crinitus
subsp. baldzhuanicus *(Lincz.) K.H. Rechinger in Pl. Syst. Evol.,* 151(3–4): 285 (1986): D. baldzhuanicus.
subsp. kermanensis *K.H. Rechinger in Pl. Syst. Evol.,* 151(3–4): 286 (1986)—Iran.
subsp. nuristanicus *(Gilli) K.H. Rechinger in Pl. Syst. Evol.,* 151(3–4): 285 (1986): D. nuristanicus.
subsp. tetralepis *(Nevski) K.H. Rechinger in Pl. Syst. Evol.,* 151(3–4): 286 (1986): D. tetralepis.
subsp. turcomanicus *(Schischk.) K.H. Rechinger in Pl. Syst. Evol.,* 151(3–4): 287 (1986): D. turcomanicus.

DIANTHUS :–
denaicus *M. Assadi in Iranian J. Bot.*, 3(1): 17 (1985)—Iran.
diversifolius *M. Assadi in Iranian J. Bot.*, 3(1): 40 (1985)—Iran.
hafezii *M. Assadi in Iranian J. Bot.*, 3(1): 23 (1985)—Iran.
×helveticorum *M. Laínz in An. Jard. Bot. Madrid,* 42(2): 549 (1985 publ. 1986)—Spain.
japonicus *C.P. Thunberg ex A. Murray in Linnaeus, Syst. Veg., ed. 14:* 417 (1784)—Japan. †
laricifolius
subsp. marizii *(Samp.) J. do Amaral Franco in Ann. Bot. Fenn.*, 23(1): 91 (1986): D. graniticus var. marizii.
orientalis
subsp. aphanoneurus *K.H. Rechinger in Pl. Syst. Evol.*, 151(3–4): 290 (1986)—Iran.
subsp. gilanicus *K.H. Rechinger in Pl. Syst. Evol.*, 151(3–4): 289 (1986)—Iran.
subsp. gorganicus *K.H. Rechinger in Pl. Syst. Evol.*, 151(3–4): 291 (1986)—Iran.
subsp. macropetalus *(Boiss.) K.H. Rechinger in Pl. Syst. Evol.*, 151(3–4): 289 (1986): D. fimbriatus var. macropetalus.
subsp. nassireddini *(Stapf) K.H. Rechinger in Pl. Syst. Evol.*, 151(3–4): 292 (1986): D. nassireddini.
subsp. obtusisquameus *(Boiss.) K.H. Rechinger in Pl. Syst. Evol.*, 151(3–4): 291 (1986): D. fimbriatus var. obtusisquameus.
subsp. stenocalyx *(Boiss.) K.H. Rechinger in Pl. Syst. Evol.*, 151(3–4): 292 (1986): D. fimbriatus var. stenocalyx.
rudbaricus *M. Assadi in Iranian J. Bot.*, 3(1): 38 (1985)—Iran.
sahandicus *M. Assadi in Iranian J. Bot.*, 3(1): 45 (1985)—Iran.
serratifolius
subsp. abbreviatus *(Heldr. ex Halácsy) A. Strid in A. Strid (ed.), Mount. Fl. Greece,* 1: 179 (1986): D. serratifolius var. abbreviatus.
vigoi *M. Laínz in An. Jard. Bot. Madrid,* 42(2): 551 (1985 publ. 1986)—Spain.
viridescens *Clem. in Atti 3 Riun. Sci. Ital.:* 520 (1841); *cf. D.J. Mabberley in Taxon,* 31(1): 70 (1982)— Locality not stated.
webbianus *Parol. ex Vis. in Atti 2 Riun. Sci. Ital.:* 180 (1841); *cf. D.J. Mabberley in Taxon,* 31(1): 70 (1982)— Locality not stated.

×**DIAPHANANGIS** hort. in Orchid Rev., 94(1112): centre page pull-out p. 8 (1986). *ORCHIDACEAE.* (Aerangis × Diaphananthe).

DICHAPETALUM (Dichapetalac.).
beckii *J. Fernández Casas in Fontqueria,* 2: 15 (1982)—Bolivia.

DICHAPETALUM :–
choristilum
var. louisii *F.J. Breteler in Agric. Univ. Wageningen Pap.*, 86(3): 32 (1986)—Gabon.
fadenii *F.J. Breteler in Agric. Univ. Wageningen Pap.*, 86(3): 33 (1986)—Kenya.
gassitae *F.J. Breteler in Agric. Univ. Wageningen Pap.*, 86(3): 35 (1986)—Gabon.
potamophilum *F.J. Breteler in Agric. Univ. Wageningen Pap.*, 86(3): 39 (1986)—Gabon, Cameroun.
prancei *J. Fernández Casas in Fontqueria,* 2: 15 (1982)—Paraguay.

DICLADANTHERA (Acanthac.).
glabra *R.M. Barker in J. Adelaide Bot. Gard.*, 9: 171 (1986)—Australia (W. Australia).

DICLIPTERA (Acanthac.).
arnhemica *R.M. Barker in J. Adelaide Bot. Gard.*, 9: 188 (1986)—Australia (N. Terr.).
australis *(Nees) R.M. Barker in J. Adelaide Bot. Gard.*, 9: 190 (1986): Brochosiphon australis.
fruticosa *K. Balkwill in Flow. Pl. Afr.*, 49(1–2): t. 1933 (1986)—South Africa (Transvaal).
miscella *R.M. Barker in J. Adelaide Bot. Gard.*, 9: 187 (1986)—Australia (W. Australia).
nasikensis *P. Lakshminarasimhan & B.D. Sharma in J. Econ. Taxon. Bot.*, 7(2): 481 (1985 publ. 1986)—India.

DICRANOPYGIUM (Cyclanthac.).
umbrophila *B.E. Hammel in Phytologia,* 60(1): 9 (1986)—Costa Rica.

DICTYONEURA (Sapindac.).
acuminata
subsp. microcarpa *J. van Dijk in Blumea,* 31(2): 445 (1986)—New Guinea (W. Irian).

DIDYMOCARPUS (Gesneriac.).
adenocalyx *W.T. Wang in Guihaia,* 6(1–2): 11 (1986)—China.

DIDYMODOXA (Urticac.).
caffra *(Thunb.) I. Friis & M. Wilmot-Dear in Bol. Soc. Brot.*, 58: 210 (1985): Urtica caffra.

DIDYMOPLEXIS (Orchidac.).
pachystomoides *(F. Muell.) L.A. Garay & W. Kittredge in Bot. Mus. Leafl. Harvard Univ.*, 30(3): (48) 182 (1985 publ. 1986): Pogonia pachystomoides.

DIGITALIS (Scrophulariac.).
×campbelliana *W. Baxt. in Loud., Hort. Brit., add. supp.2:* 630 (1839); *cf. D.J. Mabberley in Taxon,* 31(1): 70 (1982)— Locality not stated.
micrantha *Roth ex Schweigg., Enum. Pl. Hort. Bot. Regiom.:* 31 (1812); *cf. D.J. Mabberley in Taxon,* 33(3): 442 (1984)— Locality not stated.

DIGITARIA (Gramin.).
ciliata *Lag., Elenchus:* 6 (1816); *cf. D.J. Mabberley in Taxon,* 31(1): 70 (1982)—Locality not stated.
paniculata *Soderstrom ex R. McVaugh, Fl. Novo-Galiciana,* 14: 143 (1983)—Mexico.

DIMERIA (Gramin.).
jainii *P.V. Sreekumar, V.J. Nair & N.C. Nair in Curr. Sci.,* 52(6): 259 (1983)—India.

DIMETOPIA (Umbellif.).
preissiana *Bunge, Del. Sem. Hort. Bot. Dorpat.,* 1846: 3 (1846); *Ann. Sci. Nat. III,* 7: 190 (1847); *cf. D.J. Mabberley in Taxon,* 33(3): 442 (1984)— Locality not stated.

DIMORPHANTHUS (Araliac.).
japonicus *(Thunb.) Endl., Cat. Horti Vindob.,* 2: 176 (1842); *cf. D.J. Mabberley in Taxon,* 33(3): 435 (1984): Basionym not stated.

DINEBRA (Gramin.).
dura *(L.) Lag., Elenchus:* 6 (1816); *cf. D.J. Mabberley in Taxon,* 31(1): 70 (1982): Basionym not stated.

DIOSCOREA (Dioscoreac.).
basiclavicaulis *C.T. Rizzini & A. de Mattos-Filho in Rev. Brasil. Biol.,* 46(2): 317 (1986)—Brazil.
japonica *C.P. Thunberg ex A. Murray in Linnaeus, Syst. Veg., ed. 14:* 889 (1784)—Japan. †
menglaensis *H. Li in Fl. Yunnanica,* 3: 744 (1983)—China.
nanlaensis *H. Li in Fl. Yunnanica,* 3: 739 (1983)—China.
quinquelobata *C.P. Thunberg ex A. Murray in Linn.f., Syst. Veg., ed. 14:* 889 (1784)—Japan. †
septemloba *C.P. Thunberg ex A. Murray in Linnaeus, Syst. Veg., ed. 14:* 889 (1784)—Japan. †

DIOSMA (Rutac.).
acuminata *(Wendl.) C.C. Gmel., Hort. Mag. Duc. Bad. Carlsruh.:* 93 (1811); *cf. D.J. Mabberley in Taxon,* 33(3): 440 (1984): Basionym not stated.
huegeliana *Vis., Ort. Bot. Padova:* 76, 139 (1842); *cf. D.J. Mabberley in Taxon,* 33(3): 442 (1984)— Locality not stated.
umbellata *(Wendl.) Hort. Cels. ex C.C. Gmel., Hort. Mag. Duc. Bad. Carlsruh.:* 94 (1811); *cf. D.J. Mabberley in Taxon,* 33(3): 440 (1984): Basionym not stated.

DIOSPYROS (Ebenac.).
armata
var. pilosa *Z.Y. Zhang in Fl. Tsinlingensis,* 1(4): 394, 57 (1983)—China.
var. sinensis *(Hemsl.) Z.Y. Zhang in Fl. Tsinlingensis,* 1(4): 57 (1983): D. sinensis.
kaki *C.P. Thunberg ex A. Murray in Linnaeus, Syst. Veg., ed. 14:* 918 (1784)—Japan. †

DIOSPYROS :–
lotus
f. longifolia *Z.Y. Zhang in Fl. Tsinlingensis,* 1(4): 395, 60 (1983)—China.
sandwicensis
var. arachnoidea *H. St. John in Phytologia,* 59(6): 391 (1986)—Hawaiian Is.
var. degeneri *(Fosb.) H. St. John in Phytologia,* 59(6): 392 (1986): D. ferrea var. degeneri.
var. elliptica *H. St. John in Phytologia,* 59(6): 392 (1986)—Hawaiian Is.
var. frameata *H. St. John in Phytologia,* 59(6): 392 (1986)—Hawaiian Is.
var. kauaiensis *(Fosb.) H. St. John in Phytologia,* 59(6): 393 (1986): D. ferrea var. kauaiensis.
var. kokeeensis *H. St. John in Phytologia,* 59(6): 393 (1986)—Hawaiian Is.
var. lanaiensis *(Fosb.) H. St. John in Phytologia,* 59(6): 393 (1986): D. ferrea f. lanaiensis.
var. miloliiensis *H. St. John in Phytologia,* 59(6): 394 (1986)—Hawaiian Is.
var. oahuensis *H. St. John in Phytologia,* 59(6): 394 (1986)—Hawaiian Is.
var. obtusa *(Fosb.) H. St. John in Phytologia,* 59(6): 394 (1986): D. ferrea f. obtusa.
var. ovalis *H. St. John in Phytologia,* 59(6): 395 (1986)—Hawaiian Is.
var. perpubescens *H. St. John in Phytologia,* 59(6): 395 (1986)—Hawaiian Is.
var. puberula *H. St. John in Phytologia,* 59(6): 399 (1986)—Hawaiian Is.
var. pubescens *(Skottsb.) H. St. John in Phytologia,* 59(6): 395 (1986): Maba sandwicensis var. pubescens.
var. rotundata *H. St. John in Phytologia,* 59(6): 396 (1986)—Hawaiian Is.
var. rugosa *H. St. John in Phytologia,* 59(6): 397 (1986)—Hawaiian Is.
var. sclerophylla *(Fosb.) H. St. John in Phytologia,* 59(6): 397 (1986): D. ferrea f. sclerophylla.
var. toppingii *(Fosb.) H. St. John in Phytologia,* 59(6): 397 (1986): D. ferrea var. toppingii.
var. wailauensis *(Fosb.) H. St. John in Phytologia,* 59(6): 398 (1986): D. ferrea f. wailauensis.
var. wiebkei *(Fosb.) H. St. John in Phytologia,* 59(6): 398 (1986): D. ferrea f. wiebkei.

DIPLOLABELLUM (Orchidac.).
confluens *(Hand.-Mazt.) L.A. Garay & W. Kittredge in Bot. Mus. Leafl. Harvard Univ.,* 30(3): (48) 182 (1985 publ. 1986): Oreorchis patens var. confluens.

DIPLOSTEPHIUM (Compos.).
amygdalinum *Sweet, Hort. Brit., ed. 2:* 295 (1830); *cf. D.J. Mabberley in Taxon,* 30(1): 15 (1981)— Locality not stated.

DIPLOSTEPHIUM :–
humile *(Willd.) Sweet, Hort. Brit., ed. 2:* 295 (1830); *cf. D.J. Mabberley in Taxon,* 30(1): 15 (1981): Basionym not stated.
linifolium *(L.) Sweet, Hort. Brit., ed. 2:* 295 (1830); *cf. D.J. Mabberley in Taxon,* 30(1): 15 (1981): Aster linifolius.
rigidum *(Pursh) Sweet, Hort. Brit., ed. 2:* 295 (1830); *cf. D.J. Mabberley in Taxon,* 30(1): 15 (1981): Basionym not stated.

DIPLOTROPIS (Legumin.).
sect. Racemosae *H.C. de Lima in Acta Amazonica,* 15(1–2): 63 (1985 publ. 1986).

DIPODIUM (Orchidac.).
squamatum *(Forst. f.) Sm. in Rees, Cyclop.,* 39(2): Dipodium n. 2 (1819); *cf. D.J. Mabberley in Taxon,* 32(1): 82 (1983): Basionym not stated.

DIPTERACANTHUS (Acanthac.).
australasicus
subsp. corynothecus *(F. Muell. ex Benth.) R.M. Barker in J. Adelaide Bot. Gard.,* 9: 89 (1986): Ruellia corynotheca.
subsp. dalyensis *R.M. Barker in J. Adelaide Bot. Gard.,* 9: 92 (1986)—Australia (N. Terr.).
subsp. glabratus *R.M. Barker in J. Adelaide Bot. Gard.,* 9: 91 (1986)—Australia (S. Australia).
oblongatus *Nees, Del. Sem. Hort. Bot. Vratislav.,* 1840; *Linnaea 15, Lit.:* 126 (?1841); *cf. D.J. Mabberley in Taxon,* 32(1): 84 (1983)— Locality not stated.
pedunculatus *Huber, Supp. Cat. 1870:* 3 (1870–71); *Wochenschrift,* 14: 159 (1871); *cf. D.J. Mabberley in Taxon,* 34(3): 450 (1985)— Locality not stated.
schauerianus *Nees, Del. Sem. Hort. Bot. Vratislav.,* 1838; *Linnaea 13, Lit.:* 118 (?1840); *cf. D.J. Mabberley in Taxon,* 32(1): 84 (1983)— Locality not stated.
strictus *Nees, Del. Sem. Hort. Bot. Vratislav.,* 1838; *Linnaea 13, Lit.:* 119 (?1840); *cf. D.J. Mabberley in Taxon,* 32(1): 84 (1983)— Locality not stated.

DISA (Orchidac.).
lutea *H.P. Linder in Bothalia,* 15(3–4): 553 (1985), nom. nov.: Ophrys patens *L.f.*

DISCOCACTUS (Cactac.).
zehntneri
f. albispinus *(Buin. & Brederoo) Riha in Kaktusy,* 19(2): 26 (1983); *cf. Repert. Pl. Succ. (I.O.S.),* 34: 6 (1985): D. albispinus.

DISPERIS (Orchidac.).
elaphoceras *B. Verdcourt in Kew Bull.,* 41(1): 54 (1986)—Tanzania.
uzungwae *B. Verdcourt in Kew Bull.,* 41(1): 56 (1986)—Tanzania.

DISTICHIA (Juncac.).
acicularis *H. Balslev & S. Laegaard in Nordic J. Bot.,* 6(2): 151 (1986)—Ecuador.

DISTYLIUM (Hamamelidac.).
lipoense *Y.K. Li & X.M. Wang in Acta Bot. Yunnanica,* 8(3): 275 (1986)—China.

DISYNAPHIA (Compos.).
tacuarembensis *(Hieron. & Arechav.) R.M. King & H. Robinson in Phytologia,* 60(1): 80 (1986): Eupatorium tacuarembense.

DITASSA (Asclepiadac.).
conceptionis *J. Fontella Pereira in Dusenia,* 12(1): 6 (1980), nom. nov.: Calathostelma ditassoides *Fournier.*

DODECATHEON (Primulac.).
pulchellum
var. zionense *(Eastw.) S.L. Welsh in Great Basin Nat.,* 46(2): 259 (1986): D. zionense.

DOLICHOS (Legumin.).
incurvus *C.P. Thunberg ex A. Murray in Linnaeus, Syst. Veg., ed. 14:* 658 (1784)—Japan. †
lineatus *C.P. Thunberg ex A. Murray in Linnaeus, Syst. Veg., ed. 14:* 658 (1784)—Japan. †

DOLOMIAEA (Compos.).
sect. Vladimiria *(Iljin) C. Shih in Acta Phytotax. Sin.,* 24(4): 293 (1986): gen. Vladimiria.
berardioidea *(Franch.) C. Shih in Acta Phytotax. Sin.,* 24(4): 294 (1986): Jurinea edulis β. berardioidea.
denticulata *(Ling) C. Shih in Acta Phytotax. Sin.,* 24(4): 293 (1986): Vladimiria denticulata.
edulis *(Franch.) C. Shih in Acta Phytotax. Sin.,* 24(4): 294 (1986): Saussurea edulis.
forrestii *(Diels) C. Shih in Acta Phytotax. Sin.,* 24(4): 293 (1986): Jurinea forrestii.
georgii *(Anth.) C. Shih in Acta Phytotax. Sin.,* 24(4): 294 (1986): Jurinea georgii.
platylepis *(Hand.-Mazz.) C. Shih in Acta Phytotax. Sin.,* 24(4): 295 (1986): Jurinea platylepis.
salwinensis *(Hand.-Mazz.) C. Shih in Acta Phytotax. Sin.,* 24(4): 295 (1986): Jurinea salwinensis.
scabrida *(Shih & S.Y. Jin) C. Shih in Acta Phytotax. Sin.,* 24(4): 293 (1986): Vladimiria scabrida.
souliei *(Franch.) C. Shih in Acta Phytotax. Sin.,* 24(4): 294 (1986): Jurinea souliei.
var. mirabilis *(Anth.) C. Shih in Acta Phytotax. Sin.,* 24(4): 294 (1986): Jurinea mirabilis.

DONAX (Gramin.).
collinus *(Ten.) Tineo, Cat. Pl. Panorm.:* 100 (1827); *cf. D.J. Mabberley in Taxon,* 31(1): 70 (1982): Basionym not stated.

DORSTENIA (Morac.).
carautae *C.C. Berg in Proc. Kon. Nederl. Akad. Wetensch., C,* 89(2): 136 (1986)—Brazil.
cayapia
subsp. asaroides *(Hooker) C.C. Berg in Proc. Kon. Nederl. Akad. Wetensch., C,* 89(2): 143 (1986): D. asaroides.
subsp. paraguariensis *(Hassler) C.C. Berg in Proc. Kon. Nederl. Akad. Wetensch., C,* 89(2): 143 (1986): D. cayapia f. paraguariensis.
subsp. vitifolia *(Gardner) C.C. Berg in Proc. Kon. Nederl. Akad. Wetensch., C,* 89(2): 143 (1986): D. vitifolia.
ramosa
subsp. dolichocaula *(Pilger) C.C. Berg in Proc. Kon. Nederl. Akad. Wetensch., C,* 89(2): 143 (1986): D. dolichocaula.
uxpanapana *C.C. Berg & T. Wendt in Proc. Kon. Nederl. Akad. Wetensch., C,* 89(2): 129 (1986)—Mexico.

DORVALIA Hoffmanns., Preiss-Verz.: 21 (1833); cf. D.J. Mabberley in Taxon, 33(3): 434 (1984). *ONAGRACEAE.*

DORYCNIUM (Legumin.).
strictum *(Fischer & C. Meyer) P. Lassen in Willdenowia,* 16(1): 109 (1986): Lotus strictus.

DOWNINGIA (Campanulac.).
concolor
subsp. brevior *(McVaugh) R.M. Beauchamp in Phytologia,* 59(7): 438 (1986): D. concolor var. brevior.

DRABA (Crucif.).
densifolia
var. apiculata *(C.L. Hitchc.) S.L. Welsh in Great Basin Nat.,* 46(2): 255 (1986): D. apiculata.
heterocoma
subsp. archipelagi *(O.E. Schulz) K.P. Buttler in A. Strid (ed.), Mount. Fl. Greece,* 1: 315 (1986): D. heterocoma var. archipelagi.
hispanica
subsp. djurdjurae *(Batt.) W. Greuter in Willdenowia,* 15(2): 418 (1986): D. hispanica var. djurdjurae.
kassii *S.L. Welsh in Great Basin Nat.,* 46(2): 264 (1986)—U.S.A. (Utah).
lasiocarpa
subsp. dolichostyla *(O.E. Schulz) K.P. Buttler in A. Strid (ed.), Mount. Fl. Greece,* 1: 311 (1986): D. parnassica var. dolichostyla.
linearifolia *L.L. Lou & T.Y. Cheo in Fl. Xizangica,* 2: 346 (1985)—Xizang.

DRABA :–
oligosperma
var. juniperina *(Dorn) S.L. Welsh in Great Basin Nat.,* 46(2): 255 (1986): D. juniperina.
strasseri *W. Greuter in Willdenowia,* 15(2): 418 (1986)—Greece.

DRACAENA (Dracaenac.).
boscii *Hort. Cels. ex Zeyher, Verz. Saemm. Gewaech. Schwezingen:* 69 (1818); *cf. D.J. Mabberley in Taxon,* 33(3): 442 (1984)—Locality not stated.
rubra *(Kunth) Parmentier, Cat. Pl. Rar.:* 11 (1853); *cf. D.J. Mabberley in Taxon,* 33(3): 442 (1984): Cordyline rubra.

DRACOCEPHALUM (Labiat.).
butkovii *L.S. Krasovskaya in L.S. Krasovskaya & I.G. Levichev, Fl. Chatkal'skogo Zapovednika:* 170 (1986)—U.S.S.R. (Middle Asia).

DRACONIA (Compos.).
albicerata *(Krasch.) J. Soják in Čas. Nár. Muz. (Prague),* 152(1): 20 (1983): Artemisia albicerata.
bargusinensis *(Spreng.) J. Soják in Čas. Nár. Muz. (Prague),* 152(1): 20 (1983): Artemisia bargusinensis.
campestris *(L.) J. Soják in Čas. Nár. Muz. (Prague),* 152(1): 20 (1983): Artemisia campestris.
subsp. alpina *(DC.) J. Soják in Čas. Nár. Muz. (Prague),* 152(1): 20 (1983): Artemisia campestris β var. alpina.
subsp. borealis *(Pall.) J. Soják in Čas. Nár. Muz. (Prague),* 152(1): 20 (1983): Artemisia borealis.
subsp. bottnica *(Lundstr. ex Kindb.) J. Soják in Čas. Nár. Muz. (Prague),* 152(1): 20 (1983): Artemisia campestris subsp. bottnica.
subsp. glutinosa *(DC.) J. Soják in Čas. Nár. Muz. (Prague),* 152(1): 20 (1983): Artemisia glutinosa.
subsp. maritima *(Arcang.) J. Soják in Čas. Nár. Muz. (Prague),* 152(1): 20 (1983): Artemisia campestris γ subsp. maritima.
capillaris *(Thunb.) J. Soják in Čas. Nár. Muz. (Prague),* 152(1): 20 (1983): Artemisia capillaris.
commutata *(Bess.) J. Soják in Čas. Nár. Muz. (Prague),* 152(1): 20 (1983): Artemisia commutata.
daghestanica *(Krasch. & Por.) J. Soják in Čas. Nár. Muz. (Prague),* 152(1): 20 (1983): Artemisia daghestanica.
demissa *(Krasch.) J. Soják in Čas. Nár. Muz. (Prague),* 152(1): 20 (1983): Artemisia demissa.
desertorum *(Spreng.) J. Soják in Čas. Nár. Muz. (Prague),* 152(1): 20 (1983): Artemisia desertorum.

DRACONIA :–
dimoana *(M. Pop.) J. Soják in Čas. Nár. Muz. (Prague),* 152(1): 20 (1983): Artemisia dimoana.
dracunculiformis *(Krasch.) J. Soják in Čas. Nár. Muz. (Prague),* 152(1): 20 (1983): Artemisia dracunculiformis.
dracunculus *(L.) J. Soják in Čas. Nár. Muz. (Prague),* 152(1): 20 (1983): Artemisia dracunculus.
glauca *(Pall. ex Willd.) J. Soják in Čas. Nár. Muz. (Prague),* 152(1): 20 (1983): Artemisia glauca.
halodendron *(Turcz. ex Besser) J. Soják in Čas. Nár. Muz. (Prague),* 152(1): 20 (1983): Artemisia halodendron.
henriettae *(Krasch.) J. Soják in Čas. Nár. Muz. (Prague),* 152(1): 20 (1983): Artemisia henriettae.
japonica *(Thunb.) J. Soják in Čas. Nár. Muz. (Prague),* 152(1): 20 (1983): Artemisia japonica.
kelleri *(Krasch.) J. Soják in Čas. Nár. Muz. (Prague),* 152(1): 20 (1983): Artemisia kelleri.
kuschakewiczii *(C. Winkl.) J. Soják in Čas. Nár. Muz. (Prague),* 152(1): 20 (1983): Artemisia kuschakewiczii.
ledebouriana *(Bess.) J. Soják in Čas. Nár. Muz. (Prague),* 152(1): 20 (1983): Artemisia ledebouriana.
limosa *(Koidz.) J. Soják in Čas. Nár. Muz. (Prague),* 152(1): 20 (1983): Artemisia limosa.
lipskyi *(P. Polj.) J. Soják in Čas. Nár. Muz. (Prague),* 152(1): 20 (1983): Artemisia lipskyi.
littoricola *(Kitam.) J. Soják in Čas. Nár. Muz. (Prague),* 152(1): 20 (1983): Artemisia littoricola.
macilenta *(Maxim.) J. Soják in Čas. Nár. Muz. (Prague),* 152(1): 20 (1983): Artemisia campestris var. macilenta.
pamirica *(C. Winkl.) J. Soják in Čas. Nár. Muz. (Prague),* 152(1): 20 (1983): Artemisia pamirica.
pannosa *(Krasch.) J. Soják in Čas. Nár. Muz. (Prague),* 152(1): 20 (1983): Artemisia pannosa.
pycnorhiza *(Ledeb.) J. Soják in Čas. Nár. Muz. (Prague),* 152(1): 20 (1983): Artemisia pycnorhiza.
quinqueloba *(Trautv.) J. Soják in Čas. Nár. Muz. (Prague),* 152(1): 20 (1983): Artemisia quinqueloba.
salsoloides *(Willd.) J. Soják in Čas. Nár. Muz. (Prague),* 152(1): 20 (1983): Artemisia salsoloides.
saposhnikovii *(Krasch. ex Poljakov) J. Soják in Čas. Nár. Muz. (Prague),* 152(1): 21 (1983): Artemisia saposhnikovii.
scoparia *(Waldst. & Kit.) J. Soják in Čas. Nár. Muz. (Prague),* 152(1): 21 (1983): Artemisia scoparia.

DRACONIA :–
songarica *(Schrenk) J. Soják in Čas. Nár. Muz. (Prague),* 152(1): 21 (1983): Artemisia songarica.
tomentella *(Trautv.) J. Soják in Čas. Nár. Muz. (Prague),* 152(1): 21 (1983): Artemisia tomentella.
trautvetteriana *(Bess.) J. Soják in Čas. Nár. Muz. (Prague),* 152(1): 21 (1983): Artemisia trautvetteriana.
tschernieviana *(Bess.) J. Soják in Čas. Nár. Muz. (Prague),* 152(1): 21 (1983): Artemisia tschernieviana.

DRACONTIUM (Arac.).
aricuaisanum *G.S. Bunting in Phytologia,* 60(5): 301 (1986)—Venezuela.
changuango *G.S. Bunting in Phytologia,* 60(5): 302 (1986)—Venezuela.
regelianum *(Engl.) J. Bogner in Aroideana,* 8(3): 78 (1985 publ. 1986): Echidnium regelianum.

DRACOSCIADIUM O.M. Hilliard & B.L. Burtt in Notes Roy. Bot. Gard. Edinburgh, 43(2): 220 (1986). *UMBELLIFERAE.*
italae *O.M. Hilliard & B.L. Burtt in Notes Roy. Bot. Gard. Edinburgh,* 43(2): 223 (1986)—South Africa (Natal).
saniculifolium *O.M. Hilliard & B.L. Burtt in Notes Roy. Bot. Gard. Edinburgh,* 43(2): 225 (1986)—South Africa (Natal).

DRACULA (Orchidac.).
brangeri *C.A. Luer in Orchideeën,* 48(2): 47 (1986)—Colombia.
marsupialis *C.A. Luer & A.C. Hirtz in Orchidee,* 37(1): 25 (1986)—Ecuador.

DRACUNCULUS (Compos.).
sect. Campestres *Krasch. ex V.P. Amel'chenko in I.M. Krasnoborov & T.A. Safonova (eds.), Novoe Fl. Sibiri:* 241 (1986), without type.
sect. Scopariae *Krasch. ex V.P. Amel'chenko in I.M. Krasnoborov & T.A. Safonova (eds.), Novoe Fl. Sibiri:* 241 (1986), without type.

DREPANOSTACHYUM (Gramin.).
jainiana *(Das & Pal) R.B. Majumder in Bull. Bot. Surv. India,* 25(1–4): 235 (1983 publ. 1985): Chimonobambusa jainiana.
ludianense *(Yi & R.S. Wang) P.C. Keng in J. Bamboo Res.,* 5(2): 35 (1986): Ampelocalamus ludianensis.
melicoideum *P.C. Keng in J. Bamboo Res.,* 5(2): 35 (1986)—China.
naibunensis *(Hayata) P.C. Keng in J. Bamboo Res.,* 5(2): 32 (1986): Arundinaria naibunensis.
suberecta *(Munro) R.B. Majumder in Bull. Bot. Surv. India,* 25(1–4): 236 (1983 publ. 1985): Arundinaria suberecta.

DRYADODAPHNE (Monimiac.).
crassa *Schodde ex W.R. Philipson in Fl. Males.,* 10(2): 265 (1986)—New Guinea (Papua New Guinea).

DRYADODAPHNE :–

novoguineensis

var. macra *Schodde ex W.R. Philipson in Fl. Males.*, 10(2): 266 (1986)—New Guinea (Papua New Guinea).

subsp. occidentalis *Schodde ex W.R. Philipson in Fl. Males.*, 10(2): 267 (1986)—New Guinea (W. Irian).

DRYPETES (Euphorbiac.).

bhattacharyae *T. Chakrabarty in J. Econ. Taxon. Bot.*, 7(2): 453 (1985 publ. 1986), as 'bhattacharyai'—Nicobar Is.

lasiogyna

subsp. affinis *(Pax & Hoffm.) P.S. Green in J. Arnold Arbor.*, 67(1): 111 (1986): D. affinis.

longipes *X.H. Song in J. Nanjing Inst. Forest.*, 1986(1): 74 (1986)—China.

DUSCHEKIA (Betulac.).

alnobetula

subsp. suaveolens *(Requien) J. Holub in Preslia*, 58(4): 302 (1986): Alnus suaveolens.

DYAKIA E.A. Christenson in Orchid Dig., 50(2): 63 (1986). *ORCHIDACEAE.*

hendersoniana *(Reichb. f.) E.A. Christenson in Orchid Dig.*, 50(2): 63 (1986): Saccolabium hendersonianum.

E

ECHEANDIA (Anthericac.).

altipratensis *R.W. Cruden in Phytologia*, 59(6): 373 (1986) Guatemala.

breedlovei *R.W. Cruden in Phytologia*, 59(6): 374 (1986)—Mexico.

campechiana *R.W. Cruden in Phytologia*, 59(6): 378 (1986)—Mexico.

chiapensis *R.W. Cruden in Phytologia*, 59(6): 375 (1986)—Mexico.

ciliata *(Kunth) R.W. Cruden in Phytologia*, 59(6): 380 (1986): Phalangium ciliatum.

formosa *(Weatherby) R.W. Cruden in Phytologia*, 59(6): 380 (1986): E. macrocarpa var. formosa.

luteola *R.W. Cruden in Phytologia*, 59(6): 377 (1986)—Mexico, Belize.

matudae *R.W. Cruden in Phytologia*, 59(6): 376 (1986)—Mexico, Guatemala, Salvador.

molinae *R.W. Cruden in Phytologia*, 59(6): 377 (1986)—Guatemala.

petenensis *R.W. Cruden in Phytologia*, 59(6): 376 (1986)—Guatemala, Belize.

pittieri *R.W. Cruden in Phytologia*, 59(6): 379 (1986)—Panama, Colombia.

ramosissima *(Presl) R.W. Cruden in Phytologia*, 59(6): 380 (1986): Phalangium ramosissimum.

ECHEANDIA :–

skinneri *(Baker) R.W. Cruden in Phytologia*, 59(6): 380 (1986): Anthericum skinneri.

vestita *(Baker) R.W. Cruden in Phytologia*, 59(6): 380 (1986): Anthericum vestitum.

williamsii *R.W. Cruden in Phytologia*, 59(6): 373 (1986)—Honduras, Guatemala.

ECHINACANTHUS (Acanthac.).

flaviflorus *H.S. Lo & D. Fang in Acta Bot. Yunnanica*, 7(2): 141 (1985)—China.

longipes *H.S. Lo & D. Fang in Acta Bot. Yunnanica*, 7(2): 138 (1985)—China.

longzhouensis *H.S. Lo in Acta Bot. Yunnanica*, 7(2): 140 (1985)—China.

ECHINACEA (Compos.).

serotina *(Sweet) D. Don ex G. Don f. in Loud., Hort. Brit., ed. 2:* 587 (1832); *cf. D.J. Mabberley in Taxon*, 30(1): 16 (1981): Basionym not stated.

ECHINOCHLOA (Gramin.).

crus-pavonis

var. austrojaponensis *(Ohwi) S.L. Dai in Acta Sci. Nat. Univ. Sunyatseni*, 1986(3): 78 (1986), without basionym ref.: Basionym not stated.

var. breviseta *(Neilr.) S.L. Dai in Acta Sci. Nat. Univ. Sunyatseni*, 1986(3): 78 (1986), without basionym ref.: Basionym not stated.

var. praticola *(Ohwi) S.L. Dai in Acta Sci. Nat. Univ. Sunyatseni*, 1986(3): 78 (1986), without basionym ref.: Basionym not stated.

formosensis *(Ohwi) S.L. Dai in Acta Sci. Nat. Univ. Sunyatseni*, 1986(3): 78 (1986), without basionym ref.: Basionym not stated.

jaliscana *R. McVaugh, Fl. Novo-Galiciana*, 14: 153 (1983)—Mexico.

ECHINOSPARTUM (Legumin.).

boissieri

subsp. webbii *(Spach) B.E. Smythies in Englera*, 3(3): 880 (Feb. 1986): Genista webbii.

subsp. webbii *(Spach) W. Greuter & Burdet in Willdenowia*, 16(1): 109 (8 Aug. 1986): Genista webbii.

ECLIPTA (Compos.).

longifolia *Schrad., Ind. Sem. Hort. Acad. Gott.*, 1831: 3 (1831); *Linnaea 8, Lit.:* 23 (1834); *cf. D.J. Mabberley in Taxon*, 32(1): 84 (1983)— Locality not stated.

oederi *(Murr.) Weigel, Hort. Gryph.:* 33 (1782); *cf. D.J. Mabberley in Taxon*, 33(3): 442 (1984): Cotula oederi.

patula *Schrad., Ind. Sem. Hort. Acad. Gott.*, 1831: 3 (1831); *Linnaea 8, Lit.:* 23 (1834); *cf. D.J. Mabberley in Taxon*, 32(1): 84 (1983)— Locality not stated.

patula *(DC.) Endl., Cat. Horti Vindob.*, 1: 335 (1842); *cf. D.J. Mabberley in Taxon*, 33(3): 435 (1984): Basionym not stated.

EDGEWORTHIA (Thymelaeac.).
eriosolenoides *Feng & S.C. Huang in Acta Bot. Yunnanica,* 7(3): 281 (1985)—China.

EHRETIA (Boraginac.).
abyssinica
var. silvatica *(Guerke) H. Riedl in Linz. Biol. Beitr.,* 17(2): 313 (1985): E. silvatica.

ELAEAGNUS (Elaeagnac.).
conferta
var. maculata *(Servettaz) D. Basu in J. Econ. Taxon. Bot.,* 7(2): 427 (1985 publ. 1986): E. arborea var. maculata.
crispa *C.P. Thunberg ex A. Murray in Linnaeus, Syst. Veg., ed. 14:* 163 (1784)—Japan. †
glabra *C.P. Thunberg ex A. Murray in Linnaeus, Syst. Veg., ed. 14:* 164 (1784)—Japan. †
kānaii
var. osmastonii *C.L. Malhotra & D. Basu in Bull. Bot. Surv. India,* 26(1–2): 121 (1984 publ. 1985)—India.
multiflora *C.P. Thunberg ex A. Murray in Linnaeus, Syst. Veg., ed. 14:* 163 (1784)—Japan. †
pungens *C.P. Thunberg ex A. Murray in Linnaeus, Syst. Veg., ed. 14:* 164 (1784)—Japan. †
umbellata *C.P. Thunberg ex A. Murray in Linnaeus, Syst. Veg., ed. 14:* 164 (1784)—Japan. †
xizangensis *C.Y. Chang in Bull. Bot. Res. North-East. Forest. Inst.,* 6(2): 74 (1986)—Xizang.

ELAEOCARPUS (Elaeocarpac.).
capuronii *C. Tirel in Fl. Madagascar,* Fam. 125: 25 (1985)—Madagascar.
corallococcus *C. Tirel in Fl. Madagascar,* Fam. 125: 28 (1985)—Madagascar.
occidentalis *C. Tirel in Fl. Madagascar,* Fam. 125: 21 (1985)—Madagascar.
palembanicus
var. leptomischus *(Ridley) B.C. Stone in Fed. Mus. J. (Kuala Lumpur),* 26(1): 83 (1981): E. leptomischus.
perrieri *C. Tirel in Fl. Madagascar,* Fam. 125: 17 (1985)—Madagascar.

ELATOSTEMA (Urticac.).
crispulum *W.T. Wang in Acta Bot. Yunnanica,* 7(3): 295 (1985)—China.
hekouense *W.T. Wang in Acta Bot. Yunnanica,* 7(3): 296 (1985)—China.
microtrichum *W.T. Wang in Acta Bot. Yunnanica,* 7(3): 299 (1985)—China.
minutifurfuraceum *W.T. Wang in Acta Bot. Yunnanica,* 7(3): 293 (1985)—China.
stigmatosum *W.T. Wang in Acta Bot. Yunnanica,* 7(3): 297 (1985)—China.
yangbiense *W.T. Wang in Acta Bot. Yunnanica,* 7(3): 294 (1985)—China.

ELATTOSTACHYS (Sapindac.).
megalantha *S.T. Reynolds in Fl. Australia,* 25: 199, 70 (1985)—Australia (Queensland).
microcarpa *S.T. Reynolds in Fl. Australia,* 25: 199, 72 (1985)—Australia (Queensland).

ELEGIA (Restionac.).
atratiflora *E.E. Esterhuysen in Bothalia,* 15(3–4): 420 (1985)—South Africa (Cape Province).
caespitosa *E.E. Esterhuysen in Bothalia,* 15(3–4): 421 (1985)—South Africa (Cape Province).
fucata *E.E. Esterhuysen in Bothalia,* 15(3–4): 423 (1985)—South Africa (Cape Province).
grandispicata *H.P. Linder in Bothalia,* 15(3–4): 424 (1985)—South Africa (Cape Province).

ELEOCHARIS (Cyperac.).
tuvinica *S.V. Bubnova in Bot. Zhurn.,* 71(10): 1405 (1986)—U.S.S.R. (Siberia, Middle Asia).

ELETTARIA (Zingiberac.).
multiflora *(Ridley) R.M. Smith in Notes Roy. Bot. Gard. Edinburgh,* 43(3): 462 (1986): Elettariopsis multiflora.

ELLEANTHUS (Orchidac.).
teotepecensis *M.A. Soto Arenas in Orquídea (México),* 10(1): 169 (1986)—Mexico.

ELODEA (Hypericac.).
formosa *Ten., Cat. Ort. Bot. Nap.:* 93 (1845); *cf. D.J. Mabberley in Taxon,* 32(1): 84 (1983)— Locality not stated.

ELYMUS (Gramin.).
scabrus
var. plurinervis *(Vickery) B.K. Simon in Austrobaileya,* 2(3): 242 (1986): Agropyron scabrum var. plurinerve.

ELYTRIGIA (Gramin.).
repens
subsp. lolioides *(Kar. & Kir.) Á. Löve in Taxon,* 35(1): 198 (1986): Triticum lolioides.

EMILIA (Compos.).
sect. Spathulatae *(Muschl.) C. Jeffrey in Kew Bull.,* 41(4): 911 (1986): Senecio sect. Spathulati.
abyssinica *(Sch. Bip. ex A. Rich.) C. Jeffrey in Kew Bull.,* 41(4): 911 (1986): Senecio abyssinicus.
var. macroglossa *C. Jeffrey in Kew Bull.,* 41(4): 911 (1986)—Tanzania.
ambifaria *(S. Moore) C. Jeffrey in Kew Bull.,* 41(4): 919 (1986): Othonna ambifaria.
aurita *C. Jeffrey in Kew Bull.,* 41(4): 915 (1986)—Tanzania.
baberka *(Hutch.) C. Jeffrey in Kew Bull.,* 41(4): 918 (1986): Senecio baberka.
bellioides *(Chiov.) C. Jeffrey in Kew Bull.,* 41(4): 913 (1986): Senecio bellioides.

EMILIA :-
brachycephala *(R.E. Fr.) C. Jeffrey in Kew Bull.,* 41(4): 918 (1986): Senecio brachycephalus.
cenioides *C. Jeffrey in Kew Bull.,* 41(4): 914 (1986)—Tanzania.
coloniaria *(S. Moore) C. Jeffrey in Kew Bull.,* 41(4): 918 (1986): Senecio coloniarius.
crispata *C. Jeffrey in Kew Bull.,* 41(4): 916 (1986)—Tanzania.
cryptantha *C. Jeffrey in Kew Bull.,* 41(4): 915* (1986)—Tanzania, Uganda.
decipiens *C. Jeffrey in Kew Bull.,* 41(4): 917 (1986)—Tanzania, Mozambique, Malawi.
discifolia *(Oliv.) C. Jeffrey in Kew Bull.,* 41(4): 912 (1986): Senecio discifolius.
emilioides *(Sch. Bip.) C. Jeffrey in Kew Bull.,* 41(4): 918 (1986): Senecio emilioides.
fallax *(Mattf.) C. Jeffrey in Kew Bull.,* 41(4): 918 (1986): Senecio fallax.
flaccida *C. Jeffrey in Kew Bull.,* 41(4): 916 (1986)—Tanzania.
fugax *C. Jeffrey in Kew Bull.,* 41(4): 916 (1986)—Tanzania, Zambia.
gossweileri *(S. Moore) C. Jeffrey in Kew Bull.,* 41(4): 918 (1986): Crassocephalum gossweileri.
helianthella *C. Jeffrey in Kew Bull.,* 41(4): 911 (1986)—Tanzania.
hiernii *C. Jeffrey in Kew Bull.,* 41(4): 918 (1986), nom. nov.: Othonna gracilis *Hiern.*
hockii *(De Wild. & Muschl.) C. Jeffrey in Kew Bull.,* 41(4): 912 (1986): Senecio hockii.
homblei *(De Wild.) C. Jeffrey in Kew Bull.,* 41(4): 918 (1986): Senecio homblei.
irregularibracteata *(De Wild.) C. Jeffrey in Kew Bull.,* 41(4): 918 (1986): Senecio irregularibracteatus.
juncea
var. iringensis *C. Jeffrey in Kew Bull.,* 41(4): 914 (1986)—Tanzania.
kilwensis *C. Jeffrey in Kew Bull.,* 41(4): 916 (1986)—Tanzania.
kivuensis *(Muschl.) C. Jeffrey in Kew Bull.,* 41(4): 913 (1986): Senecio kivuensis.
leptocephala *(Mattf.) C. Jeffrey in Kew Bull.,* 41(4): 919 (1986): Senecio leptocephalus.
leucantha *C. Jeffrey in Kew Bull.,* 41(4): 914 (1986)—Tanzania, Zambia.
limosa *(O. Hoffm.) C. Jeffrey in Kew Bull.,* 41(4): 917 (1986): Senecio limosus.
longifolia *C. Jeffrey in Kew Bull.,* 41(4): 915 (1986)—Uganda, Tanzania.
longipes *C. Jeffrey in Kew Bull.,* 41(4): 915 (1986)—Tanzania.
longiramea *(S. Moore) C. Jeffrey in Kew Bull.,* 41(4): 919 (1986): Crassocephalum longirameum.
lopollensis *(Hiern) C. Jeffrey in Kew Bull.,* 41(4): 919 (1986): Senecio lopollensis.
marlothiana *(O. Hoffm.) C. Jeffrey in Kew Bull.,* 41(4): 919 (1986): Senecio marlothianus.
micrura *C. Jeffrey in Kew Bull.,* 41(4): 916 (1986)—Tanzania.

EMILIA :-
myriocephala *C. Jeffrey in Kew Bull.,* 41(4): 917 (1986)—Tanzania.
newtonii *(O. Hoffm.) C. Jeffrey in Kew Bull.,* 41(4): 919 (1986): Oligothrix newtonii.
pammicrocephala *(S. Moore) C. Jeffrey in Kew Bull.,* 41(4): 917 (1986): Senecio pammicrocephalus.
pseudactis *C. Jeffrey in Kew Bull.,* 41(4): 919 (1986), nom. nov.: Pseudactis emilioides *S. Moore.*
rigida *C. Jeffrey in Kew Bull.,* 41(4): 916 (1986)—Tanzania.
simulans *C. Jeffrey in Kew Bull.,* 41(4): 916 (1986)—Tanzania.
somalensis *(S. Moore) C. Jeffrey in Kew Bull.,* 41(4): 912 (1986): Euryops somalensis.
tenellula *(S. Moore) C. Jeffrey in Kew Bull.,* 41(4): 919 (1986): Senecio tenellulus.
tenera *(O. Hoffm.) C. Jeffrey in Kew Bull.,* 41(4): 917 (1986): Senecio tener.
tenuipes *C. Jeffrey in Kew Bull.,* 41(4): 914 (1986)—Tanzania.
tenuis *C. Jeffrey in Kew Bull.,* 41(4): 915 (1986)—Tanzania.
tessmannii *(Mattf.) C. Jeffrey in Kew Bull.,* 41(4): 918 (1986): Senecio tessmannii.
transvaalensis *(Bolus) C. Jeffrey in Kew Bull.,* 41(4): 919 (1986): Senecio transvaalensis.
tricholepis *C. Jeffrey in Kew Bull.,* 41(4): 912 (1986)—Kenya.
ukambensis *(O. Hoffm.) C. Jeffrey in Kew Bull.,* 41(4): 912 (1986): Senecio ukambensis.
ukingensis *(O. Hoffm.) C. Jeffrey in Kew Bull.,* 41(4): 915 (1986): Senecio ukingensis.
xyridopsis *(O. Hoffm.) C. Jeffrey in Kew Bull.,* 41(4): 919 (1986): Oligothrix xyridopsis.

EMINIUM (Arac.).
spiculatum
subsp. negevense *J. Koach & N. Feinbrun-Dothan in N. Feinbrun-Dothan, Fl. Palaestina,* 4: 397, 336 (1986)—Israel, Sinai Peninsula.

EMPETRUM (Empetrac.).
scoticum *hort. in Hort. Soc. Lond., Cat. Fr.:* 46 (1826); *cf. D.J. Mabberley in Taxon,* 33(3): 442 (1984)— Locality not stated.

ENCEPHALARTOS (Zamiac.).
regalis *Bull, Cat. 1889:* 4 (tab.), 8 (1889); *cf. D.J. Mabberley in Taxon,* 34(3): 456 (1985)— Locality not stated.

ENCHEIRIDION (Orchidac.).
sanfordii *(Jonsson) K. Senghas in Schlechter Orchideen,* 1(16–18): 1066 (1986): Microcoelia sanfordii.

ENCYCLIA (Orchidac.).
arminii *(Rchb. f.) G. Carnevali F.C. & I. Ramírez in Ernstia,* 36: 9 (1986): Epidendrum arminii.
brachychila *(Lindl.) G. Carnevali F.C. & I. Ramírez in Ernstia,* 36: 9 (1986): Epidendrum brachychilum.

ENCYCLIA :-
bradfordii *(Griseb.) G. Carnevali F.C. & I. Ramírez in Ernstia,* 36: 9 (1986): Epidendrum bradfordii.
dickensoniana *(Withner) F. Hamer in Ic. Pl. Trop.*, 13 (Orch. Nicaragua, pt. 6): t. 1215 (1985): Epidendrum dickensonianum.
garciana *(Garay & Dunst.) G. Carnevali F.C. & I. Ramírez in Ernstia,* 36: 9 (1986): Epidendrum garcianum.
lindenii *(Lindl.) G. Carnevali F.C. & I. Ramírez in Ernstia,* 36: 9 (1986): Epidendrum lindenii.
purpurochilum *(Barb. Rodr.) J.A. Fowlie in Orchid Dig.*, 50(4): 149 (1986), with incorrect basionym ref.: Epidendrum purpurachylum.
rhombilabia *S. Rosillo de Velasco in Orquídea (México)*, 10(1): 145 (1986)—Mexico.
sceptra *(Lindl.) G. Carnevali F.C. & I. Ramírez in Ernstia,* 36: 9 (1986): Epidendrum sceptrum.
tigrina *(Linden ex Lindl.) G. Carnevali F.C. & I. Ramírez in Ernstia,* 36: 9 (1986): Epidendrum tigrinum.

ENDOSAMARA R. Geesink, Scala Millettiearum (Leiden Bot. Ser., 8): 93 (1984). *LEGUMINOSAE.*
racemosa *(Roxb.) R. Geesink, Scala Millettiearum (Leiden Bot. Ser., 8):* 93 (1984): Robinia racemosa.

ENKIANTHUS (Ericac.).
calophyllus *T.Z. Hsu in Acta Bot. Yunnanica,* 7(2): 151 (1985)—China.
tubulatus *P.C. Tam in Bull. Bot. Res. North-East. Forest. Inst.*, 6(4): 125 (1986)—China.

ENTEROLOBIUM (Legumin.).
barnebianum *A.L. Mesquita & M.F. da Silva in Acta Amazonica,* 14(1–2 Suppl.): 153 (1984 publ. 1986)—Colombia, Brazil, Peru.

EPICRANTHES (Orchidac.).
abbrevilabia *(Carr) L.A. Garay & W. Kittredge in Bot. Mus. Leafl. Harvard Univ.*, 30(3): (48) 182 (1985 publ. 1986): Bulbophyllum abbrevilabium.
adangensis *(Seidenf.) L.A. Garay & W. Kittredge in Bot. Mus. Leafl. Harvard Univ.*, 30(3): (48) 182 (1985 publ. 1986): Bulbophyllum adangense.
cheiropetalum *(Ridl.) L.A. Garay & W. Kittredge in Bot. Mus. Leafl. Harvard Univ.*, 30(3): (49) 183 (1985 publ. 1986): Bulbophyllum cheiropetalum.
chlororhopalon *(Schltr.) L.A. Garay & W. Kittredge in Bot. Mus. Leafl. Harvard Univ.*, 30(3): (49) 183 (1985 publ. 1986): Bulbophyllum chlororhopalon.
cimicina *(J.J. Verm.) L.A. Garay & W. Kittredge in Bot. Mus. Leafl. Harvard Univ.*, 30(3): (49) 183 (1985 publ. 1986): Bulbophyllum cimicinum.

EPICRANTHES :-
conchophylla *(J.J. Sm.) L.A. Garay & W. Kittredge in Bot. Mus. Leafl. Harvard Univ.*, 30(3): (49) 183 (1985 publ. 1986): Bulbophyllum conchophyllum.
corneri *(Carr) L.A. Garay & W. Kittredge in Bot. Mus. Leafl. Harvard Univ.*, 30(3): (49) 183 (1985 publ. 1986): Bulbophyllum corneri.
decarhopalon *(Schltr.) L.A. Garay & W. Kittredge in Bot. Mus. Leafl. Harvard Univ.*, 30(3): (49) 183 (1985 publ. 1986): Bulbophyllum decarhopalon.
flavofimbriata *(J.J. Sm.) L.A. Garay & W. Kittredge in Bot. Mus. Leafl. Harvard Univ.*, 30(3): (49) 183 (1985 publ. 1986): Bulbophyllum flavofimbriatum.
haniffii *(Carr) L.A. Garay & W. Kittredge in Bot. Mus. Leafl. Harvard Univ.*, 30(3): (50) 184 (1985 publ. 1986): Bulbophyllum haniffii.
heterorhopalon *(Schltr.) L.A. Garay & W. Kittredge in Bot. Mus. Leafl. Harvard Univ.*, 30(3): (50) 184 (1985 publ. 1986): Bulbophyllum heterorhopalon.
hexarhopalon *(Schltr.) L.A. Garay & W. Kittredge in Bot. Mus. Leafl. Harvard Univ.*, 30(3): (50) 184 (1985 publ. 1986): Bulbophyllum hexarhopalon.
hirudinifera *(J.J. Verm.) L.A. Garay & W. Kittredge in Bot. Mus. Leafl. Harvard Univ.*, 30(3): (50) 184 (1985 publ. 1986): Bulbophyllum hirudiniferum.
johannulii *(J.J. Verm.) L.A. Garay & W. Kittredge in Bot. Mus. Leafl. Harvard Univ.*, 30(3): (50) 184 (1985 publ. 1986): Bulbophyllum johannulii.
macrorhopalon *(Schltr.) L.A. Garay & W. Kittredge in Bot. Mus. Leafl. Harvard Univ.*, 30(3): (50) 184 (1985 publ. 1986): Bulbophyllum macrorhopalon.
mobilifilum *(Carr) L.A. Garay & W. Kittredge in Bot. Mus. Leafl. Harvard Univ.*, 30(3): (50) 184 (1985 publ. 1986): Bulbophyllum mobilifilum.
octorhopalon *(Seidenf.) L.A. Garay & W. Kittredge in Bot. Mus. Leafl. Harvard Univ.*, 30(3): (50) 184 (1985 publ. 1986): Bulbophyllum octorhopalon.
papillosofilum *(Carr) L.A. Garay & W. Kittredge in Bot. Mus. Leafl. Harvard Univ.*, 30(3): (50) 184 (1985 publ. 1986): Bulbophyllum papillosofilum.
phymatum *(J.J. Verm.) L.A. Garay & W. Kittredge in Bot. Mus. Leafl. Harvard Univ.*, 30(3): (50) 184 (1985 publ. 1986): Bulbophyllum phymatum.
psilorhopalon *(Schltr.) L.A. Garay & W. Kittredge in Bot. Mus. Leafl. Harvard Univ.*, 30(3): (51) 185 (1985 publ. 1986): Bulbophyllum psilorhopalon.
rigidifilum *(J.J. Sm.) L.A. Garay & W. Kittredge in Bot. Mus. Leafl. Harvard Univ.*, 30(3): (51) 185 (1985 publ. 1986): Bulbophyllum rigidifilum.

EPICRANTHES :–
stenorhopalon *(Schltr.) L.A. Garay & W. Kittredge in Bot. Mus. Leafl. Harvard Univ.,* 30(3): (51) 185 (1985 publ. 1986): Bulbophyllum stenorhopalon.
trirhopalon *(Schltr.) L.A. Garay & W. Kittredge in Bot. Mus. Leafl. Harvard Univ.,* 30(3): (51) 185 (1985 publ. 1986): Bulbophyllum trirhopalon.
undecifila *(J.J. Sm.) L.A. Garay & W. Kittredge in Bot. Mus. Leafl. Harvard Univ.,* 30(3): (51) 185 (1985 publ. 1986): Bulbophyllum undecifilum.
vesiculosa *(J.J. Sm.) L.A. Garay & W. Kittredge in Bot. Mus. Leafl. Harvard Univ.,* 30(3): (51) 185 (1985 publ. 1986): Bulbophyllum vesiculosum.

EPIDENDRUM (Orchidac.).
aromaticum *(Sw.) Parmentier, Cat. Pl. Enghien:* 29 (1818); *cf. D.J. Mabberley in Taxon,* 33(3): 442 (1984): Vanilla aromatica.
glumarum *F. Hamer & Garay in Ic. Pl. Trop.,* 13 (Orch. Nicaragua, pt. 6): t. 1219 (1985)—Nicaragua.
siphonosepaloides *T. Hashimoto in Ann. Tsukuba Bot. Gard.,* 4: 1 (1986)—Peru.
teres *C.P. Thunberg ex A. Murray in Linnaeus, Syst. Veg., ed. 14:* 818 (1784)—Japan. †

×**EPIGLOTTIS** hort. in Orchid Rev., 94(1109): centre page pull-out p. 8 (1986). *ORCHIDACEAE.* (Epidendrum × Scaphyglottis).

EPILOBIUM (Onagrac.).
komarovii *P.N. Ovchinnikov in Dokl. Akad. Nauk Tadzh. SSR,* 23(11): 671 (1980)—U.S.S.R. (Middle Asia).
sericeum *Bernh., Sel. Sem. Hort. Erfurt.,* 1837; *Linnaea 12, Lit.:* 76 (1839); *cf. D.J. Mabberley in Taxon,* 32(1): 85 (1983)—Locality not stated.

EPIPACTIS (Orchidac.).
helleborine
subsp. neerlandica *(Vermeulen) K.P. Buttler in Willdenowia,* 16(1): 115 (1986): E. helleborine var. neerlandica.
×heterogama *M. Bayer in Mitt. Arbeitskr. Heim. Orch. Baden-Württemberg,* 18(2): 196 (1986)—Germany. (E. atrorubens × muelleri).
×reinekei *M. Bayer in Mitt. Arbeitskr. Heim. Orch. Baden-Württemberg,* 18(2): 202 (1986)—Germany. (E. helleborine × muelleri).

EPIPHYLLUM (Cactac.).
crenatum *(Lindl.) G. Don f. in Loud., Encycl. Pl., new ed.:* 1378 (1855); *cf. D.J. Mabberley in Taxon,* 30(1): 16 (1981): Basionym not stated.

ERAGROSTIS (Gramin.).
conertii *W. Lobin in Willdenowia,* 16(1): 143 (1986)—Cape Verde Is.
pectinacea
var. miserrima *(Fourn.) J.R. Reeder in Phytologia,* 60(2): 154 (1986): E. purshii var. miserrima.
petrensis *S.A. Renvoize & H.M. Longhi-Wagner in Kew Bull.,* 41(1): 71 (1986)—Brazil.

ERANTHEMUM (Acanthac.).
velutinum *Bull, Cat.,* 225: 8 (1886); *cf. D.J. Mabberley in Taxon,* 34(3): 456 (1985)—Locality not stated.

EREMOPHILA (Myoporac.).
compressa *R.J. Chinnock in Nuytsia,* 5(3): 391 (1986)—Australia (W. Australia).
lactea *R.J. Chinnock in Nuytsia,* 5(3): 393 (1986)—Australia (W. Australia).
nivea *R.J. Chinnock in Nuytsia,* 5(3): 395 (1986)—Australia (W. Australia).
verticillata *R.J. Chinnock in Nuytsia,* 5(3): 396 (1986)—Australia (W. Australia).

ERIANTHUS (Gramin.).
ecklonii *Nees in Nees & Schauer, Ind. Sem. Hort. Bot. Vratislav.,* 1835; *Ann. Sci. Nat. II,* 6: 103 (1836); *cf. D.J. Mabberley in Taxon,* 33(3): 442 (1984)— Locality not stated.

ERICA (Ericac.).
declinata *Hort. Herrenh. ex Hornem., Hort. Hafn., Suppl.:* 44 (1819); *cf. D.J. Mabberley in Taxon,* 33(3): 437 (1984)— Locality not stated.
woodii
subsp. platyura *O.M. Hilliard & B.L. Burtt in Notes Roy. Bot. Gard. Edinburgh,* 43(2): 199 (1986)—South Africa (Natal).

ERICINELLA (Ericac.).
hillburttii *E.G.H. Oliver in Bothalia,* 16(1): 46 (1986)—South Africa (Cape Province).

ERIGERON (Compos.).
campanensis *H. Valdebenito, T.K. Lowrey & T.F. Stuessy in Brittonia,* 38(1): 1 (1986)—Chile.
japonicus *C.P. Thunberg ex A. Murray in Linnaeus, Syst. Veg., ed. 14:* 754 (1784)—Japan. †
scandens *C.P. Thunberg ex A. Murray in Linnaeus, Syst. Veg., ed. 14:* 754 (1784)—Japan. †
scopulinus *G.L. Nesom & V.D. Roth in J. Ariz. Acad. Sci.,* 16(2): 39 (1981)—U.S.A. (Arizona, New Mexico).
semperflorens *Roezl ex Haage & Schmidt, Belgique horticole,* 22: 83 (1872); *Bull, Retail List,* 67: 10 (1872); *cf. D.J. Mabberley in Taxon,* 34(3): 454 (1985)— Locality not stated.
zothecinus *S.L. Welsh in Great Basin Nat.,* 46(2): 262 (1986)—U.S.A. (Utah).

ERIOCAULON (Eriocaulac.).
distichoides *J.M. Mangen in Blumea,* 31(2): 487 (1986)—New Guinea (W. Irian).

ERIOCHLOA (Gramin.).
nelsonii
var. papillosa *R.B. Shaw ex R. McVaugh, Fl. Novo-Galiciana,* 14: 178 (1983)—Mexico.

ERIOGONUM (Polygonac.).
batemanii
var. eremicum *(Reveal) S.L. Welsh in Great Basin Nat.,* 44(4): 529 (1984): E. eremicum.
var. ostlundii *(Jones) S.L. Welsh in Great Basin Nat.,* 44(4): 529 (1984): E. ostlundii.
brevicaule
var. desertorum *(Maguire) S.L. Welsh in Great Basin Nat.,* 44(4): 531 (1984): E. chrysocephalum subsp. desertorum.
var. ephedroides *(Reveal) S.L. Welsh in Great Basin Nat.,* 44(4): 531 (1984): E. ephedroides.
var. loganum *(A. Nels.) S.L. Welsh in Great Basin Nat.,* 44(4): 531 (1984): E. loganum.
var. nanum *(Reveal) S.L. Welsh in Great Basin Nat.,* 44(4): 531 (1984): E. nanum.
var. promiscuum *S.L. Welsh in Great Basin Nat.,* 44(4): 531 (1984)—U.S.A. (Utah).
var. viridulum *(Reveal) S.L. Welsh in Great Basin Nat.,* 44(4): 532 (1984): E. viridulum.
corymbosum
var. cronquistii *(Reveal) S.L. Welsh in Great Basin Nat.,* 44(4): 534 (1984): E. cronquistii.
var. humivagans *(Reveal) S.L. Welsh in Great Basin Nat.,* 44(4): 534 (1984): E. humivagans.
var. hylophilum *(Reveal & Brotherson) S.L. Welsh in Great Basin Nat.,* 44(4): 534 (1984): E. hylophilum.
var. smithii *(Reveal) S.L. Welsh in Great Basin Nat.,* 44(4): 535 (1984): E. smithii.
lonchophyllum
var. saurinum *(Reveal) S.L. Welsh in Great Basin Nat.,* 44(4): 540 (1984): E. saurinum.
nummulare
var. ammophilum *(Reveal) S.L. Welsh in Great Basin Nat.,* 44(4): 541 (1984): E. ammophilum.
racemosum
var. coccineum *(J.T. Howell) S.L. Welsh in Great Basin Nat.,* 44(4): 544 (1984): E. zionis var. coccineum.
var. zionis *(J.T. Howell) S.L. Welsh in Great Basin Nat.,* 44(4): 543 (1984): E. zionis.
spathulatum
var. natum *(Reveal) S.L. Welsh in Great Basin Nat.,* 44(4): 545 (1984): E. natum.

ERIOGONUM :–
suffrutescens *Huber, Cat. Gén. 1870 & 1871:* 3 (1870); *cf. D.J. Mabberley in Taxon,* 34(3): 453 (1985)— Locality not stated.

ERIOSEMA (Legumin.).
acuminatum *(Eckl. & Zeyh.) C.H. Stirton in Bothalia,* 16(1): 13 (1986): Desmodium squarrosum var. acuminatum.
latifolium *(Benth. ex Harv.) C.H. Stirton in Bothalia,* 16(1): 20 (1986): E. squarrosum var. latifolium.
luteopetalum *C.H. Stirton in Bothalia,* 16(1): 14 (1986)—South Africa (Natal).
rossii *C.H. Stirton in Bothalia,* 16(1): 18 (1986)—South Africa (Natal, Cape Province).
umtamvunense *C.H. Stirton in Bothalia,* 16(1): 16 (1986)—South Africa (Cape Province, Natal).

ERIOSOLENA (Thymelaeac.).
composita
var. szemaoensis *Y.Y. Qian in Acta Bot. Yunnanica,* 7(2): 136 (1985)—China.

ERISMA (Vochysiac.).
blancoa *L. Marcano Berti in Pittieria,* 13: 12 (1986)—Venezuela.
silvae *L. Marcano Berti in Pittieria,* 13: 15 (1986)—Brazil.

ERITRICHIUM (Boraginac.).
deqinense *W.T. Wang in Bull. Bot. Res. North-East. Forest. Inst.,* 6(3): 94 (1986)—China.
kangdingense *W.T. Wang in Bull. Bot. Res. North-East. Forest. Inst.,* 6(3): 92 (1986)—China.

ERODIUM (Geraniac.).
absinthoides
subsp. balcanicum *(Micevski) W. Greuter & H.M. Burdet in W. Greuter, H.M. Burdet & G. Long (eds.), Med-checklist,* 3: 249 (1986): E. absinthoides var. balcanicum.
alpinum
subsp. balcanicum *(Micevsky) W. Greuter & H.M. Burdet in W. Greuter, H.M. Burdet & G. Long (eds.), Med-checklist,* 3: xxix (1986), without basionym ref.: Basionym not stated.
melanostigma *Mart., Hort. Acad. Erlangen:* 139 (1814); *cf. D.J. Mabberley in Taxon,* 33(3): 442 (1984)— Locality not stated.

ERUCARIA (Crucif.).
rostrata *(Boiss.) A.W. Hill ex W. Greuter & Burdet in Willdenowia,* 15(2): 419 (1986): Didesmus rostratus.

×ERYDIUM hort. in Orchid Rev., 94(1115): centre page pull-out p. 10 (1986). *ORCHIDACEAE.* (Erycina × Oncidium).

ERYNGIUM (Umbellif.).
aristulatum
subsp. parishii *(Coulter & Rose) R.M. Beauchamp in Phytologia,* 59(7): 437 (1986): E. parishii.

ERYTHRINA (Legumin.).
×dyeri *E.F. Hennessy in Bothalia,* 16(1): 48 (1986)—South Africa (Natal, Cape Province). (E. caffra × lysistemon).
poianthes *Brot. ex Tilloch & Taylor, Phil. Mag.,* 61: 465 (June 1823); *cf. D.J. Mabberley in Taxon,* 30(1): 16 (1981)— Locality not stated.

ERYTHROCOCCA (Euphorbiac.).
sanjensis *A. Radcliffe-Smith in Kew Bull.,* 41(4): 963 (1986)—Tanzania.

ERYTHRORHIPSALIS (Cactac.).
burchellii *(B. & R.) Volgin in Vestn. Moskovsk. Univ. Ser. 16, Biol.,* 36(3): 19 (1981); *cf. Repert. Pl. Succ. (I.O.S.),* 36: 7 (1985 publ. 1986): Rhipsalis burchellii.
campos-portoana *(Loefgren) Volgin in Vestn. Moskovsk. Univ. Ser. 16, Biol.,* 36(3): 19 (1981); *cf. Repert. Pl. Succ. (I.O.S.),* 36: 7 (1985 publ. 1986): Rhipsalis campos-portoana.
cereuscula *(Haw.) Volgin in Vestn. Moskovsk. Univ. Ser. 16, Biol.,* 36(3): 19 (1981); *cf. Repert. Pl. Succ. (I.O.S.),* 36: 7 (1985 publ. 1986): Rhipsalis cereuscula.
cribrata *(Lem.) Volgin in Vestn. Moskovsk. Univ. Ser. 16, Biol.,* 36(3): 19 (1981); *cf. Repert. Pl. Succ. (I.O.S.),* 36: 7 (1985 publ. 1986): Hariota cribrata.

ERYTHROXYLUM (Erythroxylac.).
bezerrae *T. Plowman in Brittonia,* 38(3): 196 (1986)—Brazil.
indicum *(DC.) W. Elliot, Fl. Andhirica:* 46 (1859); *cf. D.J. Mabberley in Taxon,* 32(1): 87 (1983): Basionym not stated.
ligustrinum
var. carajasense *T. Plowman in Acta Amazonica,* 14(1–2 Suppl.): 122 (1984 publ. 1986)—Brazil.
nelson-rosae *T. Plowman in Acta Amazonica,* 14(1–2 Suppl.): 124 (1984 publ. 1986)—Brazil.
pauferrense *T. Plowman in Brittonia,* 38(3): 193 (1986)—Brazil.
schunkei *T. Plowman in Acta Amazonica,* 14(1–2 Suppl.): 127 (1984 publ. 1986)—Peru.
simonis *T. Plowman in Brittonia,* 38(3): 189 (1986)—Brazil.
tianguanum *T. Plowman in Brittonia,* 38(3): 198 (1986)—Brazil.
tucuruiense *T. Plowman in Acta Amazonica,* 14(1–2 Suppl.): 131 (1984 publ. 1986)—Brazil.

ESCHSCHOLZIA (Papaverac.).
lemmonii
subsp. kernensis *(Munz) C. Clark in Madroño,* 33(3): 225 (1986): E. caespitosa subsp. kernensis.

ESCOBARIA (Cactac.).
guadalupensis *S. Brack & K.D. Heil in Cact. Succ. J. (U.S.A.),* 58(4): 165 (1986)—U.S.A. (Texas).

ETLINGERA (Zingiberac.).
angustifolia *(Val.) R.M. Smith in Notes Roy. Bot. Gard. Edinburgh,* 43(2): 243 (1986): Geanthus angustifolius.
araneosa *(Bak.) R.M. Smith in Notes Roy. Bot. Gard. Edinburgh,* 43(2): 243 (1986): Amomum araneosum.
australasica *(R.M. Smith) R.M. Smith in Notes Roy. Bot. Gard. Edinburgh,* 43(2): 243 (1986): Achasma australasica.
brachychila *(Ridley) R.M. Smith in Notes Roy. Bot. Gard. Edinburgh,* 43(2): 243 (1986): Hornstedtia brachychila.
brevilabris *(Val.) R.M. Smith in Notes Roy. Bot. Gard. Edinburgh,* 43(2): 243 (1986): Achasma brevilabrum.
bromeliopsis *(Val.) R.M. Smith in Notes Roy. Bot. Gard. Edinburgh,* 43(2): 243 (1986): Geanthus bromeliopsis.
calycina *(Val.) R.M. Smith in Notes Roy. Bot. Gard. Edinburgh,* 43(2): 244 (1986): Geanthus calycinus.
caudiculata *(Ridley) R.M. Smith in Notes Roy. Bot. Gard. Edinburgh,* 43(2): 244 (1986): Phaeomeria caudiculata.
cevuga *(Seem.) R.M. Smith in Notes Roy. Bot. Gard. Edinburgh,* 43(2): 244 (1986): Amomum cevuga.
dekockii *(Val.) R.M. Smith in Notes Roy. Bot. Gard. Edinburgh,* 43(2): 244 (1986): Geanthus dekockii.
densiuscula *(Val.) R.M. Smith in Notes Roy. Bot. Gard. Edinburgh,* 43(2): 244 (1986): Geanthus densiusculus.
diepenhorstii *(Teysm. & Binn.) R.M. Smith in Notes Roy. Bot. Gard. Edinburgh,* 43(2): 244 (1986): Eletteria diepenhorstii.
elatior *(Jack) R.M. Smith in Notes Roy. Bot. Gard. Edinburgh,* 43(2): 244 (1986): Alpinia elatior.
fimbriobracteata *(K. Schum.) R.M. Smith in Notes Roy. Bot. Gard. Edinburgh,* 43(2): 245 (1986): Amomum fimbriobracteatum.
foetens *(Bl.) R.M. Smith in Notes Roy. Bot. Gard. Edinburgh,* 43(2): 245 (1986): Elettaria foetens.
goliathensis *(Val.) R.M. Smith in Notes Roy. Bot. Gard. Edinburgh,* 43(2): 245 (1986): Geanthus goliathensis.
gracilis *(Val.) R.M. Smith in Notes Roy. Bot. Gard. Edinburgh,* 43(2): 245 (1986): Nicolaia gracilis.

ETLINGERA :-

grandiflora *(Val.) R.M. Smith in Notes Roy. Bot. Gard. Edinburgh,* 43(2): 245 (1986): Geanthus grandiflorus.

grandiligulata *(K. Schum.) R.M. Smith in Notes Roy. Bot. Gard. Edinburgh,* 43(2): 245 (1986): Amomum grandiligulatum.

harmandii *(Gagnep.) R.M. Smith in Notes Roy. Bot. Gard. Edinburgh,* 43(2): 245 (1986): Amomum harmandii.

hemisphaerica *(Bl.) R.M. Smith in Notes Roy. Bot. Gard. Edinburgh,* 43(2): 245 (1986): Elettaria hemisphaerica.

heyniana *(Val.) R.M. Smith in Notes Roy. Bot. Gard. Edinburgh,* 43(2): 246 (1986): Nicolaia heyniana.

labellosa *(K. Schum.) R.M. Smith in Notes Roy. Bot. Gard. Edinburgh,* 43(2): 246 (1986): Amomum labellosum.

latifolia *(Val.) R.M. Smith in Notes Roy. Bot. Gard. Edinburgh,* 43(2): 246 (1986): Geanthus latifolius.

linguiformis *(Roxb.) R.M. Smith in Notes Roy. Bot. Gard. Edinburgh,* 43(2): 246 (1986): Alpinia linguiforme.

longipetala *(Val.) R.M. Smith in Notes Roy. Bot. Gard. Edinburgh,* 43(2): 247 (1986): Geanthus longipetalus.

longipetiolata *(Burtt & Smith) R.M. Smith in Notes Roy. Bot. Gard. Edinburgh,* 43(2): 247 (1986): Geanthus longipetiolatus.

loroglossa *(Gagnep.) R.M. Smith in Notes Roy. Bot. Gard. Edinburgh,* 43(2): 247 (1986): Amomum loroglossum.

lorzingii *(Val.) R.M. Smith in Notes Roy. Bot. Gard. Edinburgh,* 43(2): 247 (1986): Nicolaia lorzingii.

maingayi *(Bak.) R.M. Smith in Notes Roy. Bot. Gard. Edinburgh,* 43(2): 247 (1986): Amomum maingayi.

metriocheilos *(Griff.) R.M. Smith in Notes Roy. Bot. Gard. Edinburgh,* 43(2): 247 (1986): Achasma metriocheilos.

minor *(Ridley) R.M. Smith in Notes Roy. Bot. Gard. Edinburgh,* 43(2): 247 (1986): Phaeomeria minor.

moluccana *(K. Schum.) R.M. Smith in Notes Roy. Bot. Gard. Edinburgh,* 43(2): 248 (1986): Phaeomeria moluccana.

muluensis *R.M. Smith in Notes Roy. Bot. Gard. Edinburgh,* 43(3): 455 (1986)—Borneo (Sarawak).

nasuta *(K. Schum.) R.M. Smith in Notes Roy. Bot. Gard. Edinburgh,* 43(2): 248 (1986): Amomum nasutum.

var. reticulata *R.M. Smith in Notes Roy. Bot. Gard. Edinburgh,* 43(3): 446 (1986)—Borneo (Sarawak).

parva *(Val.) R.M. Smith in Notes Roy. Bot. Gard. Edinburgh,* 43(2): 248 (1986): Geanthus parvus.

pauciflora *(Ridley) R.M. Smith in Notes Roy. Bot. Gard. Edinburgh,* 43(2): 248 (1986): Hornstedtia pauciflora.

ETLINGERA :-

pavieana *(Pierre ex Gagnep.) R.M. Smith in Notes Roy. Bot. Gard. Edinburgh,* 43(2): 248 (1986): Amomum pavieanum.

peekelii *(Val.) R.M. Smith in Notes Roy. Bot. Gard. Edinburgh,* 43(2): 248 (1986): Nicolaia peekelii.

philippinensis *(Ridley) R.M. Smith in Notes Roy. Bot. Gard. Edinburgh,* 43(2): 248 (1986): Hornstedtia philippinensis.

polyantha *(Val.) R.M. Smith in Notes Roy. Bot. Gard. Edinburgh,* 43(2): 248 (1986): Geanthus polyanthus.

pubescens *(Burtt & Smith) R.M. Smith in Notes Roy. Bot. Gard. Edinburgh,* 43(2): 248 (1986): Geanthus pubescens.

punicea *(Roxb.) R.M. Smith in Notes Roy. Bot. Gard. Edinburgh,* 43(2): 249 (1986): Alpinia punicea.

pyramidosphaera *(K. Schum.) R.M. Smith in Notes Roy. Bot. Gard. Edinburgh,* 43(2): 249 (1986): Phaeomeria pyramidosphaera.

rosea *B.L. Burtt & R.M. Smith in Notes Roy. Bot. Gard. Edinburgh,* 43(2): 240 (1986), nom. nov.: Amomum roseum *K. Schum.*

sanguinea *(Ridley) R.M. Smith in Notes Roy. Bot. Gard. Edinburgh,* 43(2): 249 (1986): Hornstedtia sanguinea.

sericea *(Ridley) R.M. Smith in Notes Roy. Bot. Gard. Edinburgh,* 43(2): 249 (1986): Nicolaia sericea.

sessilanthera *R.M. Smith in Notes Roy. Bot. Gard. Edinburgh,* 43(2): 240 (1986)—Borneo (Sarawak).

solaris *(Bl.) R.M. Smith in Notes Roy. Bot. Gard. Edinburgh,* 43(2): 249 (1986): Elettaria solaris.

subterranea *(Holtt.) R.M. Smith in Notes Roy. Bot. Gard. Edinburgh,* 43(2): 250 (1986): Achasma subterraneum.

subulicalyx *(Val.) R.M. Smith in Notes Roy. Bot. Gard. Edinburgh,* 43(2): 250 (1986): Nicolaia subulicalyx.

sulphurea *(R.N. Parker) R.M. Smith in Notes Roy. Bot. Gard. Edinburgh,* 43(2): 250 (1986): Hornstedtia sulphurea.

triorgyalis *(Bak.) R.M. Smith in Notes Roy. Bot. Gard. Edinburgh,* 43(2): 250 (1986): Amomum triorgyale.

velutina *(Ridley) R.M. Smith in Notes Roy. Bot. Gard. Edinburgh,* 43(2): 250 (1986): Hornstedtia velutina.

venusta *(Ridley) R.M. Smith in Notes Roy. Bot. Gard. Edinburgh,* 43(2): 250 (1986): Hornstedtia venusta.

versteegii *(Val.) R.M. Smith in Notes Roy. Bot. Gard. Edinburgh,* 43(2): 250 (1986): Geanthus versteegii.

vestita *(Val.) R.M. Smith in Notes Roy. Bot. Gard. Edinburgh,* 43(2): 250 (1986): Geanthus vestitus.

walang *(Bl.) R.M. Smith in Notes Roy. Bot. Gard. Edinburgh,* 43(2): 251 (1986): Donacodes walang.

ETLINGERA :-
xanthoparyphe *(K. Schum.) R.M. Smith in Notes Roy. Bot. Gard. Edinburgh,* 43(2): 251 (1986): Amomum xanthoparyphe.
yunnanensis *(Wu & Senjen) R.M. Smith in Notes Roy. Bot. Gard. Edinburgh,* 43(2): 251 (1986): Achasma yunnanensis.

EUCALYPTUS (Myrtac.).
brevipes *M.I.H. Brooker in Nuytsia,* 5(3): 365 (1986)—Australia (W. Australia).
celastroides
subsp. virella *M.I.H. Brooker in Nuytsia,* 5(3): 359 (1986)—Australia (W. Australia).
ceracea *M.I.H. Brooker & C.C. Done in Nuytsia,* 5(3): 382 (1986)—Australia (W. Australia).
chlorophylla *M.I.H. Brooker & C.C. Done in Nuytsia,* 5(3): 389 (1986)—Australia (W. Australia).
delegatensis
subsp. tasmaniensis *D.J. Boland in Austral. Forest Res.,* 15(2): 177 (1985)—Tasmania.
erectifolia *M.I.H. Brooker & S.D. Hopper in Nuytsia,* 5(3): 351 (1986)—Australia (W. Australia).
ferriticola *M.I.H. Brooker & W.B. Edgecombe in Nuytsia,* 5(3): 373 (1986)—Australia (W. Australia).
lateritica *M.I.H. Brooker & S.D. Hopper in Nuytsia,* 5(3): 346 (1986)—Australia (W. Australia).
pilbarensis *M.I.H. Brooker & W.B. Edgecombe in Nuytsia,* 5(3): 376 (1986)—Australia (W. Australia).
rupestris *M.I.H. Brooker & C.C. Done in Nuytsia,* 5(3): 385 (1986)—Australia (W. Australia).
suberea *M.I.H. Brooker & S.D. Hopper in Nuytsia,* 5(3): 343 (1986)—Australia (W. Australia)
willisii
subsp. falciformis *M.R. Newnham, P.Y. Ladiges & T. Whiffin in Austral. J. Bot.,* 34(3): 348 (1986)—Australia (Victoria).
yilgarnensis *(Maiden) M.I.H. Brooker in Nuytsia,* 5(3): 366 (1986): E. gracilis var. yilgarnensis.

EUCOMIS (Hyacinthac.).
striata *(Ker) Ait. f., Epit. Hort. Kew.:* 367 (1814); *cf. D.J. Mabberley in Taxon,* 31(1): 70 (1982): Basionym not stated.

EUCROSIA (Amaryllidac.).
stricklandii
var. montana *A.W. Meerow in Phytologia,* 60(2): 101 (1986)—Ecuador.

EUDIANTHE (Caryophyllac.).
coeli-rosa *(L.) Endl., Cat. Horti Vindob.,* 2: 342 (1842); *cf. D.J. Mabberley in Taxon,* 33(3): 435 (1984): Basionym not stated.

EUGENIA (Myrtac.).
inirebensis *P.E. Sánchez Vindas in Phytologia,* 61(2): 127 (1986)—Mexico.
mozomboensis *P.E. Sánchez Vindas in Phytologia,* 61(2): 126 (1986)—Mexico.

EULOPHIA (Orchidac.).
ucbii *C.L. Malhotra & B. Balodi in Bull. Bot. Surv. India,* 26(1–2): 92 (1984 publ. 1985)—India.

EUONYMUS (Celastrac.).
maackii
f. salicifolia *T. Chen in Bull. Bot. Res. North-East. Forest. Inst.,* 6(2): 159 (1986)—China.
occidentalis
subsp. parishii *(Trelease) R.M. Beauchamp in Phytologia,* 59(7): 438 (1986): E. parishii.
tobira *C.P. Thunberg ex A. Murray in Linnaeus, Syst. Veg., ed. 14:* 238 (1784)—Japan. †

EUPATORIASTRUM (Compos.).
chlorostylum *B.L. Turner in Phytologia,* 59(5): 323 (1986)—Mexico.

EUPATORIUM (Compos.).
japonicum *C.P. Thunberg ex A. Murray in Linnaeus, Syst. Veg., ed. 14:* 737 (1784)—Japan. †
tanacetifolium
f. albiflorum *N.I. Matzenbacher in Comun. Mus. Ciênc. PUCRGS Sér. Bot.,* 30–39: 118 (1985), without holotype—Brazil.

EUPHORBIA (Euphorbiac.).
amygdaloides
subsp. heldreichii *(Orph. ex Boiss.) B. Aldén in A. Strid (ed.), Mount. Fl. Greece,* 1: 575 (1986): E. heldreichii.
begoniifolia *Lehm., Sem. Hort. Bot. Hamb.,* 1823: 8 (1824); *Linnaea 3, Lit.:* 9 (1828); *cf. D.J. Mabberley in Taxon,* 32(1): 85 (1983)—Locality not stated.
carinata *(Haw.) Salm-Dyck, Index Pl. Succ. Hort. Dyck.:* 24 (1822); *cf. D.J. Mabberley in Taxon,* 30(1): 16 (1981): Basionym not stated.
griffithii
var. bhutanica *(Fischer) D.G. Long in Notes Roy. Bot. Gard. Edinburgh,* 44(1): 167 (1986): E. sikkimensis subsp. bhutanica.
kansui *Liou ex S.B. Ho in Fl. Tsinlingensis,* 1(3): 450, 162 (1981)—China.
lumbricalis *L.C. Leach in S. Afr. J. Bot.,* 52(4): 369 (1986)—South Africa (Cape Province).
luteoviridis *D.G. Long in Notes Roy. Bot. Gard. Edinburgh,* 44(1): 163 (1986), as 'luteo-viridis'—Sikkim, Nepal, India, Xizang.

EUPHORBIA :–
×martinii
nothosubsp. cornubiensis *(A. Radcliffe-Smith) A. Radcliffe-Smith in Taxon,* 35(2): 349 (1986): E. × cornubiensis. (E. amygdaloides × characias subsp. wulfenii).
minima *Mart., Hort. Acad. Erlangen:* 105 (1814); *cf. D.J. Mabberley in Taxon,* 33(3): 442 (1984)— Locality not stated.
mira *L.C. Leach in S. Afr. J. Bot.,* 52(1): 10 (1986)—South Africa (Cape Province).
myrsinites
subsp. rechingeri *(Greuter) B. Aldén in A. Strid (ed.), Mount. Fl. Greece,* 1: 570 (1986): E. rechingeri.
padifolia *(L.) Salm-Dyck, Index Pl. Succ. Hort. Dyck.:* 24 (1822); *cf. D.J. Mabberley in Taxon,* 30(1): 16 (1981): Basionym not stated.
pandurata *Huber, Cat. Gén. 1871 & 1872:* 3 (1871); *cf. D.J. Mabberley in Taxon,* 34(3): 450 (1985)— Locality not stated.
parvicyathophora *W. Rauh in Cact. Succ. J. (U.S.A.),* 58(4): 143 (1986)—Madagascar.
petaloides *G. Don f. in Loud., Hort. Brit.:* 192 (1830); *cf. D.J. Mabberley in Taxon,* 31(1): 70 (1982)— Locality not stated.
talaina *A. Radcliffe-Smith in Kew Bull.,* 41(2): 319 (1986)—Pakistan.

EUTREMA (Crucif.).
obliquum *K.C. Kuan & An in Fl. Xizangica,* 2: 399 (1985)—Xizang.

EVAX (Compos.).
libyaca *S.A. Alavi in Fl. Libya,* 107: 51 (1983)—Libya.

EVOLVULUS (Convolvulac.).
yunnanensis *S.H. Huang in Acta Phytotax. Sin.,* 24(1): 17 (1986)—China.

EVOSMA (Loganiac.).
latifolia *(Aubl.) Sterler, Hort. Nymph.:* 46 (1821); *cf. D.J. Mabberley in Taxon,* 33(3): 442 (1984): Coussapoa latifolia.

EVRARDIA (Orchidac.).
asraoa *J. Joseph & N.R. Abbareddy in Bull. Bot. Surv. India,* 25(1–4): 232 (1983 publ. 1985)—India.

EXACUM (Gentianac.).
sutaepense
f. gracile *(Toyokuni) H. Toyokuni in Acta Phytotax. Geobot.,* 36(4–6): 124 (1985): E. sutaepense var. gracile.

F

FABA (Legumin.).
bona
subsp. equina *(Pers.) J. Soják in Čas. Nár. Muz. (Prague),* 152(1): 21 (1983): Vicia faba β var. equina.
subsp. minuta *(Alef.) J. Soják in Čas. Nár. Muz. (Prague),* 152(1): 21 (1983): Faba vulgaris var. minuta.

FACHEIROA (Cactac.).
subgen. Zehntnerella *(Britton & Rose) P.J. Braun & E. Esteves Pereira in Kakt. And. Sukk.,* 37(3): 56 (1986): gen. Zehntnerella.
estevesii *P.J. Braun in Kakt. And. Sukk.,* 37(4): 79 (1986)—Brazil.

FADOGIA (Rubiac.).
audruana *M. Fay, J.-P. Lebrun & A.L. Stork in Bull. Mus. Nation. Hist. Nat., B, Adansonia, Ser. 4,* 8(1): 57 (1986)—Central African Republic.

FALCARIA (Umbellif.).
agrestis *(Hoffm.) Sweet, Hort. Brit., ed. 2:* 245 (1830); *cf. D.J. Mabberley in Taxon,* 31(1): 70 (1982): Basionym not stated.

FARGESIA (Gramin.).
obliqua *T.P. Yi in Acta Bot. Yunnanica,* 8(1): 48 (1986)—China.

FARINOPSIS J. Chrtek & J. Soják in Čas. Nár. Muz. (Prague), 153(1): 10 (1984). *ROSACEAE.*
salesoviana *(Steph.) J. Chrtek & J. Soják in Čas. Nár. Muz. (Prague),* 153(1): 10 (1984): Potentilla salesoviana.

FEDIA (Valerianac.).
dentata *(L.) Bernh., Syst. Verz.:* 24 (1800); *cf. D.J. Mabberley in Taxon,* 33(3): 443 (1984): Basionym not stated.
rugulosa *Hornem., Hort. Hafn.:* 950 (1815); *cf. D.J. Mabberley in Taxon,* 33(3): 437 (1984)— Locality not stated.

FERNANDOA (Bignoniac.).
guangxiensis *D.D. Tao in Acta Phytotax. Sin.,* 24(2): 149 (1986)—China.

FEROCACTUS (Cactac.).
piliferus *(Lemaire ex Ehrenberg) G. Unger in Kakt. And. Sukk.,* 37(2): 45 (1986): Echinocactus piliferus.
f. flavispinus *(hort. ex Schelle) G. Unger in Kakt. And. Sukk.,* 37(2): 45 (1986): Echinocactus pilosus flavispinus.
var. stainesii *(Salm-Dyck) G. Unger in Kakt. And. Sukk.,* 37(2): 45 (1986): Echinocactus pilosus var. stainesii.

FERULA (Umbellif.).
hexiensis *K.M. Shen in Acta Phytotax. Sin.,* 24(4): 314 (1986)—China.

FERULAGO (Umbellif.).
sylvatica
subsp. confusa *(Velen.) P. Hartvig in A. Strid (ed.), Mount. Fl. Greece,* 1: 713 (1986): F. confusa.

FESTUCA (Gramin.).
ser. Ovinae *M. Pawlus in Fragm. Flor. Geobot.,* 29(2): 228 (1983 publ. 1985).
ser. Psammophilae *M. Pawlus in Fragm. Flor. Geobot.,* 29(2): 259 (1983 publ. 1985).
ser. Trachyphyllae *M. Pawlus in Fragm. Flor. Geobot.,* 29(2): 246 (1983 publ. 1985).
ser. Valesiacae *M. Pawlus in Fragm. Flor. Geobot.,* 29(2): 276 (1983 publ. 1985).
acamptophylla *(St.-Yves) E.B. Alekseev in Bot. Zhurn.,* 71(8): 1113 (1986): F. abyssinica subsp. acamptophylla.
brevipaleata *(St.-Yves) E.B. Alekseev in Bot. Zhurn.,* 71(8): 1113 (1986): F. camusiana var. brevipaleata.
chodatiana *(St.-Yves) E.B. Alekseev in Bot. Zhurn.,* 71(8): 1113 (1986): F. camusiana subsp. chodatiana.
claytonii *E.B. Alekseev in Bot. Zhurn.,* 71(8): 1117 (1986)—Kenya.
dracomontana *H.P. Linder in Bothalia,* 16(1): 59 (1986)—Lesotho, South Africa (Natal, Transvaal).
exaristata *E.B. Alekseev in Bot. Zhurn.,* 71(8): 1116 (1986)—Lesotho.
font-queri *(Litard.) A.M. Romo in Willdenowia,* 16(1): 115 (1986): F. rubra subsp. font-queri.
hedbergii *E.B. Alekseev in Bot. Zhurn.,* 71(8): 1113 (1986), nom. nov.: F. abyssinica var. supina *Pilg. ex St.-Yves.*
×jierru *H.S. Nava in Fontqueria,* 7: 24 (1985)—Spain. (F. gautieri × picoeuropeana).
macra *(Stapf) E.B. Alekseev in Bot. Zhurn.,* 71(8): 1116 (1986): F. caprina var. macra.
misera *C.P. Thunberg ex A. Murray in Linnaeus, Syst. Veg., ed. 14:* 119 (1784)—Japan. †
pauciflora *C.P. Thunberg ex A. Murray in Linnaeus, Syst. Veg., ed. 14:* 119 (1784)—Japan. †
picoeuropeana *H.S. Nava in Fontqueria,* 7: 23 (1985)—Spain.
richardii *E.B. Alekseev in Bot. Zhurn.,* 71(8): 1109 (1986)—Ethiopia.
rigidiuscula *E.B. Alekseev in Bot. Zhurn.,* 71(8): 1111 (1986)—Cameroun.
tarraconensis *(Litard.) A.M. Romo in Willdenowia,* 16(1): 115 (1986): F. ovina var. tarraconensis.
vaginata
var. aristata *M. Pawlus in Fragm. Flor. Geobot.,* 29(2): 275 (1983 publ. 1985)—Poland.

FESTUCELLA E.B. Alekseev in Byull. Mosk. Obshch. Ispȳt. Prir., Biol., 90(5): 104 (1985). *GRAMINEAE.*

FESTUCELLA :–
eriopoda *(Vickery) E.B. Alekseev in Byull. Mosk. Obshch. Ispȳt. Prir., Biol.,* 90(5): 104 (1985): Festuca eriopoda.

FIBIGIA (Crucif.).
lunarioides *Sweet, Hort. Brit.:* 467 (1827); *cf. D.J. Mabberley in Taxon,* 31(1): 70 (1982)—Locality not stated.

FICINIA (Cyperac.).
crinita *(Poiret) B.L Burtt in Notes Roy. Bot. Gard. Edinburgh,* 43(3): 358 (1986): Cyperus crinitus.
filiculmea *B.L. Burtt in Notes Roy. Bot. Gard. Edinburgh,* 43(3): 358 (1986)—South Africa (Natal, Cape Province).
nana *B.L. Burtt in Notes Roy. Bot. Gard. Edinburgh,* 43(3): 360 (1986)—South Africa (Natal), Lesotho.
undosa *B.L. Burtt in Notes Roy. Bot. Gard. Edinburgh,* 43(3): 360 (1986)—South Africa (Natal), Lesotho.

FICUS (Morac.).
subsect. Caulocarpae *(Mildbraed & Burret) C.C. Berg in Proc. Kon. Nederl. Akad. Wetensch., C,* 89(2): 125 (1986): F. sect. Caulocarpae.
subsect. Chlamydodorae *(Mildbraed & Burret) C.C. Berg in Proc. Kon. Nederl. Akad. Wetensch., C,* 89(2): 123 (1986): F. sect. Chlamydodorae.
subsect. Crassicostae *(Mildbraed & Burret) C.C. Berg in Proc. Kon. Nederl. Akad. Wetensch., C,* 89(2): 124 (1986): F. sect. Crassicostae.
subsect. Cyathistipulae *(Mildbraed & Burret) C.C. Berg in Proc. Kon. Nederl. Akad. Wetensch., C,* 89(2): 125 (1986): F. sect. Cyathistipulae.
subsect. Platyphyllae *(Mildbraed & Burret) C.C. Berg in Proc. Kon. Nederl. Akad. Wetensch., C,* 89(2): 122 (1986): F. sect. Platyphyllae.
ampana *C.C. Berg in Bull. Mus. Nation. Hist. Nat., B, Adansonia, Ser. 4,* 8(1): 33 (1986)—Madagascar.
antandronarum *(H. Perrier) C.C. Berg in Bull. Mus. Nation. Hist. Nat., B, Adansonia, Ser. 4,* 8(1): 45 (1986): F. pyrifolia var. antandronarum.
subsp. bernardii *C.C. Berg in Bull. Mus. Nation. Hist. Nat., B, Adansonia, Ser. 4,* 8(1): 46 (1986)—Comoro Archipelago.
aripuanensis *C.C. Berg & F. Kooy in Acta Amazonica,* 14(1–2 Suppl.): 195 (1984 publ. 1986)—Brazil.
blepharophylla *M. Vázquez Avila in Acta Amazonica,* 14(1–2 Suppl.): 197 (1984 publ. 1986)—Brazil.
cabur *Buch. ex J.E. Smith in A. Rees, Cyclop.,* 14: Ficus n. 47 (1810)—Nepal.
cremersii *C.C. Berg in Acta Amazonica,* 14(1–2 Suppl.): 199 (1984 publ. 1986)—French Guiana.

FICUS :–
galactophora *Ten., Cat. Ort. Bot. Nap.:* 86 (1845); *cf. D.J. Mabberley in Taxon,* 30(1): 16 (1981)— Locality not stated.
humbertii *C.C. Berg in Bull. Mus. Nation. Hist. Nat., B, Adansonia, Ser. 4,* 8(1): 37 (1986)—Madagascar.
infrafoliacea *Buch. ex J.E. Smith in A. Rees, Cyclop.,* 14: Ficus n. 31 (1810)—Nepal.
insipida
subsp. scabra *C.C. Berg in Acta Amazonica,* 14(1–2 Suppl.): 201 (1984 publ. 1986)—French Guiana, Brazil, Surinam, Guyana, Venezuela, Colombia.
jacobii *M. Vázquez Avila in Acta Amazonica,* 14(1–2 Suppl.): 201 (1984 publ. 1986)—Peru.
karthalensis *C.C. Berg in Bull. Mus. Nation. Hist. Nat., B, Adansonia, Ser. 4,* 8(1): 30 (1986)—Comoro Archipelago.
lauretana *M. Vázquez Avila in Acta Amazonica,* 14(1–2 Suppl.): 203 (1984 publ. 1986)—Brazil, Colombia, Peru.
leiophylla *C.C. Berg in Acta Amazonica,* 14(1–2 Suppl.): 205 (1984 publ. 1986)—French Guiana, Brazil.
macrophylla *Roxb. & Buch. ex J.E. Smith in A. Rees, Cyclop.,* 14: Ficus n. 32 (1810)—East Indies.
subsp. columnaris *(C. Moore) P.S. Green in J. Arnold Arbor.,* 67(1): 112 (1986): F. columnaris.
madagascariensis *C.C. Berg in Bull. Mus. Nation. Hist. Nat., B, Adansonia, Ser. 4,* 8(1): 34 (1986)—Madagascar.
pachyclada
subsp. arborea *(H. Perrier) C.C. Berg in Bull. Mus. Nation. Hist. Nat., B, Adansonia, Ser. 4,* 8(1): 20 (1986): F. soroceoides var. arborea.
piresiana *M. Vázquez Avila & C.C. Berg in Acta Amazonica,* 14(1–2 Suppl.): 207 (1984 publ. 1986)—Brazil, French Guiana.
reflexa
subsp. aldabrensis *(Baker) C.C. Berg in Bull. Mus. Nation. Hist. Nat., B, Adansonia, Ser. 4,* 8(1): 49 (1986): F. aldabrensis.
subsp. sechellensis *(Baker) C.C. Berg in Bull. Mus. Nation. Hist. Nat., B, Adansonia, Ser. 4,* 8(1): 49 (1986): F. rubra var. sechellensis.
repens *Roxb. ex J.E. Smith in A. Rees, Cyclop.,* 14: Ficus n. 100 (1810)—East Indies.
roraimensis *C.C. Berg in Acta Amazonica,* 14(1–2 Suppl.): 209 (1984 publ. 1986)—Brazil.
sarmentosa *Buch. ex J.E. Smith in A. Rees, Cyclop.,* 14: Ficus n. 45 (1810)—Nepal.
semicordata *Buch. ex J.E. Smith in A. Rees, Cyclop.,* 14: Ficus n. 71 (1810)—Nepal.
subincisa *Buch. ex J.E. Smith in A. Rees, Cyclop.,* 14: Ficus n. 91 (1810)—Nepal.
vittata *M. Vázquez Avila in Acta Amazonica,* 14(1–2 Suppl.): 211 (1984 publ. 1986)—Peru, Colombia, Ecuador.

FILAGO (Compos.).
asterisciflora *(Lam.) Sweet, Hort. Brit., ed. 2:* 294 (1830); *cf. D.J. Mabberley in Taxon,* 31(1): 70 (1982): Basionym not stated.

FISSISTIGMA (Annonac.).
guinanense *Y. Wan in Bull. Bot. Res. North-East. Forest. Inst.,* 6(1): 145 (1986)—China.

FLEISCHMANNIA (Compos.).
harlingii *R.M. King & H. Robinson in Phytologia,* 60(1): 81 (1986)—Ecuador.

FLICKINGERIA (Orchidac.).
puncticulosa *(J.J. Sm.) J.J. Wood in Kew Bull.,* 41(4): 819 (1986): Dendrobium puncticulosum.

FLOSMUTISIA J. Cuatrecasas in An. Jard. Bot. Madrid, 42(2): 415 (1985 publ. 1986). *COMPOSITAE.*
paramicola *J. Cuatrecasas in An. Jard. Bot. Madrid,* 42(2): 417 (1985 publ. 1986)—Colombia.

FLOURENSIA (Compos.).
cajabambensis *M.O. Dillon in Brittonia,* 38(1): 32 (1986)—Peru.

FLUEGGEA (Euphorbiac.).
virosa
subsp. himalaica *D.G. Long in Notes Roy. Bot. Gard. Edinburgh,* 44(1): 167 (1986)—Bhutan, Nepal, Sikkim, India, Assam, Burma.

FRIEDRICHSTHALIA (Boraginac.).
schimperi *Bunge, Del. Sem. Hort. Bot. Dorpat.,* 1843: 7 (1843); *cf. D.J. Mabberley in Taxon,* 32(1): 85 (1983)— Locality not stated.

FRITILLARIA (Liliac.).
amoena *C.Y. Yang in Bull. Bot. Res. North-East. Forest. Inst.,* 6(1): 173 (1986)—Xinjiang.
anhuiensis
var. albiflora *S.C. Chen & S.F. Yin in Acta Bot. Yunnanica,* 7(3): 307 (1985)—China.
austroanhuiensis *Y.K. Yang & J.K. Wu in Acta Bot. Bor.-Occid. Sin.,* 5(1): 44 (1985)—China.
chuanganensis *Y.K. Yang & J.K. Wu in Acta Bot. Bor.-Occid. Sin.,* 5(1): 28 (1985), with two types cited—China.
chuanganensis
var. glabra *(P.Y. Li) Y.K. Yang & J.K. Wu in Acta Bot. Bor.-Occid. Sin.,* 5(1): 30 (1985): F. cirrhosa f. glabra.
cirrhosa
var. brevistigma *Y.K. Yang & J.K. Wu in Acta Bot. Bor.-Occid. Sin.,* 5(1): 36 (1985)—China.
var. dingriensis *Y.K. Yang & J.Z. Zhang in Acta Bot. Bor.-Occid. Sin.,* 5(1): 34 (1985)—Xizang.

FRITILLARIA :–

var. jilongensis *Y.K. Yang & Gesan in Acta Bot. Bor.-Occid. Sin.*, 5(1): 34 (1985), with two types cited—Xizang.

delavayi

var. banmaensis *Y.K. Yang & J.K. Wu in Acta Bot. Bor.-Occid. Sin.*, 5(1): 24 (1985)—China.

duilongdeqingensis *Y.K. Yang & Gesan in Acta Bot. Bor.-Occid. Sin.*, 5(1): 30 (1985)—Xizang.

gansuensis *S.C. Chen & G.D. Yu in Acta Bot. Yunnanica*, 7(2): 148 (1985)—China.

huangshanensis *Y.K. Yang & C.J. Wu in Acta Bot. Bor.-Occid. Sin.*, 5(1): 41 (1985)—China.

lixianensis *Y.K. Yang & J.K. Wu in Acta Bot. Bor.-Occid. Sin.*, 5(1): 25 (1985)—China.

ningguoensis *S.C. Chen & S.F. Yin in Acta Bot. Yunnanica*, 7(3): 306 (1985)—China.

pingwuensis *Y.K. Yang & J.K. Wu in Acta Bot. Bor.-Occid. Sin.*, 5(1): 26 (1985)—China.

przewalskii

var. discolor *Y.K. Yang & Y.S. Zhou in Acta Bot. Bor.-Occid. Sin.*, 5(1): 40 (1985)—China.

f. emacula *Y.K. Yang & J.K. Wu in Acta Bot. Bor.-Occid. Sin.*, 5(1): 40 (1985)—China.

var. longistigma *Y.K. Yang & J.K. Wu in Acta Bot. Bor.-Occid. Sin.*, 5(1): 36 (1985)—China.

var. tessellata *Y.K. Yang & Y.S. Zhou in Acta Bot. Bor.-Occid. Sin.*, 5(1): 39 (1985)—China.

qingchuanensis *Y.K. Yang & J.K. Wu in Acta Bot. Bor.-Occid. Sin.*, 5(1): 25 (1985)—China.

taipaiensis

var. ningxiaensis *Y.K. Yang & J.K. Wu in Acta Bot. Bor.-Occid. Sin.*, 5(1): 32 (1985)—China.

var. wanyuanensis *Y.K. Yang & J.K. Wu in Acta Bot. Bor.-Occid. Sin.*, 5(1): 32 (1985), with two types cited—China.

var. zhouquensis *S.C. Chen & G.D. Yu in Acta Bot. Yunnanica*, 7(2): 149 (1985)—China.

xinyuanensis *Y.K. Yang & J.K. Wu in Acta Bot. Bor.-Occid. Sin.*, 5(1): 40 (1985)—Xinjiang.

xizangensis *Y.K. Yang & Gesan in Acta Bot. Bor.-Occid. Sin.*, 5(1): 22 (1985), with two types cited—Xizang.

yuzhongensis *G.D. Yu & Y.S. Zhou in Acta Bot. Yunnanica*, 7(2): 146 (1985)—China.

zhufenensis *Y.K. Yang & J.Z. Zhang in Acta Bot. Bor.-Occid. Sin.*, 5(1): 21 (1985)—Xizang.

FRONDARIA C.A. Luer, Ic. Pleurothallid. I-Syst. Pleurothallidinae (Monogr. Syst. Bot. Missouri Bot. Gard. 15): 29 (1986). *ORCHIDACEAE.*

caulescens *(Lindl.) C.A. Luer, Ic. Pleurothallid. I-Syst. Pleurothallidinae (Monogr. Syst. Bot. Missouri Bot. Gard. 15):* 29 (1986): Pleurothallis caulescens.

FUIRENA (Cyperac.).

obcordata *P.L. Forbes in S. Afr. J. Bot.*, 52(3): 238 (1986), sp. nov. without latin descr.—South Africa (Natal, Transvaal), Mozambique, Namibia.

FUMARIA (Papaverac.).

barnolae

subsp. algerica *(Hausskn.) M. Lidén in Op. Bot.*, 88: 50 (1986): F. agraria var. algerica.

capitata *M. Lidén in Op. Bot.*, 88: 69 (1986)—Algeria.

erostrata *(Pugsley) M. Lidén in Op. Bot.*, 88: 56 (1986): F. agraria var. erostrata.

judaica

subsp. amarysia *(Boiss. & Heldr.) M. Lidén in Op. Bot.*, 88: 49 (1986): F. amarysia.

subsp. insignis *(Pugsley) M. Lidén in Op. Bot.*, 88: 48 (1986): F. judaica var. insignis.

macrocarpa

subsp. cyrenaica *M. Lidén in Op. Bot.*, 88: 48 (1986)—Libya.

macrosepala

subsp. macrocarpa *M. Lidén in Op. Bot.*, 88: 63 (1986)—Morocco.

subsp. megasepala *(Pau) M. Lidén in Op. Bot.*, 88: 64 (1986): F. megasepala.

subsp. obscura *(Pugsley) M. Lidén in Op. Bot.*, 88: 64 (1986): F. macrosepala var. obscura.

mairei

subsp. saxicola *M. Lidén in Op. Bot.*, 88: 70 (1986)—Algeria.

officinalis

subsp. cilicica *(Hausskn.) M. Lidén in Op. Bot.*, 88: 83 (1986): F. cilicica.

ouezzanensis

subsp. ramosa *M. Lidén in Op. Bot.*, 88: 75 (1986)—Morocco.

platycarpa *M. Lidén in Op. Bot.*, 88: 87 (1986)—Algeria.

rifana *M. Lidén in Op. Bot.*, 88: 76 (1986)—Morocco.

rupestris

subsp. bifrons *(Pugsley) M. Lidén in Op. Bot.*, 88: 60 (1986): F. bifrons.

subsp. calycina *M. Lidén in Op. Bot.*, 88: 59 (1986)—Morocco.

subsp. micranthifolia *(Pugsley) M. Lidén in Op. Bot.*, 88: 60 (1986): F. micranthifolia.

subsp. pulchra *M. Lidén in Op. Bot.*, 88: 58 (1986)—Morocco.

schleicheri

subsp. microcarpa *(Hausskn.) M. Lidén in Op. Bot.*, 88: 91 (1986): F. schleicheri var. microcarpa.

sepium

subsp. rugosa *M. Lidén in Op. Bot.*, 88: 75 (1986)—Morocco.

skottsbergii *M. Lidén in Op. Bot.*, 88: 57 (1986)—Morocco.

G

GAERTNERA (Rubiac.).
godefroyana *E. Jacob de Cordemoy, Fl. L'Ile de la Réunion:* 469 (1895)—Réunion.
laxiflora *E. Jacob de Cordemoy, Fl. L'Ile de la Réunion:* 470 (1895)—Réunion.

GAGEA (Liliac.).
anonyma *K.H. Rechinger in Pl. Syst. Evol.,* 153(3–4): 288 (1986)—Iran, Afghanistan.
chloroneura *K.H. Rechinger in Pl. Syst. Evol.,* 153(3–4): 290 (1986)—Afghanistan.
czatkalica *I.G. Levichev in L.S. Krasovskaya & I.G. Levichev, Fl. Chatkal'skogo Zapovednika:* 155 (1986)—U.S.S.R. (Middle Asia).
grey-wilsonii *K.H. Rechinger in Pl. Syst. Evol.,* 153(3–4): 290 (1986)—Iran.
humicola *I.G. Levichev in L.S. Krasovskaya & I.G. Levichev, Fl. Chatkal'skogo Zapovednika:* 158 (1986)—U.S.S.R. (Middle Asia).
ignota *I.G. Levichev in L.S. Krasovskaya & I.G. Levichev, Fl. Chatkal'skogo Zapovednika:* 160 (1986)—U.S.S.R. (Middle Asia).
olgae
f. dilatata *I.G. Levichev in L.S. Krasovskaya & I.G. Levichev, Fl. Chatkal'skogo Zapovednika:* 163 (1986)—U.S.S.R. (Middle Asia).
persica
var. kashmiriensis *(Turrill) S. Dasgupta & D.B. Deb in J. Bombay Nat. Hist. Soc.,* 83(1): 87 (1986): G. kashmiriensis.
rupicola *I.G. Levichev in L.S. Krasovskaya & I.G. Levichev, Fl. Chatkal'skogo Zapovednika:* 163 (1986)—U.S.S.R. (Middle Asia).
siphonantha *K.H. Rechinger in Pl. Syst. Evol.,* 153(3–4): 289 (1986)—Afghanistan.
staintonii *K.H. Rechinger in Pl. Syst. Evol.,* 153(3–4): 289 (1986)—Pakistan, Afghanistan.
toppinii *S. Dasgupta & D.B. Deb in J. Bombay Nat. Hist. Soc.,* 83(1): 93, 164 (1986)—Pakistan.
wendelboi *K.H. Rechinger in Pl. Syst. Evol.,* 153(3–4): 291 (1986)—Iran.

GAGNEBINA (Legumin.).
commersoniana
var. aldabrensis *G.P. Lewis & Ph. Guinet in Kew Bull.,* 41(2): 465 (1986)—Aldabra Archipelago.
var. calcicola *(R. Vig.) G.P. Lewis & Ph. Guinet in Kew Bull.,* 41(2): 465 (1986): G. commersoniana subsp. calcicola.
myriophylla *(Baker) G.P. Lewis & Ph. Guinet in Kew Bull.,* 41(2): 469 (1986): Dichrostachys myriophylla.
pervilleana *(Baill.) G.P. Lewis & Ph. Guinet in Kew Bull.,* 41(2): 469 (1986): Desmanthus pervilleanus.

GALINSOGA (Compos.).
spellenbergii *B.L. Turner in Phytologia,* 59(2): 91 (1986)—Mexico.

GALIUM (Rubiac.).
mexicanum
var. asperrimum *(Gray) L.C. Higgins & S.L. Welsh in Great Basin Nat.,* 46(2): 260 (1986): G. asperrimum.
yunnanense *H. Hara & C.Y. Wu in J. Jap. Bot.,* 61(3): 74 (1986)—China.

GALTONIA (Hyacinthac.).
regalis *O.M. Hilliard & B.L. Burtt in Notes Roy. Bot. Gard. Edinburgh,* 43(3): 369 (1986)—South Africa (Natal).

GAMOCHAETA (Compos.).
monticola *M.O. Dillon & A. Sagástegui-Alva in Phytologia,* 59(4): 227 (1986)—Peru.

GAMOPLEXIS Falconer ex Lindl. in Gdner's Chron., 1847: 103 (13 Feb. 1847) et in Athenaeum, 1847(1009): 232 (27 Feb. 1847); cf. D.J. Mabberley in Taxon, 31(1): 70 (1982). *ORCHIDACEAE.*

GARCIBARRIGOA J. Cuatrecasas in Caldasia, 15(71–75): 6 (1986). *COMPOSITAE.*
telembina *(Cuatr.) J. Cuatrecasas in Caldasia,* 15(71–75): 7 (1986): Senecio telembinus.

GARDENIA (Rubiac.).
imperialis
subsp. physophylla *(K. Schum.) L. Pauwels in Bull. Soc. Roy. Bot. Belg.,* 118(2): 114 (1985): Randia physophylla.
vogelii
var. seretii *(De Wild.) L. Pauwels in Bull. Soc. Roy. Bot. Belg.,* 118(2): 111 (1985): Randia seretii.

GARDNERIA (Loganiac.).
linifolia *C.Y. Wu & S.Y. Pao in Fl. Yunnanica,* 3: 446 (1983)—China.

GAUSSIA (Palmae).
gomez-pompae *(Quero) H.J. Quero in Syst. Bot.,* 11(1): 153 (1986): Opsiandra gomez-pompae.
maya *(Cook) H.J. Quero & R.W. Read in Syst. Bot.,* 11(1): 152 (1986): Opsiandra maya.

GAYA (Malvac.).
monosperma *(K. Schum.) A. Krapovickas in Bol. Soc. Argent. Bot.,* 24(1–2): 205 (1985): Abutilon monospermum.

GAZANIA (Compos.).
pectinata *(Thunb.) Hartweg, Hort. Carlsruh.:* 126 (1825); *cf. D.J. Mabberley in Taxon,* 33(3): 442 (1984): Basionym not stated.

GAZANIOPSIS C. Huber, Trade Cat.: 4 (Oct. 1880). *COMPOSITAE.*

GAZANIOPSIS :–
stenophylla *C. Huber, Trade Cat.:* 4 (Oct. 1880)—Mexico, U.S.A. (California).

GELIDOCALAMUS (Gramin.).
longiinternodus *T.H. Wen & S.C. Chen in J. Bamboo Res.*, 5(2): 24 (1986)—China.

GENIANTHUS (Asclepiadac.).
horei *M.K. Vasudeva Rao in J. Econ. Taxon. Bot.*, 6(3): 732 (1985)—Nicobar Is.

GENISTA (Legumin.).
baetica
subsp. pumila *(Debeaux & Reverchon ex Hervier) J. Fernández Casas in Fontqueria,* 7: 17 (1985): G. baetica var. pumila.
hassertiana
subsp. glabrata *(Markgraf) W. Greuter in Willdenowia,* 15(2): 428 (1986): G. hassertiana var. glabrata.
mugronensis
subsp. rigidissima *(Vierh.) J. Fernández Casas in Fontqueria,* 7: 18 (1985): G. rigidissima.
pollinii *Spreng. ex Poll., Cat. Pl. Hort. Bot. Veron.*, 1814: 18 (1814); *cf. D.J. Mabberley in Taxon,* 33(3): 442 (1984)— Locality not stated.
sagittalis
subsp. delphinensis *(Verlot) W. Greuter in Willdenowia,* 15(2): 428 (1986): G. delphinensis.
subsp. undulata *(Ern) W. Greuter in Willdenowia,* 15(2): 428 (1986): Genistella sagittalis subsp. undulata.
tridentata
subsp. cantabrica *(Spach) W. Greuter in Willdenowia,* 15(2): 428 (1986): G. cantabrica.
subsp. lasiantha *(Spach) W. Greuter in Willdenowia,* 15(2): 428 (1986): G. lasiantha.
subsp. riphaea *(Pau & Font Quer) W. Greuter in Willdenowia,* 15(2): 429 (1986): Genistella riphaea.
subsp. stenoptera *(Spach) W. Greuter in Willdenowia,* 15(2): 429 (1986): G. stenoptera.
versicolor
subsp. pumila *(Debeaux & Reverchon ex Hervier) J. Fernández Casas in Fontqueria,* 8: 31 (1985): G. baetica var. pumila.

GENLISEA (Lentibulariac.).
pallida *E. Fromm-Trinta & P. Taylor in Bradea,* 4(27): 177 (1985)—Zambia, Angola.

GENTIANA (Gentianac.).
magellensis *(Vaccari ex Ronniger) F. Tammaro in Arch. Bot. Biogeogr. Ital.*, 62(1–2): 54 (1986): G. verna f. magellensis.

GENTIANA :–
scabra
f. alborosea *N. Yonezawa in J. Phytogeogr. Taxon.*, 33(2): 81 (1985)—Japan.
verna
subsp. arctica *(Grossh.) V.G. Sergienko, Fl. poluostrova Kanin:* 85 (1986): G. arctica.

GEOCHARIS (Zingiberac.).
fusiformis *(Ridley) R.M. Smith in Notes Roy. Bot. Gard. Edinburgh,* 43(3): 458 (1986): Amomum fusiforme.
var. borneensis *R.M. Smith in Notes Roy. Bot. Gard. Edinburgh,* 43(3): 458 (1986)—Borneo (Sabah).

GEONOMA (Palmae).
chlamydostachys *G. Galeano-Garcés in Principes,* 30(2): 71 (1986)—Colombia.
trigona *(Ruiz & Pavon) A.H. Gentry in Ann. Missouri Bot. Gard.*, 73(1): 161 (1986): Carludovica trigona.

GEOSTACHYS (Zingiberac.).
angustifolia *K. Larsen in Nordic J. Bot.*, 6(1): 31 (1986)—Thailand.

×**GERAGEUM** J. Soják in Čas. Nár. Muz. (Prague), 152(1): 21 (1983). *ROSACEAE.* (Geum × Parageum).
sudeticum *(Tausch) J. Soják in Čas. Nár. Muz. (Prague),* 152(1): 21 (1983): Geum sudeticum.

GERANIUM (Geraniac.).
angustilobum *Z.M. Tan in Bull. Bot. Res. North-East. Forest. Inst.*, 6(2): 48 (1986)—China.
franchetii
var. glandulosum *Z.M. Tan in Bull. Bot. Res. North-East. Forest. Inst.*, 6(2): 53 (1986)—China.
meiguense *Z.M. Tan in Bull. Bot. Res. North-East. Forest. Inst.*, 6(2): 47 (1986)—China.
platyrenifolium *Z.M. Tan in Bull. Bot. Res. North-East. Forest. Inst.*, 6(2): 52 (1986)—China.
pseudofarreri *Z.M. Tan in Bull. Bot. Res. North-East. Forest. Inst.*, 6(2): 50 (1986), as 'pseudo-farreri'—China.
terminale *Z.M. Tan in Bull. Bot. Res. North-East. Forest. Inst.*, 6(2): 53 (1986)—China.
wilfordii
var. glandulosum *Z.M. Tan in Bull. Bot. Res. North-East. Forest. Inst.*, 6(2): 56 (1986)—China.
yuexiense *Z.M. Tan in Bull. Bot. Res. North-East. Forest. Inst.*, 6(2): 55 (1986)—China.

GERARDIA (Scrophulariac.).
japonica *C.P. Thunberg ex A. Murray in Linnaeus, Syst. Veg., ed. 14:* 553 (1784)—Japan. †

GETHYLLIS (Amaryllidac.).
barkerae *D. Müller-Doblies in Willdenowia,* 15(2): 466 (1986)—South Africa (Cape Province).
subsp. paucifolia *D. Müller-Doblies in Willdenowia,* 15(2): 467 (1986)—South Africa (Cape Province).
britteniana
subsp. bruynsii *D. Müller-Doblies in Willdenowia,* 15(2): 467 (1986)—South Africa (Cape Province).
subsp. herrei *(L. Bolus) D. Müller-Doblies in Willdenowia,* 15(2): 467 (1986): G. herrei.
cavidens *D. Müller-Doblies in Willdenowia,* 15(2): 467 (1986)—South Africa (Cape Province).
ciliaris
subsp. longituba *(L. Bolus) D. Müller-Doblies in Willdenowia,* 15(2): 467 (1986): G. longituba.
fimbriatula *D. Müller-Doblies in Willdenowia,* 15(2): 467 (1986)—South Africa (Cape Province).
gregoriana *D. Müller-Doblies in Willdenowia,* 15(2): 468 (1986)—South Africa (Cape Province).
hallii *D. Müller-Doblies in Willdenowia,* 15(2): 468 (1986)—South Africa (Cape Province).
heinzeana *D. Müller-Doblies in Willdenowia,* 15(2): 468 (1986)—South Africa (Cape Province).
kaapensis *D. Müller-Doblies in Willdenowia,* 15(2): 468 (1986)—South Africa (Cape Province).
lata
subsp. orbicularis *D. Müller-Doblies in Willdenowia,* 15(2): 468 (1986)—South Africa (Cape Province).
marginata *D. Müller-Doblies in Willdenowia,* 15(2): 469 (1986)—South Africa (Cape Province).
oligophylla *D. Müller-Doblies in Willdenowia,* 15(2): 469 (1986)—South Africa (Cape Province).
oliverorum *D. Müller-Doblies in Willdenowia,* 15(2): 469 (1986)—South Africa (Cape Province).
pectinata *D. Müller-Doblies in Willdenowia,* 15(2): 469 (1986)—South Africa (Cape Province).
roggeveldensis *D. Müller-Doblies in Willdenowia,* 15(2): 470 (1986)—South Africa (Cape Province).
transkarooica *D. Müller-Doblies in Willdenowia,* 15(2): 470 (1986)—South Africa (Cape Province).
uteana *D. Müller-Doblies in Willdenowia,* 15(2): 470 (1986)—South Africa (Cape Province).

GEUM (Rosac.).
japonicum *C.P. Thunberg ex A. Murray in Linnaeus, Syst. Veg., ed. 14:* 480 (1784)—Japan. †

GIBASIS (Commelinac.).
karwinskyana
subsp. palmeri *D.R. Hunt in Kew Bull.,* 41(1): 124 (1986)—Mexico.
linearis
subsp. rhodantha *(Torrey) D.R. Hunt in Kew Bull.,* 41(1): 126 (1986): Tradescantia rhodantha.
triflora *(Martens & Galeotti) D.R. Hunt in Kew Bull.,* 41(1): 121 (1986): Tradescantia triflora.
venustula
subsp. peninsulae *D.R. Hunt in Kew Bull.,* 41(1): 125 (1986), nom. nov.: Tradescantia heterophylla *T.S. Brandegee.*

GILIA (Polemoniac.).
densiflora *(Benth.) Endl., Cat. Horti Vindob.,* 2: 69 (1842); *cf. D.J. Mabberley in Taxon,* 33(3): 435 (1984): Basionym not stated.
involucrata *(Ruiz & Pavon) Endl., Cat. Horti Vindob.,* 2: 69 (1842); *cf. D.J. Mabberley in Taxon,* 33(3): 435 (1984): Navarretia involucrata.

GLADIOLUS (Iridac.).
gueinzii *Kunze, Hort. Univ. Lit. Lips.,* 1844; *Linnaea,* 18: 510 (1845); *cf. D.J. Mabberley in Taxon,* 32(1): 85 (1983)— Locality not stated.
loteniensis *O.M. Hilliard & B.L. Burtt in Notes Roy. Bot. Gard. Edinburgh,* 43(2): 206 (1986)—South Africa (Natal).

GLOBBA (Zingiberac.).
japonica *C.P. Thunberg ex A. Murray in Linnaeus, Syst. Veg., ed. 14:* 73 (1784)—Japan. †

GLOCHIDION (Euphorbiac.).
airyshawii *N.P. Balakrishnan & T. Chakrabarty in Bull. Bot. Surv. India,* 25(1–4): 220 (1983 publ. 1985)—Andaman Is.
bhutanicum *D.G. Long in Notes Roy. Bot. Gard. Edinburgh,* 44(1): 169 (1986)—Bhutan.

GLOMERA (Orchidac.).
pendulata *(P. Royen) J.J. Wood in Kew Bull.,* 41(4): 819 (1986): Agrostophyllum pendulatum.

GLOSSOGYNE (Compos.).
tannensis *(Sprengel) P.J. Garnock-Jones in Taxon,* 35(1): 125 (1986): Coreopsis tannensis.

GLYCINE (Legumin.).
minus *Lag., Elenchus:* 8 (1816); *cf. D.J. Mabberley in Taxon,* 31(1): 70 (1982)—Locality not stated.

GLYCINE :–
monophylla *(Vent.) Parmentier, Cat. Pl. Enghien:* 37 (1818); *cf. D.J. Mabberley in Taxon,* 33(3): 442 (1984): Basionym not stated.
soja
var. albiflora *P.Y. Fu & Y.A. Chen in Bull. Bot. Res. North-East. Forest. Inst.,* 6(2): 119 (1986)—China.
f. angustifolia *P.Y. Fu & Y.A. Chen in Bull. Bot. Res. North-East. Forest. Inst.,* 6(2): 120 (1986)—China.
villosa *C.P. Thunberg ex A. Murray in Linnaeus, Syst. Veg., ed. 14:* 659 (1784)—Japan. †

GLYCOSMIS (Rutac.).
chlorosperma
var. bidiensis *B.C. Stone in Proc. Acad. Nat. Sci. Philadelphia,* 137(2): 3 (1985)—Borneo (Sarawak).
var. lindleyana *(Swingle) B.C. Stone in Proc. Acad. Nat. Sci. Philadelphia,* 137(2): 3 (1985): G. lindleyana.
var. macrorachis *(King) B.C. Stone in Proc. Acad. Nat. Sci. Philadelphia,* 137(2): 3 (1985): G. pentaphylla subvar. macrorachis.
var. paraphyllophora *B.C. Stone in Proc. Acad. Nat. Sci. Philadelphia,* 137(2): 3 (1985)—Malaya, Thailand.
collina *B.C. Stone in Proc. Acad. Nat. Sci. Philadelphia,* 137(2): 4 (1985)—Malaya.
cyanocarpa
var. erythrocarpoides *B.C. Stone in Proc. Acad. Nat. Sci. Philadelphia,* 137(2): 7 (1985)—Sumatra.
var. larsenii *B.C. Stone in Proc. Acad. Nat. Sci. Philadelphia,* 137(2): 8 (1985)—Thailand.
var. philippinensis *B.C. Stone in Proc. Acad. Nat. Sci. Philadelphia,* 137(2): 7 (1985)—Philippine Is.
var. rubiginosa *(Ridl.) B.C. Stone in Proc. Acad. Nat. Sci. Philadelphia,* 137(2): 7 (1985): G. rubiginosa.
var. wirawanii *B.C. Stone in Proc. Acad. Nat. Sci. Philadelphia,* 137(2): 6 (1985)—Lesser Sunda Is.
gracilis *B.C. Stone in Proc. Acad. Nat. Sci. Philadelphia,* 137(2): 10 (1985)—Vietnam.
longisepala *B.C. Stone in Proc. Acad. Nat. Sci. Philadelphia,* 137(2): 11 (1985)—Borneo (Sarawak).
mauritiana
var. fuscescens *(Kurz) B.C. Stone in Proc. Acad. Nat. Sci. Philadelphia,* 137(2): 14 (1985): G. trifoliata var. fuscescens.
var. hainanensis *B.C. Stone in Proc. Acad. Nat. Sci. Philadelphia,* 137(2): 14 (1985)—Hainan.
var. tomentosa *(Ridley) B.C. Stone in Proc. Acad. Nat. Sci. Philadelphia,* 137(2): 14 (1985): G. rupestris var. tomentosa.

GLYCOSMIS :–
parviflora
var. obtusa *(Miq.) B.C. Stone in Proc. Acad. Nat. Sci. Philadelphia,* 137(2): 15 (1985): G. obtusa.
puberula
var. craibii *(Tanaka) B.C. Stone in Proc. Acad. Nat. Sci. Philadelphia,* 137(2): 18 (1985): G. craibii.
var. eberhardtii *(Tanaka) B.C. Stone in Proc. Acad. Nat. Sci. Philadelphia,* 137(2): 18 (1985): G. sapindoides var. eberhardtii.
var. subsessilis *(Craib) B.C. Stone in Proc. Acad. Nat. Sci. Philadelphia,* 137(2): 18 (1985): G. subsessilis.
sapindoides
var. australiensis *B.C. Stone in Proc. Acad. Nat. Sci. Philadelphia,* 137(2): 18 (1985)—Australia (W. Australia).
var. microphylla *B.C. Stone in Proc. Acad. Nat. Sci. Philadelphia,* 137(2): 18 (1985)—Lesser Sunda Is.
stenura *B.C. Stone in Proc. Acad. Nat. Sci. Philadelphia,* 137(2): 19 (1985)—Malaya.
tetracronia *B.C. Stone in Proc. Acad. Nat. Sci. Philadelphia,* 137(2): 21 (1985), nom. nov.: Tetracronia cymosa *Pierre.*
trichanthera
var. kelantanica *(Stone) B.C. Stone in Proc. Acad. Nat. Sci. Philadelphia,* 137(2): 22 (1985): G. calcicola var. kelantanica.
var. sumatrana *B.C. Stone in Proc. Acad. Nat. Sci. Philadelphia,* 137(2): 22 (1985)—Sumatra.

GLYCYRRHIZA (Legumin.).
eurycarpa *P.C. Li in Acta Bot. Bor.-Occid. Sin.,* 4(2): 118 (1984)—Xinjiang, China.
yunnanensis *Cheng f. & L.K. Tai ex P.C. Li in Acta Bot. Bor.-Occid. Sin.,* 4(2): 117 (1984)—China.

GNAPHALIUM (Compos.).
dealbatum
var. luteofuscum *(Webb) W. Lobin in Courier Forschungsinst. Senckenberg,* 81: 117 (1986): G. luteo-fuscum.
japonicum *C.P. Thunberg ex A. Murray in Linnaeus, Syst. Veg., ed. 14:* 749 (1784)—Japan. †
sandwicensium
var. hawaiiense *(Degener & Sherff) W.L. Wagner, D.R. Herbst & S.H. Sohmer in Occas. Pap. Bernice P. Bishop Mus.,* 26: 115 (1986): G. hawaiiense.

GNIDIA (Thymelaeac.).
renniana *O.M. Hilliard & B.L. Burtt in Notes Roy. Bot. Gard. Edinburgh,* 43(2): 218 (1986)—South Africa (Natal).

GODETIA (Onagrac.).
roseoalba *(Bernh.) J.W. Loud., Lad. Fl. Gard. Orn. Ann.:* 53 (1840); *cf. D.J. Mabberley in Taxon,* 32(1): 87 (1983): Basionym not stated.

×GOMOGLOSSUM hort. in Orchid Rev., 94(1114): centre page pull-out p. 8 (1986). *ORCHIDACEAE.* (Gomesa × Odontoglossum).

GONATOPUS (Arac.).
clavatus *S.J. Mayo in Fl. Trop. E. Afr., Arac.:* 11 (1985)—Tanzania, Mozambique, Malawi.

GONIOTHALAMUS (Annonac.).
australis *L.W. Jessup in Austrobaileya,* 2(3): 224 (1986)—Australia (Queensland).

GONOCARPUS (Haloragac.).
effusus *A.E. Orchard in Nuytsia,* 5(3): 331 (1986)—Australia (Queensland).
ephemerus *A.E. Orchard in Nuytsia,* 5(3): 327 (1986)—Australia (W. Australia).
ericifolius *A.E. Orchard in Nuytsia,* 5(3): 329 (1986)—Australia (W. Australia).
hispidus *A.E. Orchard in Nuytsia,* 5(3): 329 (1986)—Australia (W. Australia).
urceolatus *A.E. Orchard in Nuytsia,* 5(3): 332 (1986)—Australia (Queensland).

GORDONIA (Theac.).
axillaris *(Roxb. ex Ker) Endl., Cat. Horti Vindob.,* 2: 365 (1842); *cf. D.J. Mabberley in Taxon,* 33(3): 435 (1984): Basionym not stated.

GOULDOCHLOA J. Valdés R., C.W. Morden & S.L. Hatch in Syst. Bot., 11(1): 112 (1986). *GRAMINEAE.*
curvifolia *J. Valdés R., C.W. Morden & S.L. Hatch in Syst. Bot.,* 11(1): 112 (1986)—Mexico.

GRAPHARDISIA (Myrsinac.).
coibana *C.L. Lundell in Phytologia,* 59(6): 429 (1986)—Panama.
nicaraguensis *C.L. Lundell in Phytologia,* 59(6): 429 (1986)—Nicaragua.
obtusata *C.L. Lundell in Phytologia,* 59(6): 430 (1986)—Panama.
oxyphylla *C.L. Lundell in Phytologia,* 59(6): 430 (1986)—Panama.
purpurea *C.L. Lundell in Phytologia,* 59(6): 431 (1986)—Panama.
seranoana *C.L. Lundell in Phytologia,* 59(6): 431 (1986)—Panama.
skutchii *(Morton) C.L. Lundell in Phytologia,* 59(6): 432 (1986): Ardisia skutchii.
ustupoana *C.L. Lundell in Phytologia,* 59(6): 432 (1986)—Panama.

GRAPTOPETALUM (Crassulac.).
paraguayense
subsp. bernalense *M. Kimnach & R. Moran in Cact. Succ. J. (U.S.A.),* 58(2): 54 (1986)—Mexico.

GREENEA (Rubiac.).
bahadurii *R.C. Gaur & R. Dayal in Indian J. Forest.,* 8(4): 323 (1985 publ. 1986)—India.

GREENWOODIA P. Burns-Balogh in Orquídea (México), 10(1): 1 (1986). *ORCHIDACEAE.*
sawyeri *(Standl. & L.O. Wms.) P. Burns-Balogh in Orquídea (México),* 10(1): 1 (1986): Spiranthes sawyeri.

GREVILLEA (Proteac.).
acacioides *C. Gardner ex D.J. McGillivray, New Names Grevillea (Proteac.):* 1 (1986)—Australia (W. Australia).
acerata *D.J. McGillivray, New Names Grevillea (Proteac.):* 1 (1986)—Australia (New South Wales).
acrobotrya
subsp. uniforma *D.J. McGillivray, New Names Grevillea (Proteac.):* 1 (1986)—Australia (W. Australia).
adenotricha *D.J. McGillivray, New Names Grevillea (Proteac.):* 1 (1986)—Australia (New South Wales).
aneura *D.J. McGillivray, New Names Grevillea (Proteac.):* 1 (1986)—Australia (W. Australia).
arenaria
subsp. montana *(R. Br.) D.J. McGillivray, New Names Grevillea (Proteac.):* 1 (1986): G. montana.
baileyana *D.J. McGillivray, New Names Grevillea (Proteac.):* 2 (1986), nom. nov.: Kermadecia pinnatifida *F.M. Bailey.*
barklyana
subsp. macleayana *D.J. McGillivray, New Names Grevillea (Proteac.):* 2 (1986)—Australia (New South Wales).
batrachioides *F. Muell. ex D.J. McGillivray, New Names Grevillea (Proteac.):* 2 (1986)—Australia (?W. Australia).
baueri
subsp. asperula *D.J. McGillivray, New Names Grevillea (Proteac.):* 2 (1986)—Australia (New South Wales).
beadleana *D.J. McGillivray, New Names Grevillea (Proteac.):* 2 (1986)—Australia (New South Wales).
beardiana *D.J. McGillivray, New Names Grevillea (Proteac.):* 2 (1986)—Australia (W. Australia).
bedggoodiana *J.H. Willis ex D.J. McGillivray, New Names Grevillea (Proteac.):* 3 (1986)—Australia (Victoria).
benthamiana *D.J. McGillivray, New Names Grevillea (Proteac.):* 3 (1986)—Australia (N. Terr.).
buxifolia
subsp. phylicoides *(R. Br.) D.J. McGillivray, New Names Grevillea (Proteac.):* 3 (1986): G. phylicoides.
subsp. sphacelata *(R. Br.) D.J. McGillivray, New Names Grevillea (Proteac.):* 3 (1986): G. sphacelata.
byrnesii *D.J. McGillivray, New Names Grevillea (Proteac.):* 3 (1986)—Australia (N. Terr.).

GREVILLEA :-
cagiana *D.J. McGillivray, New Names Grevillea (Proteac.):* 3 (1986)—Australia (W. Australia).
christiniae *D.J. McGillivray, New Names Grevillea (Proteac.):* 4 (1986)—Australia (W. Australia).
concinna
subsp. lemanniana *(Meisn.) D.J. McGillivray, New Names Grevillea (Proteac.):* 4 (1986): G. lemanniana.
curviloba *D.J. McGillivray, New Names Grevillea (Proteac.):* 4 (1986), nom. nov.: G. vestita var. angustata *Meisn.*
decipiens *D.J. McGillivray, New Names Grevillea (Proteac.):* 4 (1986)—Australia (W. Australia).
didymobotrya
subsp. involuta *D.J. McGillivray, New Names Grevillea (Proteac.):* 4 (1986)—Australia (W. Australia).
diffusa
subsp. evansiana *(McKee) D.J. McGillivray, New Names Grevillea (Proteac.):* 4 (1986): G. evansiana.
subsp. filipendula *D.J. McGillivray, New Names Grevillea (Proteac.):* 4 (1986)—Australia (New South Wales).
disjuncta
subsp. dolichopoda *D.J. McGillivray, New Names Grevillea (Proteac.):* 5 (1986)—Australia (W. Australia).
diversifolia
subsp. subtersericata *D.J. McGillivray, New Names Grevillea (Proteac.):* 5 (1986)—Australia (N. Terr.).
drummondii
subsp. centristigma *D.J. McGillivray, New Names Grevillea (Proteac.):* 5 (1986)—Australia (W. Australia).
subsp. pimeleoides *(W.V. Fitzg.) D.J. McGillivray, New Names Grevillea (Proteac.):* 5 (1986): G. pimeleoides.
dryandri
subsp. dasycarpa *D.J. McGillivray, New Names Grevillea (Proteac.):* 5 (1986)—Australia (N. Terr.).
eriostachya
subsp. excelsior *(Diels) D.J. McGillivray, New Names Grevillea (Proteac.):* 5 (1986): G. excelsior.
formosa *D.J. McGillivray, New Names Grevillea (Proteac.):* 6 (1986)—Australia (N. Terr.).
georgeana *D.J. McGillivray, New Names Grevillea (Proteac.):* 6 (1986)—Australia (W. Australia).
glabrata
subsp. dissectifolia *D.J. McGillivray, New Names Grevillea (Proteac.):* 6 (1986)—Australia (W. Australia).
subsp. ornithopoda *(Meisn.) D.J. McGillivray, New Names Grevillea (Proteac.):* 6 (1986): G. ornithopoda.

GREVILLEA :-
goodii
subsp. decora *(Domin) D.J. McGillivray, New Names Grevillea (Proteac.):* 6 (1986): G. decora.
subsp. pluricaulis *D.J. McGillivray, New Names Grevillea (Proteac.):* 6 (1986)—Australia (N. Terr.).
granulosa *D.J. McGillivray, New Names Grevillea (Proteac.):* 7 (1986)—Australia (W. Australia).
hakeoides
subsp. commutata *(F. Muell.) D.J. McGillivray, New Names Grevillea (Proteac.):* 7 (1986): G. commutata.
subsp. stenophylla *(W.V. Fitzg.) D.J. McGillivray, New Names Grevillea (Proteac.):* 7 (1986): G. stenophylla.
iaspicula *D.J. McGillivray, New Names Grevillea (Proteac.):* 7 (1986)—Australia (New South Wales).
infecunda *D.J. McGillivray, New Names Grevillea (Proteac.):* 7 (1986)—Australia (Victoria).
integrifolia
subsp. biformis *(Meisn.) D.J. McGillivray, New Names Grevillea (Proteac.):* 7 (1986): Grevillea biformis.
subsp. ceratocarpa *(Diels) D.J. McGillivray, New Names Grevillea (Proteac.):* 7 (1986): G. ceratocarpa.
subsp. incrassata *(Diels) D.J. McGillivray, New Names Grevillea (Proteac.):* 8 (1986): G. incrassata.
subsp. shuttleworthiana *(Meisn.) D.J. McGillivray, New Names Grevillea (Proteac.):* 8 (1986): G. shuttleworthiana.
kenneallyi *D.J. McGillivray, New Names Grevillea (Proteac.):* 8 (1986)—Australia (W. Australia).
leptopoda *D.J. McGillivray, New Names Grevillea (Proteac.):* 8 (1986)—Australia (W. Australia).
leucoclada *D.J. McGillivray, New Names Grevillea (Proteac.):* 8 (1986)—Australia (W. Australia).
linsmithii *D.J. McGillivray, New Names Grevillea (Proteac.):* 8 (1986)—Australia (Queensland).
lissopleura *D.J. McGillivray, New Names Grevillea (Proteac.):* 9 (1986)—Australia (W. Australia).
longicuspis *D.J. McGillivray, New Names Grevillea (Proteac.):* 9 (1986)—Australia (N. Terr.).
lullfitzii *D.J. McGillivray, New Names Grevillea (Proteac.):* 9 (1986)—Australia (W. Australia).
makinsonii *D.J. McGillivray, New Names Grevillea (Proteac.):* 9 (1986)—Australia (W. Australia).
manglesioides
subsp. papillosa *D.J. McGillivray, New Names Grevillea (Proteac.):* 9 (1986)—Australia (W. Australia).

GREVILLEA :-

maxwellii *D.J. McGillivray, New Names Grevillea (Proteac.):* 9 (1986)—Australia (W. Australia).

minutiflora *D.J. McGillivray, New Names Grevillea (Proteac.):* 10 (1986)—Australia (W. Australia).

molyneuxii *D.J. McGillivray, New Names Grevillea (Proteac.):* 10 (1986)—Australia (New South Wales).

murex *D.J. McGillivray, New Names Grevillea (Proteac.):* 10 (1986)—Australia (W. Australia).

myosodes *D.J. McGillivray, New Names Grevillea (Proteac.):* 10 (1986)—Australia (W. Australia).

nana
- *subsp.* abbreviata *D.J. McGillivray, New Names Grevillea (Proteac.):* 10 (1986)—Australia (W. Australia).

newbeyi *D.J. McGillivray, New Names Grevillea (Proteac.):* 10 (1986)—Australia (W. Australia).

obtusiflora
- *subsp.* granulifera *D.J. McGillivray, New Names Grevillea (Proteac.):* 11 (1986)—Australia (New South Wales).
- *subsp.* kedumbensis *D.J. McGillivray, New Names Grevillea (Proteac.):* 11 (1986)—Australia (New South Wales).

oldei *D.J. McGillivray, New Names Grevillea (Proteac.):* 11 (1986)—Australia (New South Wales).

pauciflora
- *subsp.* psilophylla *D.J. McGillivray, New Names Grevillea (Proteac.):* 11 (1986)—Australia (W. Australia).
- *subsp.* saxatilis *D.J. McGillivray, New Names Grevillea (Proteac.):* 11 (1986)—Australia (W. Australia).

petrophiloides
- *subsp.* magnifica *D.J. McGillivray, New Names Grevillea (Proteac.):* 11 (1986)—Australia (W. Australia).
- *subsp.* oligomera *D.J. McGillivray, New Names Grevillea (Proteac.):* 12 (1986)—Australia (W. Australia).

phillipsiana *D.J. McGillivray, New Names Grevillea (Proteac.):* 12 (1986)—Australia (W. Australia).

pilosa
- *subsp.* dissecta *D.J. McGillivray, New Names Grevillea (Proteac.):* 12 (1986)—Australia (W. Australia).

polyacida *D.J. McGillivray, New Names Grevillea (Proteac.):* 12 (1986)—Australia (N. Terr.).

prasina *D.J. McGillivray, New Names Grevillea (Proteac.):* 12 (1986)—Australia (N. Terr.).

psilantha *D.J. McGillivray, New Names Grevillea (Proteac.):* 12 (1986)—Australia (W. Australia).

rosieri *D.J. McGillivray, New Names Grevillea (Proteac.):* 13 (1986)—Australia (W. Australia).

GREVILLEA :-

roycei *D.J. McGillivray, New Names Grevillea (Proteac.):* 13 (1986)—Australia (W. Australia).

salicifolia *Endl., Cat. Horti Vindob.,* 1: 287 (1842); *cf. D.J. Mabberley in Taxon,* 33(3): 435 (1984)— Locality not stated.

sarissa
- *subsp.* anfractifolia *D.J. McGillivray, New Names Grevillea (Proteac.):* 13 (1986)—Australia (W. Australia).
- *subsp.* bicolor *D.J. McGillivray, New Names Grevillea (Proteac.):* 13 (1986)—Australia (W. Australia).
- *subsp.* rectitepala *D.J. McGillivray, New Names Grevillea (Proteac.):* 13 (1986)—Australia (W. Australia).
- *subsp.* succincta *D.J. McGillivray, New Names Grevillea (Proteac.):* 13 (1986)—Australia (W. Australia).
- *subsp.* umbellifera *(J.M. Black) D.J. McGillivray, New Names Grevillea (Proteac.):* 13 (1986): G. umbellifera.

scortechinii
- *subsp.* sarmentosa *(Blakely & McKie) D.J. McGillivray, New Names Grevillea (Proteac.):* 14 (1986): G. sarmentosa.

secunda *D.J. McGillivray, New Names Grevillea (Proteac.):* 14 (1986)—Australia (W. Australia).

speciosa
- *subsp.* dimorpha *(F. Muell.) D.J. McGillivray, New Names Grevillea (Proteac.):* 14 (1986): G. dimorpha.
- *subsp.* oleoides *(Sieber ex Schultes & Schultes f.) D.J. McGillivray, New Names Grevillea (Proteac.):* 14 (1986): G. oleoides.

spinosa *D.J. McGillivray, New Names Grevillea (Proteac.):* 14 (1986)—Australia (W. Australia).

spinosissima *D.J. McGillivray, New Names Grevillea (Proteac.):* 14 (1986)—Australia (W. Australia).

subtiliflora *D.J. McGillivray, New Names Grevillea (Proteac.):* 14 (1986)—Australia (W. Australia).

tetrapleura *D.J. McGillivray, New Names Grevillea (Proteac.):* 15 (1986)—Australia (W. Australia).

thelemanniana
- *subsp.* delta *D.J. McGillivray, New Names Grevillea (Proteac.):* 15 (1986)—Australia (W. Australia).
- *subsp.* fililoba *D.J. McGillivray, New Names Grevillea (Proteac.):* 15 (1986)—Australia (W. Australia).
- *subsp.* hirtella *(Benth.) D.J. McGillivray, New Names Grevillea (Proteac.):* 15 (1986): G. pinaster var. hirtella.
- *subsp.* obtusifolia *(Meisn.) D.J. McGillivray, New Names Grevillea (Proteac.):* 15 (1986): G. obtusifolia.
- *subsp.* pinaster *(Meisn.) D.J. McGillivray, New Names Grevillea (Proteac.):* 15 (1986): G. pinaster.

GREVILLEA :–
subsp. preissii *(Meisn.) D.J. McGillivray, New Names Grevillea (Proteac.):* 15 (1986): G. preissii.
uncinulata
subsp. florida *D.J. McGillivray, New Names Grevillea (Proteac.):* 15 (1986)—Australia (W. Australia).
velutinella *D.J. McGillivray, New Names Grevillea (Proteac.):* 16 (1986)—Australia (W. Australia).
versicolor *D.J. McGillivray, New Names Grevillea (Proteac.):* 16 (1986)—Australia (N. Terr.).
vestita
subsp. isopogoides *D.J. McGillivray, New Names Grevillea (Proteac.):* 16 (1986)—Australia (W. Australia).
whiteana *D.J. McGillivray, New Names Grevillea (Proteac.):* 16 (1986)—Australia (Queensland).
wickhamii
subsp. aprica *D.J. McGillivray, New Names Grevillea (Proteac.):* 16 (1986)—Australia (N. Terr.).
willisii
subsp. pachylostyla *D.J. McGillivray, New Names Grevillea (Proteac.):* 16 (1986)—Australia (Victoria).
wittweri *D.J. McGillivray, New Names Grevillea (Proteac.):* 16 (1986)—Australia (W. Australia).

GRIFFINIA (Amaryllidac.).
liboniana *Morren in Ann. Soc. Roy. Agr. Bot. Gand,* 1: 143 (1845); *cf. D.J. Mabberley in Taxon,* 30(1): 16 (1981)— Locality not stated.

GRINDELIA (Compos.).
glutinosa *(Cav.) Mart., Hort. Acad. Erlangen:* 182 (1814); *cf. D.J. Mabberley in Taxon,* 33(3): 442 (1984): Basionym not stated.

GUZMANIA (Bromeliac.).
scherzeriana
var. albiflora *W. Rauh in J. Bromeliad Soc.,* 36(4): 164 (1986)—Ecuador.

GYMNACRANTHERA (Myristicac.).
farquhariana
var. eugeniifolia *(A.DC.) R.T.A. Schouten in Blumea,* 31(2): 480 (1986): Myristica eugeniifolia.
var. paniculata *(A.DC.) R.T.A. Schouten in Blumea,* 31(2): 481 (1986): Myristica paniculata.
var. zippeliana *(Miq.) R.T.A. Schouten in Blumea,* 31(2): 482 (1986): Myristica zippeliana.
maliliensis *R.T.A. Schouten in Blumea,* 31(2): 467 (1986)—Celebes.
ocellata *R.T.A. Schouten in Blumea,* 31(2): 469 (1986)—Borneo (Sabah, Sarawak, Brunei, Kalimantan).

GYMNOCACTUS John & Riha in Kaktusy, 19(1): 22 (1981), contrary to Art. 64.1 ICBN 1983; cf. Repert. Pl. Succ. (I.O.S.), 34: 7 (1985). *CACTACEAE.*

GYMNOCALYCIUM (Cactac.).
ferox *(Backeb.) Slaba in Kaktusy,* 20(4): 80 (1984), basionym not validly publ.; *cf. Repert. Pl. Succ. (I.O.S.),* 35: 4 (1985): G. hybopleurum var. ferox.
var. ferocior *(Backeb.) Slaba in Kaktusy,* 20(4): 83 (1984), basionym not validly publ.; *cf. Repert. Pl. Succ. (I.O.S.),* 35: 4 (1985): G. hybopleurum var. ferocior.

GYMNOSPORIA (Celastrac.).
wightiana *(Babu) R.S. Rao, Fl. Goa, Diu, Daman, Dadra & Nagarhaveli (Fl. India, Ser. 2),* 1: 76 (1985): Maytenus wightiana.

GYNURA (Compos.).
campanulata *C. Jeffrey in Kew Bull.,* 41(4): 929 (1986)—Kenya.

H

HABENARIA (Orchidac.).
zothecina *L.C. Higgins & S.L. Welsh in Great Basin Nat.,* 46(2): 259 (1986)—U.S.A. (Utah).

HAGENBACHIA (Haemodorac.).
panamensis *(Standley) R.W. Cruden in Phytologia,* 59(6): 380 (1986): Anthericum panamense.

HALIMODENDRON (Legumin.).
triflorum *(L'Hérit.) Sweet, Hort. Brit., ed. 2:* 144 (1830); *cf. D.J. Mabberley in Taxon,* 31(1): 71 (1982): Basionym not stated.

HALORAGIS (Haloragac.).
acutangula
f. stellata *A.E. Orchard in Nuytsia,* 5(3): 335 (1986)—Australia (W. Australia, S. Australia).
prostrata
subsp. coquana *A.E. Orchard in Nuytsia,* 5(3): 335 (1986)—Cook Is.
subsp. tuvaluensis *A.E. Orchard in Nuytsia,* 5(3): 336 (1986)—Vanuatu.

HAPALOCHILUS (Orchidac.).

acanthoglossus *(Schltr.) L.A. Garay & W. Kittredge in Bot. Mus. Leafl. Harvard Univ.*, 30(3): (51) 185 (1985 publ. 1986): Bulbophyllum acanthoglossum.

algidus *(Ridl.) L.A. Garay & W. Kittredge in Bot. Mus. Leafl. Harvard Univ.*, 30(3): (51) 185 (1985 publ. 1986): Bulbophyllum algidum.

alkmaarensis *(J.J. Sm.) L.A. Garay & W. Kittredge in Bot. Mus. Leafl. Harvard Univ.*, 30(3): (52) 186 (1985 publ. 1986): Bulbophyllum alkmaarense.

alticola *(Schltr.) L.A. Garay & W. Kittredge in Bot. Mus. Leafl. Harvard Univ.*, 30(3): (52) 186 (1985 publ. 1986): Bulbophyllum alticola.

arfakensis *(J.J. Sm.) L.A. Garay & W. Kittredge in Bot. Mus. Leafl. Harvard Univ.*, 30(3): (52) 186 (1985 publ. 1986): Bulbophyllum arfakense.

aristilabris *(J.J. Sm.) L.A. Garay & W. Kittredge in Bot. Mus. Leafl. Harvard Univ.*, 30(3): (52) 186 (1985 publ. 1986): Bulbophyllum aristilabre.

aureoapex *(Schltr.) L.A. Garay & W. Kittredge in Bot. Mus. Leafl. Harvard Univ.*, 30(3): (52) 186 (1985 publ. 1986): Bulbophyllum aureoapex.

brevis *(Schltr.) L.A. Garay & W. Kittredge in Bot. Mus. Leafl. Harvard Univ.*, 30(3): (52) 186 (1985 publ. 1986): Bulbophyllum breve.

callipes *(J.J. Sm.) L.A. Garay & W. Kittredge in Bot. Mus. Leafl. Harvard Univ.*, 30(3): (52) 186 (1985 publ. 1986): Bulbophyllum callipes.

caudatipetalum *(J.J. Sm.) L.A. Garay & W. Kittredge in Bot. Mus. Leafl. Harvard Univ.*, 30(3): (52) 186 (1985 publ. 1986): Bulbophyllum caudatipetalum.

chrysochilus *(Schltr.) L.A. Garay & W. Kittredge in Bot. Mus. Leafl. Harvard Univ.*, 30(3): (52) 186 (1985 publ. 1986): Bulbophyllum chrysochilum.

chrysoglossus *(Schltr.) L.A. Garay & W. Kittredge in Bot. Mus. Leafl. Harvard Univ.*, 30(3): (52) 186 (1985 publ. 1986): Bulbophyllum chrysoglossum.

collinus *(Schltr.) L.A. Garay & W. Kittredge in Bot. Mus. Leafl. Harvard Univ.*, 30(3): (52) 186 (1985 publ. 1986): Bulbophyllum collinum.

coloratus *(J.J. Sm.) L.A. Garay & W. Kittredge in Bot. Mus. Leafl. Harvard Univ.*, 30(3): (53) 187 (1985 publ. 1986): Bulbophyllum coloratum.

concavibasalis *(van Royen) L.A. Garay & W. Kittredge in Bot. Mus. Leafl. Harvard Univ.*, 30(3): (53) 187 (1985 publ. 1986): Bulbophyllum concavibasalis.

concolor *(J.J. Sm.) L.A. Garay & W. Kittredge in Bot. Mus. Leafl. Harvard Univ.*, 30(3): (53) 187 (1985 publ. 1986): Bulbophyllum concolor.

HAPALOCHILUS :-

cruciatus *(J.J. Sm.) L.A. Garay & W. Kittredge in Bot. Mus. Leafl. Harvard Univ.*, 30(3): (53) 187 (1985 publ. 1986): Bulbophyllum cruciatum.

cucullatus *(Schltr.) L.A. Garay & W. Kittredge in Bot. Mus. Leafl. Harvard Univ.*, 30(3): (53) 187 (1985 publ. 1986): Bulbophyllum cucullatum.

cuniculiformis *(J.J. Sm.) L.A. Garay & W. Kittredge in Bot. Mus. Leafl. Harvard Univ.*, 30(3): (53) 187 (1985 publ. 1986): Bulbophyllum cuniculiforme.

decurvulus *(Schltr.) L.A. Garay & W. Kittredge in Bot. Mus. Leafl. Harvard Univ.*, 30(3): (53) 187 (1985 publ. 1986): Bulbophyllum decurvulum.

dolichoglottis *(Schltr.) L.A. Garay & W. Kittredge in Bot. Mus. Leafl. Harvard Univ.*, 30(3): (53) 187 (1985 publ. 1986): Bulbophyllum dolichoglottis.

fasciatus *(Schltr.) L.A. Garay & W. Kittredge in Bot. Mus. Leafl. Harvard Univ.*, 30(3): (53) 187 (1985 publ. 1986): Bulbophyllum fasciatum.

fibrinus *(J.J. Sm.) L.A. Garay & W. Kittredge in Bot. Mus. Leafl. Harvard Univ.*, 30(3): (53) 187 (1985 publ. 1986): Bulbophyllum fibrinum.

formosus *(Schltr.) L.A. Garay & W. Kittredge in Bot. Mus. Leafl. Harvard Univ.*, 30(3): (54) 188 (1985 publ. 1986): Bulbophyllum formosum.

frustrans *(J.J. Sm.) L.A. Garay & W. Kittredge in Bot. Mus. Leafl. Harvard Univ.*, 30(3): (54) 188 (1985 publ. 1986): Bulbophyllum frustrans.

geniculifer *(J.J. Sm.) L.A. Garay & W. Kittredge in Bot. Mus. Leafl. Harvard Univ.*, 30(3): (54) 188 (1985 publ. 1986): Bulbophyllum geniculiferum.

gobiensis *(Schltr.) L.A. Garay & W. Kittredge in Bot. Mus. Leafl. Harvard Univ.*, 30(3): (54) 188 (1985 publ. 1986): Bulbophyllum gobiense.

holochilus *(J.J. Sm.) L.A. Garay & W. Kittredge in Bot. Mus. Leafl. Harvard Univ.*, 30(3): (54) 188 (1985 publ. 1986): Bulbophyllum holochilum.

humilis *(Schltr.) L.A. Garay & W. Kittredge in Bot. Mus. Leafl. Harvard Univ.*, 30(3): (54) 188 (1985 publ. 1986): Bulbophyllum humile.

inclinatus *(J.J. Sm.) L.A. Garay & W. Kittredge in Bot. Mus. Leafl. Harvard Univ.*, 30(3): (54) 188 (1985 publ. 1986): Bulbophyllum inclinatum.

jadunae *(Schltr.) L.A. Garay & W. Kittredge in Bot. Mus. Leafl. Harvard Univ.*, 30(3): (54) 188 (1985 publ. 1986): Bulbophyllum jadunae.

jensenii *(J.J. Sm.) L.A. Garay & W. Kittredge in Bot. Mus. Leafl. Harvard Univ.*, 30(3): (54) 188 (1985 publ. 1986): Bulbophyllum jensenii.

HAPALOCHILUS :-

kelelensis *(Schltr.) L.A. Garay & W. Kittredge in Bot. Mus. Leafl. Harvard Univ.*, 30(3): (54) 188 (1985 publ. 1986): Bulbophyllum kelelense.

kermesinus *(Ridl.) L.A. Garay & W. Kittredge in Bot. Mus. Leafl. Harvard Univ.*, 30(3): (54) 188 (1985 publ. 1986): Bulbophyllum kermesinum.

leontoglossus *(Schltr.) L.A. Garay & W. Kittredge in Bot. Mus. Leafl. Harvard Univ.*, 30(3): (55) 189 (1985 publ. 1986): Bulbophyllum leontoglossum.

leucorhodus *(Schltr.) L.A. Garay & W. Kittredge in Bot. Mus. Leafl. Harvard Univ.*, 30(3): (55) 189 (1985 publ. 1986): Bulbophyllum leucorhodum.

longilabris *(Schltr.) L.A. Garay & W. Kittredge in Bot. Mus. Leafl. Harvard Univ.*, 30(3): (55) 189 (1985 publ. 1986): Bulbophyllum longilabre.

melinoglossus *(Schltr.) L.A. Garay & W. Kittredge in Bot. Mus. Leafl. Harvard Univ.*, 30(3): (55) 189 (1985 publ. 1986): Bulbophyllum melinoglossum.

microrhombos *(Schltr.) L.A. Garay & W. Kittredge in Bot. Mus. Leafl. Harvard Univ.*, 30(3): (55) 189 (1985 publ. 1986): Bulbophyllum microrhombos.

monosema *(Schltr.) L.A. Garay & W. Kittredge in Bot. Mus. Leafl. Harvard Univ.*, 30(3): (55) 189 (1985 publ. 1986): Bulbophyllum monosema.

mutatus *(J.J. Sm.) L.A. Garay & W. Kittredge in Bot. Mus. Leafl. Harvard Univ.*, 30(3): (55) 189 (1985 publ. 1986): Bulbophyllum mutatum.

mystrochilus *(Schltr.) L.A. Garay & W. Kittredge in Bot. Mus. Leafl. Harvard Univ.*, 30(3): (55) 189 (1985 publ. 1986): Bulbophyllum mystrochilum.

novae-hiberniae *(Schltr.) L.A. Garay & W. Kittredge in Bot. Mus. Leafl. Harvard Univ.*, 30(3): (55) 189 (1985 publ. 1986): Bulbophyllum novae-hiberniae.

olorinus *(J.J. Sm.) L.A. Garay & W. Kittredge in Bot. Mus. Leafl. Harvard Univ.*, 30(3): (55) 189 (1985 publ. 1986): Bulbophyllum olorinum.

oxyanthus *(Schltr.) L.A. Garay & W. Kittredge in Bot. Mus. Leafl. Harvard Univ.*, 30(3): (56) 190 (1985 publ. 1986): Bulbophyllum oxyanthum.

pemae *(Schltr.) L.A. Garay & W. Kittredge in Bot. Mus. Leafl. Harvard Univ.*, 30(3): (56) 190 (1985 publ. 1986): Bulbophyllum pemae.

quadricaudatus *(J.J. Sm.) L.A. Garay & W. Kittredge in Bot. Mus. Leafl. Harvard Univ.*, 30(3): (56) 190 (1985 publ. 1986): Bulbophyllum quadricaudatum.

rectilabris *(J.J. Sm.) L.A. Garay & W. Kittredge in Bot. Mus. Leafl. Harvard Univ.*, 30(3): (56) 190 (1985 publ. 1986): Bulbophyllum rectilabre.

HAPALOCHILUS :-

reflexus *L.A. Garay & W. Kittredge in Bot. Mus. Leafl. Harvard Univ.*, 30(3): (56) 190 (1985 publ. 1986), nom. nov.: Bulbophyllum rhynchoglossum *Schltr. 1912 not 1910.*

scaphosepalum *(Ridl.) L.A. Garay & W. Kittredge in Bot. Mus. Leafl. Harvard Univ.*, 30(3): (56) 190 (1985 publ. 1986): Bulbophyllum scaphosepalum.

schizopetalum *(L.O. Wms.) L.A. Garay & W. Kittredge in Bot. Mus. Leafl. Harvard Univ.*, 30(3): (56) 190 (1985 publ. 1986): Bulbophyllum schizopetalum.

scitulus *(Ridl.) L.A. Garay & W. Kittredge in Bot. Mus. Leafl. Harvard Univ.*, 30(3): (56) 190 (1985 publ. 1986): Bulbophyllum scitulum.

scyphochilus *(Schltr.) L.A. Garay & W. Kittredge in Bot. Mus. Leafl. Harvard Univ.*, 30(3): (56) 190 (1985 publ. 1986): Bulbophyllum scyphochilum.

speciosus *(Schltr.) L.A. Garay & W. Kittredge in Bot. Mus. Leafl. Harvard Univ.*, 30(3): (56) 190 (1985 publ. 1986): Bulbophyllum speciosum.

stabilis *(J.J. Sm.) L.A. Garay & W. Kittredge in Bot. Mus. Leafl. Harvard Univ.*, 30(3): (57) 191 (1985 publ. 1986): Bulbophyllum stabile.

stellula *(Ridl.) L.A. Garay & W. Kittredge in Bot. Mus. Leafl. Harvard Univ.*, 30(3): (57) 191 (1985 publ. 1986): Bulbophyllum stellula.

stenophyton *L.A. Garay & W. Kittredge in Bot. Mus. Leafl. Harvard Univ.*, 30(3): (57) 191 (1985 publ. 1986), nom. nov.: Bulbophyllum stenophyllum *Schltr.*

stictanthus *(Schltr.) L.A. Garay & W. Kittredge in Bot. Mus. Leafl. Harvard Univ.*, 30(3): (57) 191 (1985 publ. 1986): Bulbophyllum stictanthum.

striatus *L.A. Garay & W. Kittredge in Bot. Mus. Leafl. Harvard Univ.*, 30(3): (57) 191 (1985 publ. 1986), nom. nov.: Bulbophyllum pulchrum *Schltr.*

torricellensis *(Schltr.) L.A. Garay & W. Kittredge in Bot. Mus. Leafl. Harvard Univ.*, 30(3): (57) 191 (1985 publ. 1986): Bulbophyllum torricellense.

trachyglossus *(Schltr.) L.A. Garay & W. Kittredge in Bot. Mus. Leafl. Harvard Univ.*, 30(3): (57) 191 (1985 publ. 1986): Bulbophyllum trachyglossum.

trigonocarpus *(Schltr.) L.A. Garay & W. Kittredge in Bot. Mus. Leafl. Harvard Univ.*, 30(3): (57) 191 (1985 publ. 1986): Bulbophyllum trigonocarpum.

warianus *(Schltr.) L.A. Garay & W. Kittredge in Bot. Mus. Leafl. Harvard Univ.*, 30(3): (57) 191 (1985 publ. 1986): Bulbophyllum warianum.

xanthoacron *(J.J. Sm.) L.A. Garay & W. Kittredge in Bot. Mus. Leafl. Harvard Univ.*, 30(3): (57) 191 (1985 publ. 1986): Bulbophyllum xanthoacron.

HAPALOCHILUS :–
xanthophaeus *(Schltr.) L.A. Garay & W. Kittredge in Bot. Mus. Leafl. Harvard Univ.,* 30(3): (58) 192 (1985 publ. 1986): Bulbophyllum xanthophaeum.

HAPLOPHYLLUM (Rutac.).
sect. Indehiscentes *(Engl.) C.C. Townsend in Hooker's Icon. Pl.,* 40(1–3): 12 (1986): Ruta sect. Indehiscentes.
linifolium
subsp. africanum *C.C. Townsend in Hooker's Icon. Pl.,* 40(1–3): t. 3961, p. 3 (1986), nom. nov.: H. linifolium var. sexovulatum *Litardière & Maire.*

HARTIGHSEA A. Jussieu in Mirb. & Cass. apud Guill. in Bull. Sci. Nat. Geol., 23: 237 (1830); cf. D.J. Mabberley in Taxon, 31(1): 66 (1982). *MELIACEAE.*

HARVEYA (Scrophulariac.).
leucopharynx *O.M. Hilliard & B.L. Burtt in Notes Roy. Bot. Gard. Edinburgh,* 43(3): 379 (1986)—South Africa (Natal, Cape Province).
pulchra *O.M. Hilliard & B.L. Burtt in Notes Roy. Bot. Gard. Edinburgh,* 43(3): 381 (1986)—South Africa (Natal, Cape Province), Lesotho.
silvatica *O.M. Hilliard & B.L. Burtt in Notes Roy. Bot. Gard. Edinburgh,* 43(3): 382 (1986)—South Africa (Natal, Transvaal), Swaziland.

HASTINGSIA (Liliac.).
atropurpurea *R.W. Becking in Madroño,* 33(3): 175 (1986)—U.S.A. (Oregon).

HAWORTHIA (Aloeac.).
ser. Attenuatae *(Pilbeam) Rowley, Haworthia Drawings J.T. Bates:* 14 (1985); *cf. Repert. Pl. Succ. (I.O.S.),* 36: 5 (1985 publ. 1986): H. subsect. Attenuatae.
ser. Fusiformes *(Bark.) Rowley, Haworthia Drawings J.T. Bates:* 14 (1985); *cf. Repert. Pl. Succ. (I.O.S.),* 36: 5 (1985 publ. 1986): H. sect Fusiformes.
ser. Limpidae *(A. Berger) Rowley, Haworthia Drawings J.T. Bates:* 13 (1985); *cf. Repert. Pl. Succ. (I.O.S.),* 36: 5 (1985 publ. 1986): H. sect. Limpidae.
ser. Obtusatae *(A. Berger) Rowley, Haworthia Drawings J.T. Bates:* 14 (1985); *cf. Repert. Pl. Succ. (I.O.S.),* 36: 5 (1985 publ. 1986): H. sect. Obtusatae.
ser. Proliferae *(Pilbeam) Rowley, Haworthia Drawings J.T. Bates:* 14 (1985); *cf. Repert. Pl. Succ. (I.O.S.),* 36: 5 (1985 publ. 1986): H. subsect. Proliferae.
ser. Scabrae *(A. Berger) Rowley, Haworthia Drawings J.T. Bates:* 14 (1985); *cf. Repert. Pl. Succ. (I.O.S.),* 36: 5 (1985 publ. 1986): H. sect. Scabrae.

HAWORTHIA :–
ser. Turgidae *(Pilbeam) Rowley, Haworthia Drawings J.T. Bates:* 14 (1985); *cf. Repert. Pl. Succ. (I.O.S.),* 36: 5 (1985 publ. 1986): H. subsect. Turgidae.
fasciata
var. browniana *(v. Pölln.) C.L. Scott, Genus Haworthia:* 21 (1985); *cf. Repert. Pl. Succ. (I.O.S.),* 36: 5 (1985 publ. 1986): H. browniana.
multilineata *(G.G. Smith) C.L. Scott, Genus Haworthia:* 134 (1985); *cf. Repert. Pl. Succ. (I.O.S.),* 36: 5 (1985 publ. 1986): H. retusa var. multilineata.

HEDYCHIUM (Zingiberac.).
convexum *S.Q. Tong in Acta Bot. Yunnanica,* 8(1): 41 (1986)—China.
pauciflorum *S.Q. Tong in Acta Bot. Yunnanica,* 8(1): 43 (1986)—China.
yungjiangense *S.Q. Tong in Acta Bot. Yunnanica,* 8(1): 40 (1986)—China.

HEDYOTIS (Rubiac.).
devicolamensis *D.B. Deb & R. Dutta in J. Bombay Nat. Hist. Soc.,* 82(3): 619 (1985 publ. 1986)—India.

HEDYSARUM (Legumin.).
alpinum
subsp. boreoeuropaeum *B.A. Yurtsev in Arktich. Fl. SSSR,* 9(2): 182 (1986), as 'boreo-europaeum'—U.S.S.R. (European Russia).
americanum
subsp. philoscium *(A. Nels.) B.A. Yurtsev in Arktich. Fl. SSSR,* 9(2): 153 (1986): H. philoscium.
argyreum *W. Greuter & Burdet in Willdenowia,* 16(1): 109 (1986), nom. nov.: H. argentatum *Maire.*
caudatum *C.P. Thunberg ex A. Murray in Linnaeus, Syst. Veg., ed. 14:* 675 (1784)—Japan. †
dentatoalatum *K.T. Fu in Fl. Tsinlingensis,* 1(3): 448, 73 (1981), as 'dentato-alatum'—China.
hedysaroides
subsp. austrosibiricum *(B. Fedtsch.) B.A. Yurtsev in Arktich. Fl. SSSR,* 9(2): 157 (1986): H. austrosibiricum.
subsp. tschuktschorum *B.A. Yurtsev in Arktich. Fl. SSSR,* 9(2): 182, 159 (1986)—U.S.S.R. (Soviet Far East, Siberia), Alaska.
incanum *C.P. Thunberg ex A. Murray in Linnaeus, Syst. Veg., ed. 14:* 677 (1784)—Japan. †
microphyllum *C.P. Thunberg ex A. Murray in Linnaeus, Syst. Veg., ed. 14:* 675 (1784)—Japan. †
pilosum *C.P. Thunberg ex A. Murray in Linnaeus, Syst. Veg., ed. 14:* 675 (1784)—Japan. †

HEDYSARUM :–
polybotrys
var. robustum *K.T. Fu in Fl. Tsinlingensis,* 1(3): 447, 72 (1981)—China.
racemosum *C.P. Thunberg ex A. Murray in Linnaeus, Syst. Veg., ed. 14:* 675 (1784)—Japan. †
sericeum *C.P. Thunberg ex A. Murray in Linnaeus, Syst. Veg., ed. 14:* 675 (1784)—Japan. †
striatum *C.P. Thunberg ex A. Murray in Linnaeus, Syst. Veg., ed. 14:* 675 (1784)—Japan. †
taipeicum *(Hand.-Mazz.) K.T. Fu in Fl. Tsinlingensis,* 1(3): 72 (1981): H. esculentum var. taipeicum.
tomentosum *C.P. Thunberg ex A. Murray in Linnaeus, Syst. Veg., ed. 14:* 675 (1784)—Japan. †
virgatum *C.P. Thunberg ex A. Murray in Linnaeus, Syst. Veg., ed. 14:* 675 (1784)—Japan. †

HELENIUM (Compos.).
varium *Schrad., Ind. Sem. Hort. Acad. Gott.,* 1830: 3 (1830); *cf. D.J. Mabberley in Taxon,* 32(1): 85 (1983)— Locality not stated.

HELIANTHEMUM (Cistac.).
canum *(L.) Hornem., Hort. Hafn.:* 496 (1815); *cf. D.J. Mabberley in Taxon,* 33(3): 437 (1984): Basionym not stated.
juliae *W. Wildpret de la Torre in Vieraea,* 16(1–2): 361 (1986)—Canary Is.
pourretii *M.E. Timbal-Lagrave in Bull. Soc. Sci. Phys. Natur. Toulouse,* 2: 65 (1875)—France.
rosmarinifolium *Hort. Erlang. ex Mart., Hort. Acad. Erlangen:* 114 (1814); *cf. D.J. Mabberley in Taxon,* 33(3): 442 (1984)—Locality not stated.

HELIANTHUS (Compos.).
grandiflorus *Wender., Ind. Sem. Hort. Acad. Marb.,* 1835; *Linnaea 11, Lit.:* 92 (1837); *cf. D.J. Mabberley in Taxon,* 32(1): 85 (1983)—Locality not stated.
parviflorus *Hornem., Hort. Hafn.:* 842 (1815); *cf. D.J. Mabberley in Taxon,* 33(3): 437 (1984)— Locality not stated.

HELICHRYSUM (Compos.).
evansii *O.M. Hilliard in Notes Roy. Bot. Gard. Edinburgh,* 43(2): 198 (1986), nom. nov.: H. alticolum var. montanum *Bolus.*
polioides *B.L. Burtt in Notes Roy. Bot. Gard. Edinburgh,* 43(2): 233 (1986)—Malawi.

HELICIA (Proteac.).
recurva *D.B. Foreman in Muelleria,* 6(3–4): 193 (1986)—Australia (Queensland).

HELIOTROPIUM (Boraginac.).
bacciferum
subsp. erosum *(Lehmann) H. Riedl in Linz. Biol. Beitr.,* 17(2): 299 (1985): H. erosum.

HELIOTROPIUM :–
balansae *H. Riedl in Linz. Biol. Beitr.,* 17(2): 295 (1985)—Morocco.

HELLEBORUS (Ranunculac.).
lividus
subsp. corsicus *(Briquet) P.F. Yeo in Taxon,* 35(1): 157 (1986): H. trifolius subsp. corsicus.
nothosubsp. sternii *(Turrill) P.F. Yeo in Taxon,* 35(1): 157 (1986): H. × sternii.

HEMEROCALLIS (Hemerocallidac.).
cordata *C.P. Thunberg ex A. Murray in Linnaeus, Syst. Veg., ed. 14:* 339 (1784)—Japan. †
japonica *C.P. Thunberg ex A. Murray in Linnaeus, Syst. Veg., ed. 14:* 339 (1784)—Japan. †

HEMIGENIA (Labiat.).
conferta *B.J. Conn in Muelleria,* 6(3–4): 259 (1986)—Australia (W. Australia).

HENRIETTEA (Melastomatac.).
squamulosa *(Cogn.) W.S. Judd in Brittonia,* 38(3): 240 (1986): Calycogonium squamulosum.

HERACLEUM (Umbellif.).
dissectifolium *K.T. Fu in Fl. Tsinlingensis,* 1(3): 464, 431 (1981)—China.
egrissicum *R.I. Gagnidze in Soobshch. Akad. Nauk Gruz. SSR,* 101(1): 117 (1981)—U.S.S.R. (Caucasus).
moellendorffii
var. paucivittatum *R.H. Shan & T.S. Wang in Acta Phytotax. Sin.,* 24(4): 316 (1986)—China.
var. sagenifolium *K.T. Fu in Fl. Tsinlingensis,* 1(3): 464, 431 (1981)—China.

HERMANNIA (Sterculiac.).
cordata *(E. Phillips) B. de Winter in Notes Roy. Bot. Gard. Edinburgh,* 43(3): 401 (1986): Mahernia cordata.

HERMINIUM (Orchidac.).
alpinum *(Jacq.) Sweet, Hort. Brit.:* 382 (1827); *cf. D.J. Mabberley in Taxon,* 31(1): 71 (1982): Basionym not stated.

HERSCHELIANTHE (Orchidac.).
barbata *(L.f.) N.C. Anthony in Bothalia,* 15(3–4): 554 (1985): Orchis barbata.
forficaria *(H. Bol.) N.C. Anthony in Bothalia,* 15(3–4): 554 (1985): Disa forficaria.
lugens
var. nigrescens *(Linder) N.C. Anthony in Bothalia,* 15(3–4): 554 (1985): Herschelia lugens var. nigrescens.
newdigateae *(L. Bol.) N.C. Anthony in Bothalia,* 15(3–4): 554 (1985): Disa newdigateae.
schlechteriana *(H. Bol.) N.C. Anthony in Bothalia,* 15(3–4): 554 (1985): Disa schlechteriana.

HERSCHELIANTHE :–
spathulata
subsp. tripartita *(Lindl.) N.C. Anthony in Bothalia,* 15(3–4): 554 (1985): Disa tripartita.

HESPERANTHA (Iridac.).
alborosea *O.M. Hilliard & B.L. Burtt in Notes Roy. Bot. Gard. Edinburgh,* 43(3): 421 (1986)—South Africa (Natal).
baurii
subsp. formosa *O.M. Hilliard & B.L. Burtt in Notes Roy. Bot. Gard. Edinburgh,* 43(3): 430 (1986)—Lesotho, South Africa (Orange Free State, Natal).
curvula *O.M. Hilliard & B.L. Burtt in Notes Roy. Bot. Gard. Edinburgh,* 43(3): 416 (1986)—South Africa (Natal).
glareosa *O.M. Hilliard & B.L. Burtt in Notes Roy. Bot. Gard. Edinburgh,* 43(3): 424 (1986)—South Africa (Natal, Transvaal, Orange Free State), Lesotho.
hutchingsiae *O.M. Hilliard & B.L. Burtt in Notes Roy. Bot. Gard. Edinburgh,* 43(3): 414 (1986)—South Africa (Cape Province).
ingeliensis *O.M. Hilliard & B.L. Burtt in Notes Roy. Bot. Gard. Edinburgh,* 43(3): 424 (1986)—South Africa (Natal, Cape Province).
pubinervia *O.M. Hilliard & B.L. Burtt in Notes Roy. Bot. Gard. Edinburgh,* 43(3): 419 (1986)—South Africa (Natal).
scopulosa *O.M. Hilliard & B.L. Burtt in Notes Roy. Bot. Gard. Edinburgh,* 43(3): 417 (1986)—South Africa (Natal).

HESPERIS (Crucif.).
torulosa *Hort. ex Lag., Elenchus:* 9 (1816); *cf. D.J. Mabberley in Taxon,* 31(1): 71 (1982)— Locality not stated.

HETERANTHERA (Pontederiac.).
multiflora *(Grisebach) C.N. Horn in Phytologia,* 59(4): 290 (1986): H. reniformis var. multiflora.

HETEROPSIS (Arac.).
spruceana
var. robusta *G.S. Bunting in Phytologia,* 60(5): 303 (1986)—Venezuela.
steyermarkii *G.S. Bunting in Phytologia,* 60(5): 305 (1986)—Venezuela.
tenuispadix *G.S. Bunting in Phytologia,* 60(5): 306 (1986)—Venezuela.

HEXASTYLIS (Aristolochiac.).
rhombiformis *L.L. Gaddy in Brittonia,* 38(1): 82 (1986)—U.S.A. (N. Carolina).

HIBISCUS (Malvac.).
amazonicus *P.A. Fryxell in Acta Amazonica,* 14(1–2 Suppl.): 101 (1984 publ. 1986)— Brazil.
chrysanthus *Bull, Retail List,* 199: 13 (1884); *cf. D.J. Mabberley in Taxon,* 34(3): 456 (1985)— Locality not stated.

HIBISCUS :–
hoshiarpurensis *T.K. Paul & M.P. Nayar in Bull. Bot. Surv. India,* 25(1–4): 188 (1983 publ. 1985)—India.
leiospermus *K.T. Fu & C.C. Fu in Fl. Tsinlingensis,* 1(3): 454, 290 (1981)—China.
pacificus *Nakai ex Y. Jotani & H. Ohba in J. Jap. Bot.,* 61(4): 99 (1986)—Volcano Is.
spinulosus *Huber, Cat. Gén. 1871 & 1872:* 4 (1871); *cf. D.J. Mabberley in Taxon,* 34(3): 450 (1985)— Locality not stated.

HIMERANTHUS (Solanac.).
runcinatus *(Lam.) Endl., Cat. Horti Vindob.,* 2: 83 (1842); *cf. D.J. Mabberley in Taxon,* 33(3): 435 (1984): Jaborosa runcinata.

HIPPOCREPIS (Legumin.).
scorpioides
var. nevadensis *(Uhrova) P. Lassen in Willdenowia,* 16(1): 111 (1986): H. comosa var. nevadensis.
unisiliquosa
subsp. biflora *(Sprengel) O. de Bolòs & J. Vigo, Fl. Països Catalans:* 642 (1984): H. biflora.

HIRTELLA (Chrysobalanac.).
margae *G.T. Prance in Proc. Kon. Nederl. Acad. Wetensch., C,* 89(1): 111 (1986)— Surinam, Guyana.
racemosa
var. hispida *G.T. Prance in Proc. Kon. Nederl. Acad. Wetensch., C,* 89(1): 113 (1986)—Brazil, French Guiana.

HOLMGRENANTHE W.J. Elisens, Monogr. Maurandyinae (Scrophulariac.-Antirrhin.) (Syst. Bot. Monog., 5): 54 (1985). *SCROPHULARIACEAE.*
petrophila *(Coville & C. Morton) W.J. Elisens, Monogr. Maurandyinae (Scrophulariac.-Antirrhin.) (Syst. Bot. Monog., 5):* 54 (1985): Maurandia petrophila.

HOLOGRAPHIS (Acanthac.).
websteri *T.F. Daniel in Southwest. Nat.,* 31(2): 169 (1986)—Mexico.

HOLSTIANTHUS J.A. Steyermark in Ann. Missouri Bot. Gard., 73(2): 495 (1986). *RUBIACEAE.*
barbigularis *J.A. Steyermark in Ann. Missouri Bot. Gard.,* 73(2): 495 (1986)—Venezuela.

HOMALOMENA (Arac.).
engleri *J. Bogner in Aroideana,* 8(3): 75 (1985 publ. 1986), nom. nov.: Diandriella novoguineensis *Engl.*

HOOKEROCHLOA E.B. Alekseev in Byull. Mosk. Obshch. Ispўt. Prir., Biol., 90(5): 106 (1985). *GRAMINEAE.*
hookeriana *(F. Muell. ex Hook. f.) E.B. Alekseev in Byull. Mosk. Obshch. Ispўt. Prir., Biol.,* 90(5): 106 (1985): Festuca hookeriana.

HOPEA (Dipterocarpac.).
modesta *(A.DC.) A.J.G.J. Kostermans in Ceylon J. Sci., Biol. Sci.,* 15(1–2): 47 (1982): H. jucunda var. modesta.

HORDEUM (Gramin.).
fuegianum *R. von Bothmer, N. Jacobsen & R.B. Jørgensen in Nordic J. Bot.,* 6(4): 404 (1986)—Tierra del Fuego, Chile.

HORNSCHUCHIA (Annonac.).
polyantha *P.J.M. Maas in Proc. Kon. Nederl. Akad. Wetensch., C,* 89(3): 258 (1986)—Brazil.

HORSFIELDIA (Myristicac.).
affinis *W.J.J.O. de Wilde in Gard. Bull. Singapore,* 38(2): 217 (1985)—Borneo (Sarawak, Brunei, Kalimantan).
amplomontana *W.J.J.O. de Wilde in Gard. Bull. Singapore,* 39(1): 34 (1986)—Borneo (Sabah).
androphora *W.J.J.O. de Wilde in Gard. Bull. Singapore,* 39(1): 32 (1986)—Borneo (Sarawak, Sabah).
atjehensis *W.J.J.O. de Wilde in Gard. Bull. Singapore,* 38(2): 186 (1985)—Sumatra.
borneensis *W.J.J.O. de Wilde in Gard. Bull. Singapore,* 39(1): 27 (1986)—Borneo (Sarawak, Sabah, Kalimantan).
coriacea *W.J.J.O. de Wilde in Gard. Bull. Singapore,* 39(1): 50 (1986)—Celebes.
disticha *W.J.J.O. de Wilde in Gard. Bull. Singapore,* 39(1): 10 (1986)—Borneo (Brunei).
endertii *W.J.J.O. de Wilde in Gard. Bull. Singapore,* 39(1): 24 (1986)—Borneo (Kalimantan, Sarawak, Sabah).
glabra
var. javanica *W.J.J.O. de Wilde in Gard. Bull. Singapore,* 39(1): 59 (1986)—Java.
var. oviflora *W.J.J.O. de Wilde in Gard. Bull. Singapore,* 39(1): 59 (1986)—Java.
gracilis *W.J.J.O. de Wilde in Gard. Bull. Singapore,* 38(2): 211 (1985)—Borneo (Sarawak).
hirtiflora *W.J.J.O. de Wilde in Gard. Bull. Singapore,* 39(1): 1 (1986)—Sumatra.
laticostata *(Sinclair) W.J.J.O. de Wilde in Gard. Bull. Singapore,* 39(1): 15 (1986): H. brachiata var. laticostata.
macilenta *W.J.J.O. de Wilde in Gard. Bull. Singapore,* 39(1): 13 (1986)—Borneo (Sabah, Sarawak), Malaya, Sumatra.
nervosa *W.J.J.O. de Wilde in Gard. Bull. Singapore,* 39(1): 16 (1986)—Borneo (Sarawak).
obscura *W.J.J.O. de Wilde in Gard. Bull. Singapore,* 39(1): 42 (1986)—Borneo (Kalimantan, ?Sarawak).
obtusa *W.J.J.O. de Wilde in Gard. Bull. Singapore,* 39(1): 9 (1986)—Borneo (Sarawak).

HORSFIELDIA :–
pachyrachis *W.J.J.O. de Wilde in Gard. Bull. Singapore,* 39(1): 5 (1986)—Borneo (Kalimantan).
pallidicaula *W.J.J.O. de Wilde in Gard. Bull. Singapore,* 38(2): 191 (1985)—Borneo (Sarawak, Sabah, ?Kalimantan).
var. macrocarya *W.J.J.O. de Wilde in Gard. Bull. Singapore,* 38(2): 193 (1985)—Borneo (Sarawak).
var. microcarya *(Sinclair) W.J.J.O. de Wilde in Gard. Bull. Singapore,* 38(2): 193 (1985): H. bracteosa var. microcarya.
polyspherula
var. maxima *W.J.J.O. de Wilde in Gard. Bull. Singapore,* 39(1): 22 (1986)—Borneo (Sarawak, Sabah, Kalimantan).
var. sumatrana *(Miq.) W.J.J.O. de Wilde in Gard. Bull. Singapore,* 39(1): 20 (1986): Myristica glabra var. sumatrana.
pulcherrima *W.J.J.O. de Wilde in Gard. Bull. Singapore,* 38(2): 206 (1985)—Malaya, Sumatra.
punctata *W.J.J.O. de Wilde in Gard. Bull. Singapore,* 39(1): 37 (1986)—Malaya.
sessilifolia *W.J.J.O. de Wilde in Gard. Bull. Singapore,* 38(2): 201 (1985)—Borneo (Sarawak).
sparsa *W.J.J.O. de Wilde in Gard. Bull. Singapore,* 38(2): 194 (1985)—Malaya, Thailand, Singapore, Sumatra.
splendida *W.J.J.O. de Wilde in Gard. Bull. Singapore,* 38(2): 213 (1985)—Borneo (Sarawak, Brunei, Sabah, Kalimantan).
sterilis *W.J.J.O. de Wilde in Gard. Bull. Singapore,* 38(2): 224 (1985)—Borneo (Sabah).
subalpina
subsp. kinabaluensis *W.J.J.O. de Wilde in Gard. Bull. Singapore,* 39(1): 41 (1986)—Borneo (Sabah, ?Sarawak).
sucosa
subsp. bifissa *W.J.J.O. de Wilde in Gard. Bull. Singapore,* 38(2): 190 (1985)—Borneo (Kalimantan, Sabah).
tenuifolia *(Sinclair) W.J.J.O. de Wilde in Gard. Bull. Singapore,* 39(1): 11 (1986): H. polyspherula var. tenuifolia.
triandra *W.J.J.O. de Wilde in Gard. Bull. Singapore,* 38(2): 195 (1985)—Sumatra.
tristis *W.J.J.O. de Wilde in Gard. Bull. Singapore,* 38(2): 197 (1985)—Borneo (Sarawak, Kalimantan), Sumatra.
xanthina
subsp. macrophylla *W.J.J.O. de Wilde in Gard. Bull. Singapore,* 39(1): 47 (1986)—Borneo (Sabah, Sarawak).

HOUSSAYANTHUS A.T. Hunziker in Kurtziana, 11: 17 (1978). *SAPINDACEAE.*
fiebrigii *(Barkley) A.T. Hunziker in Kurtziana,* 11: 20 (1978): Urvillea fiebrigii.
incanus *(Radlk.) M.S. Ferrucci in Candollea,* 41(1): 218 (1986): Serjania incana.

HOUSSAYANTHUS :–
macrolophus *(Radlk.) A.T. Hunziker in Kurtziana,* 11: 18 (1978): Cardiospermum? macrolophum.

HOUSTONIA (Rubiac.).
greenei *(A. Gray) E.E. Terrell in Phytologia,* 59(1): 79 (1985): Oldenlandia greenei.
nigricans
var. floridana *(Standley) E.E. Terrell in Phytologia,* 59(1): 79 (1985): H. floridana.
var. pulvinata *(Small) E.E. Terrell in Phytologia,* 59(1): 79 (1985): H. pulvinata.
rosea *(Rafinesque) E.E. Terrell in Rhodora,* 88(855): 395 (1986): Hedyotis rosea.

HOUTTUYNIA (Saururac.).
cordata *C.P. Thunberg in Kong. Vetensk. Akad. Handl., Stockholm,* 4: 149, 152, t. 5 (1783), as 'Houtuynia'—Japan.
f. polypetaloidea *T. Yamazaki in J. Jap. Bot.,* 61(10): 310 (1986)—Cultivation.
f. viridis *J. Ohara in J. Phytogeogr. Taxon.,* 33(2): 72 (1985)—Japan.

HOVENIA (Rhamnac.).
dulcis
var. acerba *(Lindl.) G. Sen Gupta & B. Safui in Bull. Bot. Surv. India,* 26(1–2): 55 (1984 publ. 1985): H. acerba.

HUODENDRON (Styracac.).
decorticatum *C.Y. Wu in Fl. Yunnanica,* 3: 418 (1983)—China.
tibeticum
var. denticulatum *C.Y. Wu in Fl. Yunnanica,* 3: 416 (1983)—China.

HYALOLAENA (Umbellif.).
viridiflora *E.V. Klyuikov in Bot. Zhurn.,* 71(9): 1271 (1986)—U.S.S.R. (Middle Asia).

HYALOSERIS (Compos.).
longicephala *B.L. Turner in Phytologia,* 59(5): 317 (1986)—Bolivia.

HYDRANGEA (Hydrangeac.).
arborescens
f. carnea *(Raf.) L.J. Uttal in Sida,* 11(3): 352 (1986): H. vulgaris var. carnea.

HYDROSTEMMA Wall. in Taylor & Phillips, Phil. Mag., n.s., 1: 454 (June 1827); cf. D.J. Mabberley in Taxon, 31(1): 68 (1982). *NYMPHAEACEAE.*

HYMENOCALLIS (Amaryllidac.).
acutifolia *(Herb.) Sweet, Hort. Brit., ed. 2:* 513 (1830); *cf. D.J. Mabberley in Taxon,* 30(1): 16 (1981): Basionym not stated.
incaica *P. Ravenna in Nordic J. Bot.,* 6(4): 463 (1986), with two types cited—Peru.

HYMENOCARPOS (Legumin.).
cornicina *(L.) P. Lassen in Willdenowia,* 16(1): 111 (1986): Anthyllis cornicina.

HYMENOCARPOS :–
hamosus *(Desf.) P. Lassen in Willdenowia,* 16(1): 111 (1986): Anthyllis hamosa.
lotoides *(L.) P. Lassen in Willdenowia,* 16(1): 111 (1986): Anthyllis lotoides.

HYMENOLEPIS (Compos.).
cynopus *K. Bremer & M. Källersjö in Nordic J. Bot.,* 5(6): 519 (1985 publ. 1986)—South Africa (Cape Province).
dentata *(DC.) M. Källersjö in Nordic J. Bot.,* 5(6): 534 (1985 publ. 1986): Oligodora dentata.
gnidioides *(S. Moore) M. Källersjö in Nordic J. Bot.,* 5(6): 534 (1985 publ. 1986): Phaeocephalus gnidioides.
indivisa *(Harv.) M. Källersjö in Nordic J. Bot.,* 5(6): 534 (1985 publ. 1986): Athanasia indivisa.
speciosa *(Hutch.) M. Källersjö in Nordic J. Bot.,* 5(6): 534 (1985 publ. 1986): Athanasia speciosa.

HYMENOLOBUS (Crucif.).
revelieri *(Jordan) S. Brullo in Boll. Sedute Accad. Gioenia Sci. Nat. Catania,* 15(320): 227 (1982): Hutchinsia revelieri.
subsp. sommieri *(Pamp.) S. Brullo in Boll. Sedute Accad. Gioenia Sci. Nat. Catania,* 15(320): 227 (1982): Hutchinsia procumbens f. sommieri.

HYMENORCHIS (Orchidac.).
serrulata *(Hallé) L.A. Garay in Bot. Mus. Leafl. Harvard Univ.,* 30(3): (58) 192 (1985 publ. 1986): Saccolabium serrulatum.

HYPERICUM (Guttif.).
subsect. Platyadenum *N.K.B. Robson in Notes Roy. Bot. Gard. Edinburgh,* 43(2): 256 (1986).
subsect. Stenadenum *N.K.B. Robson in Notes Roy. Bot. Gard. Edinburgh,* 43(2): 257 (1986).
davisii *N.K.B. Robson in Notes Roy. Bot. Gard. Edinburgh,* 43(2): 263 (1986)—Turkey, U.S.S.R. (Caucasus), Iran.
elongatum
subsp. apiculatum *N.K.B. Robson in Notes Roy. Bot. Gard. Edinburgh,* 43(2): 261 (1986)—Turkey.
subsp. callithyrsum *(Cosson) N.K.B. Robson in Notes Roy. Bot. Gard. Edinburgh,* 43(2): 262 (1986): H. callithyrsum.
subsp. microcalycinum *(Boiss. & Heldr.) N.K.B. Robson in Notes Roy. Bot. Gard. Edinburgh,* 43(2): 261 (1986): H. microcalycinum.
subsp. racemosum *(Kuntze) N.K.B. Robson in Notes Roy. Bot. Gard. Edinburgh,* 43(2): 261 (1986): H. hyssopifolium var. racemosum.
subsp. tymphresteum *(Boiss. & Spruner) N.K.B. Robson in Notes Roy. Bot. Gard. Edinburgh,* 43(2): 262 (1986): H. tymphresteum.

HYPERICUM :–
erectum *C.P. Thunberg ex A. Murray in Linnaeus, Syst. Veg., ed. 14:* 702 (1784)—Japan. †
hookerianum
var. dentatum *S.N. Biswas in Bull. Bot. Surv. India,* 25(1–4): 195 (1983 publ. 1985)—India, Sikkim.
japonicum *C.P. Thunberg ex A. Murray in Linnaeus, Syst. Veg., ed. 14:* 702 (1784)—Japan. †
patulum *C.P. Thunberg ex A. Murray in Linnaeus, Syst. Veg., ed. 14:* 700 (1784)—Japan. †
rumeliacum
subsp. apollinis *(Boiss. & Heldr.) N.K.B. Robson & A. Strid in A. Strid (ed.), Mount. Fl. Greece,* 1: 606 (1986): H. apollinis.
sorgerae *N.K.B. Robson in Notes Roy. Bot. Gard. Edinburgh,* 43(2): 264 (1986)—Turkey.
supinum *Clusius ex Vis. in Atti 2 Riun. Sci. Ital.:* 175 (1841) *et Racc.:* 37 (1845); *cf. D.J. Mabberley in Taxon,* 31(1): 71 (1982)—Locality not stated.

HYPODISCUS (Restionac.).
laevigatus *(Kunth) H.P. Linder in Bothalia,* 15(3–4): 489 (1985): Boeckhia laevigata.
montanus *E.E. Esterhuysen in Bothalia,* 15(3–4): 489 (1985)—South Africa (Cape Province).
procurrens *E.E. Esterhuysen in Bothalia,* 15(3–4): 490 (1985)—South Africa (Cape Province).
squamosus *E.E. Esterhuysen in Bothalia,* 15(3–4): 492 (1985)—South Africa (Cape Province).

HYPOESTES (Acanthac.).
floribunda
var. cinerea *R.M. Barker in J. Adelaide Bot. Gard.,* 9: 214 (1986)—Australia (N. Terr., Queensland).
var. neoguineensis *R.M. Barker in J. Adelaide Bot. Gard.,* 9: 211 (1986)—New Guinea (Papua New Guinea).
var. suaveolens *(Gardner) R.M. Barker in J. Adelaide Bot. Gard.,* 9: 223 (1986): H. suaveolens.
var. varia *R.M. Barker in J. Adelaide Bot. Gard.,* 9: 216 (1986)—Australia (N. Terr., W. Australia).
var. velutina *R.M. Barker in J. Adelaide Bot. Gard.,* 9: 209 (1986)—Australia (Queensland).
var. yorkensis *R.M. Barker in J. Adelaide Bot. Gard.,* 9: 212 (1986)—Australia (Queensland).
sparsiflora *R.M. Barker in J. Adelaide Bot. Gard.,* 9: 197 (1986)—Australia (W. Australia).

HYPOXIS (Hypoxidac.).
spicata *C.P. Thunberg ex A. Murray in Linnaeus, Syst. Veg., ed. 14:* 326 (1784)—Japan. †

HYPTIS (Labiat.).
sect. Pachyphyllae *(Epling) R.M. Harley in Kew Bull.,* 41(4): 996 (1986): H. subsect. Pachyphyllae.
carvalhoi *R.M. Harley in Kew Bull.,* 41(1): 141 (1986)—Brazil.
hagei *R.M. Harley in Kew Bull.,* 41(1): 145 (1986)—Brazil.
huberi *R.M. Harley in Kew Bull.,* 41(1): 147 (1986)—Venezuela.
imbricatiformis *R.M. Harley in Kew Bull.,* 41(4): 1000 (1986)—Brazil.
rondonica *R.M. Harley in Kew Bull.,* 41(1): 141 (1986)—Brazil.

HYSSARIA A.A. Kolakovskiĭ in Soobshch. Akad. Nauk Gruz. SSR, 103(1): 151 (1981). *CAMPANULACEAE.*
lehmanniana *(Bunge) A.A. Kolakovskiĭ in Soobshch. Akad. Nauk Gruz. SSR,* 103(1): 151 (1981): Campanula lehmanniana.

I

IBERIS (Crucif.).
saxatilis
subsp. pseudosaxatilis *(Emberger) M. Moreno & M. Velasco in Taxon,* 35(1): 93 (1986): I. sempervirens subsp. pseudosaxatilis.

ILEX (Aquifoliac.).
ser. Rivulares *Y.K. Li in Acta Phytotax. Sin.,* 24(5): 399 (1986).
blancheana *W.S. Judd in Rhodora,* 88(856): 436 (1986)—Hispaniola (Haiti).
crenata *C.P. Thunberg ex A. Murray in Linnaeus, Syst. Veg., ed. 14:* 168 (1784)—Japan. †
emarginata *C.P. Thunberg ex A. Murray in Linnaeus, Syst. Veg., ed. 14:* 168 (1784)—Japan. †
fargesii
var. angustifolia *C.Y. Chang in Fl. Tsinlingensis,* 1(3): 451, 197 (1981)—China.
var. brevifolia *S. Andrews in Kew Mag.,* 3(3): 134 (1986)—China.
subsp. melanotricha *(Merr.) S. Andrews in Kew Mag.,* 3(3): 134 (1986): I. melanotricha.
var. parvifolia *(S.Y. Hu) S. Andrews in Kew Mag.,* 3(3): 134 (1986): I. franchetiana var. parvifolia.

ILEX :–
integra *C.P. Thunberg ex A. Murray in Linnaeus, Syst. Veg., ed. 14:* 168 (1784)—Japan. †
japonica *C.P. Thunberg ex A. Murray in Linnaeus, Syst. Veg., ed. 14:* 168 (1784)—Japan. †
latifolia *C.P. Thunberg ex A. Murray in Linnaeus, Syst. Veg., ed. 14:* 168 (1784)—Japan. †
rivularis *Y.K. Li in Acta Phytotax. Sin.*, 24(5): 399 (1986)—China.
rotunda *C.P. Thunberg ex A. Murray in Linnaeus, Syst. Veg., ed. 14:* 168 (1784)—Japan. †
serrata *C.P. Thunberg ex A. Murray in Linnaeus, Syst. Veg., ed. 14:* 168 (1784)—Japan. †
ternatiflora *(C.H. Wright) R.A. Howard in Brittonia,* 38(1): 15 (1986): Quiina ternatiflora.
walsinghamii *R.A. Howard in Brittonia,* 38(1): 14 (1986), nom. nov.: I. wrightii *Loes.*

ILLIGERA (Hernandiac.).
gammiei *M.P. Nayar. & G.S. Giri in Bull. Bot. Surv. India,* 25(1–4): 249 (1983 publ. 1985)—Assam.

IMBRALYX R. Geesink, Scala Millettiearum (Leiden Bot. Ser., 8): 95 (1984). *LEGUMINOSAE.*
albiflorus *(Prain) R. Geesink, Scala Millettiearum (Leiden Bot. Ser., 8):* 95 (1984): Millettia albiflora.

INDIGOFERA (Legumin.).
ammoxylum *(DC.) R.M. Polhill in Kew Bull.,* 41(2): 343 (1986): Bremontiera ammoxylum.
glandulosa
var. sykesii *Baker ex B.K. Vijay Kumar & N. Ramayya in Curr. Sci.,* 52(9): 429 (1983), isonym of I. glandulosa var. sykesii Baker 1876—India.
pseudoevansii *O.M. Hilliard & B.L. Burtt in Notes Roy. Bot. Gard. Edinburgh,* 43(2): 208 (1986)—South Africa (Natal).
santapaui *M. Sanjappa in Bull. Bot. Surv. India,* 25(1–4): 202 (1983 publ. 1985)—India.
suffruticosa
var. canescens *(J.A. Schmidt) W. Lobin in Courier Forschungsinst. Senckenberg,* 81: 118 (1986): I. anil var. canescens.
tenuipes *R.M. Polhill in Kew Bull.,* 41(2): 345 (1986)—Madagascar, Réunion.
trita
var. marginulata *(Grah. ex Wt. & Arn.) M. Sanjappa in Bull. Bot. Surv. India,* 26(1–2): 118 (1984 publ. 1985): I. marginulata.
var. purandharensis *M. Sanjappa in Bull. Bot. Surv. India,* 26(1–2): 117 (1984 publ. 1985)—India.

INDOCALAMUS (Gramin.).
pumilus *Q.H. Dai & C.F. Huang in Acta Phytotax. Sin.,* 24(5): 394 (1986)—Cultivation.

INULA (Compos.).
dubia *C.P. Thunberg ex A. Murray in Linnaeus, Syst. Veg., ed. 14:* 767 (1784)—Japan. †
japonica *C.P. Thunberg ex A. Murray in Linnaeus, Syst. Veg., ed. 14:* 767 (1784)—Japan. †
f. giraldii *(Diels) J.Q. Fu in Fl. Tsinlingensis,* 1(5): 207 (1985): I. giraldii.
macrophylla *C.C. Gmel., Hort. Mag. Duc. Bad. Carlsruh.:* 145 (1811); *cf. D.J. Mabberley in Taxon,* 33(3): 440 (1984)— Locality not stated.

INULANTHERA M. Källersjö in Nordic J. Bot., 5(6): 539 (1985 publ. 1986). *COMPOSITAE.*
brownii *(Hochreut.) M. Källersjö in Nordic J. Bot.,* 5(6): 540 (1985 publ. 1986), without basionym ref.: Athanasia brownii.
calva *(Hutch.) M. Källersjö in Nordic J. Bot.,* 5(6): 540 (1985 publ. 1986): Athanasia calva.
coronopifolia *(Harv.) M. Källersjö in Nordic J. Bot.,* 5(6): 540 (1985 publ. 1986): Athanasia coronopifolia.
dregeana *(DC.) M. Källersjö in Nordic J. Bot.,* 5(6): 540 (1985 publ. 1986): Hymenolepis dregeana.
leucoclada *(DC.) M. Källersjö in Nordic J. Bot.,* 5(6): 540 (1985 publ. 1986): Hymenolepis leucoclada.
montana *(Wood) M. Källersjö in Nordic J. Bot.,* 5(6): 540 (1985 publ. 1986), incorrect basionym page (10 in error for 171): Athanasia montana.
nuda *M. Källersjö in Nordic J. Bot.,* 5(6): 540 (1985 publ. 1986), nom. nov.: Pentzia schistostephioides *Taylor.*
schistostephioides *(Hiern) M. Källersjö in Nordic J. Bot.,* 5(6): 540 (1985 publ. 1986): Athanasia schistostephioides.
thodei *(Bolus) M. Källersjö in Nordic J. Bot.,* 5(6): 540 (1985 publ. 1986): Athanasia thodei.
tridens *(Oliv.) M. Källersjö in Nordic J. Bot.,* 5(6): 540 (1985 publ. 1986): Athanasia tridens.

IONIDIUM (Violac.).
marcucii *E.N. Bancroft in Jam. Phys. J.,* 2: 338 (1835); *cf. D.J. Mabberley in Taxon,* 30(1): 10 (1981)— ?Jamaica.

IPOMOEA (Convolvulac.).
antonschmidii *R.W. Johnson in Austrobaileya,* 2(3): 217 (1986)—Australia (Queensland).
argillicola *R.W. Johnson in Austrobaileya,* 2(3): 217 (1986)—Australia (Queensland, N. Terr.).

IPOMOEA :–
classeniana *Huber, Cat. Gén. 1870 & 1871:* 7 (1870); *cf. D.J. Mabberley in Taxon,* 34(3): 452 (1985)— Locality not stated.
fulva *Bertol., Hort. Bot. Bonon.,* 1826: 5 (1826) *et Giorn. Arcad.,* 29: 343 (1826); *cf. D.J. Mabberley in Taxon,* 32(1): 82 (1983)— Locality not stated.
oculata *Huber, Cat. Gén. 1871 & 1872:* 3 (1871); *cf. D.J. Mabberley in Taxon,* 34(3): 451 (1985)— Locality not stated.
papillosa *Bertol., Hort. Bot. Bonon.,* 1826: 5 (1826) *et Giorn. Arcad.,* 29: 343 (1826); *cf. D.J. Mabberley in Taxon,* 32(1): 82 (1983)— Brazil.
polpha *R.W. Johnson in Austrobaileya,* 2(3): 220 (1986)—Australia (Queensland).
rotundifolia *(Schrank) Germann, Verz. Pfl. Bot. Gart. Kais. Univ. Dorpat,* 1807: 75 (1807); *cf. D.J. Mabberley in Taxon,* 32(1): 85 (1983): Convolvulus rotundifolius.
saintronanensis *R.W. Johnson in Austrobaileya,* 2(3): 222 (1986)—Australia (Queensland).
truncata *Roezl in Deut. Gärtn.-Zeit.,* 1881; *Belgique horticole,* 33: 135 (1883); *cf. D.J. Mabberley in Taxon,* 34(3): 454 (1985)— Locality not stated.

IPOMOPSIS (Polemoniac.).
aggregata
subsp. collina *(Greene) D.H. Wilken & S.T. Allard in Syst. Bot.,* 11(1): 11 (1986): Callisteris collina.

IRIS (Iridac.).
lortetii
var. samariae *(Dinsmore) N. Feinbrun-Dothan, Fl. Palaestina,* 4: 121 (1986): I. samariae.
missuriensis *Mertens, Hort. Leeuven.,* 1840; *cf. D.J. Mabberley in Taxon,* 33(3): 442 (1984)— Locality not stated.
pariensis *S.L. Welsh in Great Basin Nat.,* 46(2): 256 (1986)—U.S.A. (Utah).

ISACHNE (Gramin.).
veldkampii *K.G. Bhat & C.R. Nagendran in Curr. Sci.,* 52(6): 258 (1983)—India.

ISATIS (Crucif.).
tinctoria
subsp. athoa *(Boiss.) K. Papanicolaou in A. Strid (ed.), Mount. Fl. Greece,* 1: 238 (1986): I. athoa.

ISCHYROLEPIS (Restionac.).
affinis *E.E. Esterhuysen in Bothalia,* 15(3–4): 401 (1985)—South Africa (Cape Province).
arida *(Pillans) H.P. Linder in Bothalia,* 15(3–4): 402 (1985): Restio aridus.
caespitosa *E.E. Esterhuysen in Bothalia,* 15(3–4): 402 (1985)—South Africa (Cape Province).
capensis *(L.) H.P. Linder in Bothalia,* 15(3–4): 402 (1985): Schoenus capensis.
cincinnata *(Mast.) H.P. Linder in Bothalia,* 15(3–4): 403 (1985): Restio cincinnatus.

ISCHYROLEPIS :–
coactilis *(Mast.) H.P. Linder in Bothalia,* 15(3–4): 403 (1985): Restio coactilis.
curvibracteata *E.E. Esterhuysen in Bothalia,* 15(3–4): 403 (1985)—South Africa (Cape Province).
curviramis *(Kunth) H.P. Linder in Bothalia,* 15(3–4): 404 (1985): Restio curviramis.
distracta *(Mast.) H.P. Linder in Bothalia,* 15(3–4): 404 (1985): Restio distractus.
duthieae *(Pillans) H.P. Linder in Bothalia,* 15(3–4): 404 (1985): Restio duthieae.
eleocharis *(Mast.) H.P. Linder in Bothalia,* 15(3–4): 404 (1985): Restio eleocharis.
esterhuyseniae *(Pillans) H.P. Linder in Bothalia,* 15(3–4): 404 (1985): Restio esterhuyseniae.
feminea *E.E. Esterhuysen in Bothalia,* 15(3–4): 404 (1985)—South Africa (Cape Province).
fraterna *(Kunth) H.P. Linder in Bothalia,* 15(3–4): 405 (1985): Restio fraternus.
fuscidula *(Pillans) H.P. Linder in Bothalia,* 15(3–4): 405 (1985): Restio fuscidulus.
gaudichaudiana *(Kunth) H.P. Linder in Bothalia,* 15(3–4): 405 (1985): Restio gaudichaudianus.
gossypina *(Mast.) H.P. Linder in Bothalia,* 15(3–4): 405 (1985): Restio gossypinus.
helenae *(Mast.) H.P. Linder in Bothalia,* 15(3–4): 406 (1985): Restio helenae.
hystrix *(Mast.) H.P. Linder in Bothalia,* 15(3–4): 406 (1985): Restio hystrix.
karooica *E.E. Esterhuysen in Bothalia,* 15(3–4): 406 (1985)—South Africa (Cape Province).
laniger *(Kunth) H.P. Linder in Bothalia,* 15(3–4): 406 (1985): Restio laniger.
leptoclados *(Mast.) H.P. Linder in Bothalia,* 15(3–4): 407 (1985): Restio leptoclados.
longiaristata *Pillans ex H.P. Linder in Bothalia,* 15(3–4): 407 (1985)—South Africa (Cape Province).
macer *(Kunth) H.P. Linder in Bothalia,* 15(3–4): 407 (1985): Restio macer.
marlothii *(Pillans) H.P. Linder in Bothalia,* 15(3–4): 407 (1985): Restio marlothii.
monanthos *(Mast.) H.P. Linder in Bothalia,* 15(3–4): 407 (1985): Restio monanthos.
nana *E.E. Esterhuysen in Bothalia,* 15(3–4): 409 (1985)—South Africa (Cape Province).
nubigena *E.E. Esterhuysen in Bothalia,* 15(3–4): 409 (1985)—South Africa (Cape Province).
ocreata *(Kunth) H.P. Linder in Bothalia,* 15(3–4): 410 (1985): Restio ocreatus.
paludosa *(Pillans) H.P. Linder in Bothalia,* 15(3–4): 410 (1985): Restio paludosus.
papillosa *E.E. Esterhuysen in Bothalia,* 15(3–4): 410 (1985)—South Africa (Cape Province).
pratensis *E.E. Esterhuysen in Bothalia,* 15(3–4): 411 (1985)—South Africa (Cape Province).
pygmaea *(Pillans) H.P. Linder in Bothalia,* 15(3–4): 413 (1985): Restio pygmaeus.

ISCHYROLEPIS :–
rivula *E.E. Esterhuysen in Bothalia,* 15(3–4): 413 (1985)—South Africa (Cape Province).
rottboellioides *(Kunth) H.P. Linder in Bothalia,* 15(3–4): 413 (1985): Restio rottboellioides.
sabulosa *(Pillans) H.P. Linder in Bothalia,* 15(3–4): 413 (1985): Restio sabulosus.
schoenoides *(Kunth) H.P. Linder in Bothalia,* 15(3–4): 413 (1985): Restio schoenoides.
setiger *(Kunth) H.P. Linder in Bothalia,* 15(3–4): 413 (1985): Restio setiger.
sieberi *(Kunth) H.P. Linder in Bothalia,* 15(3–4): 414 (1985): Restio sieberi.
sporadica *E.E. Esterhuysen in Bothalia,* 15(3–4): 414 (1985)—South Africa (Cape Province).
tenuissima *(Kunth) H.P. Linder in Bothalia,* 15(3–4): 415 (1985): Restio tenuissimus.
triflora *(Rottb.) H.P. Linder in Bothalia,* 15(3–4): 415 (1985): Restio triflorus.
unispicata *H.P. Linder in Bothalia,* 15(3–4): 415 (1985)—South Africa (Cape Province).
vilis *(Kunth) H.P. Linder in Bothalia,* 15(3–4): 417 (1985): Restio vilis.
virgea *(Mast.) H.P. Linder in Bothalia,* 15(3–4): 417 (1985): Restio virgeus.
wallichii *(Mast.) H.P. Linder in Bothalia,* 15(3–4): 417 (1985): Restio wallichii.
wittebergensis *E.E. Esterhuysen in Bothalia,* 15(3–4): 417 (1985)—South Africa (Cape Province).

ISOCOMA (Compos.).
acradenia
subsp. eremophila *(Greene) R.M. Beauchamp in Phytologia,* 59(7): 437 (1986): I. eremophila.
veneta
var. furfuracea *(Greene) R.M. Beauchamp in Phytologia,* 59(7): 437 (1986): Biglovia furfuracea.
var. oxyphylla *(Greene) R.M. Beauchamp in Phytologia,* 59(7): 438 (1986): I. oxyphylla.

ISODON (Labiat.).
lophanthoides *(Ham. ex D. Don) B.N. Mehrotra & B.S. Aswal in J. Econ. Taxon. Bot.,* 6(3): 612 (1985), without basionym ref.: Hyssopus lophanthoides.

ISOGLOSSA (Acanthac.).
eranthemoides *(F. Muell.) R.M. Barker in J. Adelaide Bot. Gard.,* 9: 229 (1986): Justicia eranthemoides.

ISOLEPIS (Cyperac.).
angelica *B.L. Burtt in Notes Roy. Bot. Gard. Edinburgh,* 43(3): 362 (1986)—Lesotho, South Africa (Cape Province).
costata
var. macra *(Boeck.) B.L. Burtt in Notes Roy. Bot. Gard. Edinburgh,* 43(3): 363 (1986): Scirpus macer.

ISOLEPIS :–
pellocolea *B.L. Burtt in Notes Roy. Bot. Gard. Edinburgh,* 43(3): 363 (1986)—South Africa (Natal).

ISOMETRUM (Gesneriac.).
sect. Pachysiphon *K.Y. Pan in Acta Bot. Yunnanica,* 8(1): 26 (1986).
crenatum *K.Y. Pan in Acta Bot. Yunnanica,* 8(1): 27 (1986)—China.
lancifolium *(Franch.) K.Y. Pan in Acta Bot. Yunnanica,* 8(1): 30 (1986): Didissandra lancifolium.
var. mucronatum *K.Y. Pan in Acta Bot. Yunnanica,* 8(1): 30 (1986)—China.
var. tsingchengshanicum *W.T. Wang & K.Y. Pan in Acta Bot. Yunnanica,* 8(1): 30 (1986)—China.
pinnatilobatum *K.Y. Pan in Acta Bot. Yunnanica,* 8(1): 34 (1986)—China.
sichuanicum *K.Y. Pan in Acta Bot. Yunnanica,* 8(1): 33 (1986)—China.
villosum *K.Y. Pan in Acta Bot. Yunnanica,* 8(1): 31 (1986)—China.

ITEADAPHNE (Laurac.).
caudata *(Nees) H.W. Li in Acta Bot. Yunnanica,* 7(2): 132 (1985): Daphnidium caudatum.

ITYSA P. Ravenna in Nordic J. Bot., 6(5): 582 (1986). *IRIDACEAE.*
gardneri *(Bak.) P. Ravenna in Nordic J. Bot.,* 6(5): 582 (1986): Calydorea gardneri.
venezolensis *P. Ravenna in Nordic J. Bot.,* 6(5): 584 (1986)—Venezuela.

IXOCA (Caryophyllac.).
intonsa *(Greuter & Melzheimer) J. Holub in Preslia,* 58(4): 302 (1986): Silene intonsa.
pusilla
subsp. albanica *(K. Malý) J. Holub in Preslia,* 58(4): 302 (1986): Heliosperma albanica.
widderi *(Kofol-Seliger & T. Wraber) J. Holub in Preslia,* 58(4): 302 (1986): Silene veselskyi subsp. widderi.

IXORA (Rubiac.).
sect. Microthamnus *A.M. Homolle ex M. Guédès in Phytologia,* 60(4): 246 (1986).
beddomei *T. Husain & S.R. Paul in Candollea,* 41(1): 87 (1986)—India.
bemangidiensis *M. Guédès in Phytologia,* 60(4): 246 (1986)—Madagascar.
foliicalyx *M. Guédès in Phytologia,* 60(4): 250 (1986)—Madagascar.
littoralis *(A.M. Hom. ex J. Arènes) M. Guédès in Phytologia,* 60(4): 248 (1986): Thouarsiora littoralis.
mercaraica *T. Husain & S.R. Paul in Candollea,* 41(1): 88 (1986)—India.
oournuensis *J. Florence in Bull. Mus. Nation. Hist. Nat., B, Adansonia, Ser. 4,* 8(1): 10 (1986)—Marquesas Is.
reducta *Drake ex M. Guédès in Phytologia,* 60(4): 245 (1986)—Madagascar.

IXORA :–
sambiranensis *A.M. Homolle ex M. Guédès in Phytologia,* 60(4): 245 (1986)—Madagascar.
trimera *M. Guédès in Phytologia,* 60(4): 250 (1986)—Madagascar.

J

JALCOPHILA M.O. Dillon & A. Sagástegui Alva in Brittonia, 38(2): 162 (1986). *COMPOSITAE.*
ecuadorensis *M.O. Dillon & A. Sagástegui Alva in Brittonia,* 38(2): 165 (1986)—Ecuador.
peruviana *M.O. Dillon & A. Sagástegui Alva in Brittonia,* 38(2): 163 (1986)—Peru.

JALTOMATA (Solanac.).
glandulosa *(Miers) R. Castillo & R.E. Schultes in Rhodora,* 88(854): 292 (1986): Saracha glandulosa.
vestita *(Miers) R. Castillo & R.E. Schultes in Rhodora,* 88(854): 292 (1986): Saracha vestita.

JASARUM G.S. Bunting in Acta Bot. Venezuel., 10(1–4): 264 (1975). *ARACEAE.*

JASMINUM (Oleac.).
cordifolium
subsp. andamanicum *S.K. Srivastava & S.L. Kapoor in Bull. Bot. Surv. India,* 25(1–4): 219 (1983 publ. 1985)—Andaman Is.
fluminense
subsp. gratissimum *(Deflers) P.S. Green in Kew Bull.,* 41(2): 417 (1986): J. gratissimum.
grandiflorum
subsp. floribundum *(R. Br. ex Fresen.) P.S. Green in Kew Bull.,* 41(2): 414 (1986): J. floribundum.
laurifolium *Roxb. ex Hornem., Hort. Hafn., Suppl.:* 112 (1819); *cf. D.J. Mabberley in Taxon,* 33(3): 437 (1984)— Locality not stated.
yuanjiangense *P.Y. Bai in Acta Bot. Yunnanica,* 7(4): 421 (1985)—China.

JATROPHA (Euphorbiac.).
tuberosa *W. Elliot, Fl. Andhirica:* 85 (1859); *cf. D.J. Mabberley in Taxon,* 32(1): 87 (1983)— Locality not stated.

JEDDA J.R. Clarkson in Austrobaileya, 2(3): 203 (1986). *THYMELAEACEAE.*
multicaulis *J.R. Clarkson in Austrobaileya,* 2(3): 203 (1986)—Australia (Queensland).

JEJOSEPHIA A. Nageswara Rao & K.J. Mani in J. Econ. Taxon. Bot., 7(1): 217 (1985). *ORCHIDACEAE.*

JEJOSEPHIA :–
pusilla *(Joseph & H. Deka) A. Nageswara Rao & K.J. Mani in J. Econ. Taxon. Bot.,* 7(1): 217 (1985): Trias pusilla.

JUNCUS (Juncac.).
acutus
subsp. littoralis *(C.A. Mey.) N. Feinbrun-Dothan, Fl. Palaestina,* 4: 143 (1986): J. littoralis.
mollifolius *O.M. Hilliard & B.L. Burtt in Notes Roy. Bot. Gard. Edinburgh,* 43(3): 367 (1986)—South Africa (Natal), Lesotho.
orchonicus *V.S. Novikov in Byull. Mosk. Obshch. Ispўt. Prir., Biol.,* 90(5): 110 (1985)—Mongolia, U.S.S.R. (Siberia).
yoskisukei *A.K. Goel in J. Econ. Taxon. Bot.,* 7(1): 208 (1985), nom. nov.: J. albescens *Satake.*

JUNGIA (Labiat.).
splendens *(Ker-Gawl.) J. Soják in Čas. Nár. Muz. (Prague),* 152(1): 21 (1983): Salvia splendens.

JUSTICIA (Acanthac.).
cuneata
subsp. hoerleiniana *(P.G. Mey.) K. Immelman in Bothalia,* 16(1): 40 (1986): J. hoerleiniana.
subsp. latifolia *(Nees) K. Immelman in Bothalia,* 16(1): 40 (1986): Gendarussa orchioides var. latifolia.
japonica *C.P. Thunberg ex A. Murray in Linnaeus, Syst. Veg., ed. 14:* 63 (1784)—Japan. †
orchioides
subsp. glabrata *K. Immelman in Bothalia,* 16(1): 39 (1986)—South Africa (Cape Province).
parvibracteata *K. Immelman in Bothalia,* 16(1): 39 (1986)—South Africa (Cape Province).
petiolaris
subsp. bowiei *(C.B. Cl.) K. Immelman in Bothalia,* 16(1): 40 (1986): J. bowiei.
subsp. incerta *(C.B. Cl.) K. Immelman in Bothalia,* 16(1): 40 (1986): J. incerta.
protracta
subsp. rhodesiana *(S. Moore) K. Immelman in Bothalia,* 16(1): 40 (1986): J. rhodesiana.
ramosissima *Roxb. ex Hornem., Hort. Hafn., Suppl.:* 112 (1819); *cf. D.J. Mabberley in Taxon,* 33(3): 437 (1984)— Locality not stated.
scabra *Vahl, Enum.,* 1: 120 (1804); *cf. D.J. Mabberley in Taxon,* 32(1): 81 (1983)— Locality not stated.
sulphuriflora *M. Hedrén in Nordic J. Bot.,* 6(3): 303 (1986)—Tanzania.
ukagurensis *M. Hedrén in Nordic J. Bot.,* 6(3): 301 (1986)—Tanzania.
zarucchii *D.C. Wasshausen in Acta Amazonica,* 14(1–2 Suppl.): 145 (1984 publ. 1986)—Brazil.

K

KAEMPFERIA (Zingiberac.).
latifolia *Donn ex Hornem., Hort. Hafn.:* 6 (1813); *cf. D.J. Mabberley in Taxon,* 33(3): 437 (1984)— Locality not stated.

KALOPANAX (Araliac.).
septemlobus
var. pilosus *Q.Z. Yu in Acta Bot. Bor.-Occid. Sin.,* 5(3): 234 (1985), without type—China.

KAMETTIA (Apocynac.).
caryophyllata *(Roxb.) D.H. Nicolson & C.R. Suresh in Taxon,* 35(2): 354 (1986): Echites caryophyllata.

KARNATAKA P.K. Mukherjee & L. Constance in Brittonia, 38(2): 145 (1986). *UMBELLIFERAE.*
benthamii *(C.B. Clarke) P.K. Mukherjee & L. Constance in Brittonia,* 38(2): 145 (1986): Schultzia ? benthamii.

KEDARNATHA P.K. Mukherjee & L. Constance in Brittonia, 38(2): 147 (1986). *UMBELLIFERAE.*
sanctuarii *P.K. Mukherjee & L. Constance in Brittonia,* 38(2): 147 (1986)—India.

KEETIA (Rubiac.).
acuminata *(De Wild.) D.M. Bridson in Kew Bull.,* 41(4): 985 (1986): Plectronia acuminata.
angustifolia *D.M. Bridson in Kew Bull.,* 41(4): 971 (1986)—Rwanda, Burundi, Uganda.
carmichaelii *D.M. Bridson in Kew Bull.,* 41(4): 984 (1986)—Tanzania.
cornelia *(Cham. & Schlecht.) D.M. Bridson in Kew Bull.,* 41(4): 985 (1986): Canthium cornelia.
ferruginea *D.M. Bridson in Kew Bull.,* 41(4): 977 (1986)—Tanzania.
foetida *(Hiern) D.M. Bridson in Kew Bull.,* 41(4): 981 (1986): Canthium foetidum.
gracilis *(Hiern) D.M. Bridson in Kew Bull.,* 41(4): 986 (1986): Canthium gracile.
gueinzii *(Sond.) D.M. Bridson in Kew Bull.,* 41(4): 970 (1986): Canthium gueinzii.
hispida *(Benth.) D.M. Bridson in Kew Bull.,* 41(4): 986 (1986): Canthium hispidum.
inaequilatera *(Hutch. & Dalz.) D.M. Bridson in Kew Bull.,* 41(4): 987 (1986): Canthium inaequilaterum.
koritschoneri *D.M. Bridson in Kew Bull.,* 41(4): 984 (1986)—Tanzania.
leucantha *(K. Krause) D.M. Bridson in Kew Bull.,* 41(4): 987 (1986): Plectronia leucantha.
mannii *(Hiern) D.M. Bridson in Kew Bull.,* 41(4): 988 (1986): Canthium mannii.
molundensis *(K. Krause) D.M. Bridson in Kew Bull.,* 41(4): 976 (1986): Plectronia molundensis.

KEETIA :–
var. macrostipulata *(De Wild.) D.M. Bridson in Kew Bull.,* 41(4): 976 (1986): Plectronia macrostipulata.
multiflora *(Schum. & Thonn.) D.M. Bridson in Kew Bull.,* 41(4): 988 (1986): Psychotria multiflora.
procteri *D.M. Bridson in Kew Bull.,* 41(4): 981 (1986)—Tanzania.
purpurascens *(Bullock) D.M. Bridson in Kew Bull.,* 41(4): 981 (1986): Canthium purpurascens.
purseglovei *D.M. Bridson in Kew Bull.,* 41(4): 972 (1986)—Uganda, Cameroun, Zaïre.
ripae *(De Wild.) D.M. Bridson in Kew Bull.,* 41(4): 989 (1986): Plectronia ripae.
rubens *(Hiern) D.M. Bridson in Kew Bull.,* 41(4): 989 (1986): Canthium rubens.
rufivillosa *(Robyns ex Hutch. & Dalz.) D.M. Bridson in Kew Bull.,* 41(4): 990 (1986): Canthium rufivillosum.
rwandensis *D.M. Bridson in Kew Bull.,* 41(4): 990 (1986)—Rwanda.
tenuiflora *(Hiern) D.M. Bridson in Kew Bull.,* 41(4): 982 (1986): Canthium tenuiflorum.
venosa *(Oliv.) D.M. Bridson in Kew Bull.,* 41(4): 974 (1986): Plectronia venosa.
venosissima *(Hutch. & Dalz.) D.M. Bridson in Kew Bull.,* 41(4): 991 (1986): Canthium venosissima.
zanzibarica *(Klotzsch) D.M. Bridson in Kew Bull.,* 41(4): 979 (1986): Canthium zanzibaricum.
subsp. cornelioides *(De Wild.) D.M. Bridson in Kew Bull.,* 41(4): 979 (1986): Plectronia cornelioides.
subsp. gentilii *(De Wild.) D.M. Bridson in Kew Bull.,* 41(4): 980 (1986): Plectronia gentilii.

KELLERIA (Thymelaeac.).
lyallii *(Hook.f.) S. Berggren in Minneskr. Fisiogr. Sällsk. Lund,* 8: 20 (1878): Drapetes lyallii.

KERNERA (Crucif.).
auriculata *(Lam.) Sweet, Hort. Brit.:* 467 (1827); *cf. D.J. Mabberley in Taxon,* 31(1): 71 (1982): Basionym not stated.
saxatilis *(L.) Sweet, Hort. Brit.,* 467 (1827); *cf. D.J. Mabberley in Taxon,* 31(1): 71 (1982): Basionym not stated.

KHAYA A. Jussieu in Mirb. & Cass. apud Guill. in Bull. Sci. Nat. Geol., 23: 238 (1830); cf. D.J. Mabberley in Taxon, 31(1): 66 (1982). *MELIACEAE.*

KICKXIA (Scrophulariac.).
elatine
subsp. bombycina *(Boiss. & Bl.) J. Chrtek in Cas. Nár. Muz. (Prague),* 153(2): 99 (1984): Linaria bombycina.
subsp. caucasica *(Mussin) J. Chrtek in Čas. Nár. Muz. (Prague),* 153(2): 99 (1984): Linaria caucasica.

KIELMEYERA (Guttif.).
albopunctata *N. Saddi in Rodriguésia,* 36(60): 61 (1984)—Brazil.
decipiens *N. Saddi in Rodriguésia,* 36(60): 60 (1984)—Brazil.
divergens *N. Saddi in Rodriguésia,* 36(60): 61 (1984)—Brazil.
insignis *N. Saddi in Rodriguésia,* 36(60): 60 (1984)—Brazil.
occhioniana *N. Saddi in Rodriguésia,* 36(60): 62 (1984)—Brazil.
paranaensis *N. Saddi in Rodriguésia,* 36(60): 43 (1984)—Brazil.
rizziniana *N. Saddi in Rodriguésia,* 36(60): 60 (1984)—Brazil.
rufotomentosa *N. Saddi in Rodriguésia,* 36(60): 62 (1984)—Brazil.
sigillata *N. Saddi in Rodriguésia,* 36(60): 63 (1984)—Brazil.

KITAGAWIA M.G. Pimenov in Bot. Zhurn., 71(7): 943 (1986). *UMBELLIFERAE.*
baicalensis *(Redow. ex Willd.) M.G. Pimenov in Bot. Zhurn.,* 71(7): 944 (1986): Selinum baicalense.
eryngiifolia *(Kom.) M.G. Pimenov in Bot. Zhurn.,* 71(7): 948 (1986): Peucedanum eryngiifolium.
komarovii *M.G. Pimenov in Bot. Zhurn.,* 71(7): 948 (1986), nom. nov.: Peucedanum elegans *Kom.*
litoralis *(Worosch. & Gorovoi) M.G. Pimenov in Bot. Zhurn.,* 71(7): 948 (1986): Peucedanum litorale.
terebinthacea *(Fisch. ex Spreng.) M.G. Pimenov in Bot. Zhurn.,* 71(7): 944 (1986): Selinum terebinthaceum.
subsp. trichootheca *M.G. Pimenov in Bot. Zhurn.,* 71(7): 947 (1986)—U.S.S.R. (Soviet Far East), Mongolia, China, Sakhalin Is., Kuril Is., Japan.

KLEINIA (Compos.).
subgen. Notonia *(DC.) C. Jeffrey in Kew Bull.,* 41(4): 926 (1986): gen. Notonia.
abyssinica
var. hildebrandtii *(Vatke) C. Jeffrey in Kew Bull.,* 41(4): 928 (1986): Notonia hildebrandtii.
breviflora *C. Jeffrey in Kew Bull.,* 41(4): 925 (1986)—Kenya, Tanzania.
carnosa *(Ait.) Endl., Cat. Horti Vindob.,* 1: 378 (1842); *cf. D.J. Mabberley in Taxon,* 33(3): 435 (1984): Basionym not stated.
dolichocoma *C. Jeffrey in Kew Bull.,* 41(4): 925 (1986)—Kenya, Somali Republic.
gregorii *(S. Moore) C. Jeffrey in Kew Bull.,* 41(4): 927 (1986): Notonia gregorii.
implexa *(Bally) C. Jeffrey in Kew Bull.,* 41(4): 926 (1986): Senecio implexus.
leptophylla *C. Jeffrey in Kew Bull.,* 41(4): 927 (1986)—Kenya, Ethiopia.
mweroensis *(Bak.) C. Jeffrey in Kew Bull.,* 41(4): 927 (1986): Senecio mweroensis.

KLEINIA :–
oligondonta *C. Jeffrey in Kew Bull.,* 41(4): 926 (1986)—Tanzania, Kenya.
patriciae *C. Jeffrey in Kew Bull.,* 41(4): 927 (1986)—Tanzania.
petraea *(R.E. Fr.) C. Jeffrey in Kew Bull.,* 41(4): 926 (1986): Notonia petraea.
picticaulis *(Bally) C. Jeffrey in Kew Bull.,* 41(4): 927 (1986): Senecio picticaulis.
scottioides *C. Jeffrey in Kew Bull.,* 41(4): 925 (1986)—Kenya, Uganda.
vermicularis *C. Jeffrey in Kew Bull.,* 41(4): 928 (1986)—Tanzania.

×**KNAPPARA** hort. in Orchid Rev., 94(1109): centre page pull-out p. 8 (1986). *ORCHIDACEAE.* (Ascocentrum × Rhynchostylis × Vanda × Vandopsis).

KNAUTIA (Dipsacac.).
hybrida *(All.) Hort. ex Lag., Elenchus:* 9 (1816); *cf. D.J. Mabberley in Taxon,* 31(1): 71 (1982): Basionym not stated.

KNEIFFIA (Onagrac.).
gracilis *(Fisch. & Mey.) Bartl., Ind. Sem. Hort. Acad. Gott.,* 1837: 4 (1837); *cf. D.J. Mabberley in Taxon,* 32(1): 85 (1983): Basionym not stated.
serotina *(Lehm.) Bartl., Ind. Sem. Hort. Acad. Gott.,* 1837: 4 (1837); *cf. D.J. Mabberley in Taxon,* 32(1): 85 (1983): Oenothera serotina.

KNIPHOFIA (Asphodelac.).
media *(Ker) Endl., Cat. Horti Vindob.,* 1: 110 (1842); *cf. D.J. Mabberley in Taxon,* 33(3): 435 (1984): Basionym not stated.

KOBRESIA (Cyperac.).
helanshanica *W.Z. Di & M.J. Zhong in Acta Bot. Bor.-Occid. Sin.,* 5(4): 311 (1985)—China.

KOELERIA (Gramin.).
calarashica *M.G. Kalenichenko in Bot. Zhurn.,* 71(1): 91 (1986)—U.S.S.R. (Moldavia).
subspicata *(L.) Schrad. ex Mart., Hort. Acad. Erlangen:* 40 (1814); *cf. D.J. Mabberley in Taxon,* 33(3): 442 (1984): Basionym not stated.
taurica *M.G. Kalenichenko in Bot. Zhurn.,* 71(1): 91 (1986)—U.S.S.R. (Crimea).

KONIGA (Crucif.).
nummularia *(Ehrbg.) Bunge, Del. Sem. Hort. Bot. Dorpat.,* 1837; *Linnaea 12, Lit.:* 72 (?1839); *cf. D.J. Mabberley in Taxon,* 32(1): 85 (1983): Basionym not stated.

KOPSIA (Apocynac.).
deverrei *L. Allorge in Phytologia,* 59(2): 93 (1986)—Malaya.

KORTHALSELLA (Viscac.).
japonica
var. fasciculata *(Van Tiegh.) H.S. Kiu in Fl. Yunnanica,* 3: 374 (1983): Bifaria fasciculata.

KUBITZKIA H. van der Werff in Taxon, 35(1): 165 (1986), nom. nov.: gen. Systemonodaphne Mez. (Laurac.).
macrantha *(Kostermans) H. van der Werff in Taxon,* 35(1): 165 (1986): Systemonodaphne macrantha.
mezii *(Kostermans) H. van der Werff in Taxon,* 35(1): 165 (1986): Systemonodaphne mezii.

KUHLHASSELTIA (Orchidac.).
halconensis *J.J. Wood in Kew Bull.,* 41(4): 814 (1986), nom. nov.: Herpysma merrillii *Ames.*

KUHNIA (Compos.).
maximilianii *Sinning, Del. Sem. Hort. Bot. Bonnen.,* 1838: 2, 3 (1938); *cf. D.J. Mabberley in Taxon,* 32(1): 85 (1983)— Locality not stated.

KUNDMANNIA (Umbellif.).
pastinacifolia *(Bertol.) Sweet, Hort. Brit., ed. 2:* 247 (1830); *cf. D.J. Mabberley in Taxon,* 31(1): 71 (1982): Basionym not stated.

L

LACTUCA (Compos.).
inermis
var. myriocephala *(Dethier) A. Lawalrée in Fl. Afr. Centr., Spermatophytes,* Compos. 66 (1986): L. capensis var. myriocephala.

LAELIA (Orchidac.).
hookeri *Schouw, Forteg. Kjöb. Bot. Hav. Pl.:* 25 (1847); *cf. D.J. Mabberley in Taxon,* 33(3): 439 (1984)— Locality not stated.

LAGENANDRA (Arac.).
dewitii *W. Crusio & A. de Graaf in Aqua Planta,* 1986(2): 57 (1986)—Sri Lanka.

LAGENARIA (Cucurbitac.).
natalensis *Huber, Cat. Graines 1867:* 9 (1867); *cf. D.J. Mabberley in Taxon,* 34(3): 452 (1985)— Locality not stated.

LAGUNAEA (Malvac.).
lobata *(Murr.) Schreber ex Spreng., Bot. Gart. Univ. Halle:* 52 (1800); *cf. D.J. Mabberley in Taxon,* 31(1): 71 (1982): Basionym not stated.

LAMIUM (Labiat.).
×conradiae *G. Bosc in Candollea,* 41(1): 60 (1986), without type—Corsica. (L. corsicum × garganicum subsp. laevigatum).

LAMPRANTHUS (Aizoac.).
deltoides *(L.) Glen ex D.O. Wijnands, Botany Commelins:* 148 (1983): Mesembryanthemum deltoides.
deltoides *(L.) H.F. Glen in Bothalia,* 16(1): 55 (1986): Mesembryanthemum deltoides.
glaucus
subsp. aureus *(L.) Glen ex D.O. Wijnands, Botany Commelins:* 149 (1983), without basionym ref.: Basionym not stated.

LAMPROCONUS (Bromeliac.).
maidifolia *(C. Morr.) Jacob-Makoy, Prix Cour. Pl. Nouv.,* 120: 23 (1879); *cf. D.J. Mabberley in Taxon,* 34(3): 456 (1985): Basionym not stated.

LANGLOISIA (Polemoniac.).
setosissima
subsp. punctata *(Coville) S. Timbrook in Madroño,* 33(3): 169 (1986): Gilia setosissima var. punctata.

LANTANA (Verbenac.).
ruiz-teranii *S. López-Palacios & Steyermark in Ernstia,* 36: 10 (1986)—Venezuela.
urticiformis *Huber, Cat. Gén. 1871 & 1872:* 4 (1871); *cf. D.J. Mabberley in Taxon,* 34(3): 451 (1985)— Brazil.

LAPORTEA (Urticac.).
peduncularis
subsp. latidens *I. Friis in Bol. Soc. Brot.,* 58: 206 (1985)—Mozambique, South Africa (Natal).

LAPPULA (Boraginac.).
semnanensis *H. Riedl & M. Iranshahr in Linz. Biol. Beitr.,* 17(2): 327 (1985)—Iran.

LARIX (Pinac.).
communis *Lindl. in Loud., Encycl. Pl.:* 806 (1829); *cf. D.J. Mabberley in Taxon,* 31(1): 71 (1982)— Locality not stated.

LASERPITIUM (Umbellif.).
siler
subsp. laeve *(Halácsy) P. Hartvig in A. Strid (ed.), Mount. Fl. Greece,* 1: 732 (1986): L. garganicum var. laeve.

LASIADENIA (Thymelaeac.).
ottohuberi *T. Plowman & L.I. Nevling in Brittonia,* 38(2): 114 (1986)—Venezuela.

LASIOCARYUM (Boraginac.).
munroi
var. diffusum *(Brand.) B.N. Mehrotra & B.S. Aswal in J. Econ. Taxon. Bot.,* 6(3): 608 (1985), without basionym ref.: Microcaryum diffusum.

LATHYRUS (Legumin.).
anhuiensis *Y.J. Zhu & R.X. Meng in Acta Phytotax. Sin.*, 24(5): 402 (1986)—China.
davidii
var. roseus *C.W. Chang in Fl. Tsinlingensis*, 1(3): 450, 102 (1981)—China.
glandulosus *S.L. Broich in Madroño*, 33(2): 136 (1986)—U.S.A. (California).
italicus *Juss. ex Spreng., Bot. Gart. Univ. Halle:* 53 (1800); *cf. D.J. Mabberley in Taxon,* 32(1): 85 (1983)— Locality not stated.
japonicus
subsp. pubescens *A.A. Korobkov in Arktich. Fl. SSSR,* 9(2): 173 (1986)—U.S.S.R. (European Russia, Soviet Far East), Canada (Labrador, N.W. Terr., Ontario), Alaska, Newfoundland, Greenland, Aleutian Is., etc.
pratensis
subsp. lusseri *(Heer ex Koch) J. Soják in Čas. Nár. Muz. (Prague),* 152(1): 21 (1983): L. lusseri.
subsp. velutinus *(DC.) J. Soják in Čas. Nár. Muz. (Prague),* 152(1): 21 (1983): L. pratensis β var. velutinus.
tataricus *Schreb. ex Spreng., Bot. Gart. Univ. Halle:* 54 (1800); *cf. D.J. Mabberley in Taxon,* 32(1): 85 (1983)— Locality not stated.

LAUROCERASUS (Rosac.).
zippeliana
var. crassistyla *(Card.) T.T. Yü & L.T. Lu in Fl. Reipubl. Popul. Sin.*, 38: 118 (1986): Prunus macrophylla var. crassistyla.

LAURUS (Laurac.).
glauca *C.P. Thunberg ex A. Murray in Linnaeus, Syst. Veg., ed. 14:* 383 (1784)—Japan. †
lucida *C.P. Thunberg ex A. Murray in Linnaeus, Syst. Veg., ed. 14:* 384 (1784)—Japan. †
pedunculata *C.P. Thunberg ex A. Murray in Linnaeus, Syst. Veg., ed. 14:* 383 (1784)—Japan. †
umbellata *C.P. Thunberg ex A. Murray in Linnaeus, Syst. Veg., ed. 14:* 384 (1784)—Japan. †

LAVANDULA (Labiat.).
sect. Dentata *M. Suárez-Cervera & J.A. Seoane-Camba in An. Jard. Bot. Madrid,* 42(2): 402 (1985 publ. 1986).

LAVAUXIA (Onagrac.).
triloba *(Nutt.) Bartl., Ind. Sem. Hort. Acad. Gott.*, 1837: 4 (1837); *cf. D.J. Mabberley in Taxon,* 32(1): 85 (1983): Oenothera triloba.

LAVRANIA D.C.H. Plowes in Cact. Succ. J. (U.S.A.), 58(3): 122 (1986). *ASCLEPIADACEAE.*
haagnerae *D.C.H. Plowes in Cact. Succ. J. (U.S.A.),* 58(3): 123 (1986)—Namibia.

LAXMANNIA (Anthericac.).
compacta *J.G. Conran & P.I. Forster in Austrobaileya,* 2(3): 248 (1986)—Australia (Queensland, New South Wales).

LEMURELLA (Orchidac.).
sarcanthoides *(Schltr.) K. Senghas in Schlechter Orchideen,* 1(16–18): 1018 (1986): Oeoniella sarcanthoides.

LENS (Legumin.).
odemensis *G. Ladizinsky in Notes Roy. Bot. Gard. Edinburgh,* 43(3): 489 (1986)—Israel, Turkey, East Aegean Is.

LEOCHILUS (Orchidac.).
hagsateri *M.W. Chase in Syst. Bot.*, 11(1): 244 (1986)—Mexico.
macrantherus *(Hook.) Liebm. in Schouw, Forteg. Kjöb. Bot. Hav. Pl.:* 27 (1847); *cf. D.J. Mabberley in Taxon,* 33(3): 439 (1984): Basionym not stated.
puertoricensis *M.W. Chase in Syst. Bot.*, 11(1): 242 (1986)—Puerto Rico.

LEONTODON (Compos.).
asperrimus *(Willd.) Endl., Cat. Horti Vindob.*, 1: 410 (1842); *cf. D.J. Mabberley in Taxon,* 33(3): 435 (1984): Basionym not stated.

LEONURUS (Labiat.).
comosus *Koen. ex Rottb., Pl. Hort. Univ. Rar. Prog.:* 18 (1773); *cf. D.J. Mabberley in Taxon,* 33(3): 438 (1984)— Locality not stated.
malebaricus *Koen. ex Rottb., Pl. Hort. Univ. Rar. Prog.:* 20 (1773); *cf. D.J. Mabberley in Taxon,* 33(3): 438 (1984)— Locality not stated.

LEOPOLDIA (Hyacinthac.).
deserticola *(Rech. f.) N. Feinbrun-Dothan, Fl. Palaestina,* 4: 70 (1986): Muscari deserticolum.
longipes
subsp. negevensis *N. Feinbrun-Dothan & Danin in N. Feinbrun-Dothan, Fl. Palaestina,* 4: 397, 71 (1986)—Israel, Sinai Peninsula.

LEPANTHES (Orchidac.).
subgen. Brachycladium *C.A. Luer, Ic. Pleurothallid. I-Syst. Pleurothallidinae (Monogr. Syst. Bot. Missouri Bot. Gard. 15):* 31 (1986), nom. nov.: L. sect. Brachyclada *Rchb. f.*
sect. Caprimulginae *C.A. Luer, Ic. Pleurothallid. I-Syst. Pleurothallidinae (Monogr. Syst. Bot. Missouri Bot. Gard. 15):* 34 (1986).
subgen. Draconanthes *C.A. Luer, Ic. Pleurothallid. I-Syst. Pleurothallidinae (Monogr. Syst. Bot. Missouri Bot. Gard. 15):* 32 (1986).

LEPANTHES :–
sect. Felinae *C.A. Luer, Ic. Pleurothallid. I-Syst. Pleurothallidinae (Monogr. Syst. Bot. Missouri Bot. Gard. 15):* 34 (1986).
subgen. Marsipanthes *C.A. Luer, Ic. Pleurothallid. I-Syst. Pleurothallidinae (Monogr. Syst. Bot. Missouri Bot. Gard. 15):* 34 (1986).
sect. Marsipanthes *C.A. Luer, Ic. Pleurothallid. I-Syst. Pleurothallidinae (Monogr. Syst. Bot. Missouri Bot. Gard. 15):* 34 (1986).
ankistra *C.A. Luer & R.L. Dressler in Orquideologia,* 16(3): 12 (1986)—Panama.
antilocapra *C.A. Luer & R.L. Dressler in Orquideologia,* 16(3): 15 (1986)—Panama.
arachnion *C.A. Luer & R.L. Dressler in Orquideologia,* 16(3): 6 (1986)—Panama.
asoma *C.A. Luer & A.C. Hirtz in Orchidee,* 37(5): 213 (1986)—Ecuador.
biappendiculata *C.A. Luer in Phytologia,* 59(7): 444 (1986), as 'biappandiculata'—Venezuela.
calocodon *C.A. Luer in Phytologia,* 59(7): 445 (1986)—Venezuela.
erepsis *C.A. Luer & A.C. Hirtz in Orchidee,* 37(5): 215 (1986)—Ecuador.
glochidea *C.A. Luer in Phytologia,* 59(7): 446 (1986)—Venezuela.
navicularis *C.A. Luer in Phytologia,* 59(7): 447 (1986)—Venezuela.
odobenella *C.A. Luer & A.C. Hirtz in Orchidee,* 37(5): 217 (1986)—Ecuador.
ophiostele *C.A. Luer in Phytologia,* 59(7): 448 (1986)—Venezuela.
pantomima *C.A. Luer & R.L. Dressler in Orquideologia,* 16(3): 3 (1986)—Panama.
pectinata *C.A. Luer in Phytologia,* 59(7): 449 (1986)—Venezuela.
pecunialis *C.A. Luer & A.C. Hirtz in Orchidee,* 37(5): 219 (1986)—Ecuador.
scolex *C.A. Luer in Phytologia,* 59(7): 450 (1986)—Venezuela.
truncata *C.A. Luer & R.L. Dressler in Orquideologia,* 16(3): 9 (1986)—Panama.

LEPANTHOPSIS (Orchidac.).
pristis *C.A. Luer & R. Escobar R. in Orquideologia,* 16(3): 27 (1986)—Colombia.

LEPIDIUM (Crucif.).
aletes
var. integrifolium *(Thell.) O. Boelcke in Parodiana,* 4(1): 58 (1986): L. calycinum var. integrifolium.
gracile *(Chod. & Hassler) O. Boelcke in Parodiana,* 4(1): 36 (1986): L. bonariense var. gracile.
gussonii *Schrad., Ind. Sem. Hort. Acad. Gott.,* 1833: 4 (1833); *cf. D.J. Mabberley in Taxon,* 32(1): 85 (1983)— Locality not stated.
stuckertianum *(Thell.) O. Boelcke in Parodiana,* 4(1): 40 (1986): L. bonariense var. stuckertianum.

LEPIGONUM (Caryophyllac.).
mollugineum *(Lag.) Bartl., Ind. Sem. Hort. Acad. Gott.,* 1838: 4 (1838); *cf. D.J. Mabberley in Taxon,* 32(1): 85 (1983): Basionym not stated.
segetale *(L.) Bartl., Ind. Sem. Hort. Acad. Gott.,* 1839: 5 (1839); *cf. D.J. Mabberley in Taxon,* 32(1): 85 (1983): Basionym not stated.

LEPTOCORYPHIUM (Gramin.).
villaregalis *R. McVaugh & R. Guzmán in R. McVaugh, Fl. Novo-Galiciana,* 14: 218 (1983)—Mexico.

LEPTODERMIS (Rubiac.).
diffusa
var. umbellata *(Batal.) Z.Y. Zhang in Fl. Tsinlingensis,* 1(5): 10 (1985): L. umbellata.

LEPTOMISCHUS (Rubiac.).
parviflorus *H.S. Lo in Bull. Bot. Res. North-East. Forest. Inst.,* 6(4): 49 (1986)—China, Vietnam.

LEPTOPUS (Euphorbiac.).
chinensis
var. pubescens *(Hutch.) S.B. Ho in Fl. Tsinlingensis,* 1(3): 168 (1981): Andrachne capillipes var. pubescens.

LEPTOSIPHON (Polemoniac.).
multiflorus *Haage & Schmidt, Belgique horticole,* 22: 85 (1872); *Bull, Retail List,* 67: 11 (1872); *cf. D.J. Mabberley in Taxon,* 34(3): 454 (1985)— Locality not stated.

LESPEDEZA (Legumin.).
frutescens *Hornem., Hort. Hafn.:* 699 (1815); *cf. D.J. Mabberley in Taxon,* 33(3): 437 (1984)— Locality not stated.
thunbergii
f. nivea *(S. Akiyama & H. Ohba) H. Ohashi in J. Jap. Bot.,* 61(4): 124 (1986): L. patens f. nivea.
var. patens *(Nakai) H. Ohashi in J. Jap. Bot.,* 61(4): 124 (1986): L. patens.
f. patens *(Nakai) H. Ohashi in J. Jap. Bot.,* 61(4): 124 (1986): L. patens.

LESTIBUDESIA (Amaranthac.).
paniculata *R. Br. ex Sweet, Hort. Suburb. Lond.:* 48 (1818); *cf. D.J. Mabberley in Taxon,* 30(1): 16 (1981)— Locality not stated.

LETHIA P. Ravenna in Nordic J. Bot., 6(5): 585 (1986). *IRIDACEAE.*
umbellata *(Klatt) P. Ravenna in Nordic J. Bot.,* 6(5): 587 (1986): Herbertia umbellata.

LEUCANTHEMOPSIS (Compos.).
trifurcata *(Desf.) S.A. Alavi in Fl. Libya,* 107: 168 (1983): Chrysanthemum trifurcatum.

LEUCAS (Labiat.).
indica *(L.) Ait. f., Hort. Kew. ed. 2,* 3: 409 (1811); *cf. D.J. Mabberley in Taxon,* 33(3): 439 (1984): Leonurus indicus.
martinicensis *(Jacq.) R. Br. in Ait., Hort. Kew. ed. 2, 3:* 409 (1811); *cf. D.J. Mabberley in Taxon,* 30(1): 16 (1981): Basionym not stated.

LEUCOPOGON (Epacridac.).
cochlearifolius *A. Strid in Willdenowia,* 16(1): 171 (1986)—Australia (W. Australia).
conchifolius *A. Strid in Willdenowia,* 16(1): 169 (1986)—Australia (W. Australia).
infuscatus *A. Strid in Willdenowia,* 16(1): 173 (1986)—Australia (W. Australia).
lloydiorum *A. Strid in Willdenowia,* 16(1): 174 (1986)—Australia (W. Australia).

LIBANOTHAMNUS (Compos.).
occultus
var. salomonii *J. Cuatrecasas & López-Figueiras in Phytologia,* 61(1): 51 (1986)—Venezuela.

LIBANOTIS (Umbellif.).
laserpitifolia *(Palibin) K.T. Fu in Fl. Tsinlingensis,* 1(3): 411 (1981): Seseli laserpitifolium.
wannienchun *K.T. Fu in Fl. Tsinlingensis,* 1(3): 458, 409 (1981)—China.

LICANIA (Chrysobalanac.).
granvillei *G.T. Prance in Proc. Kon. Nederl. Acad. Wetensch., C,* 89(1): 114 (1986)—French Guiana, Brazil, Colombia.

LICARIA (Laurac.).
sarapiquensis *B.E. Hammel in J. Arnold Arbor.,* 67(1): 124 (1986)—Costa Rica.

×**LICHTARA** hort. in Orchid Rev., 94(1112): centre page pull-out p. 8 (1986). *ORCHIDACEAE.* (Doritis × Gastrochilus × Phalaenopsis).

LIGULARIA (Compos.).
sibirica
subsp. arctica *(Pojark.) V.G. Sergienko, Fl. poluostrova Kanin:* 99 (1986): L. arctica.

LIGUSTICUM (Umbellif.).
gyirongense *R.H. Shan & H.T. Chang in Acta Phytotax. Sin.,* 24(4): 315 (1986)—Xizang.
sinense
var. alpinum *Shan ex K.T. Fu in Fl. Tsinlingensis,* 1(3): 461, 419 (1981)—China.

LIGUSTRUM (Oleac.).
acutissimum
var. glabrum *Z.Y. Zhang in Fl. Tsinlingensis,* 1(4): 395, 91 (1983)—China.
latifolium *C.P. Thunberg ex A. Murray in Linnaeus, Syst. Veg., ed. 14:* 56 (1784)—Japan.

LILIUM (Liliac.).
sect. Asteridium *S.G. Haw, Lilies of China:* 159 (1986).
sect. Dimorphophyllum *S.G. Haw, Lilies of China:* 159 (1986).
bakerianum
subsp. sempervivoideum *(Lévl.) D.R. McKean in Notes Roy. Bot. Gard. Edinburgh,* 44(1): 185 (1986): Lilium sempervivoidium.
brevistylum *(S.Y. Liang) S.Y. Liang in Acta Bot. Yunnanica,* 8(1): 52 (1986): L. nanum var. brevistylum.
habaense *F.T. Wang & T. Tang in Acta Bot. Yunnanica,* 8(1): 51 (1986)—China.
japonicum *C.P. Thunberg ex A. Murray in Linnaeus, Syst. Veg., ed. 14:* 324 (1784)—Japan. †
jinfushanense *L.J. Peng & B.N. Wang in Acta Bot. Yunnanica,* 8(2): 225 (1986)—China.
ningnanense *J.M. Xu in Bull. Bot. Res. North-East. Forest. Inst.,* 6(2): 68 (1986)—China.
pinifolium *L.J. Peng in Acta Bot. Yunnanica,* 7(3): 317 (1985)—China.

LIMACIOPSIS (Menispermac.).
valida *(Diels) H.S. Lo in Fl. Yunnanica,* 3: 240 (1983): Pachygone valida.

LIMNANTHES (Limnanthac.).
gracilis
subsp. parishii *(Jepson) R.M. Beauchamp in Phytologia,* 59(7): 440 (1986): L. versicolor var. parishii.

LIMNOPHILA (Scrophulariac.).
brownii *B.S. Wannan in Taxon,* 35(3): 597 (1986)—Australia (Queensland).

LIMODORUM (Orchidac.).
abortivum
var. trabutianum *(Batt.) C. Raynaud, Orchid. Maroc.* 97 (1985): L. trabutianum.
striatum *C.P. Thunberg ex A. Murray in Linnaeus, Syst. Veg., ed. 14:* 816 (1784)—Japan. †

LIMONIUM (Plumbaginac.).
binervosum
subsp. anglicum *M.J. Ingrouille in Bot. J. Linn. Soc.,* 92(3): 188 (1986)—Great Britain.
var. aurigniense *M.J. Ingrouille in Bot. J. Linn. Soc.,* 92(3): 190 (1986)—Channel Is.
subsp. cantianum *M.J. Ingrouille in Bot. J. Linn. Soc.,* 92(3): 188 (1986)—Great Britain.
subsp. mutatum *M.J. Ingrouille in Bot. J. Linn. Soc.,* 92(3): 189 (1986)—Great Britain.
subsp. sarniense *M.J. Ingrouille in Bot. J. Linn. Soc.,* 92(3): 190 (1986)—Channel Is.

LIMONIUM :–
var. sarniense *M.J. Ingrouille in Bot. J. Linn. Soc.*, 92(3): 190 (1986)—Channel Is.
subsp. saxonicum *M.J. Ingrouille in Bot. J. Linn. Soc.*, 92(3): 189 (1986)—Great Britain.
var. sercquense *M.J. Ingrouille in Bot. J. Linn. Soc.*, 92(3): 191 (1986)—Channel Is.
britannicum *M.J. Ingrouille in Bot. J. Linn. Soc.*, 92(3): 199 (1986)—Great Britain.
subsp. celticum *M.J. Ingrouille in Bot. J. Linn. Soc.*, 92(3): 203 (1986)—Great Britain.
var. celticum *M.J. Ingrouille in Bot. J. Linn. Soc.*, 92(3): 203 (1986)—Great Britain.
subsp. coombense *M.J. Ingrouille in Bot. J. Linn. Soc.*, 92(3): 201 (1986)—Great Britain.
var. coombense *M.J. Ingrouille in Bot. J. Linn. Soc.*, 92(3): 202 (1986)—Great Britain.
var. grandicaule *M.J. Ingrouille in Bot. J. Linn. Soc.*, 92(3): 202 (1986)—Great Britain.
var. kelseyanum *M.J. Ingrouille in Bot. J. Linn. Soc.*, 92(3): 201 (1986)—Great Britain.
var. pharense *M.J. Ingrouille in Bot. J. Linn. Soc.*, 92(3): 203 (1986)—Great Britain.
subsp. transcanalis *M.J. Ingrouille in Bot. J. Linn. Soc.*, 92(3): 202 (1986)—Great Britain.
dodartiforme *M.J. Ingrouille in Bot. J. Linn. Soc.*, 92(3): 208 (1986)—Great Britain.
loganicum *M.J. Ingrouille in Bot. J. Linn. Soc.*, 92(3): 205 (1986), as 'longanicum'—Great Britain.
paradoxum
var. mutabile *M.J. Ingrouille in Bot. J. Linn. Soc.*, 92(3): 192 (1986)—Great Britain.
parvum *M.J. Ingrouille in Bot. J. Linn. Soc.*, 92(3): 204 (1986)—Great Britain.
procerum *(C.E. Salmon) M.J. Ingrouille in Bot. J. Linn. Soc.*, 92(3): 193 (1986): L. occidentale var. procerum.
subsp. cambrense *M.J. Ingrouille in Bot. J. Linn. Soc.*, 92(3): 198 (1986)—Great Britain.
var. cornubiense *M.J. Ingrouille in Bot. J. Linn. Soc.*, 92(3): 197 (1986)—Great Britain.
subsp. devoniense *M.J. Ingrouille in Bot. J. Linn. Soc.*, 92(3): 197 (1986)—Great Britain.
var. hibernicum *M.J. Ingrouille in Bot. J. Linn. Soc.*, 92(3): 195 (1986)—Ireland.
var. medium *M.J. Ingrouille in Bot. J. Linn. Soc.*, 92(3): 195 (1986)—Great Britain.
var. paramedium *M.J. Ingrouille in Bot. J. Linn. Soc.*, 92(3): 197 (1986)—Great Britain.
var. wessexense *M.J. Ingrouille in Bot. J. Linn. Soc.*, 92(3): 197 (1986)—Great Britain.

LIMONIUM :–
recurvum
var. donegalense *M.J. Ingrouille in Bot. J. Linn. Soc.*, 92(3): 213 (1985)—Ireland.
subsp. humile *(Girard) M.J. Ingrouille in Bot. J. Linn. Soc.*, 92(3): 213 (1986): Statice dodartii var. humile.
var. humile *(Girard) M.J. Ingrouille in Bot. J. Linn. Soc.*, 92(3): 213 (1986): Statice dodartii var. humile.
var. kerryense *M.J. Ingrouille in Bot. J. Linn. Soc.*, 92(3): 212 (1986)—Ireland.
subsp. portlandicum *M.J. Ingrouille in Bot. J. Linn. Soc.*, 92(3): 211 (1986)—Great Britain, Ireland.
var. portlandicum *M.J. Ingrouille in Bot. J. Linn. Soc.*, 92(3): 212 (1986)—Great Britain.
var. pseudoparadoxum *M.J. Ingrouille in Bot. J. Linn. Soc.*, 92(3): 214 (1986)—Ireland.
subsp. pseudotranswallianum *M.J. Ingrouille in Bot. J. Linn. Soc.*, 92(3): 212 (1986)—Ireland.
var. recurviforme *M.J. Ingrouille in Bot. J. Linn. Soc.*, 92(3): 212 (1986)—Great Britain.
todaroanum *F.M. Raimondo & S. Pignatti in Webbia*, 39(2): 417 (1986)—Sicily.

LIMOSELLA (Scrophulariac.).
inflata *O.M. Hilliard & B.L. Burtt in Notes Roy. Bot. Gard. Edinburgh*, 43(2): 213 (1986)—South Africa (Natal, Cape Province, Orange Free State), Lesotho, Botswana.
vesiculosa *O.M. Hilliard & B.L. Burtt in Notes Roy. Bot. Gard. Edinburgh*, 43(2): 215 (1986)—Lesotho, South Africa (Natal, Cape Province).

LINARIA (Scrophulariac.).
benitoi *J. Fernández Casas in Fontqueria*, 2: 29 (1982)—Spain.
dianthifolia *Henckel, Adumb. Pl. Hort. Hal.*: 11 (1806); *cf. D.J. Mabberley in Taxon*, 32(1): 85 (1983)— Locality not stated.
italica *Trev., Ind. Sem. Wratislav. App. II:* 2 (1820); *cf. D.J. Mabberley in Taxon*, 31(1): 71 (1982)— Locality not stated.
michauxii
var. parviflora *J. Léonard in Bull. Jard. Bot. Nation. Belg.*, 56(1–2): 224 (1986)—Iran.
orbensis *J.L. Carretero & H. Boira in An. Jard. Bot. Madrid*, 42(2): 411 (1985 publ. 1986)—Spain.

LINDERA (Laurac.).
thomsonii
var. velutina *(Forr.) L.C. Wang in Acta Bot. Bor.-Occid. Sin.*, 6(1): 64 (1986): L. stychnifolia var. velutina.

LINDERNIA (Scrophulariac.).
japonica *C.P. Thunberg ex A. Murray in Linnaeus, Syst. Veg., ed. 14:* 567 (1784)—Japan. †

LINDERNIA :-
taishanensis *F.Z. Li in Bull. Bot. Res. North-East. Forest. Inst.*, 6(1): 169 (1986)—China.

LINGNANIA (Gramin.).
chungii
var. barbellata *Q.H. Dai in Acta Phytotax. Sin.*, 24(5): 395 (1986)—China.
longianthera *G.A. Fu in J. Bamboo Res.*, 5(2): 41 (1986)—Hainan.

LINUM (Linac.).
capitatum
subsp. serrulatum *(Bertol.) P. Hartvig in A. Strid (ed.), Mount. Fl. Greece,* 1: 556 (1986): L. serrulatum.
flavum
subsp. albanicum *(Janchen) P. Hartvig in A. Strid (ed.), Mount. Fl. Greece,* 1: 556 (1986): L. albanicum.
olympicum
subsp. athoum *P. Hartvig in A. Strid (ed.), Mount. Fl. Greece,* 1: 563 (1986)—Greece.

LIPARIS (Orchidac.).
vestita
subsp. seidenfadenii *S. Misra in Nordic J. Bot.*, 6(1): 26 (1986)—India.

LIPOCARPHA (Cyperac.).
perspicua *S.S. Hooper in Kew Bull.*, 41(2): 424 (1986)—Angola.
reddyi *S.S. Hooper in Kew Bull.*, 41(2): 427 (1986)—India.

LIRIOPE (Convallariac.).
spicata
var. humilis *F.Z. Li in Bull. Bot. Res. North-East. Forest. Inst.*, 6(1): 170 (1986)—China.

LISTERA (Orchidac.).
nandadeviensis *P.K. Hajra in Bull. Bot. Surv. India*, 25(1–4): 181 (1983 publ. 1985)—India.

LITHODORA (Boraginac.).
lusitanica *(Samp.) J. Holub in Preslia,* 58(4): 302 (1986): Lithospermum lusitanicum.

LITHOSPERMUM (Boraginac.).
lanceolatum *(Pursh) Sweet, Hort. Brit., ed. 2:* 374 (1830); *cf. D.J. Mabberley in Taxon,* 30(1): 16 (1981): Basionym not stated.

LITSEA (Laurac.).
glutinosa
var. brachyphylla *(Hand.-Mazz.) L.C. Wang in Acta Bot. Bor.-Occid. Sin.*, 6(1): 64 (1986): L. sebifera var. brachyphylla.
involucrata *(Roxb.) Hereman, Paxt. Bot. Dict.:* 612 (1868); *cf. D.J. Mabberley in Taxon,* 32(1): 87 (1983): Basionym not stated.

LITTAEA (Agavac.).
juncea *E. Morr., Belg. Hort.*, 16: 343 (1866); *cf. D.J. Mabberley in Taxon,* 34(3): 456 (1985)— Locality not stated.

LLOYDIA (Liliac.).
yanyuanensis *S.Y. Liang in Acta Bot. Yunnanica,* 8(2): 227 (1986)—China.

LOBADIUM (Anacardiac.).
aromaticum *(Ait.) Sweet, Hort. Brit.:* 98 (1826); *cf. D.J. Mabberley in Taxon,* 30(1): 16 (1981): Rhus aromatica.

LOBELIA (Campanulac.).
blanda *(G. Don f.) Endl., Cat. Horti Vindob.*, 1: 437 (1842); *cf. D.J. Mabberley in Taxon,* 33(3): 435 (1984): Basionym not stated.

LOBIVIA (Cactac.).
shaferi
var. albiflora *(Rausch) E. Herzog in Kakt./Sukk.*, 19(1–2): 9 (1985); *cf. Repert. Pl. Succ. (I.O.S.),* 36: 8 (1985 publ. 1986): L. aurea var. albiflora.
subsp. aurea *(B. & R.) E. Herzog in Kakt./Sukk.*, 19(1–2): 6 (1985); *cf. Repert. Pl. Succ. (I.O.S.),* 36: 8 (1985 publ. 1986): Echinopsis aurea.
f. cylindracea *(Backeb.) E. Herzog in Kakt./Sukk.*, 19(1–2): 3 (1985); *cf. Repert. Pl. Succ. (I.O.S.),* 36: 7 (1985 publ. 1986): L. cylindracea.
subsp. fallax *(Oehme) E. Herzog in Kakt./Sukk.*, 19(1–2): 14 (1985); *cf. Repert. Pl. Succ. (I.O.S.),* 36: 8 (1985 publ. 1986): L. fallax.
var. flaviflora *(Spegazzini) E. Herzog in Kakt./Sukk.*, 19(1–2): 3 (1985), nom. inval. (art. 33.2); *cf. Repert. Pl. Succ. (I.O.S.),* 36: 8 (1985 publ. 1986): Cereus huascha var. flaviflora.
subsp. leucomalla *(Wessner) E. Herzog in Kakt./Sukk.*, 19(1–2): 10 (1985); *cf. Repert. Pl. Succ. (I.O.S.),* 36: 8 (1985 publ. 1986): L. leucomalla.
f. luteiflora *(Backeb.) E. Herzog in Kakt./Sukk.*, 19(1–2): 7 (1985), basionym not validly publ.; *cf. Repert. Pl. Succ. (I.O.S.),* 36: 7 (1985 publ. 1986): Pseudolobivia luteiflora.
var. quinesensis *(Rausch) E. Herzog in Kakt./Sukk.*, 19(1–2): 11 (1985); *cf. Repert. Pl. Succ. (I.O.S.),* 36: 8 (1985 publ. 1986): Echinopsis aurea var. quinesensis.
subsp. rubriflora *E. Herzog in Kakt./Sukk.*, 19(1–2): 13 (1985), nom. inval. (art. 33.2); *cf. Repert. Pl. Succ. (I.O.S.),* 36: 8 (1985 publ. 1986)— Locality not stated.

LOESELIASTRUM (Brand) S. Timbrook in Madroño, 33(3): 170 (1986): Langloisia sect. Loeseliastrum (Polemoniac.).
matthewsii *(A. Gray) S. Timbrook in Madroño,* 33(3): 171 (1986): Loeselia matthewsii.

LOESELIASTRUM :-
schottii *(Torr.) S. Timbrook in Madroño,* 33(3): 172 (1986): Navarretia schottii.

LOLIUM (Gramin.).
rigidum
subsp. negevense *N. Feinbrun-Dothan, Fl. Palaestina,* 4: 397, 232 (1986)—Israel, Sinai Peninsula.

LOMANDRA (Xanthorrhoeac.).
sect. Capitatae *(G. Don) A.T. Lee in Fl. Australia,* 46: 223 (1986): Xerotes sect. Capitateae.
ser. Sparsiflorae *(Benth.) A.T. Lee in Fl. Australia,* 46: 223 (1986): Xerotes ser. Sparsiflorae.
micrantha
subsp. teretifolia *J. Everett in Fl. Australia,* 46: 222, 131 (1986)—Australia (W. Australia, Victoria).
subsp. tuberculata *J. Everett in Fl. Australia,* 46: 222, 131 (1986)—Australia (S. Australia, New South Wales, Victoria).
multiflora
subsp. dura *(F. Muell.) T.D. Macfarlane in Fl. Australia,* 46: 224 (1986): Xerotes dura.
nana *(A. Lee) A.T. Lee in Fl. Australia,* 46: 223 (1986): L. glauca subsp. nana.
teres *T.D. Macfarlane in Fl. Australia,* 46: 224, 106 (1986)—Australia (Queensland).

LOMATIUM (Umbellif.).
cookii *J.S. Kagan in Madroño,* 33(1): 71 (1986)—U.S.A. (Oregon).
kingii *(Wats.) A. Cronquist in Great Basin Nat.,* 46(2): 254 (1986): Peucedanum kingii.
var. alpinum *(Wats.) A. Cronquist in Great Basin Nat.,* 46(2): 255 (1986): Peucedanum graveolens var. alpinum.
scabrum
var. tripinnatum *S. Goodrich in Great Basin Nat.,* 46(1): 99 (1986)—U.S.A. (Utah, Arizona).

LOMATOGONIUM (Gentianac.).
perenne *T.N. He & S.W. Liu ex J.X. Yang in Fl. Tsinlingensis,* 1(4): 396, 122 (1983)—China.

LOMELOSIA (Dipsacac.).
graminifolia
subsp. arizagae *(Uribe & Alejandre) A.M. Romo in Willdenowia,* 15(2): 421 (1986): Scabiosa graminifolia subsp. arizagae.

LONICERA (Caprifoliac.).
japonica *C.P. Thunberg ex A. Murray in Linnaeus, Syst. Veg., ed. 14:* 216 (1784)—Japan. †

LOPHOCHLAENA (Gramin.).
oregona *(Chase) P.P.H. But in Madroño,* 33(2): 146 (1986): Pleuropogon oregonus.

LOPHOSPERMUM D. Don ex R. Taylor, Phil. Mag., 67: 222 (31 Mar. 1826); cf. D.J. Mabberley in Taxon, 30(1): 16 (1981). *SCROPHULARIACEAE.*
sect. Rhodochiton *(Zuccarini ex Otto & A. Dietrich) W.J. Elisens, Monogr. Maurandyinae (Scrophulariac.-Antirrhin.) (Syst. Bot. Monog., 5):* 78 (1985): gen. Rhodochiton.
breedlovei *W.J. Elisens, Monogr. Maurandyinae (Scrophulariac.-Antirrhin.) (Syst. Bot. Monog., 5):* 77 (1985)—Mexico.
chiapense *W.J. Elisens, Monogr. Maurandyinae (Scrophulariac.-Antirrhin.) (Syst. Bot. Monog., 5):* 73 (1985)—Mexico.
hintonii *W.J. Elisens, Monogr. Maurandyinae (Scrophulariac.-Antirrhin.) (Syst. Bot. Monog., 5):* 81 (1985)—Mexico.
nubicola *W.J. Elisens, Monogr. Maurandyinae (Scrophulariac.-Antirrhin.) (Syst. Bot. Monog., 5):* 83 (1985)—Mexico, Guatemala.
purpurascens *W.J. Elisens, Monogr. Maurandyinae (Scrophulariac.-Antirrhin.) (Syst. Bot. Monog., 5):* 78 (1985)—Mexico.
turneri *W.J. Elisens, Monogr. Maurandyinae (Scrophulariac.-Antirrhin.) (Syst. Bot. Monog., 5):* 71 (1985)—Mexico, Guatemala.

LORICARIA (Compos.).
ollgaardii *M.O. Dillon & A. Sagástegui-Alva in Phytologia,* 59(4): 228 (1986)—Ecuador.
thyrsoidea *(Cuatr.) M.O. Dillon & A. Sagástegui-Alva in Phytologia,* 59(4): 230 (1986): L. thuyoides var. thyrsoidea.

LOTUS (Legumin.).
benoistii *(Maire) P. Lassen in Willdenowia,* 16(1): 111 (1986): Benedictella benoistii.
canescens *Sieb. ex Trev., Ind. Sem. Wratislav. App. III:* 2 (1821); *cf. D.J. Mabberley in Taxon,* 32(1): 85 (1983)— Crete.
conjugatus
subsp. requienii *(Sang.) W. Greuter in Willdenowia,* 16(1): 112 (1986): L. requienii.
diffusus *Spreng., Nachtr. Bot. Gart. Univ. Halle:* 27 (1801); *cf. D.J. Mabberley in Taxon,* 31(1): 71 (1982)— Locality not stated.

LOUISEANIA (Rosac.).
pedunculata *(Pall.) G.V. Eremin & A.A. Yushev in Byull. Vses. Ord. Lenina Inst. Rast. N.I. Vavilova,* 147: 31 (1985): Amygdalus pedunculata.

LUDWIGIA (Onagrac.).
glandulosa
subsp. brachycarpa *(Torrey & Gray) C.I. Peng in Ann. Missouri Bot. Gard.,* 73(2): 490 (1986): L. cylindrica β. brachycarpa.
linifolia *(Vahl) R.S. Rao, Fl. Goa, Diu, Daman, Dadra & Nagarhaveli (Fl. India, Ser. 2),* 1: 179 (1985): Jussiaea linifolia.

LUETZELBURGIA (Legumin.).
andrade-limae *H.C. de Lima in An. Soc. Bot. Brasil, XXXIV Congr. Nac. Bot.,* 2: 191 (1984)—Brazil.

×**LUISTYLIS** hort. in Orchid Rev., 94(1110): centre page pull-out p. 8 (1986). *ORCHIDACEAE.* (Luisia × Rhynchostylis).

LUPINUS (Legumin.).
argenteus
var. fulvomaculatus *(Payson) R.C. Barneby in Great Basin Nat.,* 46(2): 257 (1986): L. fulvomaculatus.
var. moabensis *S.L. Welsh in Great Basin Nat.,* 46(2): 262 (1986)—U.S.A. (Utah).
var. palmeri *(Wats.) R.C. Barneby in Great Basin Nat.,* 46(2): 257 (1986): L. palmeri.
densiflorus
subsp. lacteus *(Kellogg) R.M. Beauchamp in Phytologia,* 59(7): 439 (1986): L. lacteus.
excubitus
subsp. austromontanus *(Heller) R.M. Beauchamp in Phytologia,* 59(7): 439 (1986): L. austromontanus.
subsp. medius *(Jepson) R.M. Beauchamp in Phytologia,* 59(7): 439 (1986): L. albifrons var. medius.
polyphyllus
var. ammophilus *(Greene) R.C. Barneby in Great Basin Nat.,* 46(2): 257 (1986): L. ammophilus.
var. humicola *(A. Nels.) R.C. Barneby in Great Basin Nat.,* 46(2): 257 (1986): L. humicola.

LUVUNGA (Rutac.).
minutiflora *B.C. Stone in Proc. Acad. Nat. Sci. Philadelphia,* 137(2): 221 (1985)—Borneo (Sabah).

LYCHNIS (Caryophyllac.).
apetala
var. kingii *(Wats.) S.L. Welsh in Great Basin Nat.,* 46(2): 255 (1986): L. kingii.
coronata *C.P. Thunberg ex A. Murray in Linnaeus, Syst. Veg., ed. 14:* 435 (1784)—Japan. †

LYCIUM (Solanac.).
japonicum *C.P. Thunberg ex A. Murray in Linnaeus, Syst. Veg., ed. 14:* 228 (1784)—Japan. †

LYSIMACHIA (Primulac.).
japonica *C.P. Thunberg ex A. Murray in Linnaeus, Syst. Veg., ed. 14:* 196 (1784)—Japan. †

LYSIONOTUS (Gesneriac.).
sangzhiensis *W.T. Wang in Guihaia,* 6(3): 164 (1986)—China.

M

MAACKIA (Legumin.).
hwashanensis *W.T. Wang ex C.W. Chang in Fl. Tsinlingensis,* 1(3): 444, 17 (1981)—China.

MABEA (Euphorbiac.).
angularis *G. den Hollander in Proc. Kon. Nederl. Akad. Wetensch., C,* 89(2): 147 (1986)—Brazil, Guyana.
montana
subsp. biglandulosa *(Müller-Argoviensis) G. den Hollander in Proc. Kon. Nederl. Akad. Wetensch., C,* 89(2): 149 (1986): M. biglandulosa.
subsp. lucida *(Pax & Hoffmann) G. den Hollander in Proc. Kon. Nederl. Akad. Wetensch., C,* 89(2): 150 (1986): M. lucida.
setulosa *(Müller-Argoviensis) G. den Hollander in Proc. Kon. Nederl. Akad. Wetensch., C,* 89(2): 153 (1986): M. occidentalis var. setulosa.
speciosa
subsp. concolor *(Müller-Argoviensis) G. den Hollander in Proc. Kon. Nederl. Akad. Wetensch., C,* 89(2): 154 (1986): M. piriri var. concolor.

MABRYA W.J. Elisens, Monogr. Maurandyinae (Scrophulariac.-Antirrhin.) (Syst. Bot. Monog., 5): 57 (1985). *SCROPHULARIACEAE.*
acerifolia *(Pennell) W.J. Elisens, Monogr. Maurandyinae (Scrophulariac.-Antirrhin.) (Syst. Bot. Monog., 5):* 58 (1985): Maurandya acerifolia.
coccinea *(I.M. Johnston) W.J. Elisens, Monogr. Maurandyinae (Scrophulariac.-Antirrhin.) (Syst. Bot. Monog., 5):* 63 (1985): Maurandya coccinea.
erecta *(Hemsley) W.J. Elisens, Monogr. Maurandyinae (Scrophulariac.-Antirrhin.) (Syst. Bot. Monog., 5):* 65 (1985): Maurandia erecta.
geniculata *(Robinson & Fernald) W.J. Elisens, Monogr. Maurandyinae (Scrophulariac.-Antirrhin.) (Syst. Bot. Monog., 5):* 60 (1985): Maurandia geniculata.
subsp. flaviflora *(I.M. Johnston) W.J. Elisens, Monogr. Maurandyinae (Scrophulariac.-Antirrhin.) (Syst. Bot. Monog., 5):* 62 (1985): Maurandya flaviflora.
subsp. lanata *W.J. Elisens, Monogr. Maurandyinae (Scrophulariac.-Antirrhin.) (Syst. Bot. Monog., 5):* 63 (1985)—Mexico.
rosei *(Munz) W.J. Elisens, Monogr. Maurandyinae (Scrophulariac.-Antirrhin.) (Syst. Bot. Monog., 5):* 64 (1985): Maurandya rosei.

MACARANGA (Euphorbiac.).
capensis
var. kilimandscharica *(Pax) I. Friis & M.G. Gilbert in Kew Bull.*, 41(1): 68 (1986): M. kilimandscharica.

MACHAERANTHERA (Compos.).
asteroides
var. glandulosa *B.L. Turner in Phytologia*, 60(1): 77 (1986)—U.S.A. (Arizona).
var. lagunensis *(Keck) B.L. Turner in Phytologia*, 60(1): 77 (1986): M. lagunensis.
bigelovii
var. commixta *(Greene) B.L. Turner in Phytologia*, 60(1): 77 (1986): M. commixta.
canescens
var. ambigua *B.L. Turner in Phytologia*, 60(1): 77 (1986)—U.S.A. (Arizona).
var. aristata *(Eastwood) B.L. Turner in Phytologia*, 60(1): 78 (1986): Aster canescens var. aristatus.
var. nebraskana *B.L. Turner in Phytologia*, 60(1): 78 (1986)—U.S.A. (Nebraska).
var. sessiliflora *(Nutt.) B.L. Turner in Phytologia*, 60(1): 78 (1986): Dieteria sessiliflora.
var. shastensis *(A. Gray) B.L. Turner in Phytologia*, 60(1): 79 (1986): M. shastensis.
var. ziegleri *(Munz) B.L. Turner in Phytologia*, 60(1): 79 (1986): M. canescens subsp. ziegleri.

MACHAERIUM (Legumin.).
nicaraguense *V.E. Rudd in Phytologia*, 60(2): 93 (1986)—Nicaragua, Honduras.

MACLURA (Morac.).
sect. Plecospermum *(Trécul) C.C. Berg in Proc. Kon. Nederl. Akad. Wetensch., C*, 89(3): 245 (1986): gen. Plecospermum.
andamanica *(J.D. Hooker) C.C. Berg in Proc. Kon. Nederl. Akad. Wetensch., C*, 89(3): 245 (1986): Plecospermum andamanicum.
plumiera *G. Don f. in Loud., Hort. Brit.*: 380 (1830); *cf. D.J. Mabberley in Taxon*, 30(1): 16 (1981)— Locality not stated.
spinosa *(Willdenow) C.C. Berg in Proc. Kon. Nederl. Akad. Wetensch., C*, 89(3): 245 (1986): Trophis spinosa.

MACROPANAX (Araliac.).
meghalayensis *K. Haridasan & R.R. Rao, Forest Fl. Meghalaya:* 434 (1985)—India.

MAGNOLIA (Magnoliac.).
odoratissima *Y.W. Law & R.Z. Zhou in Bull. Bot. Res. North-East. Forest. Inst.*, 6(2): 139 (1986)—China.
officinalis
var. pubescens *C.Y. Deng in J. Nanjing Inst. Forest.*, 1986(1): 145 (1986)—China.

MAHONIA (Berberidac.).
repens *(Lindl.) G. Don f. in Loud., Hort. Brit.*: 476 (1830); *cf. D.J. Mabberley in Taxon*, 30(1): 16 (1981): Basionym not stated.

MAIANTHEMUM (Convallariac.).
amoenum *(H.L. Wendl.) J.V. LaFrankie in Taxon*, 35(3): 588 (1986): Smilacina amoena.
atropurpureum *(Franchet) J.V. LaFrankie in Taxon*, 35(3): 588 (1986): Tovaria atropurpurea.
dahuricum *(Fisch. & Mey.) J.V. LaFrankie in Taxon*, 35(3): 588 (1986): Smilacina dahurica.
flexuosum *(Bertol.) J.V. LaFrankie in Taxon*, 35(3): 588 (1986): Smilacina flexuosa.
formosanum *(Hayata) J.V. LaFrankie in Taxon*, 35(3): 588 (1986): Smilacina formosana.
forrestii *(W.W. Smith) J.V. LaFrankie in Taxon*, 35(3): 588 (1986), basionym as 'Smilacina forestii': Tovaria forrestii.
fuscum *(Wall.) J.V. LaFrankie in Taxon*, 35(3): 588 (1986): Smilacina fusca.
gigas *(Woodson) J.V. LaFrankie in J. Arnold Arbor.*, 67(4): 402 (1986): Smilacina gigas.
var. crassipes *(Standley & Steyerm.) J.V. LaFrankie in J. Arnold Arbor.*, 67(4): 404 (1986): Smilacina crassipes.
henryi *(Baker) J.V. LaFrankie in Taxon*, 35(3): 588 (1986): Oligobotrya henryi.
hondoense *(Ohwi) J.V. LaFrankie in Taxon*, 35(3): 588 (1986): Smilacina hondoensis.
japonicum *(A. Gray) J.V. LaFrankie in Taxon*, 35(3): 588 (1986): Smilacina japonica.
lichiangense *(W.W. Smith) J.V. LaFrankie in Taxon*, 35(3): 589 (1986), basionym as 'Smilacina lichiangensis': Tovaria lichiangensis.
macrophyllum *(Martens & Galeotti) J.V. LaFrankie in Taxon*, 35(3): 588 (1986): Smilacina macrophylla.
monteverdense *J.V. LaFrankie in J. Arnold Arbor.*, 67(4): 409 (1986)—Costa Rica, Nicaragua.
oleraceum *(Baker) J.V. LaFrankie in Taxon*, 35(3): 589 (1986): Tovaria oleracea.
paludicolum *J.V. LaFrankie in Amer. J. Bot.*, 73(9): 1258 (1986)—Costa Rica.
paniculatum *(Martens & Galeotti) J.V. LaFrankie in Taxon*, 35(3): 588 (1986): Smilacina paniculata.
purpureum *(Wall.) J.V. LaFrankie in Taxon*, 35(3): 589 (1986): Smilacina purpureum.
racemosum
subsp. amplexicaule *(Nutt.) J.V. LaFrankie in J. Arnold Arbor.*, 67(4): 418 (1986): Smilacina amplexicaulis.
robustum *(Makino & Honda) J.V. LaFrankie in Taxon*, 35(3): 589 (1986): Smilacina robusta.
salvinii *(Baker) J.V. LaFrankie in Taxon*, 35(3): 588 (1986): Tovaria salvinii.

MAIANTHEMUM :–
scilloideum *(Martens & Galeotti) J.V. LaFrankie in Taxon,* 35(3): 588 (1986): Smilacina scilloidea.
septifolium *J.V. LaFrankie in J. Arnold Arbor.,* 67(4): 427 (1986)—Mexico.
tatsiense *(Franchet) J.V. LaFrankie in Taxon,* 35(3): 589 (1986): Tovaria tatsiensis.
tubiferum *(Batalin) J.V. LaFrankie in Taxon,* 35(3): 589 (1986): Smilacina tubifera.
yesoense *(Franchet & Savatier) J.V. LaFrankie in Taxon,* 35(3): 589 (1986): Tovaria yesoensis.

×**MAILAMAIARA** hort. in Orchid Rev., 94(1109): centre page pull-out p. 8 (1986). *ORCHIDACEAE.* (Cattleya × Diacrium × Laelia × Schomburgkia).

MAJORANA (Labiat.).
smyrnaea *(L.) T. Nees, Pl. Offic. Suppl. I,* (?1831) *et Nees & Eberm., Handb. Med. Bot.,* 2: 572 (1831); *cf. D.J. Mabberley in Taxon,* 31(1): 71 (1982): Basionym not stated.

MALAISIA (Morac.).
scandens
subsp. megacarpa *P.S. Green in J. Arnold Arbor.,* 67(1): 113 (1986)—Lord Howe Is.

MALLEA A. Jussieu in Mirb. & Cass. apud Guill. in Bull. Sci. Nat. Geol., 23: 236 (1830); cf. D.J. Mabberley in Taxon, 31(1): 66 (1982). *MELIACEAE.*

MALLOTUS (Euphorbiac.).
resinosus
var. subramanyamii *(Ellis) T. Chakrabarty in J. Econ. Taxon. Bot.,* 6(3): 704 (1985): M. subramanyamii.
subramanyamii *J.L. Ellis in Bull. Bot. Surv. India,* 25(1–4): 199 (1983 publ. 1985)—India.

MALMEA (Annonac.).
anomala *(R.E. Fries) P.J.M. Maas in Proc. Kon. Nederl. Akad. Wetensch., C,* 89(3): 260 (1986): Cremastosperma anomalum.

MALVA (Malvac.).
aurantiaco-rubra *Huber, Cat. Gén. 1870 & 1871:* 3 (1870); *cf. D.J. Mabberley in Taxon,* 34(3): 453 (1985)— Locality not stated.
cuneata *P. Baltasar Merino, Algunas Pl. Rar. ... de La Guardia (Pontevedra):* 31 (1895)—Spain. †
lagascae *Capelli, Cat. Stirp. Reg. Hort. Bot. Taur.:* 36 (1821); *cf. D.J. Mabberley in Taxon,* 32(1): 85 (1983)— Locality not stated.
paxtonii *G. Don f. in Loud., Encycl. Pl., new ed.:* 1424 (1855); *cf. D.J. Mabberley in Taxon,* 30(1): 16 (1981)— Locality not stated.

MAMMILLARIA (Cactac.).
ser. Megastigmatae *T.M.W. Neutelings in Succulenta,* 65(1): 3 (1986).
albicans
f. slevinii *(Britton & Rose) T.M.W. Neutelings in Succulenta,* 65(1): 5 (1986), without basionym ref.: Basionym not stated.
berkiana *A.B. Lau in Kakt. And. Sukk.,* 37(2): 33 (1986)—Mexico.
coronata *G. Don f. in Loud., Hort. Brit.:* 194 (1830); *cf. D.J. Mabberley in Taxon,* 30(1): 16 (1981)— Locality not stated.
dioica
f. angelensis *(Craig) T.M.W. Neutelings in Succulenta,* 65(1): 4 (1986), without basionym ref.: M. angelensis.
var. armillata *(K. Brandegee) T.M.W. Neutelings in Succulenta,* 65(1): 4 (1986), without basionym ref.: M. armillata.
var. capensis *(Gates) T.M.W. Neutelings in Succulenta,* 65(1): 4 (1986), without basionym ref.: Basionym not stated.
var. cerralboa *(Britton & Rose) T.M.W. Neutelings in Succulenta,* 65(1): 4 (1986), without basionym ref.: Basionym not stated.
f. estebanensis *(Lindsay) T.M.W. Neutelings in Succulenta,* 65(1): 4 (1986), without basionym ref.: M. estebanensis.
f. incerta *(Parish) T.M.W. Neutelings in Succulenta,* 65(1): 4 (1986), without basionym ref.: M. incerta.
var. neopalmeri *(Coulter) T.M.W. Neutelings in Succulenta,* 65(1): 4 (1986), without basionym ref.: Basionym not stated.
f. phitauiana *(Baxter) T.M.W. Neutelings in Succulenta,* 65(1): 4 (1986), without basionym ref.: Basionym not stated.
f. verhaertiana *(Bödeker) T.M.W. Neutelings in Succulenta,* 65(1): 4 (1986), without basionym ref.: M. verhaertiana.
goodridgii
var. blossfeldiana *(Bödeker) T.M.W. Neutelings in Succulenta,* 65(1): 5 (1986), without basionym ref.: M. blossfeldiana.
var. bullardiana *(Gates) T.M.W. Neutelings in Succulenta,* 65(1): 5 (1986), without basionym ref.: Basionym not stated.
var. hutchinsoniana *(Gates) T.M.W. Neutelings in Succulenta,* 65(1): 5 (1986), without basionym ref.: Basionym not stated.
var. louisae *(Lindsay) T.M.W. Neutelings in Succulenta,* 65(1): 5 (1986), without basionym ref.: M. louisae.
f. shurlyana *(Gates) T.M.W. Neutelings in Succulenta,* 65(1): 5 (1986), without basionym ref.: M. shurlyana.
guelzowiana
var. robustior *R. Wolf in Kakt. And. Sukk.,* 37(12): 257 (1986)—Mexico.

MAMMILLARIA :-
mazatlanensis
var. occidentalis *(Britton & Rose) T.M.W. Neutelings in Succulenta,* 65(1): 4 (1986), without basionym ref.: Basionym not stated.
f. patonii *(Bravo) T.M.W. Neutelings in Succulenta,* 65(1): 4 (1986), without basionym ref.: Basionym not stated.
f. sinalensis *(Craig) T.M.W. Neutelings in Succulenta,* 65(1): 4 (1986), without basionym ref.: M. occidentalis var. sinalensis.
microcarpa
f. auricarpa *(Marshall) T.M.W. Neutelings in Succulenta,* 65(1): 3 (1986), without basionym ref.: M. microcarpa var. auricarpa.
f. gueldemanniana *(Backeberg) T.M.W. Neutelings in Succulenta,* 65(1): 4 (1986), without basionym ref.: M. gueldemanniana.
f. oliviae *(Orcutt) T.M.W. Neutelings in Succulenta,* 65(1): 4 (1986), without basionym ref.: M. oliviae.
var. sheldonii *(Britton & Rose) T.M.W. Neutelings in Succulenta,* 65(1): 3 (1986), without basionym ref.: Basionym not stated.
f. swinglei *(Britton & Rose) T.M.W. Neutelings in Succulenta,* 65(1): 4 (1986), without basionym ref.: Basionym not stated.
thornberi
var. yaquensis *(Craig) T.M.W. Neutelings in Succulenta,* 65(1): 4 (1986), without basionym ref.: M. yaquensis.

MANIHOT (Euphorbiac.).
cannabina *Sweet, Hort. Brit., ed. 2:* 458 (1830); *cf. D.J. Mabberley in Taxon,* 30(1): 16 (1981)— Locality not stated.
neusana *N.M.A. Nassar in Ciênc. Cult.,* 38(2): 340 (1986)—Brazil.

MAPANIA (Cyperac.).
herrerae *J. Gómez Laurito in Phytologia,* 60(1): 73 (1986)—Costa Rica.

MARISCUS (Cyperac.).
drakensbergensis *P. Vorster in S. Afr. J. Bot.,* 52(2): 181 (1986)—South Africa (Natal).
patulus *Schrad., Ind. Sem. Hort. Acad. Gott.,* 1835: 4 (1835); *Linnaea 11, Lit.:* 88 (1837); *cf. D.J. Mabberley in Taxon,* 32(1): 85 (1983)— Locality not stated.
solidus *(Kunth) P. Vorster in S. Afr. J. Bot.,* 52(3): 265 (1986): Cyperus solidus.

MARSDENIA (Asclepiadac.).
guanchezii *G. Morillo in Ernstia,* 37: 10 (1986)—Venezuela.

MASDEVALLIA (Orchidac.).
subgen. Amanda *C.A. Luer, Ic. Pleurothallid. II-Syst. Masdevallia (Monogr. Syst. Bot. Missouri Bot. Gard. 16):* 10 (1986), nom. nov.: M. sect. Amandae *Rchb. f.*
subgen. Meleagris *C.A. Luer, Ic. Pleurothallid. II-Syst. Masdevallia (Monogr. Syst. Bot. Missouri Bot. Gard. 16):* 51 (1986), nom. nov.: gen. Rodrigoa *Braas.*
subgen. Pelecaniceps *C.A. Luer, Ic. Pleurothallid. II-Syst. Masdevallia (Monogr. Syst. Bot. Missouri Bot. Gard. 16):* 53 (1986), nom. nov.: gen. Luerella *Braas.*
subsect. Alaticaules *(Krzl.) C.A. Luer, Ic. Pleurothallid. II-Syst. Masdevallia (Monogr. Syst. Bot. Missouri Bot. Gard. 16):* 43 (1986): M. sect. Alaticaules.
sect. Amaluzae *C.A. Luer, Ic. Pleurothallid. II-Syst. Masdevallia (Monogr. Syst. Bot. Missouri Bot. Gard. 16):* 18 (1986).
sect. Aphanes *C.A. Luer, Ic. Pleurothallid. II-Syst. Masdevallia (Monogr. Syst. Bot. Missouri Bot. Gard. 16):* 18 (1986).
sect. Caudivolvulae *C.A. Luer, Ic. Pleurothallid. II-Syst. Masdevallia (Monogr. Syst. Bot. Missouri Bot. Gard. 16):* 21 (1986).
subsect. Dentatae *C.A. Luer, Ic. Pleurothallid. II-Syst. Masdevallia (Monogr. Syst. Bot. Missouri Bot. Gard. 16):* 48 (1986).
subsect. Durae *C.A. Luer, Ic. Pleurothallid. II-Syst. Masdevallia (Monogr. Syst. Bot. Missouri Bot. Gard. 16):* 23 (1986).
sect. Ligiae *C.A. Luer, Ic. Pleurothallid. II-Syst. Masdevallia (Monogr. Syst. Bot. Missouri Bot. Gard. 16):* 26 (1986).
sect. Mentosae *C.A. Luer, Ic. Pleurothallid. II-Syst. Masdevallia (Monogr. Syst. Bot. Missouri Bot. Gard. 16):* 38 (1986).
sect. Nidificae *C.A. Luer, Ic. Pleurothallid. II-Syst. Masdevallia (Monogr. Syst. Bot. Missouri Bot. Gard. 16):* 12 (1986).
sect. Ophioglossae *C.A. Luer, Ic. Pleurothallid. II-Syst. Masdevallia (Monogr. Syst. Bot. Missouri Bot. Gard. 16):* 15 (1986).
subsect. Oscillantes *C.A. Luer, Ic. Pleurothallid. II-Syst. Masdevallia (Monogr. Syst. Bot. Missouri Bot. Gard. 16):* 32 (1986).
sect. Pygmaeae *C.A. Luer, Ic. Pleurothallid. II-Syst. Masdevallia (Monogr. Syst. Bot. Missouri Bot. Gard. 16):* 15 (1986).
subsect. Reichenbachianae *(Woolw.) C.A. Luer, Ic. Pleurothallid. II-Syst. Masdevallia (Monogr. Syst. Bot. Missouri Bot. Gard. 16):* 48 (1986): M. sect. Reichenbachianae.
subsect. Saltatrices *(Rchb. f.) C.A. Luer, Ic. Pleurothallid. II-Syst. Masdevallia (Monogr. Syst. Bot. Missouri Bot. Gard. 16):* 36 (1986): M. sect. Saltatrices.
subgen. Teagueia *C.A. Luer, Ic. Pleurothallid. II-Syst. Masdevallia (Monogr. Syst. Bot. Missouri Bot. Gard. 16):* 53 (1986).

MASDEVALLIA :–
subsect. Tubulosae *(Rchb. f.) C.A. Luer, Ic. Pleurothallid. II-Syst. Masdevallia (Monogr. Syst. Bot. Missouri Bot. Gard. 16):* 38 (1986): M. sect. Tubulosae.
asterotricha *W. Königer in Orchidee,* 37(3): 104 (1986)—Peru.
audax *W. Königer in Orchidee,* 37(3): 106 (1986)—Peru.
condorensis *C.A. Luer & Hirtz in Lindleyana,* 1(3): 180 (1986)—Ecuador.
coriacea
subsp. bonplandii *(Rchb. f.) C.A. Luer, Ic. Pleurothallid. II-Syst. Masdevallia (Monogr. Syst. Bot. Missouri Bot. Gard. 16):* 23 (1986): M. bonplandii.
discoidea *C.A. Luer & Würstle in Lindleyana,* 1(3): 182 (1986)—Brazil.
hoeijeri *C.A. Luer & Hirtz in Lindleyana,* 1(3): 184 (1986)—Ecuador.
hubeinii *C.A. Luer & B. Würstle in Orchideeën,* 48(2): 50 (1986)—Colombia.
hystrix *C.A. Luer & A.C. Hirtz in Orchidee,* 37(3): 139 (1986)—Ecuador.
lintricula *W. Königer in Orchidee,* 37(3): 106 (1986)—Peru.
ophioglossa
subsp. grossa *(Luer) C.A. Luer, Ic. Pleurothallid. II-Syst. Masdevallia (Monogr. Syst. Bot. Missouri Bot. Gard. 16):* 15 (1986): M. grossa.
picturata
subsp. minor *(Cogn.) C.A. Luer, Ic. Pleurothallid. II-Syst. Masdevallia (Monogr. Syst. Bot. Missouri Bot. Gard. 16):* 12 (1986): M. picturata var. minor.
pileata *C.A. Luer & B. Würstle in Orchideeën,* 48(2): 50 (1986)—Colombia.
rufescens *W. Königer in Orchidee,* 37(3): 110 (1986)—Peru.
tubulosa
subsp. syringodes *(Luer & Andreetta) C.A. Luer, Ic. Pleurothallid. II-Syst. Masdevallia (Monogr. Syst. Bot. Missouri Bot. Gard. 16):* 38 (1986): M. syringodes.
valenciae *C.A. Luer & R. Escobar R. in Orquideologia,* 16(3): 36 (1986)—Colombia.
ventricularia
subsp. filaria *(Luer & Escobar) C.A. Luer, Ic. Pleurothallid. II Syst. Masdevallia (Monogr. Syst. Bot. Missouri Bot. Gard. 16):* 36 (1986): M. filaria.

MASTERSIELLA (Restionac.).
purpurea *(Pillans) H.P. Linder in Bothalia,* 15(3–4): 487 (1985): Hypolaena purpurea.
spathulata *(Pillans) H.P. Linder in Bothalia,* 15(3–4): 487 (1985): Hypolaena spathulata.

MATELEA (Asclepiadac.).
holstii *G. Morillo & G. Carnevali in Ernstia,* 40: 21 (1986)—Venezuela.

MATTHIOLA (Crucif.).
carnica *F. Tammaro in Arch. Bot. Biogeogr. Ital.,* 61(1–2): 22 (1985 publ. 1986)—Italy.

MATUCANA (Cactac.).
krahnii *(Donald) R. Bregman in Kakt. And. Sukk.,* 37(12): 253 (1986): Borzicactus krahnii.
polzii *L. Diers, J.D. Donald & E. Zecher in Kakt. And. Sukk.,* 37(6): 118 (1986)—Peru.

MAURANDYA (Scrophulariac.).
antirrhiniflora
subsp. hederifolia *(Rothmaler) W.J. Elisens, Monogr. Maurandyinae (Scrophulariac.-Antirrhin.) (Syst. Bot. Monog., 5):* 50 (1985): Maurandella hederaefolia.

MAURIA (Anacardiac.).
membranifolia *A. Barfod & L.B. Holm-Nielsen in Nordic J. Bot.,* 6(4): 423 (1986)—Ecuador.

MAUSOLEA Bunge ex D. Podlech in Fl. Iran., 158: 223 (1986). *COMPOSITAE.*
eriocarpa *(Bunge) Poljak. ex D. Podlech in Fl. Iran.,* 158: 224 (1986): Artemisia eriocarpa.

MECONOPSIS (Papaverac.).
pseudohorridula *C.Y. Wu & H. Chuang in Fl. Xizangica,* 2: 234 (1985)—Xizang.

MECRANIUM (Melastomatac.).
revolutum *J.D. Skean & W.S. Judd in Brittonia,* 38(3): 230 (1986)—Hispaniola (Haiti).

MEDICAGO (Legumin.).
citrina *(Font Quer) W. Greuter in Willdenowia,* 16(1): 112 (1986): M. arborea var. citrina.
sativa
var. integrifoliola *C.W. Chang in Fl. Tsinlingensis,* 1(3): 444, 25 (1981)—China.

MEEHANIA (Labiat.).
urticifolia
f. rosea *J. Ōhara in J. Phytogeogr. Taxon.,* 33(2): 72 (1985)—Japan.

MELALEUCA (Myrtac.).
obliqua *De Vriese, Hort. Spaarn-Berg.:* 119 (1839); *cf. D.J. Mabberley in Taxon,* 33(3): 442 (1984)— Locality not stated.

MELANOLOMA (Compos.).
atlantis *(Maire & Weiller) J. Fernández Casas & A. Susanna de la Serna in Fontqueria,* 1: 8 (1982): Centaurea atlantis.
gattefossei *(Maire) J. Fernández Casas & A. Susanna de la Serna in Fontqueria,* 1: 6 (1982): Centaurea gattefossei.
takredense *(Cosson ex Maire) J. Fernández Casas & A. Susanna de la Serna in Fontqueria,* 1: 4 (1982): Centaurea takredensis.

MELASMA (Scrophulariac.).
brevipedicellatum *S. Lisowski & R. Mielcarek in Fragm. Flor. Geobot.,* 29(2): 215 (1983 publ. 1985)—Zaïre.

MELICA (Gramin.).
latifolia *Roxb. ex Hornem., Hort. Hafn., Suppl.:* 117 (1819); *cf. D.J. Mabberley in Taxon,* 33(3): 437 (1984)— Locality not stated.

MELILOTUS (Legumin.).
hamosa *(Bieb.) Besser ex Schrad., Ind. Sem. Hort. Acad. Gott.,* 1821: 3 (1821); *cf. D.J. Mabberley in Taxon,* 32(1): 85 (1983): Basionym not stated.

MELISSITUS (Legumin.).
ruthenicus *(L.) C.W. Chang in Fl. Tsinlingensis,* 1(3): 23 (1981): Trigonella ruthenica.
var. liaosiensis *(P.Y. Fu & Y.A. Chen) X.P. Jiang in Fl. Beijing,* 1: 429 (1984), without basionym ref.: Pocockia liaosiensis.

MELOCACTUS (Cactac.).
lanssensianus *P.J. Braun in Succulenta,* 65(2): 26 (1986)—Brazil.

MELOCALAMUS (Gramin.).
indicus *R.B. Majumder in Bull. Bot. Surv. India,* 25(1–4): 236 (1983 publ. 1985)—Assam, India.

MELOCHIA (Sterculiac.).
diffusa *Colla, Hort. Rip.:* 88 (1824); *cf. D.J. Mabberley in Taxon,* 30(1): 16 (1981)—Locality not stated.

MELOLOBIUM (Legumin.).
involucratum *(Thunb.) C.H. Stirton in S. Afr. J. Bot.,* 52(4): 355 (1986), as '(Harv.)': Psoralea involucrata.

MENISPERMUM (Menispermac.).
acutum *C.P. Thunberg ex A. Murray in Linnaeus, Syst. Veg., ed. 14:* 892 (1784)—Japan. †
japonicum *C.P. Thunberg ex A. Murray in Linnaeus, Syst. Veg., ed. 14:* 892 (1784)—Japan. †
platyphyllum *(St. Hil.) Mart., Consp. Reg. Veg.:* 42 (1835); *cf. D.J. Mabberley in Taxon,* 30(1): 16 (1981): Basionym not stated.
trilobum *C.P. Thunberg ex A. Murray in Linnaeus, Syst. Veg., ed. 14:* 892 (1784)—Japan. †

MENTHA (Labiat.).
cinerea *P. Baltasar Merino, Algunas Pl. Rar. ... de La Guardia (Pontevedra):* 28 (1895)—Spain. †
pallida *Nees ex Mart., Hort. Acad. Erlangen:* 121 (1814); *cf. D.J. Mabberley in Taxon,* 33(3): 442 (1984)— Locality not stated.

MENTZELIA (Loasac.).
bartonioides *(Zucc.) Lehm., Del. Sem. Hort. Hamb.,* 1850: 6 (1850); *cf. D.J. Mabberley in Taxon,* 33(3): 442 (1984): Eucnide bartonioides.

MENTZELIA :–
shultziorum *B.A. Prigge in Great Basin Nat.,* 46(2): 361 (1986)—U.S.A. (Utah).

MENZIESIA (Ericac.).
katsumatae *M. Tashiro & H. Hatta in Bot. Mag. Tokyo,* 99(1053): 29 (1986)—Japan.
yakushimensis *M. Tashiro & H. Hatta in Bot. Mag. Tokyo,* 99(1053): 33 (1986)—Japan.

MERINTHOPODIUM (Solanac.).
vogelii *(Cuatrecasas) R. Castillo & R.E. Schultes in Rhodora,* 88(854): 292 (1986): Markea vogelii.

MERREMIA (Convolvulac.).
pavonii *(H. Hallier) D.F. Austin & G.W. Staples in J. Arnold Arbor.,* 67(2): 264 (1986): Operculina pavoni.

MERTENSIA (Boraginac.).
lanceolata
var. nivalis *(Wats.) L.C. Higgins in Great Basin Nat.,* 46(2): 255 (1986): M. paniculata var. nivalis.

MESPILUS (Rosac.).
acuminata *(Lindl.) Ohlendorff, Verz. Baum. Hamb. Bot. Gart.,* 1832: 17 (1832); *cf. D.J. Mabberley in Taxon,* 33(3): 442 (1984): Cotoneaster acuminatus.
frigida *(Wall.) Ohlendorff, Verz. Baum. Hamb. Bot. Gart.,* 1832: 18 (1832); *cf. D.J. Mabberley in Taxon,* 33:(3): 442 (1984): Cotoneaster frigidus.
japonica *C.P. Thunberg ex A. Murray in Linnaeus, Syst. Veg., ed. 14:* 466 (1784)—Japan. †
microphylla *(Lindl.) Ohlendorff, Verz. Baum. Hamb. Bot. Gart.,* 1832: 18 (1832); *cf. D.J. Mabberley in Taxon,* 33(3): 442 (1984): Cotoneaster microphyllus.

METASTELMA (Asclepiadac.).
giuliettianum *J. Fontella Pereira in Phytologia,* 59(4): 224 (1986)—Brazil.
harleyi *J. Fontella Pereira in Phytologia,* 59(4): 225 (1986)—Brazil.

METROSIDEROS (Myrtac.).
bartlettii *J.W. Dawson in New Zealand J. Bot.,* 23(4): 607 (1985 publ. 1986)—New Zealand.

MICHELIA (Magnoliac.).
angustioblonga *Y.W. Law & Y.F. Wu in Bull. Bot. Res. North-East. Forest. Inst.,* 6(2): 97 (1986)—China.
foveolata
var. cinerascens *Y.W. Law & Y.F. Wu in Bull. Bot. Res. North-East. Forest. Inst.,* 6(2): 99 (1986)—China.

MICONIA (Melastomatac.).
globuliflora
var. dominicae *R.A. Howard & E.A. Kellogg in J. Arnold Arbor.,* 67(2): 242 (1986)—Windward Is., Leeward Is., Trinidad.

MICONIA :-
var. vulcanica *(Naudin) R.A. Howard & E.A. Kellogg in J. Arnold Arbor.*, 67(2): 243 (1986): M. vulcanica.
secunda *R.A. Howard & E.A. Kellogg in J. Arnold Arbor.*, 67(2): 250 (1986)—Windward Is.

MICRANTHOCEREUS (Cactac.).
streckeri *W. van Heek & L. van Criekinge in Kakt. And. Sukk.*, 37(5): 105 (1986)—Brazil.

MICROCORYS (Labiat.).
cephalantha *B.J. Conn in Muelleria,* 6(3–4): 260 (1986)—Australia (W. Australia).
elliptica *B.J. Conn in Muelleria,* 6(3–4): 262 (1986)—Australia (N. Terr.).
wilsoniana *B.J. Conn in Muelleria,* 6(3–4): 263 (1986)—Australia (W. Australia).

MICRORRHINUM (Scrophulariac.).
pluttulum *(Rech.f.) F. Speta in Phyton (Austria),* 26(1): 51 (1986): Chaenorhinum pluttulum.

MIKANIA (Compos.).
harlingii *R.M. King & H. Robinson in Phytologia,* 60(1): 82 (1986)—Ecuador.
tehuacanensis *W.C. Holmes in Phytologia,* 59(7): 442 (1986)—Mexico.

MIKANIOPSIS (Compos.).
bambuseti *(R.E. Fr.) C. Jeffrey in Kew Bull.*, 41(4): 878 (1986): Senecio bambuseti.
cissampelina *(DC.) C. Jeffrey in Kew Bull.*, 41(4): 879 (1986): Cacalia ? cissampelina.

MILIUM (Gramin.).
album *(L.) Lag., Elenchus:* 10 (1816); *cf. D.J. Mabberley in Taxon,* 31(1): 71 (1982): Basionym not stated.
caninum *(L.) Lag., Elenchus:* 10 (1816); *cf. D.J. Mabberley in Taxon,* 31(1): 71 (1982): Agrostis canina.
capillare *(L.) Lag., Elenchus:* 10 (1816); *cf. D.J. Mabberley in Taxon,* 31(1): 71 (1982): Basionym not stated.
globosum *C.P. Thunberg ex A. Murray in Linnaeus, Syst. Veg., ed. 14:* 109 (1784)—Japan. †
hispidum *(Willd.) Lag., Elenchus:* 10 (1816); *cf. D.J. Mabberley in Taxon,* 31(1): 71 (1982): Basionym not stated.
interruptum *(L.) Lag., Elenchus:* 10 (1816); *cf. D.J. Mabberley in Taxon,* 31(1): 71 (1982): Basionym not stated.
spica-venti *(L.) Lag., Elenchus:* 10 (1816); *cf. D.J. Mabberley in Taxon,* 31(1): 71 (1982): Basionym not stated.
stoloniferum *(L.) Lag., Elenchus:* 10 (1816); *cf. D.J. Mabberley in Taxon,* 31(1): 71 (1982): Agrostis stolonifera.

MILIUSA (Annonac.).
brahei *(F. Muell.) L.W. Jessup in Austrobaileya,* 2(3): 227 (1986): Saccopetalum brahei.

MILLETTIA (Legumin.).
orissae *G. Panigrahi & S.C. Mishra in Indian J. Forest.*, 8(3): 234 (1985), nom. nov.: Robinia racemosa *Roxb.*

MILTONIOIDES (Orchidac.).
pauciflora *(L.O. Wms.) F. Hamer & Garay in Ic. Pl. Trop.*, 13 (Orch. Nicaragua, pt. 6): t. 1048a (1985): Odontoglossum pauciflorum.
schroederiana *(O'Brien) E. Lückel in Orchidee,* 37(2): 55 (1986): Miltonia schroederiana.

MIMOSA (Legumin.).
caudero *L. Cárdenas in Ernstia,* 40: 15 (1986)—Venezuela.
turneri *R.C. Barneby in Brittonia,* 38(1): 4 (1986)—U.S.A. (Texas), Mexico.

MIMULUS (Scrophulariac.).
leptanthus *(Nutt) A.L. Grant ex L.H. Bailey in Gentes Herbarum,* 1(3): 136 (1923): Diplacus leptanthus.
longiflorus *(Nutt) A.L. Grant ex L.H. Bailey in Gentes Herbarum,* 1(3): 136 (1923): Diplacus longiflorus.
roezlii *Benary ex Haage & Schmidt, Cat. 1872:* 8 (1872); *cf. D.J. Mabberley in Taxon,* 34(3): 454 (1985)— Locality not stated.

MINUARTIA (Caryophyllac.).
setacea
subsp. stojanovii *(Kitanov) A. Strid in A. Strid (ed.), Mount. Fl. Greece,* 1: 95 (1986): M. setacea var. stojanovii.

MIRABILIS (Nyctaginac.).
linearis
var. decipiens *(Standl.) S.L. Welsh in Great Basin Nat.*, 46(2): 258 (1986): Allionia decipiens.

MIRANDEA (Acanthac.).
andradenia *T.F. Daniel in Southwest. Nat.*, 31(2): 170 (1986)—Mexico.

MISCHOCARPUS (Sapindac.).
albescens *S.T. Reynolds in Fl. Australia,* 25: 200, 97 (1985)—Australia (Queensland).
australis *S.T. Reynolds in Fl. Australia,* 25: 200, 100 (1985)—Australia (Queensland, New South Wales).
macrocarpus *S.T. Reynolds in Fl. Australia,* 25: 200, 100 (1985)—Australia (Queensland).
stipitatus *S.T. Reynolds in Fl. Australia,* 25: 200, 100 (1985)—Australia (Queensland).

MIYAMAYOMENA (Compos.).
sect. Heterolepis *Ling & Y.L. Chen in Bull. Bot. Res. North-East. Forest. Inst.*, 6(2): 41 (1986).
angustifolia *(Hand.-Mazz.) Y.L. Chen in Bull. Bot. Res. North-East. Forest. Inst.*, 6(2): 43 (1986): Asteromaea angustifolia.
simplex *(Hand.-Mazz.) Y.L. Chen in Bull. Bot. Res. North-East. Forest. Inst.*, 6(2): 42 (1986): Asteromaea simplex.

MNESITHEA (Gramin.).
afraurita *(Stapf) R. de Koning & M.S.M. Sosef in Blumea,* 31(2): 290 (1986): Rottboellia afraurita.
annua *(Lazarides) R. de Koning & M.S.M. Sosef in Blumea,* 31(2): 295 (1986): Heteropholis annua.
aurita *(Steud.) R. de Koning & M.S.M. Sosef in Blumea,* 31(2): 290 (1986): Rottboellia aurita.
balansae *(Hack.) R. de Koning & M.S.M. Sosef in Blumea,* 31(2): 290 (1986): Rottboellia balansae.
benoistii *(Camus) R. de Koning & M.S.M. Sosef in Blumea,* 31(2): 287 (1986): Heteropholis benoistii.
capensis *(Stapf) R. de Koning & M.S.M. Sosef in Blumea,* 31(2): 290 (1986): Coelorachis capensis.
clarkei *(Hack.) R. de Koning & M.S.M. Sosef in Blumea,* 31(2): 290 (1986): Rottboellia clarkei.
cylindrica *(Michx.) R. de Koning & M.S.M. Sosef in Blumea,* 31(2): 290 (1986): Tripsacum cylindricum.
formosa *(R. Br.) R. de Koning & M.S.M. Sosef in Blumea,* 31(2): 288 (1986): Rottboellia formosa.
glandulosa *(Trin.) R. de Koning & M.S.M. Sosef in Blumea,* 31(2): 290 (1986): Rottboellia glandulosa.
granularis *(L.) R. de Koning & M.S.M. Sosef in Blumea,* 31(2): 295 (1986): Cenchrus granularis.
helferi *(Hook. f.) R. de Koning & M.S.M. Sosef in Blumea,* 31(2): 291 (1986): Rottboellia helferi.
impressa *(Griseb.) R. de Koning & M.S.M. Sosef in Blumea,* 31(2): 291 (1986): Rottboellia impressa.
khasiana *(Hack.) R. de Koning & M.S.M. Sosef in Blumea,* 31(2): 291 (1986): Rottboellia striata subsp. khasiana.
laevis
var. chenii *(Hsu) R. de Koning & M.S.M. Sosef in Blumea,* 31(2): 286 (1986): Thaumastochloa chenii.
var. cochinchinensis *(Lour.) R. de Koning & M.S.M. Sosef in Blumea,* 31(2): 286 (1986): Phleum cochinchinensis.
laevispica *(Keng) R. de Koning & M.S.M. Sosef in Blumea,* 31(2): 291 (1986): Rottboellia laevispica.
lepidura *(Stapf) R. de Koning & M.S.M. Sosef in Blumea,* 31(2): 291 (1986): Coelorachis lepidura.
nigrescens *(Thw.) R. de Koning & M.S.M. Sosef in Blumea,* 31(2): 287 (1986): Rottboellia nigrescens.
parodiana *(Henr.) R. de Koning & M.S.M. Sosef in Blumea,* 31(2): 291 (1986): Coelorachis parodiana.
pulcherrima *(Kunth) R. de Koning & M.S.M. Sosef in Blumea,* 31(2): 291 (1986): Ratzeburgia pulcherrima.

MNESITHEA :–
ramosa *(Fourn.) R. de Koning & M.S.M. Sosef in Blumea,* 31(2): 291 (1986): Apogonia ramosa.
rottboellioides *(R. Br.) R. de Koning & M.S.M. Sosef in Blumea,* 31(2): 291 (1986): Ischaemum rottboellioides.
rugosa *(Nutt.) R. de Koning & M.S.M. Sosef in Blumea,* 31(2): 291 (1986): Rottboellia rugosa.
selloana *(Hack.) R. de Koning & M.S.M. Sosef in Blumea,* 31(2): 292 (1986): Rottboellia selloana.
striata *(Steud.) R. de Koning & M.S.M. Sosef in Blumea,* 31(2): 292 (1986): Rottboellia striata.
subgibbosa *(Hack.) R. de Koning & M.S.M. Sosef in Blumea,* 31(2): 292 (1986): Rottboellia loricata subsp./var. subgibbosa.
sulcata *(Stapf) R. de Koning & M.S.M. Sosef in Blumea,* 31(2): 287 (1986): Peltophorus sulcatus.
tessellata *(Steud.) R. de Koning & M.S.M. Sosef in Blumea,* 31(2): 293 (1986): Rottboellia tessellata.
tuberculosa *(Nash) R. de Koning & M.S.M. Sosef in Blumea,* 31(2): 293 (1986): Manisuris tuberculosa.

MOEHRINGIA (Caryophyllac.).
intricata
subsp. castellana *J.M. Montserrat in An. Jard. Bot. Madrid,* 42(2): 548 (1985 publ. 1986)—Spain.
subsp. tejedensis *(Willk.) J.M. Montserrat in An. Jard. Bot. Madrid,* 42(2): 548 (1985 publ. 1986): M. tejedensis.

MOLLIA (Caryophyllac.).
alsinifolia *(Biv.) Guss., Cat. Pl. Hort. Reg. Bocc.:* 42 (1821); *cf. D.J. Mabberley in Taxon,* 31(1): 71 (1982): Basionym not stated.
polycarpoides *(Biv.) Guss., Cat. Pl. Hort. Reg. Bocc.:* 42 (1821); *cf. D.J. Mabberley in Taxon,* 31(1): 71 (1982): Basionym not stated.

MONACTIS (Compos.).
rhombifolia *A. Sagástegui-Alva & M.O. Dillon in Phytologia,* 61(1): 5 (1986)—Peru.

MONANTHOCITRUS (Rutac.).
bispinosa *B.C. Stone in Proc. Acad. Nat. Sci. Philadelphia,* 137(2): 217 (1985)—New Guinea (W. Irian).
paludosa *(Lauterb.) B.C. Stone in Proc. Acad. Nat. Sci. Philadelphia,* 137(2): 217 (1985): Citrus paludosa.

MONANTHOTAXIS (Annonac.).
discrepantinervia *B. Verdcourt in Kew Bull.,* 41(2): 295 (1986)—Tanzania.

MONOCERA (Elaeocarpac.).
dentata *(Forst. & Forst. f.) Endl., Cat. Horti Vindob.,* 2: 365 (1842); *cf. D.J. Mabberley in Taxon,* 33(3): 435 (1984): Basionym not stated.
reticulata *(Sm.) Endl., Cat. Horti Vindob.,* 2: 365 (1842); *cf. D.J. Mabberley in Taxon,* 33(3): 435 (1984): Elaeocarpus reticulatus.

MONTBRETIA (Iridac.).
crocata *(L.) Endl., Cat. Horti Vindob.,* 1: 166 (1842); *cf. D.J. Mabberley in Taxon,* 33(3): 435 (1984): Basionym not stated.
deusta *(Ait.) Endl., Cat. Horti Vindob.,* 1: 166 (1842); *cf. D.J. Mabberley in Taxon,* 33(3): 435 (1984): Basionym not stated.
fenestrata *(Jacq.) Endl., Cat. Horti Vindob.,* 1: 166 (1842); *cf. D.J. Mabberley in Taxon,* 33(3): 435 (1984): Basionym not stated.

MORAEA (Iridac.).
deserticola *P. Goldblatt in Ann. Missouri Bot. Gard.,* 73(1): 113 (1986)—South Africa (Cape Province).
fugax
subsp. filicaulis *(Baker) P. Goldblatt in Ann. Missouri Bot. Gard.,* 73(1): 155 (1986): M. filicaulis.
graniticola *P. Goldblatt in Ann. Missouri Bot. Gard.,* 73(1): 107 (1986)—Namibia.
hexaglottis *P. Goldblatt in Ann. Missouri Bot. Gard.,* 73(1): 108 (1986)—Namibia.
macrocarpa *P. Goldblatt in Ann. Missouri Bot. Gard.,* 73(1): 151 (1986)—South Africa (Cape Province).
pseudospicata *P. Goldblatt in Ann. Missouri Bot. Gard.,* 73(1): 111 (1986)—South Africa (Cape Province).
rigidifolia *P. Goldblatt in Ann. Missouri Bot. Gard.,* 73(1): 109 (1986)—Namibia.
worcesterensis *P. Goldblatt in Ann. Missouri Bot. Gard.,* 73(1): 115 (1986)—South Africa (Cape Province).

MORICANDIA (Crucif.).
moricandioides
subsp. cavanillesiana *(Font Quer & A. Bolós) W. Greuter & H.M. Burdet in W. Greuter, H.M. Burdet & G. Long (eds.), Med-checklist,* 3: 144 (1986): M. ramburii subsp. cavanillesiana.

MORINDA (Rubiac.).
sessiliflora *Bertol., Syll. Pl. Hort. Bot. Bonon.,* 1827: 5 (1827) *et Nov. Comm. Acad. Sci. Bonon.,* 4: 19, t. 4 (1840); *cf. D.J. Mabberley in Taxon,* 32(1): 82 (1983)— Windward Is.

MOSCHOXYLUM A. Jussieu in Mirb. & Cass. apud Guill. in Bull. Sci. Nat. Geol., 23: 237 (1830); cf. D.J. Mabberley in Taxon, 31(1): 66 (1982). *MELIACEAE.*

MOURETIA (Rubiac.).
tonkinensis
var. inaequalis *H.S. Lo in Bull. Bot. Res. North-East. Forest. Inst.,* 6(4): 48 (1986)—Vietnam.

MUNRONIA (Meliac.).
hainanensis
var. microphyllina *X.M. Chen in J. Wuhan Bot. Res.,* 4(2): 173 (1986)—China.

MURRAYA (Rutac.).
sect. Bergera *(Koenig ex L.) P.P.H. But & Y.C. Kong in Acta Phytotax. Sin.,* 24(3): 189 (1986): gen. Bergera.

MUSSAENDA (Rubiac.).
chingii *C.Y. Wu ex H.H. Hsue & H. Wu in Acta Phytotax. Sin.,* 24(3): 236 (1986)—China.
inflata *H.H. Hsue & H. Wu in Acta Phytotax. Sin.,* 24(3): 234 (1986)—China.
mollissima *C.Y. Wu ex H.H. Hsue & H. Wu in Acta Phytotax. Sin.,* 24(3): 235 (1986)—China.
multinervis *C.Y. Wu ex H.H. Hsue & H. Wu in Acta Phytotax. Sin.,* 24(3): 237 (1986)—China.
pingbianensis *C.Y. Wu ex H.H. Hsue & H. Wu in Acta Phytotax. Sin.,* 24(3): 233 (1986)—China.
pubescens
f. clematidiflora *Chun ex H.H. Hsue & H. Wu in Acta Phytotax. Sin.,* 24(3): 239 (1986)—China.

MYOGALUM (Liliac.).
coarctatum *(Jacq.) Endl., Cat. Horti Vindob.,* 1: 134 (1842); *cf. D.J. Mabberley in Taxon,* 33(3): 435 (1984): Basionym not stated.
flavescens *(Jacq.) Endl., Cat. Horti Vindob.,* 1: 134 (1842); *cf. D.J. Mabberley in Taxon,* 33(3): 435 (1984): Basionym not stated.
thyrsoides *(Jacq.) Endl., Cat. Horti Vindob.,* 1: 134 (1842); *cf. D.J. Mabberley in Taxon,* 33(3): 435 (1984): Ornithogalum thyrsoides.

MYOPORUM (Myoporac.).
glandulosum *(Spin) Spin, Cat. Jard. Bot. St. Seb.:* 18, 29 (1818); *cf. D.J. Mabberley in Taxon,* 33(3): 442 (1984): Bertolonia glandulosa.
turbinatum *R.J. Chinnock in Nuytsia,* 5(3): 398 (1986)—Australia (W. Australia).

MYOSOTIS (Boraginac.).
tenuis *(Ledeb.) Mörch, Cat. Hort. Bot. Hafn.:* 64 (1839); *cf. D.J. Mabberley in Taxon,* 33(3): 439 (1984): Basionym not stated.

MYOXANTHUS (Orchidac.).
aspasicensis *(Rchb. f.) C.A. Luer, Ic. Pleurothallid. I-Syst. Pleurothallidinae (Monogr. Syst. Bot. Missouri Bot. Gard. 15):* 38 (1986): Pleurothallis aspasicensis.

MYOXANTHUS :–
lappiformis *(Heller & L.O. Wms.) C.A. Luer, Ic. Pleurothallid. I-Syst. Pleurothallidinae (Monogr. Syst. Bot. Missouri Bot. Gard. 15):* 38 (1986): Pleurothallis lappiformis.
pan *(Luer) C.A. Luer, Ic. Pleurothallid. I-Syst. Pleurothallidinae (Monogr. Syst. Bot. Missouri Bot. Gard. 15):* 38 (1986): Pleurothallis pan.
pastacensis *(Luer) C.A. Luer, Ic. Pleurothallid. I-Syst. Pleurothallidinae (Monogr. Syst. Bot. Missouri Bot. Gard. 15):* 38 (1986): Pleurothallis pastacensis.
sempergemmatus *(Luer) C.A. Luer, Ic. Pleurothallid. I-Syst. Pleurothallidinae (Monogr. Syst. Bot. Missouri Bot. Gard. 15):* 38 (1986): Pleurothallis sempergemmata.
spilanthus *(Barb. Rodr.) C.A. Luer, Ic. Pleurothallid. I-Syst. Pleurothallidinae (Monogr. Syst. Bot. Missouri Bot. Gard. 15):* 38 (1986): Pleurothallis spilantha.
stonei *(Luer) C.A. Luer, Ic. Pleurothallid. I-Syst. Pleurothallidinae (Monogr. Syst. Bot. Missouri Bot. Gard. 15):* 38 (1986): Pleurothallis stonei.
uncinatus *(Fawc.) C.A. Luer, Ic. Pleurothallid. I-Syst. Pleurothallidinae (Monogr. Syst. Bot. Missouri Bot. Gard. 15):* 38 (1986): Pleurothallis uncinata.

MYRIACTIS (Compos.).
cuchumatanica *(Beaman & De Jong) J. Cuatrecasas in An. Jard. Bot. Madrid,* 42(2): 422 (1985 publ. 1986): Lagenophora cuchumatanica.
minuscula *(Cuatr.) J. Cuatrecasas in An. Jard. Bot. Madrid,* 42(2): 422 (1985 publ. 1986): Lagenifera minuscula.
panamensis *(Blake) J. Cuatrecasas in An. Jard. Bot. Madrid,* 42(2): 422 (1985 publ. 1986): Lagenophora panamensis.
sakirana *(Cuatr.) J. Cuatrecasas in An. Jard. Bot. Madrid,* 42(2): 422 (1985 publ. 1986): Lagenifera sakirana.
westonii *(Cuatr.) J. Cuatrecasas in An. Jard. Bot. Madrid,* 42(2): 422 (1985 publ. 1986): Lagenifera westonii.

MYRICA (Myricac.).
nagi *C.P. Thunberg ex A. Murray in Linnaeus, Syst. Veg., ed. 14:* 884 (1784)—Japan. †

MYRIOPHYLLUM (Haloragac.).
alpinum *A.E. Orchard in Brunonia,* 8(2): 208 (1986)—Australia (New South Wales, Victoria).
austropygmaeum *A.E. Orchard in Brunonia,* 8(2): 280 (1986)—Tasmania.
balladoniense *A.E. Orchard in Brunonia,* 8(2): 221 (1986)—Australia (W. Australia).
callitrichoides
subsp. striatum *A.E. Orchard in Brunonia,* 8(2): 252 (1986)—Australia (W. Australia).
costatum *A.E. Orchard in Brunonia,* 8(2): 245 (1986)—Australia (W. Australia).

MYRIOPHYLLUM :–
crispatum *A.E. Orchard in Brunonia,* 8(2): 210 (1986)—Australia (New South Wales, Queensland, Victoria, ?S. Australia, W. Australia), Tasmania.
decussatum *A.E. Orchard in Brunonia,* 8(2): 282 (1986)—Australia (W. Australia).
echinatum *A.E. Orchard in Brunonia,* 8(2): 259 (1986)—Australia (W. Australia).
gracile
var. laeve *A.E. Orchard in Brunonia,* 8(2): 235 (1986)—Australia (Queensland).
var. lineare *A.E. Orchard in Brunonia,* 8(2): 234 (1986)—Australia (New South Wales, Queensland).
implicatum *A.E. Orchard in Brunonia,* 8(2): 240 (1986)—Australia (Queensland, ?New South Wales).
limnophilum *A.E. Orchard in Brunonia,* 8(2): 252 (1986)—Australia (W. Australia).
lophatum *A.E. Orchard in Brunonia,* 8(2): 277 (1986)—Australia (New South Wales, Victoria).
muricatum *A.E. Orchard in Brunonia,* 8(2): 223 (1986)—Australia (N. Terr., Queensland).
papillosum *A.E. Orchard in Brunonia,* 8(2): 214 (1986)—Australia (New South Wales, Victoria, S. Australia, ?Queensland).
pedunculatum
subsp. longibracteolatum *(Schindler) A.E. Orchard in Brunonia,* 8(2): 273 (1986): M. longibracteolatum.
petraeum *A.E. Orchard in Brunonia,* 8(2): 247 (1986)—Australia (W. Australia).
simulans *A.E. Orchard in Brunonia,* 8(2): 203 (1986)—Australia (S. Australia, Victoria, New South Wales, Queensland), Tasmania.
striatum *A.E. Orchard in Brunonia,* 8(2): 243 (1986)—Australia (New South Wales, Queensland, Victoria).

×**MYRMECOLAELIA** (Orchidac.).
fuchsii *F. Hamer in Ic. Pl. Trop.,* 13 (Orch. Nicaragua, pt. 6): t. 1236 (1985), without latin descr.—Nicaragua. (Laelia rubescens × Myrmecophila wendlandii).

MYRTUS (Myrtac.).
laevis *C.P. Thunberg ex A. Murray in Linnaeus, Syst. Veg., ed. 14:* 461 (1784)—Japan. †
pimentoides *(DC.) T. Nees, Pl. Offic., Suppl.:* t. 89 (1833); *cf. D.J. Mabberley in Taxon,* 30(1): 16 (1981): Basionym not stated.

MYSCOLUS (Compos.).
hispanicus *(L.) Endl., Cat. Horti Vindob.,* 1: 410 (1842); *cf. D.J. Mabberley in Taxon,* 33(3): 435 (1984): Scorzonera hispanica.

MZYMTELLA A.A. Kolakovskiĭ in Soobshch. Akad. Nauk Gruz. SSR, 103(1): 150 (1981). *CAMPANULACEAE.*
sclerophylla *A.A. Kolakovskiĭ in Soobshch. Akad. Nauk Gruz. SSR,* 103(1): 151 (1981)—U.S.S.R. (Caucasus).

N

NAJAS (Najadac.).
arguta
var. podostemon *(Magnus) R.M. Lowden in Aquatic Bot.*, 24(2): 180 (1986): N. podostemon.
guadalupensis
f. floridana *(Haynes & Wentz.) R.M. Lowden in Aquatic Bot.*, 24(2): 174 (1986): N. guadalupensis var. floridana.
orientalis *L. Triest & P. Uotila in Ann. Bot. Fenn.*, 23(2): 169 (1986)—Japan, U.S.S.R. (Soviet Far East), China.

NAPEANTHUS (Gesneriac.).
bicolor *(L.O. Wms.) K. Barringer in Phytologia*, 59(5): 365 (1986): Tetranema bicolor.

NARCISSUS (Amaryllidac.).
nothosubsect Angustizettae *J. Fernández Casas in Fontqueria,* 6: 39 (1984). (N. subsect. Angustifoliae × subsect. Tazettae).
nothosect. Braxini *J. Fernández Casas in Fontqueria,* 6: 36 (1984). (N. sect. Braxireon × sect. Serotini).
nothosect. Bulbissi *J. Fernández Casas in Fontqueria,* 6: 36 (1984). (N. sect. Bulbocodii × sect. Pseudonarcissi).
nothosect. Bulbomedes *J. Fernández Casas in Fontqueria,* 6: 36 (1984). (N. sect. Bulbocodii × sect. Ganymedes).
nothosect. Bulboquillae *J. Fernández Casas in Fontqueria,* 6: 37 (1984). (N. sect. Bulbocodii × sect. Jonquillae).
nothosubsect Chlorifoliae *J. Fernández Casas in Fontqueria,* 6: 39 (1984). (N. subsect. Angustifoliae × subsect. Chloranthi).
sect. Dubii *J. Fernández Casas in Fontqueria,* 6: 39 (1984).
nothosect. Dubiquillae *J. Fernández Casas in Fontqueria,* 6: 37 (1984). (N. sect. Dubii × sect. Jonquillae).
nothosect. Ganycissi *J. Fernández Casas in Fontqueria,* 6: 37 (1984). (N. sect. Ganymedes × sect. Pseudonarcissi).
nothosect. Ganyquillae *J. Fernández Casas in Fontqueria,* 6: 37 (1984). (N. sect. Ganymedes × sect. Jonquillae).
nothosect. Joncissus *J. Fernández Casas in Fontqueria,* 6: 37 (1984). (N. sect. Jonquillae × sect. Narcissus).
nothosect. Jonissi *J. Fernández Casas in Fontqueria,* 6: 37 (1984). (N. sect. Jonquillae × sect. Pseudonarcissi).
nothosect. Jonquizettae *J. Fernández Casas in Fontqueria,* 6: 39 (1984). (N. sect. Jonquillae × sect. Tazettae).
nothosect. Narcisettae *J. Fernández Casas in Fontqueria,* 6: 39 (1984). (N. sect. Narcissus × sect. Tazettae).
nothosect. Nassi *J. Fernández Casas in Fontqueria,* 6: 39 (1984). (N. sect. Narcissus × sect. Pseudonarcissi).

NARCISSUS :–
nothosect. Pseudozettae *J. Fernández Casas in Fontqueria,* 6: 39 (1984). (N. sect. Pseudonarcissi × sect. Tazettae).
nothosect. Serozettae *J. Fernández Casas in Fontqueria,* 6: 39 (1984). (N. sect. Serotini × sect. Tazettae).
ajax *Sweet, Hort. Suburb. Lond.*: 67 (1818); *cf. D.J. Mabberley in Taxon,* 31(1): 71 (1982)—Locality not stated.
assoanus
var. pallens *(Freyn ex Willk.) J. Fernández Casas in Fontqueria,* 1: 10 (1982): N. pallens.
baeticus *J. Fernández Casas in Fontqueria,* 1: 11 (1982)—Spain.
bulbocodium
subsp. citrinus *(Baker) J. Fernández Casas in Fontqueria,* 2: 39 (1982): N. bulbocodium var. citrinus.
calcicarpetanus *J. Fernández Casas in Fontqueria,* 2: 33 (1982)—Spain.
cantabricus
subsp. hedraeanthus *(Webb & Heldr.) J. Fernández Casas in Fontqueria,* 1: 12 (1982): Corbularia hedraeantha.
capax *Salisb. ex Sweet, Hort. Suburb. Lond.*, (1818); *cf. D.J. Mabberley in Taxon,* 31(1): 71 (1982)— Locality not stated.
×carpetanus *(Barra & G. López) J. Fernández Casas in Fontqueria,* 6: 49 (1984): N. bulbocodium nothosubsp. carpetanus.
×cazorlanus *J. Fernández Casas in Fontqueria,* 2: 37 (1982)—Spain. (N. cantabricus subsp. hedraeanthus × triandrus subsp. pallidulus).
citrinus *(Baker) J. Fernández Casas in Fontqueria,* 6: 49 (1984): N. bulbocodium var. citrinus.
×consolationis *J. Fernández Casas in Fontqueria,* 2: 37 (1982)—Spain. (N. nivalis × triandrus subsp. pallidulus).
cordubensis *J. Fernández Casas in Fontqueria,* 1: 10 (1982)—Spain.
eugeniae *J. Fernández Casas in Fontqueria,* 1: 11 (1982)—Spain.
fernandesii
var. cordubensis *(Fernández Casas) J. Fernández Casas in Fontqueria,* 8: 31 (1985): N. cordubensis.
hispanicus
var. bujei *(J. Fernández Casas) J. Fernández Casas in Fontqueria,* 6: 41 (1984): N. longispathus var. bujei.
longispathus
var. bujei *J. Fernández Casas in Fontqueria,* 2: 33 (1982)—Spain.
×lopezii *J. Fernández Casas in Fontqueria,* 2: 37 (1982)—Spain. (N. bulbocodium × obvallaris).
marianicus *J. Fernández Casas in Fontqueria,* 1: 10 (1982)—Spain.
×montcaunicus *J. Fernández Casas in Fontqueria,* 2: 31 (1982)—Spain. (N. bulbocodium × eugeniae).

NARCISSUS :-
×ponsii-sorollae *J. Fernández-Casas, Exsicc. nobis distrib.*, 3: 12 (1980)—Spain. (N. pallidulus × requienii).
pseudonarcissus
subsp. leonensis *(Pugsley) J. Fernández Casas & Laínz in Fontqueria*, 6: 50 (1984): N. leonensis.
subsp. primigenius *(Fernández Suárez ex Laínz) J. Fernández Casas & Laínz in Fontqueria*, 6: 49 (1984): N. pseudonarcissus var. primigenius.
×pugsleyi *J. Fernández Casas in Fontqueria*, 6: 41 (1984)—Spain. (N. alpestris × assoanus).
radinganorum *J. Fernández Casas in Fontqueria*, 6: 41 (1984)—Spain.
redoutei *Sweet, Hort. Brit., ed. 2:* 514 (1830); *cf. D.J. Mabberley in Taxon*, 31(1): 71 (1982)— Locality not stated.
×rozeirae *J. Fernández Casas & Pérez Chiscano in Fontqueria*, 6: 43 (1984)—Spain. (N. bulbocodium × pallidulus).
segimonensis *J. Fernández Casas in Fontqueria*, 2: 31 (1982)—Spain.
×stenanthus *(Lange) J. Fernández Casas in Fontqueria*, 1: 12 (1982): N. pseudonarcissus f. stenantha. (N. confusus × nivalis).
×susannae *J. Fernández Casas, Exsicc. nobis distrib.*, 3: 13 (1980)—Spain. (N. cantabricus × pallidulus).

NARTHEX (Umbellif.).
asafoetida *Falconer ex Lindl., Gard. Chron.*, 1846: 743 (7 Nov. 1846) *et in Athenaeum*, 1846(996): 1222 (1846); *cf. D.J. Mabberley in Taxon*, 31(1): 71 (1982)— Locality not stated.

NAVARRETIA (Polemoniac.).
pusilla *Hort. Par. ex Schrad., Ind. Sem. Hort. Acad. Gott.*, 1833: 4 (1833); *cf. D.J. Mabberley in Taxon*, 33(3): 442 (1984)— Locality not stated.

NAVIA (Bromeliac.).
grafii *J.A. Steyermark & B. Holst in Ernstia*, 38: 44 (1986)—Venezuela.

NEANOTIS (Rubiac.).
hirsuta
var. glabra *(Honda) H. Hara in J. Jap. Bot.*, 61(3): 73 (1986): Oldenlandia hirsuta var. glabra.
f. glabricalycina *(Honda) H. Hara in J. Jap. Bot.*, 61(3): 72 (1986): Oldenlandia hirsuta var. glabricalycina.

NECEPSIA (Euphorbiac.).
afzelii
var. sitae *A. Bouchat & J. Léonard in Bull. Jard. Bot. Nation. Belg.*, 56(1–2): 186 (1986)—Congo - Brazzaville.
subsp. zenkeri *A. Bouchat & J. Léonard in Bull. Jard. Bot. Nation. Belg.*, 56(1–2): 186 (1986)—Cameroun, Gabon.

NECEPSIA :-
castaneifolia *(Baillon) A. Bouchat & J. Léonard in Bull. Jard. Bot. Nation. Belg.*, 56(1–2): 191 (1986): Palissya castaneifolia.
subsp. capuronii *A. Bouchat & J. Léonard in Bull. Jard. Bot. Nation. Belg.*, 56(1–2): 194 (1986)—Madagascar.
subsp. chirindica *(Radcliffe-Smith) A. Bouchat & J. Léonard in Bull. Jard. Bot. Nation. Belg.*, 56(1–2): 194 (1986): Neopalissya castaneifolia subsp. chirindica.
subsp. kimbozensis *(Radcliffe-Smith) A. Bouchat & J. Léonard in Bull. Jard. Bot. Nation. Belg.*, 56(1–2): 193 (1986): Neopalissya castaneifolia subsp. kimbozensis.
zairensis *A. Bouchat & J. Léonard in Bull. Jard. Bot. Nation. Belg.*, 56(1–2): 187 (1986)—Zaïre, Angola, Gabon, Congo - Brazzaville.
var. lujae *A. Bouchat & J. Léonard in Bull. Jard. Bot. Nation. Belg.*, 56(1–2): 190 (1986)—Zaïre.

NECTANDRA (Laurac.).
hypoleuca *B.E. Hammel in J. Arnold Arbor.*, 67(1): 126 (1986)—Costa Rica.
membranacea
subsp. cuspidata *(Nees) J.G. Rohwer in Mitt. Inst. Allg. Bot. Hamburg*, 20: 72 (1986): N. cuspidata.
mollis
subsp. laurel *(Nees) J.G. Rohwer in Mitt. Inst. Allg. Bot. Hamburg*, 20: 61 (1986): N. laurel.
subsp. oppositifolia *(Nees) J.G. Rohwer in Mitt. Inst. Allg. Bot. Hamburg*, 20: 61 (1986): N. oppositifolia.

NEMAUCHENES (Compos.).
aspera *(L.) Endl., Cat. Horti Vindob.*, 1: 423 (1842); *cf. D.J. Mabberley in Taxon*, 33(3): 435 (1984): Crepis aspera.

NEMEDRA A. Jussieu in Mirb. & Cass. apud Guill. in Bull. Sci. Nat. Geol., 23: 236 (1830); cf. D.J. Mabberley in Taxon, 31(1): 66 (1982). *MELIACEAE.*

NEMESIA (Scrophulariac.).
glabriuscula *O.M. Hilliard & B.L. Burtt in Notes Roy. Bot. Gard. Edinburgh*, 43(3): 393 (1986)—South Africa (Natal).
rupicola *O.M. Hilliard in Notes Roy. Bot. Gard. Edinburgh*, 43(3): 396 (1986)—South Africa (Natal, Transvaal, Orange Free State), Lesotho.
silvatica *O.M. Hilliard in Notes Roy. Bot. Gard. Edinburgh*, 43(3): 398 (1986)—South Africa (Natal, Cape Province).
umbonata *(Hiern) O.M. Hilliard & B.L. Burtt in Notes Roy. Bot. Gard. Edinburgh*, 43(3): 399 (1986): Diclis umbonata.

NENAX (Rubiac.).
acerosa
subsp. macrocarpa *(Eckl. & Zeyh.) C. Puff in Fl. S. Afr.,* 31(1:2): 46 (1986): Ambraria hirta var. macrocarpa.
arenicola *C. Puff in Fl. S. Afr.,* 31(1:2): 47 (1986)—South Africa (Cape Province).
cinerea *(Thunb.) C. Puff in Fl. S. Afr.,* 31(1:2): 40 (1986): Cliffortia cinerea.
coronata *C. Puff in Fl. S. Afr.,* 31(1:2): 41 (1986)—South Africa (Cape Province).
elsieae *C. Puff in Fl. S. Afr.,* 31(1:2): 43 (1986)—South Africa (Cape Province).
hirta
subsp. calciphila *C. Puff in Fl. S. Afr.,* 31(1:2): 44 (1986)—South Africa (Cape Province).
namaquensis *C. Puff in Fl. S. Afr.,* 31(1:2): 41 (1986)—South Africa (Cape Province).

NEOCUATRECASIA (Compos.).
cuzcoensis *R.M. King & H. Robinson in Phytologia,* 60(1): 83 (1986)—Peru.

NEOLEHMANNIA (Orchidac.).
curvicolumna *(A.H.& S.) F. Hamer in Ic. Pl. Trop.,* 13 (Orch. Nicaragua, pt. 6): t. 1237 (1985): Epidendrum curvicolumna.
difformis
var. angustata *T. Hashimoto in Ann. Tsukuba Bot. Gard.,* 4: 2 (1986)—Peru.

NEOLITSEA (Laurac.).
calcicola *Z.R. Xu in Acta Sci. Nat. Univ. Sunyatseni,* 1986(2): 99 (1986), without type—China.

NEOLLOYDIA (Cactac.).
laui *(Glass & Foster) E.F. Anderson in Bradleya,* 4: 22 (1986): Turbinicarpus laui.
lophophoroides *(Werderm.) E.F. Anderson in Bradleya,* 4: 22 (1986): Thelocactus lophophoroides.
mandragora *(Fric ex A. Berger) E.F. Anderson in Bradleya,* 4: 14 (1986): Echinocactus mandragora.
pseudomacrochele *(Backeb.) E.F. Anderson in Bradleya,* 4: 22 (1986): Strombocactus pseudomacrochele.
pseudopectinata *(Backeb.) E.F. Anderson in Bradleya,* 4: 15 (1986): Pelecyphora pseudopectinata.
schmiedickeana *(Boedeker) E.F. Anderson in Bradleya,* 4: 19 (1986): Echinocactus schmiedickeanus.
var. dickisoniae *(Glass & Foster) E.F. Anderson in Bradleya,* 4: 20 (1986): Turbinicarpus schmiedickeanus var. dickisoniae.
var. flaviflora *(G. Frank & A. Lau) E.F. Anderson in Bradleya,* 4: 20 (1986): Turbinicarpus flaviflorus.
var. gracilis *(Glass & Foster) E.F. Anderson in Bradleya,* 4: 19 (1986): Turbinicarpus gracilis.

NEOLLOYDIA :-
var. klinkeriana *(Backeb. & H.J. Jacobsen) E.F. Anderson in Bradleya,* 4: 20 (1986): Turbinicarpus klinkerianus.
var. macrochele *(Werderm.) E.F. Anderson in Bradleya,* 4: 20 (1986): Echinocactus macrochele.
var. schwarzii *(Shurly) E.F. Anderson in Bradleya,* 4: 20 (1986): Strombocactus schwarzii.
subterranea
var. zaragosae *(Glass & Foster) E.F. Anderson in Bradleya,* 4: 14 (1986): Gymnocactus subterraneus var. zaragosae.
valdeziana *(H. Moeller) E.F. Anderson in Bradleya,* 4: 14 (1986): Pelecyphora valdeziana.
viereckii
var. major *(Glass & Foster) E.F. Anderson in Bradleya,* 4: 15 (1986): Gymnocactus viereckii var. major.

NEOREGELIA (Bromeliac.).
coimbrae *E. Pereira & E.M.C. Leme in Rev. Brasil. Biol.,* 45(4): 631 (1985 publ. 1986)—Brazil.
gavionensis *G. Martinelli & E.M.C. Leme in J. Bromeliad Soc.,* 36(2): 71 (1986)—Brazil.
martinellii *W. Weber in Feddes Repert.,* 97(3–4): 119 (1986)—Brazil.

NEOURBANIA (Orchidac.).
nicaraguensis *F. Hamer & Garay in Ic. Pl. Trop.,* 13 (Orch. Nicaragua, pt. 6): t. 1238 (1985)—Nicaragua, Costa Rica.

NEPENTHES (Nepenthac.).
adnata *R. Tamin & M. Hotta in M. Hotta (ed.), Divers. & Dynam. Pl. Life Sumatra,* 1: 76 (1986), without latin descr. or type—Sumatra.
rafflesiana
var. longicirrhosa *R. Tamin & M. Hotta in M. Hotta (ed.), Divers. & Dynam. Pl. Life Sumatra,* 1: 93 (1986), without latin descr. or type—Sumatra.
rosulata *R. Tamin & M. Hotta in M. Hotta (ed.), Divers. & Dynam. Pl. Life Sumatra,* 1: 95 (1986), without latin descr. or type—Sumatra.
spinosa *R. Tamin & M. Hotta in M. Hotta (ed.), Divers. & Dynam. Pl. Life Sumatra,* 1: 103 (1986), without latin descr. or type—Sumatra.

NEPETA (Labiat.).
incana *C.P. Thunberg ex A. Murray in Linnaeus, Syst. Veg., ed. 14:* 529 (1784)—Japan. †
willdenowiana *G.F. Hoffm., Hort. Mosq.:* 25 (1808); *cf. D.J. Mabberley in Taxon,* 31(1): 71 (1982)— Locality not stated.

NEPHELIUM (Sapindac.).
aculeatum *P.W. Leenhouts in Blumea,* 31(2): 382 (1986)—Borneo (Sabah).

NEPHELIUM :–
cuspidatum
var. bassacense *(Pierre) P.W. Leenhouts in Blumea,* 31(2): 387 (1986): N. bassacense.
subvar. beccarianum *(Radlk.) P.W. Leenhouts in Blumea,* 31(2): 390 (1986): N. beccarianum.
subvar. dasyneurum *(Radlk.) P.W. Leenhouts in Blumea,* 31(2): 388 (1986): N. dasyneurum.
var. eriopetalum *(Miq.) P.W. Leenhouts in Blumea,* 31(2): 389 (1986): N. eriopetalum.
var. multinerve *(Radlk.) P.W. Leenhouts in Blumea,* 31(2): 389 (1986): N. multinerve.
var. ophiodes *(Radlk.) P.W. Leenhouts in Blumea,* 31(2): 390 (1986): N. ophiodes.
subvar. ophiodes *(Radlk.) P.W. Leenhouts in Blumea,* 31(2): 390 (1986): N. ophiodes.
var. robustum *(Radlk.) P.W. Leenhouts in Blumea,* 31(2): 391 (1986): N. robustum.
havilandii *P.W. Leenhouts in Blumea,* 31(2): 394 (1986)—Borneo (Sarawak).
lappaceum
var. pallens *(Hiern) P.W. Leenhouts in Blumea,* 31(2): 402 (1986): N. mutabile var. pallens.
var. xanthioides *(Radlk.) P.W. Leenhouts in Blumea,* 31(2): 403 (1986): N. xanthioides.
meduseum *P.W. Leenhouts in Blumea,* 31(2): 409 (1986)—Borneo (Sarawak, Brunei, Kalimantan).
papillatum *P.W. Leenhouts in Blumea,* 31(2): 413 (1986)—Borneo (Sabah).
ramboutan-ake *(Labill.) P.W. Leenhouts in Blumea,* 31(2): 415 (1986): Litchi ramboutan-ake.
uncinatum *Radlk. ex P.W. Leenhouts in Blumea,* 31(2): 421 (1986)—Borneo (Sabah, Sarawak, Brunei, Kalimantan), Malaya, Sumatra.

NEPHTHYTIS (Arac.).
hallaei *(Bogner) J. Bogner in Aroideana,* 8(3): 78 (1985 publ. 1986): Callopsis hallaei. †

NERIUM (Apocynac.).
coccineum *Roxb. ex Hornem., Hort. Hafn., Suppl.:* 126 (1819); *cf. D.J. Mabberley in Taxon,* 33(3): 437 (1984)— Locality not stated.

NEVILLEA (Restionac.).
singularis *E.E. Esterhuysen in Bothalia,* 15(3–4): 484 (1985)—South Africa (Cape Province).

NICOBARIODENDRON M.K. Vasudeva Rao & T. Chakrabarty in J. Econ. Taxon. Bot., 7(3): 513 (1985 publ. 1986). *CELASTRACEAE.*
sleumeri *M.K. Vasudeva Rao & T. Chakrabarty in J. Econ. Taxon. Bot.,* 7(3): 514 (1985 publ. 1986)—Nicobar Is.

NICOLASIA (Compos.).
coronata
subsp. planifolia *S. Lisowski in Bull. Jard. Bot. Nation. Belg.,* 56(1–2): 203 (1986)—Zaïre.

NICOTIANA (Solanac.).
fastigiata *Lehm ex Nees, Del. Sem. Hort. Bot. Vrat.,* 1844: 2, 3 (1844); *cf. D.J. Mabberley in Taxon,* 32(1): 85 (1983)— Locality not stated.

NIDULARIUM (Bromeliac.).
alvimii *W. Weber in Feddes Repert.,* 97(3–4): 122 (1986)—Brazil.
angraensis *E. Pereira & E.M.C. Leme in Bradea,* 4(32): 229 (1986)—Brazil.
billbergioides
f. azureum *E. Pereira & E.M.C. Leme in Bradea,* 4(32): 235 (1986)—Brazil.
correia-araujoi *E. Pereira & E.M.C. Leme in Bradea,* 4(32): 230 (1986), as 'correia-arauji'—Brazil.
lymanioides *E. Pereira & E.M.C. Leme in Bradea,* 4(32): 229 (1986)—Brazil.
microps
f. acuminatum *E. Pereira & E.M.C. Leme in Bradea,* 4(32): 226 (1986)—Brazil.
f. bicense *(Ule) E. Pereira & E.M.C. Leme in Bradea,* 4(32): 225 (1986): N. microcephalum var. bicensis.
f. elatum *E. Pereira & E.M.C. Leme in Bradea,* 4(32): 226 (1986)—Brazil.
f. pallidum *(L.B. Smith) E. Pereira & E.M.C. Leme in Bradea,* 4(32): 226 (1986): N. microps var. pallidum.
proprocerum
var. cariacicaensis *W. Weber in Feddes Repert.,* 97(3–4): 124 (1986)—Brazil.
pulcherrimum *E. Pereira & E.M.C. Leme in Bradea,* 4(32): 227 (1986)—Brazil.
selloanum *(Baker) E. Pereira & E.M.C. Leme in Bradea,* 4(32): 235 (1986): Quesnelia selloana.
simulans *E. Pereira & E.M.C. Leme in Bradea,* 4(32): 228 (1986)—Brazil.
weberi *E. Pereira & E.M.C. Leme in Bradea,* 4(32): 231 (1986)—Brazil.

NIGELLA (Ranunculac.).
atropurpurea *Huber, Cat. 1866:* 4 (1866); *cf. D.J. Mabberley in Taxon,* 34(3): 451 (1985)— Locality not stated.
nana *J.W. Loud., Lad. Fl. Gard. Orn. Ann.:* 11 (1840); *cf. D.J. Mabberley in Taxon,* 32(1): 87 (1983)— Locality not stated.

NOCCAEA (Crucif.).
zaffranii *F.K. Meyer in Willdenowia,* 15(2): 389 (1986)—Crete.

NOLTEA (Rhamnac.).
africana *(L.) Endl., Cat. Horti Vindob.,* 2: 385 (1842); *cf. D.J. Mabberley in Taxon,* 33(3): 435 (1984): Basionym not stated.

NOTHOFAGUS (Fagac.).
stylosa *C.G.G.J. van Steenis in Kew Bull.,* 41(3): 732 (1986)—New Guinea (W. Irian).

NOTHOSCORDUM (Alliac.).
gracile *(Aiton) W.T. Stearn in Taxon,* 35(2): 338 (1986): Allium gracile.

NOTICASTRUM (Compos.).
macbridei *J. Cuatrecasas in Fontqueria,* 8: 15 (1985)—Peru.

NOTOCACTUS (Cactac.).
arnostianus *K. Lisal & J. Kolarik in Internoto,* 7(1): 8 (1986)—Brazil.
gibberulus *K.H. Prestlé in Succulenta,* 65(6–7): 142 (1986)—Brazil.
graessneri
f. microdasys *P.J. Braun in Kakt. And. Sukk.,* 37(3): 57 (1986)—Brazil.
macambarensis *K.H. Prestlé in Internoto,* 7(3): 72 (1986)—Brazil.
meonacanthus *K.H. Prestlé in Internoto,* 7(2): 37 (1986)—Brazil.
ritteranus *K. Lisal & J. Kolarik in Internoto,* 7(1): 5 (1986)—Brazil.
villa-velhensis *(Backeb. & Voll) Slaba in Kaktusy,* 20(1): 7 (1984); *cf. Repert. Pl. Succ. (I.O.S.),* 35: 5 (1985): N. ottonis var. villa-velhensis.

NYMPHAEA (Nymphaeac.).
×daubenyana *W.T. Baxter ex Daubeny, Suppl. Oxf. Bot. Gard.:* 40 (1864); *cf. D.J. Mabberley in Taxon,* 33(3): 442 (1984)—Locality not stated.

NYMPHOIDES (Menyanthac.).
disperma *H.I. Aston in Muelleria,* 6(3–4): 197 (1986)—Australia (W. Australia).

O

OBERNA (Caryophyllac.).
suffrutescens *(Greuter, Mattäs & Risse) J. Holub in Preslia,* 58(4): 302 (1986): Silene vulgaris subsp. suffrutescens.

OCIMUM (Labiat.).
acutum *C.P. Thunberg ex A. Murray in Linnaeus, Syst. Veg., ed. 14:* 546 (1784)—Japan. †
crispum *C.P. Thunberg ex A. Murray in Linnaeus, Syst. Veg., ed. 14:* 546 (1784)—Japan. †
inflexum *C.P. Thunberg ex A. Murray in Linnaeus, Syst. Veg., ed. 14:* 546 (1784)—Japan. †

OCIMUM :–
nutans *Koen. ex Rottb., Pl. Hort. Univ. Rar. Prog.:* 22 (1773); *cf. D.J. Mabberley in Taxon,* 33(3): 439 (1984)— Locality not stated.
punctatum *C.P. Thunberg ex A. Murray in Linnaeus, Syst. Veg., ed. 14:* 545 (1784)—Japan. †
rugosum *C.P. Thunberg ex A. Murray in Linnaeus, Syst. Veg., ed. 14:* 546 (1784)—Japan. †
sideritidis *Koen. ex Rottb., Pl. Hort. Univ. Rar. Prog.:* 21 (1773); *cf. D.J. Mabberley in Taxon,* 33(3): 439 (1984)— Locality not stated.
virgatum *C.P. Thunberg ex A. Murray in Linnaeus, Syst. Veg., ed. 14:* 546 (1784)—Japan. †

OCOTEA (Laurac.).
camphoromoea *J.G. Rohwer in Mitt. Inst. Allg. Bot. Hamburg,* 20: 179 (1986), nom. nov.: Camphoromoea subtriplinervia *Nees.*
deflexa *J.G. Rohwer in Mitt. Inst. Allg. Bot. Hamburg,* 20: 214 (1986), nom. nov.: Oreodaphne declinata *Meissn.*
fendleri *(Meissn.) J.G. Rohwer in Mitt. Inst. Allg. Bot. Hamburg,* 20: 152 (1986): Gymnobalanus fendleri.
glaucosericea *J.G. Rohwer in Mitt. Inst. Allg. Bot. Hamburg,* 20: 144 (1986), nom. nov.: Nectandra hypoglauca *Standl. ex C.K. Allen.*
glomerata
subsp. magnifica *(O.C. Schmidt) J.G. Rohwer in Mitt. Inst. Allg. Bot. Hamburg,* 20: 159 (1986): O. magnifica.
hartshorniana *B.E. Hammel in J. Arnold Arbor.,* 67(1): 128 (1986)—Costa Rica, Ecuador.
lobbii *(Meissn.) J.G. Rohwer in Mitt. Inst. Allg. Bot. Hamburg,* 20: 113 (1986): Oreodaphne lobbii.
microneura *(Meissn.) J.G. Rohwer in Mitt. Inst. Allg. Bot. Hamburg,* 20: 130 (1986): Mespilodaphne microneura.
nitida *(Meissn.) J.G. Rohwer in Mitt. Inst. Allg. Bot. Hamburg,* 20: 160 (1986): Aydendron nitidum.
oblonga
subsp. cuprea *(Meissn.) J.G. Rohwer in Mitt. Inst. Allg. Bot. Hamburg,* 20: 151 (1986): Oreodaphne cuprea.
oblongoobovata *(Nees) J.G. Rohwer in Mitt. Inst. Allg. Bot. Hamburg,* 20: 153 (1986): Oreodaphne oblongoobovata.
odorifera *(Vell.) J.G. Rohwer in Mitt. Inst. Allg. Bot. Hamburg,* 20: 111 (1986): Laurus odorifera.
perrobusta *(C.K. Allen) J.G. Rohwer in Mitt. Inst. Allg. Bot. Hamburg,* 20: 170 (1986): Aiouea perrobusta.
peruviana *J.G. Rohwer in Mitt. Inst. Allg. Bot. Hamburg,* 20: 154 (1986), nom. nov.: Oreodaphne cuneifolia *Nees.*
producta *(C.K. Allen) J.G. Rohwer in Mitt. Inst. Allg. Bot. Hamburg,* 20: 143 (1986): Nectandra producta.

OCOTEA :–
reticularis *(Britton & Wilson) C.K. Allen ex J.G. Rohwer in Mitt. Inst. Allg. Bot. Hamburg,* 20: 126 (1986): Nectandra reticularis.
rigens *(Nees) J.G. Rohwer in Mitt. Inst. Allg. Bot. Hamburg,* 20: 215 (1986): Oreodaphne rigens.
tabacifolia *(Meissn.) J.G. Rohwer in Mitt. Inst. Allg. Bot. Hamburg,* 20: 173 (1986): Persea tabacifolia.
velutina *(Nees) J.G. Rohwer in Mitt. Inst. Allg. Bot. Hamburg,* 20: 172 (1986): Oreodaphne velutina.

OCTOMERIA (Orchidac.).
gemmula *G. Carnevali F.C. & I. Ramírez in Ernstia,* 39: 13 (1986)—Venezuela.

ODONTARRHENA (Crucif.).
muralis *(Waldst. & Kit.) Endl., Cat. Horti Vindob.,* 2: 245 (1842); *cf. D.J. Mabberley in Taxon,* 33(3): 435 (1984): Alyssum murale.

ODONTOGLOSSUM (Orchidac.).
matangense *L. Bockemühl in Orchidee,* 37(5): 209 (1986)—Ecuador.
ramosissimum
var. albomaculatum *L. Bockemühl in Orchidee,* 37(4): 161 (1986), as 'albo-maculatum'—Ecuador.
reversum *L. Bockemühl in Orchidee,* 37(5): 207 (1986)—Colombia.

ODONTONEMA (Acanthac.).
auriculatum *(Rose) T.F. Daniel in Southwest. Nat.,* 31(2): 174 (1986): Jacobinia auriculata.

OENOTHERA (Onagrac.).
subsect. Australis *W.L. Wagner & W. Dietrich in Ann. Missouri Bot. Gard.,* 73(2): 479 (1986).
sect. Contortae *W.L. Wagner in Ann. Missouri Bot. Gard.,* 73(2): 478 (1986).
sect. Eremia *W.L. Wagner in Ann. Missouri Bot. Gard.,* 73(2): 477 (1986).
sect. Ravenia *W.L. Wagner in Ann. Missouri Bot. Gard.,* 73(2): 477 (1986).
caespitosa
var. macroglottis *(Rydb.) A. Cronquist in Great Basin Nat.,* 46(2): 259 (1986): Pachylophus macroglottis.
var. navajoensis *(Wagner, Stockhouse & Klein) A. Cronquist in Great Basin Nat.,* 46(2): 259 (1986), without full basionym ref.: O. caespitosa subsp. navajoensis.
cordata *J.W. Loud., Lad. Fl. Gard. Orn. Perenn.,* 1: 167 (1843); *cf. D.J. Mabberley in Taxon,* 32(1): 87 (1983)— Locality not stated.
coryi *W.L. Wagner in Ann. Missouri Bot. Gard.,* 73(2): 475 (1986)—U.S.A. (Texas).
flava
var. acutissima *(W.L. Wagner) S.L. Welsh in Great Basin Nat.,* 46(2): 259 (1986): O. acutissima.

OENOTHERA :–
subsp. taraxacoides *(Woot. & Standl.) W.L. Wagner in Ann. Missouri Bot. Gard.,* 73(2): 479 (1986): Lavauxia taraxacoides.
gigantea *Huber, Cat. Gén. 1871 & 1872:* 4 (1871); *cf. D.J. Mabberley in Taxon,* 34(3): 453 (1985)— Locality not stated.
primiveris
var. bufonis *(Jones) A. Cronquist in Great Basin Nat.,* 46(2): 259 (1986): O. bufonis.
salicifolia *Desf. ex Lehm., Sem. Hort. Bot. Hamb.,* 1824: 20 (1824); *cf. D.J. Mabberley in Taxon,* 32(1): 85 (1983)— Locality not stated.
spectabilis *Lehm., Sem. Hort. Bot. Hamb.,* 1824: 20 (1824); *cf. D.J. Mabberley in Taxon,* 32(1): 85 (1983)— Locality not stated.

OERSTEDELLA (Orchidac.).
aberrans *(Schltr.) F. Hamer in Ic. Pl. Trop.,* 9 (Orch. Nicaragua, pt. 3): t. 891 (1983): Epidendrum aberrans.

OLAX (Olacac.).
acuminata *Wall. ex Benth. in Proc. Linn. Soc.,* 1: 89 (1840); *cf. D.J. Mabberley in Taxon,* 31(1): 71 (1982)— Locality not stated.
nana *Wall. ex Benth. in Proc. Linn. Soc.,* 1: 88 (1840); *cf. D.J. Mabberley in Taxon,* 31(1): 71 (1982)— Locality not stated.

OLDENLANDIA (Rubiac.).
drymarioides *(Standley) E.E. Terrell in Phytologia,* 59(1): 80 (1985): Houstonia drymarioides.

OLEA (Oleac.).
capensis
subsp. hochstetteri *(Baker) I. Friis & P.S. Green in Kew Bull.,* 41(1): 36 (1986): O. hochstetteri.
subsp. welwitschii *(Knobl.) I. Friis & P.S. Green in Kew Bull.,* 41(1): 36 (1986): Mayepea welwitschii.
fragrans *C.P. Thunberg ex A. Murray in Linnaeus, Syst. Veg., ed. 14:* 57 (1784)—Japan. †

OLEARIA (Compos.).
macdonnellensis *D.A. Cooke in Muelleria,* 6(3–4): 181 (1986)—Australia (N. Terr.).
tridens *D.A. Cooke in Muelleria,* 6(3–4): 182 (1986)—Australia (N. Terr.).

OLIGOPHYTON H.P. Linder in Kew Bull., 41(2): 314 (1986). *ORCHIDACEAE.*
drummondii *H.P. Linder & G. Williamson in Kew Bull.,* 41(2): 314 (1986)—Zimbabwe.

OLYRA (Gramin.).
holttumiana *T.R. Soderstrom & F.O. Zuloaga in Kew Bull.,* 41(3): 722 (1986)—Panama.

ONAGRA (Onagrac.).
elata *(Kunth) Bartl., Ind. Sem. Hort. Acad. Gott.,* 1838: 5 (1838); *cf. D.J. Mabberley in Taxon,* 32(1): 85 (1983): Oenothera elata.

ONOBROMA (Compos.).
cretica *(L.) Hornem., Hort. Hafn.*: 786 (1815); *cf. D.J. Mabberley in Taxon,* 33(3): 437 (1984): Basionym not stated.
lanata *(L.) Hornem., Hort. Hafn.*: 786 (1815); *cf. D.J. Mabberley in Taxon,* 33(3): 437 (1984): Carthamus lanatus.
mitissima *(L.) Hornem., Hort. Hafn.*: 786 (1815); *cf. D.J. Mabberley in Taxon,* 33(3): 437 (1984): Basionym not stated.

ONOBRYCHIS (Legumin.).
humilis *(L.) G. López González in An. Jard. Bot. Madrid,* 42(2): 321 (1985 publ. 1986): Hedysarum humile.
montana
subsp. macrocarpa *A. Strid in A. Strid (ed.), Mount. Fl. Greece,* 1: 537 (1986)—Greece.

ONONIS (Legumin.).
exalopecuroides *G. López González in An. Jard. Bot. Madrid,* 42(2): 322 (1985 publ. 1986)—Israel.
repens
subsp. antiquorum *(L.) W. Greuter in Willdenowia,* 16(1): 113 (1986): O. antiquorum.
subsp. arvensis *(L.) W. Greuter in Willdenowia,* 16(1): 113 (1986): O. arvensis.
subsp. diacantha *(Reichenb.) W. Greuter in Willdenowia,* 16(1): 113 (1986): O. diacantha.
subsp. leiosperma *(Boiss.) W. Greuter in Willdenowia,* 16(1): 113 (1986): O. leiosperma.
subsp. masquillierii *(Bertol.) W. Greuter in Willdenowia,* 16(1): 113 (1986): O. masquillierii.
subsp. spinosa *W. Greuter in Willdenowia,* 16(1): 113 (1986), nom. nov.: O. campestris *Koch & Ziz.*
subsp. spinosiformis *(Simonkai) W. Greuter in Willdenowia,* 16(1): 113 (1986): O. spinosiformis.

ONOPORDUM (Compos.).
bracteatum
var. arachnoideum *S. Erik & H. Sümbül in Notes Roy. Bot. Gard. Edinburgh,* 44(1): 155 (1986)—Turkey.

OPHRYS (Orchidac.).
×aetolica *H. Baumann & S. Künkele in Mitt. Arbeitskr. Heim. Orch. Baden-Württemberg,* 18(3): 507 (1986)—Greece. (O. mammosa × oestrifera).

OPHRYS :–
×alberobellensis *H. Baumann & S. Künkele in Mitt. Arbeitskr. Heim. Orch. Baden-Württemberg,* 18(3): 496 (1986)—Italy. (O. incubacea × tarentina).
×angelensis *H. Baumann & S. Künkele in Mitt. Arbeitskr. Heim. Orch. Baden-Württemberg,* 18(3): 493 (1986)—Italy. (O. incubacea × promontorii).
×antiochiana *H. Baumann & S. Künkele in Mitt. Arbeitskr. Heim. Orch. Baden-Württemberg,* 18(3): 527 (1986)—Turkey. (O. straussii × transhyrcana).
×apicula
nothosubsp. fabrei *(Bernard) H. Baumann & S. Künkele in Mitt. Arbeitskr. Heim. Orch. Baden-Württemberg,* 18(3): 416 (1986): O. × fabrei.
×apolloniensis *H. Baumann & S. Künkele in Mitt. Arbeitskr. Heim. Orch. Baden-Württemberg,* 18(3): 475 (1986)—Greece. (O. hebes × oestrifera).
×aquilana *H. Baumann & S. Künkele in Mitt. Arbeitskr. Heim. Orch. Baden-Württemberg,* 18(3): 483 (1986)—Italy. (O. holosericea × promontorii).
aveyronensis *(J.J. Wood) H. Baumann & S. Künkele in Mitt. Arbeitskr. Heim. Orch. Baden-Württemberg,* 18(3): 329 (1986): O. sphegodes subsp. aveyronensis.
aymoninii *(Breistr.) K.P. Buttler in Willdenowia,* 16(1): 115 (1986): O. insectifera subsp. aymoninii.
×azurea *H. Baumann & S. Künkele in Mitt. Arbeitskr. Heim. Orch. Baden-Württemberg,* 18(3): 516 (1986)—Italy. (O. promontorii × pseudobertolonii).
×baumanniana
nothosubsp. hierapetrae *H. Baumann & S. Künkele in Mitt. Arbeitskr. Heim. Orch. Baden-Württemberg,* 18(3): 454 (1986)—Crete. (O. doerfleri × sphegodes subsp. cretensis).
bertolonii
var. bertoloniiformis *M. Balayer in Bull. Soc. Bot. France, Lett. Bot.,* 133(3): 281 (1986)—France.
var. ferrequinoides *M. Balayer in Bull. Soc. Bot. France, Lett. Bot.,* 133(3): 281 (1986)—France.
×calabrica *H. Baumann & S. Künkele in Mitt. Arbeitskr. Heim. Orch. Baden-Württemberg,* 18(3): 523 (1986)—Italy. (O. sphegodes × tarentina).
×calenae *(O. & E. Danesch) H. Baumann & S. Künkele in Mitt. Arbeitskr. Heim. Orch. Baden-Württemberg,* 18(3): 513 (1986): O. × philippei nm. calenae.
×camarinana *H. Baumann & S. Künkele in Mitt. Arbeitskr. Heim. Orch. Baden-Württemberg,* 18(3): 490 (1986)—Sicily. (O. incubacea × lunulata).
×carica *H. Baumann & S. Künkele in Mitt. Arbeitskr. Heim. Orch. Baden-Württemberg,* 18(3): 458 (1986)—Turkey. (O. ferrum-equinum × holosericea).

OPHRYS :-
×cosana *H. Baumann & S. Künkele in Mitt. Arbeitskr. Heim. Orch. Baden-Württemberg,* 18(3): 438 (1986)—Italy. (O. bombyliflora × incubacea).
×cosentiana *H. Baumann & S. Künkele in Mitt. Arbeitskr. Heim. Orch. Baden-Württemberg,* 18(3): 406 (1986)—Italy. (O. apulica × holosericea).
nothosubsp. nociana *H. Baumann & S. Künkele in Mitt. Arbeitskr. Heim. Orch. Baden-Württemberg,* 18(3): 407 (1986)—Italy. (O. apulica × holosericea subsp. parvimaculata).
×creutzburgii *H. Baumann & S. Künkele in Mitt. Arbeitskr. Heim. Orch. Baden-Württemberg,* 18(3): 467 (1986)—Rhodes. (O. fusca × omegaifera).
×daunia *H. Baumann & S. Künkele in Mitt. Arbeitskr. Heim. Orch. Baden-Württemberg,* 18(3): 436 (1986)—Italy. (O. bombyliflora × garganica).
×dodonica *B. & E. Willing in Orchidophile,* 72: 1100 (1986)—Greece. (O. helenae × oestrifera).
×eliasii
nothosubsp. conimbricensis *(O. & E. Danesch) H. Baumann & S. Künkele in Mitt. Arbeitskr. Heim. Orch. Baden-Württemberg,* 18(3): 471 (1986): O. × eliasii nm. conimbricensis.
×ficuzzana *H. Baumann & S. Künkele in Mitt. Arbeitskr. Heim. Orch. Baden-Württemberg,* 18(3): 468 (1986)—Sicily. (O. fusca × pallida).
×fortosica *B. & E. Willing in Orchidophile,* 72: 1102 (1986)—Greece. (O. ferrum-equinum × helenae).
fuciflora
var. apiformoides *M. Balayer in Bull. Soc. Bot. France, Lett. Bot.,* 133(3): 282 (1986)—France.
var. carmeloides *M. Balayer in Bull. Soc. Bot. France, Lett. Bot.,* 133(3): 282 (1986)—France.
fusca
var. maculata *M. Balayer in Bull. Soc. Bot. France, Lett. Bot.,* 133(3): 281 (1986)—France.
subsp. minima *M. Balayer in Bull. Soc. Bot. France, Lett. Bot.,* 133(3): 281 (1986)—France.
var. rubescens *M. Balayer in Bull. Soc. Bot. France, Lett. Bot.,* 133(3): 281 (1986)—France.
×gauthieri
nothosubsp. fenarolii *(Ferlan) H. Baumann & S. Künkele in Mitt. Arbeitskr. Heim. Orch. Baden-Württemberg,* 18(3): 465 (1986): O. × fenarolii.
×gelana *H. Baumann & S. Künkele in Mitt. Arbeitskr. Heim. Orch. Baden-Württemberg,* 18(3): 492 (1986)—Sicily. (O. incubacea × oxyrrhynchos).

OPHRYS :-
gracilis *(Büel & Danesch) P. Englmaier in Abh. Zool.-Bot. Ges. Österr.,* 22: 140 (1984): O. fuciflora subsp. gracilis.
holoserica
subsp. apulica *(O. Danesch & E. Danesch) K.P. Buttler in Willdenowia,* 16(1): 115 (1986): O. fuciflora subsp. apulica.
nothosubsp. halicarnassia *H. Baumann & S. Künkele in Mitt. Arbeitskr. Heim. Orch. Baden-Württemberg,* 18(3): 478 (1986)—Turkey. (O. holosericea subsp. heterochila × subsp. holosericea).
×ioanninensis *B. & E. Willing in Orchidophile,* 72: 1100 (1986)—Greece. (O. helenae × mammosa).
×kozanica *B. & E. Willing in Orchidophile,* 72: 1098 (1986)—Greece. (O. mammosa × reinholdii).
×laconica *H. Baumann & S. Künkele in Mitt. Arbeitskr. Heim. Orch. Baden-Württemberg,* 18(3): 426 (1986)—Greece. (O. attica × tenthredinifera).
×ligariana *H. Baumann & S. Künkele in Mitt. Arbeitskr. Heim. Orch. Baden-Württemberg,* 18(3): 457 (1986)—Greece. (O. ferrum-equinum × helenae).
lutea
var. minor *(Tod.) C. Raynaud, Orchid. Maroc:* 16 (1985): Arachnites minor.
×lydia *H. Baumann & S. Künkele in Mitt. Arbeitskr. Heim. Orch. Baden-Württemberg,* 18(3): 511 (1986)—Turkey. (O. mammosa × vernixia).
nothosubsp. magnessa *H. Baumann & S. Künkele in Mitt. Arbeitskr. Heim. Orch. Baden-Württemberg,* 18(3): 512 (1986)—Greece. (O. mammosa × vernixia subsp. vernixia).
×macchiatii
nothosubsp. neokelleri *(Soó) H. Baumann & S. Künkele in Mitt. Arbeitskr. Heim. Orch. Baden-Württemberg,* 18(3): 525 (1986): O. × neokelleri.
×macrostachys
nothosubsp. liceana *(Renz & Taubenheim) H. Baumann & S. Künkele in Mitt. Arbeitskr. Heim. Orch. Baden-Württemberg,* 18(3): 443 (1986): O. × liceana.
nothosubsp. palaestina *H. Baumann & S. Künkele in Mitt. Arbeitskr. Heim. Orch. Baden-Württemberg,* 18(3): 444 (1986)—Israel. (O. bornmuelleri subsp. bornmuelleri × umbilicata subsp. umbilicata).
mammosa
subsp. transhyrcana *(Černjak.) K.P. Buttler in Willdenowia,* 16(1): 115 (1986): O. transhyrcana.
×manacorensis *H. Baumann & S. Künkele in Mitt. Arbeitskr. Heim. Orch. Baden-Württemberg,* 18(3): 410 (1986)—Italy. (O. arachnitiformis × garganica).

OPHRYS :–

×maremmae
nothosubsp. loutanensis *H. Baumann & S. Künkele in Mitt. Arbeitskr. Heim. Orch. Baden-Württemberg,* 18(3): 487 (1986)—Rhodes. (O. holosericea subsp. heterochila × tenthredinifera subsp. villosa).
nothosubsp. normanii *(J.J. Wood) H. Baumann & S. Künkele in Mitt. Arbeitskr. Heim. Orch. Baden-Württemberg,* 18(3): 486 (1986): O. × normanii.

×marmarensis *H. Baumann & S. Künkele in Mitt. Arbeitskr. Heim. Orch. Baden-Württemberg,* 18(3): 489 (1986)—Rhodes. (O. holosericea × umbilicata).

×messeniensis *H. Baumann & S. Künkele in Mitt. Arbeitskr. Heim. Orch. Baden-Württemberg,* 18(3): 514 (1986)—Greece. (O. oestrifera × tenthredinifera).

×methonensis *H. Baumann & S. Künkele in Mitt. Arbeitskr. Heim. Orch. Baden-Württemberg,* 18(3): 422 (1986)—Greece. (O. argolica × tenthredinifera).

×monopolitana *H. Baumann & S. Künkele in Mitt. Arbeitskr. Heim. Orch. Baden-Württemberg,* 18(3): 430 (1986)—Italy. (O. bertolonii × tarentina).

×montana *H. Baumann & S. Künkele in Mitt. Arbeitskr. Heim. Orch. Baden-Württemberg,* 18(3): 474 (1986)—Greece. (O. hebes × mammosa).

×montserratensis
nothosubsp. neoruppertii *(A. Camus) H. Baumann & S. Künkele in Mitt. Arbeitskr. Heim. Orch. Baden-Württemberg,* 18(3): 518 (1986): O. × neoruppertii.

×moreana *H. Baumann & S. Künkele in Mitt. Arbeitskr. Heim. Orch. Baden-Württemberg,* 18(3): 420 (1986)—Greece. (O. argolica × ferrum-equinum).

×mulierum *H. Baumann & S. Künkele in Mitt. Arbeitskr. Heim. Orch. Baden-Württemberg,* 18(3): 508 (1986)—Greece. (O. mammosa × reinholdii).

×necropolitana *H. Baumann & S. Künkele in Mitt. Arbeitskr. Heim. Orch. Baden-Württemberg,* 18(3): 452 (1986)—Sicily. (O. discors × oxyrrhynchos).

nervosa *C.P. Thunberg ex A. Murray in Linnaeus, Syst. Veg., ed. 14:* 814 (1784)—Japan. †

×nicosiana *H. Baumann & S. Künkele in Mitt. Arbeitskr. Heim. Orch. Baden-Württemberg,* 18(3): 463 (1986)—Cyprus. (O. flavomarginata × kotschyi).

×olympica *H. Baumann & S. Künkele in Mitt. Arbeitskr. Heim. Orch. Baden-Württemberg,* 18(3): 477 (1986)—Greece. (O. helenae × mammosa).

×pano-lefkaron *M. Bourdon in Orchidophile,* 71: 1056 (1986)—Cyprus, Rhodes. (O. iricolor × mammosa).

×paphosiana *H. Baumann & S. Künkele in Mitt. Arbeitskr. Heim. Orch. Baden-Württemberg,* 18(3): 500 (1986)—Cyprus. (O. kotschyi × umbilicata).

OPHRYS :–

×personei
nothosubsp. bourlieri *(Maire) H. Baumann & S. Künkele in Mitt. Arbeitskr. Heim. Orch. Baden-Württemberg,* 18(3): 505 (1986): O. × bourlieri.

×placotica *B. & E. Willing in Orchidophile,* 72: 1102 (1986)—Greece. (O. attica × sprunneri).

pseudobertolonii
subsp. bertoloniiformis *(O. & E. Danesch) H. Baumann & S. Künkele in Mitt. Arbeitskr. Heim. Orch. Baden-Württemberg,* 18(3): 367 (1986): O. bertoloniiformis.
subsp. catalaunica *(O. & E. Danesch) H. Baumann & S. Künkele in Mitt. Arbeitskr. Heim. Orch. Baden-Württemberg,* 18(3): 368 (1986): O. catalaunica.

×pugliana *H. Baumann & S. Künkele in Mitt. Arbeitskr. Heim. Orch. Baden-Württemberg,* 18(3): 481 (1986)—Italy. (O. holosericea × oxyrrhynchos).

reinholdii
f. albovirescens *B. & E. Willing in Orchidophile,* 72: 1099 (1986)—Greece.

×russonensis *H. Baumann & S. Künkele in Mitt. Arbeitskr. Heim. Orch. Baden-Württemberg,* 18(3): 425 (1986)—Greece. (O. attica × mammosa).

×saratoi
nothosubsp. gelmii *(Murr) H. Baumann & S. Künkele in Mitt. Arbeitskr. Heim. Orch. Baden-Württemberg,* 18(3): 519 (1986): O. × gelmii.

×scalana *H. Baumann & S. Künkele in Mitt. Arbeitskr. Heim. Orch. Baden-Württemberg,* 18(3): 462 (1986)—Greece. (O. ferrum-equinum × vernixia).

×sicana *H. Baumann & S. Künkele in Mitt. Arbeitskr. Heim. Orch. Baden-Württemberg,* 18(3): 502 (1986)—Sicily. (O. lunulata × oxyrrhynchos).

×sieberi *H. Baumann & S. Künkele in Mitt. Arbeitskr. Heim. Orch. Baden-Württemberg,* 18(3): 453 (1986)—Crete. (O. doerfleri × mammosa).

×sivana *H. Baumann & S. Künkele in Mitt. Arbeitskr. Heim. Orch. Baden-Württemberg,* 18(3): 446 (1986)—Crete. (O. candica × holosericea).

sphegodes
subsp. cretensis *H. Baumann & S. Künkele in Mitt. Arbeitskr. Heim. Orch. Baden-Württemberg,* 18(3): 375 (1986)—Crete.
var. garganicoides *M. Balayer in Bull. Soc. Bot. France, Lett. Bot.,* 133(3): 282 (1986)—France.
subsp. gortynia *H. Baumann & S. Künkele in Mitt. Arbeitskr. Heim. Orch. Baden-Württemberg,* 18(3): 377 (1986)—Crete.
subsp. integra *(Moggr. & Rchb. f.) H. Baumann & S. Künkele in Mitt. Arbeitskr. Heim. Orch. Baden-Württemberg,* 18(3): 380 (1986): O. insectifera subsp. integra.

OPHRYS :–
var. subaesculapiana *M. Balayer in Bull. Soc. Bot. France, Lett. Bot.*, 133(3): 282 (1986)—France.
var. subspruneriana *M. Balayer in Bull. Soc. Bot. France, Lett. Bot.*, 282 (1986)—France.
var. subtommasiniana *M. Balayer in Bull. Soc. Bot. France, Lett. Bot.*, 133(3): 282 (1986)—France.
tenthredinifera
f. lutescens *(Batt.) C. Raynaud, Orchid. Maroc:* 26 (1985): O. tenthredinifera var. lutescens.
subsp. villosa *(Desf.) H. Baumann & S. Künkele in Mitt. Arbeitskr. Heim. Orch. Baden-Württemberg,* 18(3): 384 (1986): O. villosa.
×trullana *H. Baumann & S. Künkele in Mitt. Arbeitskr. Heim. Orch. Baden-Württemberg,* 18(3): 472 (1986)—Italy. (O. garganica × tarentina).
×tuscanica *H. Baumann & S. Künkele in Mitt. Arbeitskr. Heim. Orch. Baden-Württemberg,* 18(3): 450 (1986)—Italy. (O. crabronifera × tenthredinifera).
umbilicata
subsp. rhodia *H. Baumann & S. Künkele in Mitt. Arbeitskr. Heim. Orch. Baden-Württemberg,* 18(3): 388 (1986)—Rhodes.
×venusiana *H. Baumann & S. Künkele in Mitt. Arbeitskr. Heim. Orch. Baden-Württemberg,* 18(3): 528 (1986)—Italy. (O. tarentina × tenthredinifera).
vernixia
nothosubsp. buttleri *H. Baumann & S. Künkele in Mitt. Arbeitskr. Heim. Orch. Baden-Württemberg,* 18(3): 530 (1986)—Turkey. (O. vernixia subsp. regis-ferdinandii × subsp. vernixia).
subsp. lusitanica *(O. & E. Danesch) H. Baumann & S. Künkele in Mitt. Arbeitskr. Heim. Orch. Baden-Württemberg,* 18(3): 391 (1986): O. speculum subsp. lusitanica.
×vittoriana *H. Baumann & S. Künkele in Mitt. Arbeitskr. Heim. Orch. Baden-Württemberg,* 18(3): 428 (1986)—Sicily. (O. bertolonii × lunulata).
×vogatsica *B. & E. Willing in Orchidophile,* 72: 1100 (1986)—Greece. (O. reinholdii × sphegodes).
×warwarensis *H. Baumann & S. Künkele in Mitt. Arbeitskr. Heim. Orch. Baden-Württemberg,* 18(3): 445 (1986)—Crete. (O. candica × heldreichii).
×willingiorum *H. Baumann & S. Künkele in Mitt. Arbeitskr. Heim. Orch. Baden-Württemberg,* 18(3): 501 (1986)—Cyprus. (O. levantina × umbilicata).

OPIZIA (Gramin.).
bracteata *R. McVaugh, Fl. Novo-Galiciana,* 14: 269 (1983)—Mexico.

OPLISMENUS (Gramin.).
burmannii
var. nudicaulis *(Vasey) R. McVaugh, Fl. Novo-Galiciana,* 14: 274 (1983): O. humboldtianus var. nudicaulis.

OPUNTIA (Cactac.).
ambigua *Link ex Tenore in Atti 7 Adun. Sci. Ital.:* 884 (1846); *cf. D.J. Mabberley in Taxon,* 30(1): 16 (1981)— Locality not stated.
chisosensis *(Anthony) D.J. Ferguson in Cact. Succ. J. (U.S.A.),* 58(3): 124 (1986): O. lindheimeri var. chisosensis.
erinacea
var. aurea *(Baxter) S.L. Welsh in Great Basin Nat.*, 46(2): 255 (1986): O. aurea.
virens *Tenore in Atti 7 Adun. Sci. Ital.:* 886 (1846); *cf. D.J. Mabberley in Taxon,* 30(1): 16 (1981)— Locality not stated.

×**ORCHIDACTYLA** (Orchidac.).
aquitaniensis *(G. Keller & Jeanjean) L.V. Aver'yanov in Bot. Zhurn.*, 71(1): 93 (1986): Orchis × aquitaniensis.
castellanensis *M. Balayer in Bull. Soc. Bot. France, Lett. Bot.*, 133(3): 283 (1986)—France. (Dactylorhiza sambucina var. laurentina × Orchis mascula).
kabyliensis *(G. Keller) L.V. Aver'yanov in Bot. Zhurn.*, 71(1): 93 (1986): Orchis kabyliensis.
lamarquei *(G. Keller & Jeanjean) L.V. Aver'yanov in Bot. Zhurn.*, 71(1): 93 (1986): Orchis lamarquei.
masteineri *(Ciferri & Giacomini) L.V. Aver'yanov in Bot. Zhurn.*, 71(1): 93 (1986): Orchis masteineri.
romanallens *(Ciferri & Giacomini) L.V. Aver'yanov in Bot. Zhurn.*, 71(1): 93 (1986): Orchis × romanallens.

ORCHIS (Orchidac.).
×aurunca *W. Rossi & F. Minutillo in Atti Accad. Naz. Lincei, Rend. Sci. Fis. Mat. Nat.*, 78(5): 1 (1985)—Italy. (O. pauciflora × provincialis).
×bassoulsii *M. Balayer in Bull. Soc. Bot. France, Lett. Bot.*, 133(3): 282 (1986)—France. (O. militaris × papilionacea).
champagneuxii
var. mesomelana *(Reichenb. f.) D. Tyteca in Orchidophile,* 70: 997 (1986): O. morio var. mesomelana.
coriophora
var. subsancta *M. Balayer in Bull. Soc. Bot. France, Lett. Bot.*, 133(3): 281 (1986)—France.
falcata *C.P. Thunberg ex A. Murray in Linnaeus, Syst. Veg., ed. 14:* 811 (1784)—Japan. †
japonica *C.P. Thunberg ex A. Murray in Linnaeus, Syst. Veg., ed. 14:* 811 (1784)—Japan. †

ORCHIS :–
×lauzensis *M. Balayer in Bull. Soc. Bot. France, Lett. Bot.*, 133(3): 283 (1986)—France. (O. morio × provincialis).
mascula
var. bicolor *M. Balayer in Bull. Soc. Bot. France, Lett. Bot.*, 133(3): 281 (1986)—France.
subsp. longibracteatoides *M. Balayer in Bull. Soc. Bot. France, Lett. Bot.*, 133(3): 281 (1986)—France.
morio
subsp. albanica *(Gölz & Reinh.) K.P. Buttler in Willdenowia*, 16(1): 116 (1986): O. albanica.
olbiensis
subsp. ichnusae *(Corrias) K.P. Buttler in Willdenowia*, 16(1): 116 (1986): O. mascula subsp. ichnusae.
papilionacea
subsp. expansa *(Tenore) C. Raynaud, Orchid. Maroc:* 43 (1985): O. expansa.
saccata
f. flavescens *(Soó) C. Raynaud, Orchid. Maroc:* 41 (1985): O. saccata lus. flavescens.

OREOCALAMUS (Gramin.).
armata *(Gamble) T.H. Wen in J. Bamboo Res.*, 5(2): 22 (1986): Arundinaria armata.

OREOPANAX (Araliac.).
nicaraguensis *M.J. Cannon & J.F.M. Cannon in Ann. Missouri Bot. Gard.*, 73(2): 482 (1986)—Nicaragua, Costa Rica, Panama.

OREOSCHIMPERELLA (Umbellif.).
arabiae-felicis *(C.C. Townsend) C.C. Townsend in Kew Bull.*, 41(2): 456 (1986): Trachyspermum arabiae-felicis.
var. laevis *C.C. Townsend in Kew Bull.*, 41(2): 456 (1986)—Saudi Arabia.

ORITROPHIUM (Compos.).
callacallense *J. Cuatrecasas in Fontqueria*, 8: 9 (1985)—Peru.
ollgaardii *J. Cuatrecasas in Fontqueria*, 8: 11 (1985)—Ecuador.

ORNICHIA J. Klackenberg in Bull. Mus. Nation. Hist. Nat., B, Adansonia, Ser. 4, 8(2): 196 (1986). *GENTIANACEAE.*
lancifolia *(Baker) J. Klackenberg in Bull. Mus. Nation. Hist. Nat., B, Adansonia,* Ser. 4, 8(2): 198 (1986): Chironia lancifolia.
madagascariensis *(Baker) J. Klackenberg in Bull. Mus. Nation. Hist. Nat., B, Adansonia,* Ser. 4, 8(2): 202 (1986): Chironia madagascariensis.
subsp. pubescens *(Baker) J. Klackenberg in Bull. Mus. Nation. Hist. Nat., B, Adansonia,* Ser. 4, 8(2): 205 (1986): Chironia pubescens.
trinervis *(Desr.) J. Klackenberg in Bull. Mus. Nation. Hist. Nat., B, Adansonia,* Ser. 4, 8(2): 200 (1986): Lisianthus trinervis.

ORNITHOGALUM (Hyacinthac.).
byzantinum *Strangew. ex Ten., Cat. Ort. Bot. Nap.:* 51 (1845); *cf. D.J. Mabberley in Taxon,* 32(1): 85 (1983)— Locality not stated.
japonicum *C.P. Thunberg ex A. Murray in Linnaeus, Syst. Veg., ed. 14:* 328 (1784)—Japan. †
neurostegium
subsp. eigii *(Feinbr.) N. Feinbrun-Dothan, Fl. Palaestina,* 4: 55 (1986): O. eigii.
sorgerae *H. Wittmann in Stapfia,* 13: 80 (1985)—Turkey.
spetae *H. Wittmann in Stapfia,* 13: 62 (1985)—Greece.

ORNITHOPUS (Legumin.).
coronatus *(L.) Hornem., Hort. Hafn.:* 696 (1815); *cf. D.J. Mabberley in Taxon,* 33(3): 437 (1984): Coronilla coronata.
creticus *(L.) Hornem., Hort. Hafn.:* 697 (1815); *cf. D.J. Mabberley in Taxon,* 33(3): 437 (1984): Coronilla cretica.
glaucus *(L.) Hornem., Hort. Hafn.:* 696 (1815); *cf. D.J. Mabberley in Taxon,* 33(3): 437 (1984): Basionym not stated.
junceus *(L.) Hornem., Hort. Hafn.:* 696 (1815); *cf. D.J. Mabberley in Taxon,* 33(3): 437 (1984): Coronilla juncea.
minimus *(L.) Hornem., Hort. Hafn.:* 696 (1815); *cf. D.J. Mabberley in Taxon,* 33(3): 437 (1984): Coronilla minima.
valentinus *(L.) Hornem., Hort. Hafn.:* 696 (1815); *cf. D.J. Mabberley in Taxon,* 33(3): 437 (1984): Coronilla valentina.
varius *(L.) Hornem., Hort. Hafn.:* 697 (1815); *cf. D.J. Mabberley in Taxon,* 33(3): 437 (1984): Coronilla varia.

ORONTIUM (Arac.).
japonicum *C.P. Thunberg ex A. Murray in Linnaeus, Syst. Veg., ed. 14:* 340 (1784)—Japan. †

OROSTACHYS (Crassulac.).
umbilicus *(L.) Hort. Monac. ex Schrank & Mart., Hort. Reg. Monac.:* 132 (1829); *cf. D.J. Mabberley in Taxon,* 33(3): 443 (1984): Basionym not stated.

OROYA (Cactac.).
peruviana
var. baumanii *(Knize) Slaba in Kaktusy,* 21(1): 10 (1985); *cf. Repert. Pl. Succ. (I.O.S.),* 36: 8 (1985 publ. 1986): O. baumanii.
var. citriflora *(Knize) Slaba in Kaktusy,* 21(1): 9 (1985); *cf. Repert. Pl. Succ. (I.O.S.),* 36: 8 (1985 publ. 1986): O. gibbosa var. citriflora.
var. depressa *(Rauh & Backeb.) Slaba in Kaktusy,* 21(1): 9 (1985); *cf. Repert. Pl. Succ. (I.O.S.),* 36: 8 (1985 publ. 1986): O. neoperuviana var. depressa.

OROYA :–
var. neoperuviana *(Backeb.) Slaba in Kaktusy,* 21(1): 7 (1985); *cf. Repert. Pl. Succ. (I.O.S.),* 36: 8 (1985 publ. 1986): O. neoperuviana.
subocculta
var. laxiareolata *(Rauh & Backeb.) Slaba in Kaktusy,* 21(2): 34 (1985); *cf. Repert. Pl. Succ. (I.O.S.),* 36: 8 (1985 publ. 1986): O. laxiareolata.
var. pluricentralis *(Backeb.) Slaba in Kaktusy,* 21(2): 34 (1985), basionym not validly publ.; *cf. Repert. Pl. Succ. (I.O.S.),* 36: 8 (1985 publ. 1986): O. laxiareolata var. pluricentralis.

ORTHILIA (Ericac.).
obtusata
var. xizangensis *Y.L. Chou in Bull. Bot. Res. North-East. Forest. Inst.,* 6(3): 145 (1986)—Xizang.

ORTHOPHYTUM (Bromeliac.).
alvimii *W. Weber in Feddes Repert.,* 97(3–4): 126 (1986)—Brazil.

ORTHOPOGON (Gramin.).
hirtellus *(L.) R. Br. ex Sweet, Hort. Suburb. Lond.:* 19 (June-July 1818); *cf. D.J. Mabberley in Taxon,* 30(1): 16 (1981): Basionym not stated.

OSBECKIA (Melastomatac.).
darjeelingensis *G.S. Giri & M.P. Nayar in Bull. Bot. Surv. India,* 25(1–4): 241 (1983 publ. 1985)—India, Sikkim.
lineolata
var. anamalayana *G.S. Giri & M.P. Nayar in Bull. Bot. Surv. India,* 25(1–4): 244 (1983 publ. 1985)—India.
mehrana *G.S. Giri & M.P. Nayar in Kew Bull.,* 41(2): 429 (1986)—India.

OSBERTIA (Compos.).
chihuahuana *B.L. Turner & S. Sundberg in Pl. Syst. Evol.,* 151(3–4): 233 (1986)—Mexico.

OSSICULUM P.J. Cribb & F.M. van der Laan in Kew Bull., 41(4): 823 (1986). *ORCHIDACEAE.*
aurantiacum *P.J. Cribb & F.M. van der Laan in Kew Bull.,* 41(4): 824 (1986)—Cameroun.

OSYRIS (Santalac.).
japonica *C.P. Thunberg ex A. Murray in Linnaeus, Syst. Veg., ed. 14:* 881 (1784)—Japan. †

OTATEA (Gramin.).
fimbriata *Soderstrom ex R. McVaugh, Fl. Novo-Galiciana,* 14: 280 (1983)—Mexico.

OTHOLOBIUM (Legumin.).
acuminatum *(Lam.) C.H. Stirton in S. Afr. J. Bot.,* 52(1): 2 (1986): Psoralea acuminata.
argenteum *(Thunb.) C.H. Stirton in S. Afr. J. Bot.,* 52(1): 2 (1986): Psoralea argentea.

OTHOLOBIUM :–
bolusii *(Forbes) C.H. Stirton in S. Afr. J. Bot.,* 52(1): 2 (1986): Psoralea bolusii.
bowieanum *(Harv.) C.H. Stirton in S. Afr. J. Bot.,* 52(1): 2 (1986): Psoralea bowieana.
bracteolatum *(Eckl. & Zeyh.) C.H. Stirton in S. Afr. J. Bot.,* 52(1): 2 (1986): Psoralea bracteolata.
candicans *(Eckl. & Zeyh.) C.H. Stirton in S. Afr. J. Bot.,* 52(1): 2 (1986): Psoralea candicans.
carneum *(E. Mey.) C.H. Stirton in S. Afr. J. Bot.,* 52(1): 2 (1986): Psoralea carnea.
decumbens *(Ait.) C.H. Stirton in S. Afr. J. Bot.,* 52(1): 2 (1986): Psoralea decumbens.
foliosum *(Oliv.) C.H. Stirton in S. Afr. J. Bot.,* 52(1): 2 (1986): Psoralea foliosa.
fruticans *(L.) C.H. Stirton in S. Afr. J. Bot.,* 52(1): 3 (1986): Trifolium fruticans.
hamatum *(Harv.) C.H. Stirton in S. Afr. J. Bot.,* 52(1): 3 (1986): Psoralea hamata.
heterosepalum *(Fourcade) C.H. Stirton in S. Afr. J. Bot.,* 52(1): 3 (1986): Psoralea heterosepala.
hirtum *(L.) C.H. Stirton in S. Afr. J. Bot.,* 52(1): 3 (1986): Psoralea hirta.
macradenium *(Harv.) C.H. Stirton in S. Afr. J. Bot.,* 52(1): 3 (1986): Psoralea macradenia.
mundianum *(Eckl. & Zeyh.) C.H. Stirton in S. Afr. J. Bot.,* 52(1): 3 (1986): Psoralea mundiana.
obliquum *(E. Mey.) C.H. Stirton in S. Afr. J. Bot.,* 52(1): 3 (1986): Psoralea obliqua.
parviflorum *(E. Mey.) C.H. Stirton in S. Afr. J. Bot.,* 52(1): 3 (1986): Psoralea parviflora.
polyphyllum *(Eckl. & Zeyh.) C.H. Stirton in S. Afr. J. Bot.,* 52(1): 3 (1986): Psoralea polyphylla.
polystictum *(Benth. ex Harv.) C.H. Stirton in S. Afr. J. Bot.,* 52(1): 3 (1986): Psoralea polysticta.
racemosum *(Thunb.) C.H. Stirton in S. Afr. J. Bot.,* 52(1): 3 (1986): Psoralea racemosa.
rotundifolium *(L.f.) C.H. Stirton in S. Afr. J. Bot.,* 52(1): 4 (1986): Psoralea rotundifolia.
sericeum *(Poir.) C.H. Stirton in S. Afr. J. Bot.,* 52(1): 4 (1986): Psoralea sericea.
spicatum *(L.) C.H. Stirton in S. Afr. J. Bot.,* 52(1): 4 (1986): Psoralea spicata.
stachyerum *(Eckl. & Zeyh.) C.H. Stirton in S. Afr. J. Bot.,* 52(1): 4 (1986): Psoralea stachyera.
striatum *(Thunb.) C.H. Stirton in S. Afr. J. Bot.,* 52(1): 4 (1986): Psoralea striata.
swartbergense *C.H. Stirton in S. Afr. J. Bot.,* 52(1): 4 (1986)—South Africa (Cape Province).
thomii *(Harv.) C.H. Stirton in S. Afr. J. Bot.,* 52(1): 4 (1986): Psoralea thomii.
trianthum *(E. Mey.) C.H. Stirton in S. Afr. J. Bot.,* 52(1): 4 (1986): Psoralea triantha.
uncinatum *(Eckl. & Zeyh.) C.H. Stirton in S. Afr. J. Bot.,* 52(1): 4 (1986): Psoralea uncinata.

OTHOLOBIUM :–
venustum *(Eckl. & Zeyh.) C.H. Stirton in S. Afr. J. Bot.*, 52(1): 4 (1986): Psoralea venusta.
wilmsii *(Harms) C.H. Stirton in S. Afr. J. Bot.*, 52(1): 4 (1986): Psoralea wilmsii.
zeyheri *(Harv.) C.H. Stirton in S. Afr. J. Bot.*, 52(1): 4 (1986): Psoralea zeyheri.

OTHONNA (Compos.).
gregorii *(F. Muell.) C. Jeffrey in Kew Bull.*, 41(4): 876 (1986): Senecio gregorii.

OTITES (Caryophyllac.).
cuneifolia
subsp. arenaria *(Podp.) J. Holub in Preslia*, 58(4): 302 (1986): Silene pseudo-otites var. arenaria.
velebitica *(Degen) J. Holub in Preslia*, 58(4): 302 (1986): Silene otites subsp. velebitica.

OTTONIA (Piperac.).
glaucescens *(Jacq. f.) Endl., Cat. Horti Vindob.*, 1: 225 (1842); *cf. D.J. Mabberley in Taxon*, 33(3): 435 (1984): Basionym not stated.
plantaginea *(Lam.) Endl., Cat. Horti Vindob.*, 1: 225 (1842); *cf. D.J. Mabberley in Taxon*, 33(3): 435 (1984): Basionym not stated.

OURATEA (Ochnac.).
kananariensis *C. Sastre in Caldasia*, 15(71–75): 187 (1986)—Colombia.
tumacoensis *C. Sastre in Caldasia*, 15(71–75): 189 (1986)—Colombia.

OXANDRA (Annonac.).
reticulata *P.J.M. Maas in Proc. Kon. Nederl. Akad. Wetensch., C*, 89(3): 261 (1986)—Brazil.

OXANTHERA (Rutac.).
brevipes *B.C. Stone in Proc. Acad. Nat. Sci. Philadelphia*, 137(2): 219 (1985)—New Caledonia.

OXYCEROS (Rubiac.).
malabarica *(Lam.) D.D. Tirvengadum in Ceylon J. Sci., Biol. Sci.*, 14(1–2): 4 (1981): Randia malabarica.

OXYRIA (Polygonac.).
acida *R. Br. ex Lindl. in Loud., Encycl. Pl.*: 294 (1829); *cf. D.J. Mabberley in Taxon*, 31(1): 71 (1982)— Locality not stated.

OXYTROPIS (Legumin.).
beringensis *B.A. Yurtsev in Arktich. Fl. SSSR*, 9(2): 179, 116 (1986)—U.S.S.R. (Soviet Far East).
chinglingensis *C.W. Chang in Fl. Tsinlingensis*, 1(3): 446, 68 (1981)—China.
davisii *(Welsh) B.A. Yurtsev in Arktich. Fl. SSSR*, 9(2): 97 (1986): O. campestris var. davisii.
deflexa
subsp. dezhnevii *(Jurtz.) B.A. Yurtsev in Arktich. Fl. SSSR*, 9(2): 79 (1986): O. deflexa var. dezhnevii.

OXYTROPIS :–
subsp. retrorsa *(Fern.) B.A. Yurtsev in Arktich. Fl. SSSR*, 9(2): 77 (1986): O. retrorsa.
gerzeensis *P.C. Li in Fl. Xizangica*, 2: 859 (1985)—Xizang.
hirta
var. wutuensis *C.W. Chang in Fl. Tsinlingensis*, 1(3): 445, 65 (1981)—China.
humilis *C.W. Chang in Fl. Tsinlingensis*, 1(3): 447, 68 (1981)—China.
inopinata *B.A. Yurtsev in Arktich. Fl. SSSR*, 9(2): 182, 144 (1986)—U.S.S.R. (Siberia).
kateninii *B.A. Yurtsev in Arktich. Fl. SSSR*, 9(2): 182, 146 (1986)—U.S.S.R. (Soviet Far East).
leucantha
subsp. subarctica *B.A. Yurtsev in Arktich. Fl. SSSR*, 9(2): 180, 119 (1986)—U.S.S.R. (Siberia).
subsp. tschukotcensis *B.A. Yurtsev in Arktich. Fl. SSSR*, 9(2): 180, 120 (1986)—U.S.S.R. (Soviet Far East, Siberia).
melanocalyx
var. brevidentata *C.W. Chang in Fl. Tsinlingensis*, 1(3): 445, 66 (1981)—China.
middendorffii
subsp. coerulescens *B.A. Yurtsev & Petrovsky in Arktich. Fl. SSSR*, 9(2): 181, 137 (1986)—U.S.S.R. (Soviet Far East).
subsp. jarovoji *B.A. Yurtsev in Arktich. Fl. SSSR*, 9(2): 181 (1986)—U.S.S.R. (Siberia).
subsp. submiddendorffii *B.A. Yurtsev in Arktich. Fl. SSSR*, 9(2): 181, 137 (1986), as 'sub-middendorffii'—U.S.S.R. (Soviet Far East).
subsp. trautvetteri *(Meinsh.) B.A. Yurtsev in Arktich. Fl. SSSR*, 9(2): 141 (1986): O. trautvetteri.
ochrantha
var. albopilosa *P.C. Li in Fl. Xizangica*, 2: 870 (1985)—Xizang.
ochrocephala
var. longibracteata *P.C. Li in Fl. Xizangica*, 2: 859 (1985)—Xizang.
sitaipaiensis *T.P. Wang ex C.W. Chang in Fl. Tsinlingensis*, 1(3): 446, 67 (1981)—China.
sordida
subsp. arctolenensis *B.A. Yurtsev in Arktich. Fl. SSSR*, 9(2): 179, 114 (1986)—U.S.S.R. (Siberia).
subsp. barnebyana *(Welsh) B.A. Yurtsev in Arktich. Fl. SSSR*, 9(2): 109 (1986): O. arctica var. barnebyana.
subsp. murrayi *B.A. Yurtsev in Arktich. Fl. SSSR*, 9(2): 179 (1986)—Canada (Yukon).
subsp. schamurinii *B.A. Yurtsev in Arktich. Fl. SSSR*, 9(2): 179, 115 (1986)—U.S.S.R. (Soviet Far East, Siberia).
subpodoloba *P.C. Li in Fl. Xizangica*, 2: 864 (1985)—Xizang.

OXYTROPIS :-
tichomirovii *B.A. Yurtsev in Arktich. Fl. SSSR*, 9(2): 180, 128 (1986)—U.S.S.R. (Siberia).
uschakovii *B.A. Yurtsev in Arktich. Fl. SSSR*, 9(2): 178, 104 (1986)—U.S.S.R. (Soviet Far East).
vassilczenkoi
subsp. substepposa *B.A. Yurtsev in Arktich. Fl. SSSR*, 9(2): 178, 102 (1986)—U.S.S.R. (Soviet Far East, Siberia).
vassilievii *B.A. Yurtsev in Arktich. Fl. SSSR*, 9(2): 180 (1986)—U.S.S.R. (Soviet Far East).
vasskovskyi *B.A. Yurtsev in Arktich. Fl. SSSR*, 9(2): 181, 143 (1986)—U.S.S.R. (Siberia, Soviet Far East).

P

PACHYPODIUM (Crucif.).
pannonicum *(Jacq.) Endl., Cat. Horti Vindob.*, 2: 251 (1842); *cf. D.J. Mabberley in Taxon*, 33(3): 435 (1984): Basionym not stated.

PADELLUS (Rosac.).
mahaleb
subsp. cupaniana *(Gussone) J. Soják in Čas. Nár. Muz. (Prague)*, 153(3): 171 (1984 publ. 1985): Prunus cupaniana.
subsp. simonkaii *(Pénzes) J. Soják in Čas. Nár. Muz. (Prague)*, 153(3): 171 (1984 publ. 1985): Prunus mahaleb subsp. simonkaii.

PADUS (Rosac.).
ser. Maackiopadus *(Koehne) T.T. Yü & T.C. Ku in Fl. Reipubl. Popul. Sin.*, 38: 94 (1986): Prunus subsect. Maackiopadus.
brachypoda
var. microdonta *(Koehne) T.T. Yü & T.C. Ku in Fl. Reipubl. Popul. Sin.*, 38: 99 (1986): Prunus brachypoda var. microdonta.
buergeriana *(Miq.) T.T. Yü & T.C. Ku in Fl. Reipubl. Popul. Sin.*, 38: 91 (1986): Prunus buergeriana.
obtusata *(Koehne) T.T. Yü & T.C. Ku in Fl. Reipubl. Popul. Sin.*, 38: 101 (1986): Prunus obtusata.
perulata *(Koehne) T.T. Yü & T.C. Ku in Fl. Reipubl. Popul. Sin.*, 38: 92 (1986): Prunus perulata.
racemosa
var. asiatica *(Kom.) T.T. Yü & T.C. Ku in Fl. Reipubl. Popul. Sin.*, 38: 98 (1986): P. asiatica.
stellipila *(Koehne) T.T. Yü & T.C. Ku in Fl. Reipubl. Popul. Sin.*, 38: 92 (1986): Prunus stellipila.

PAEONIA (Paeoniac.).
arborea *C.C. Gmel., Hort. Mag. Duc. Bad. Carlsruh.*: 192 (1811); *cf. D.J. Mabberley in Taxon*, 33(3): 440 (1984)— Locality not stated.

PALLENIS (Compos.).
cyrenaica *S.A. Alavi in Fl. Libya*, 107: 109 (1983)—Libya.

PALMA (Palmae).
chilensis *R. Alsop in Molina, Geog. Hist. Chile (Connecticut ed.)*: 124 (1808); *cf. D.J. Mabberley in Taxon*, 32(1): 87 (1983)—Locality not stated.

PANAX (Araliac.).
austrocaledonicus *H. Baillon, Adansonia*, 12: 152 (1878)—New Caledonia.
balansae *H. Baillon, Adansonia*, 12: 152 (1878)—New Caledonia.
microbotrys *H. Baillon, Adansonia*, 12: 152 (1878)—New Caledonia.
myriophyllus *H. Baillon, Adansonia*, 12: 152 (1878)—New Caledonia.
pancheri *H. Baillon, Adansonia*, 12: 152 (1878)—New Caledonia.
pulchellus *H. Baillon, Adansonia*, 12: 152 (1878)—New Caledonia.
suborbicularis *H. Baillon, Adansonia*, 12: 152 (1878)—New Caledonia.

PANDANUS (Pandanac.).
angustifolia *C.C. Gmel., Hort. Mag. Duc. Bad. Carlsruh.*: 193 (1811); *cf. D.J. Mabberley in Taxon*, 33(3): 440 (1984)— Locality not stated.
×nullumiae *R. Tucker in Austrobaileya*, 2(3): 294 (1986)—Australia (Queensland). (P. conicus × yalna).
yalna *R. Tucker in Austrobaileya*, 2(3): 288 (1986)—Australia (Queensland).

PANICUM (Gramin.).
johnii *S.M. Almeida in J. Bombay Nat. Hist. Soc.*, 83(1): 184 (1986)—India.
maximum
subsp. pubescens *M. Sharma in J. Econ. Taxon. Bot.*, 7(1): 106 (1985)—India.
serotinum *(Walt.) C.C. Gmel., Hort. Mag. Duc. Bad. Carlsruh.*: 194 (1811); *cf. D.J. Mabberley in Taxon*, 33(3): 440 (1984): Basionym not stated.

PAPAVER (Papaverac.).
sect. Dubia *A.D. Mikheev in Bot. Zhurn.*, 71(6): 808 (1986).
alberti *A.D. Mikheev in Bot. Zhurn.*, 71(6): 809 (1986)—U.S.S.R. (Caucasus).
argemone
subsp. davisii *J.W. Kadereit in Notes Roy. Bot. Gard. Edinburgh*, 44(1): 38 (1986)—Turkey.
subsp. meiklei *J.W. Kadereit in Notes Roy. Bot. Gard. Edinburgh*, 44(1): 38 (1986), as 'meiklii'—Cyprus.

PAPAVER :–
subsp. minus *(Boiv.) J.W. Kadereit in Notes Roy. Bot. Gard. Edinburgh,* 44(1): 39 (1986): Closterandra minor.
subsp. nigrotinctum *(Fedde) J.W. Kadereit in Notes Roy. Bot. Gard. Edinburgh,* 44(1): 37 (1986): P. nigrotinctum.
×kobayashii *G. Siniscalco Gigliano in Delpinoa,* 23–24: 68 (1981–1982 publ. 1985)—Cultivation. (P. bracteatum × somniferum).
pavoninum
subsp. ocellatum *(Woron) J.W. Kadereit in Notes Roy. Bot. Gard. Edinburgh,* 44(1): 30 (1986): P. ocellatum.
radicatum
var. pygmaeum *(Rydb.) S.L. Welsh in Great Basin Nat.,* 46(2): 259 (1986): P. pygmaeum.

PAPHIOPEDILUM (Orchidac.).
emersonii *H. Koopowitz & P. Cribb in Orchid Advocate,* 12(3): 86 (1986)—China.
glanduliferum
var. wilhelminae *(L.O. Williams) P.J. Cribb in Kew Mag.,* 3(4): 160 (1986): P. wilhelminae.

PAPYRUS (Cyperac.).
auricomus *(Sieb. ex Spreng.) Schrad., Ind. Sem. Hort. Acad. Gott.,* 1835: 5 (1835); *cf. D.J. Mabberley in Taxon,* 33(3): 443 (1984): Cyperus auricomus.

PARADERRIS (Miq.) R. Geesink, Scala Millettiearum (Leiden Bot. Ser., 8): 109 (1984): Derris sect. Paraderris (Legumin.).
cuneifolia *(Benth.) R. Geesink, Scala Millettiearum (Leiden Bot. Ser., 8):* 109 (1984): Derris cuneifolia.

PARAFESTUCA E.B. Alekseev in Byull. Mosk. Obshch. Ispȳt. Prir., Biol., 90(5): 107 (1985). *GRAMINEAE.*
albida *(Lowe) E.B. Alekseev in Byull. Mosk. Obshch. Ispȳt. Prir., Biol.,* 90(5): 108 (1985): Festuca albida.

PARALEPISTEMON J. Lejoly & S. Lisowski in Bull. Jard. Bot. Nation. Belg., 56(1–2): 196 (1986). *CONVOLVULACEAE.*
shirensis *(Oliv.) J. Lejoly & S. Lisowski in Bull. Jard. Bot. Nation. Belg.,* 56(1–2): 197 (1986): Ipomoea shirensis.

PARAMOMUM S.Q. Tong in Acta Bot. Yunnanica, 7(3): 309 (1985). *ZINGIBERACEAE.*
petaloideum *S.Q. Tong in Acta Bot. Yunnanica,* 7(3): 310 (1985)—China.

PARAPIPTADENIA (Legumin.).
zehntneri *(Harms) M.P.M. de Lima & H.C. de Lima in Rodriguésia,* 36(60): 26 (1984): Piptadenia zehntneri.

PARATHESIS (Myrsinac.).
baruana *C.L. Lundell in Phytologia,* 59(6): 433 (1986)—Panama.

PARIS (Trilliac.).
sect. Thibetica *H. Li in Bull. Bot. Res. North-East. Forest. Inst.,* 6(1): 132 (1986).
cronquistii
var. xichouensis *H. Li in Bull. Bot. Res. North-East. Forest. Inst.,* 6(1): 113 (1986)—China.
delavayi
var. ovalifolia *H. Li in Bull. Bot. Res. North-East. Forest. Inst.,* 6(1): 115 (1986)—China.
polyphylla
var. alba *H. Li & R.J. Mitchell in Bull. Bot. Res. North-East. Forest. Inst.,* 6(1): 123 (1986)—China.
f. latifolia *(Wang & Chang) H. Li in Bull. Bot. Res. North-East. Forest. Inst.,* 6(1): 125 (1986): P. polyphylla var. latifolia.
f. macrosepala *H. Li in Bull. Bot. Res. North-East. Forest. Inst.,* 6(1): 127 (1986)—China.
var. nana *H. Li in Bull. Bot. Res. North-East. Forest. Inst.,* 6(1): 123 (1986)—China.
var. pseudothibetica *H. Li in Bull. Bot. Res. North-East. Forest. Inst.,* 6(1): 126 (1986)—China.
wenxianensis *Z.X. Peng & R.N. Zhao in Acta Bot. Bor.-Occid. Sin.,* 6(2): 133 (1986)—China.

PARODIA (Cactac.).
weberioides *F. Brandt in Letzeb. Cacteefrenn,* 5(2): 29–34 (1984); *cf. Repert. Pl. Succ. (I.O.S.),* 35: 5 (1985)— Locality not stated.

PARODIOCHLOA A.M. Molina in Parodiana, 4(1): 110 (1986). *GRAMINEAE.*
trachyantha *(Phil.) A.M. Molina in Parodiana,* 4(1): 112 (1986): Koeleria trachyantha.

PARONYCHIA (Caryophyllac.).
vulgaris *Destreux, Elenchus Pl.:* 32 (1821); *cf. D.J. Mabberley in Taxon,* 32(1): 85 (1983) Locality not stated.

PARRYA (Crucif.).
arabidiflora *(DC.) Sweet, Hort. Brit., ed. 2:* 24 (1830); *cf. D.J. Mabberley in Taxon,* 31(1): 71 (1982): Basionym not stated.

PASPALUM (Gramin.).
moluccanum *Huber, Cat. 1866:* 6 (1866); *cf. D.J. Mabberley in Taxon,* 34(3): 451 (1985)— Locality not stated.
uyucensis *R.W. Pohl in Ann. Missouri Bot. Gard.,* 73(2): 501 (1986)—Honduras.
villosum *C.P. Thunberg ex A. Murray in Linnaeus, Syst. Veg., ed. 14:* 105 (1784)—Japan. †

PASSIFLORA (Passiflorac.).
bicolorata *Baxter in Loud., Encycl. new ed.:* 1234 (1855); *cf. D.J. Mabberley in Taxon,* 30(1): 16 (1981)— Locality not stated.
colombiana *L.K. Escobar in Syst. Bot.,* 11(1): 95 (1986)—Colombia.

PASSIFLORA :–
fanchonae *C. Feuillet in Candollea,* 41(1): 175 (1986)—French Guiana.
huamachucoensis *L.K. Escobar in Syst. Bot.,* 11(1): 88 (1986)—Peru.
lobata *(Killip) Hutch. ex J.M. MacDougal in Phytologia,* 60(6): 446 (1986): Tetrastylis lobata.
rufostipulata *C. Feuillet in Candollea,* 41(1): 173 (1986)—French Guiana.
rugosa
var. venezolana *L.K. Escobar in Syst. Bot.,* 11(1): 90 (1986)—Venezuela, Colombia.
runa *L.K. Escobar in Syst. Bot.,* 11(1): 90 (1986)—Peru.
trifoliata
var. tarmensis *L.K. Escobar in Syst. Bot.,* 11(1): 93 (1986)—Peru.

PATERSONIA (Iridac.).
lanata
f. calvata *D.A. Cooke in Fl. Australia,* 46: 220, 23 (1986)—Australia (W. Australia).
rudis
subsp. velutina *D.A. Cooke in Fl. Australia,* 46: 220, 22 (1986)—Australia (W. Australia).

×**PAULARA** hort. in Orchid Rev., 94(1116): centre page pull-out p. 8 (1986). *ORCHIDACEAE.* (Ascocentrum × Doritis × Phalaenopsis × Renanthera × Vanda).

PAULLINIA (Sapindac.).
japonica *C.P. Thunberg ex A. Murray in Linnaeus, Syst. Veg., ed. 14:* 380 (1784)—Japan. †

PAULOWNIA (Scrophulariac.).
tomentosa
var. glabrata *(Rehd.) S.Z. Qu in Fl. Tsinlingensis,* 1(4): 319 (1983): P. glabrata.

PAVONIA (Malvac.).
fryxelliana *F.R. Fosberg & M.-H. Sachet in Ceylon J. Sci., Biol. Sci.,* 15(1-2): 38 (1982 publ. 1985)—Sri Lanka.

PEDICULARIS (Scrophulariac.).
ser. Perpusillae *Tsoong ex T. Yamazaki in J. Jap. Bot.,* 61(3): 78 (1986).
anas
subsp. nepalensis *(Yamaz.) T. Yamazaki in J. Jap. Bot.,* 61(10): 300 (1986): P. cheilanthifolia subsp. nepalensis.
muscoides
subsp. himalayca *T. Yamazaki in J. Jap. Bot.,* 61(10): 298 (1986)—Nepal.
oxyrhyncha *T. Yamazaki in J. Jap. Bot.,* 61(10): 297 (1986)—Nepal.
paradoxa *(Prain) T. Yamazaki in J. Jap. Bot.,* 61(10): 299 (1986): P. instar var. paradoxa.
sceptrum-carolinum
var. uniflora *V.G. Sergienko, Fl. poluostrova Kanin:* 91 (1986)—U.S.S.R. (European Russia).

PEDICULARIS :–
yalungensis *T. Yamazaki in J. Jap. Bot.,* 61(3): 77 (1986)—Nepal.

PEDIOMELUM (Legumin.).
aromaticum *(Payson) S.L. Welsh in Great Basin Nat.,* 46(2): 257 (1986): Psoralea aromatica.
var. tuhyi *S.L. Welsh in Great Basin Nat.,* 46(2): 257 (1986)—U.S.A. (Utah).
epipsilum *(Barneby) S.L. Welsh in Great Basin Nat.,* 46(2): 257 (1986): Psoralea epipsila.
megalanthum
var. epipsilum *(Barneby) J.W. Grimes in Brittonia,* 38(2): 185 (1986): Psoralea epipsila.
var. retrorsum *(Rydb.) J.W. Grimes in Brittonia,* 38(2): 185 (1986): P. retrorsum.
pariense *(Welsh & Atwood) J.W. Grimes in Brittonia,* 38(2): 185 (1986): Psoralea pariensis.
pariense *(Welsh & Atwood) S.L. Welsh in Great Basin Nat.,* 46(2): 257 (1986): Psoralea pariensis.

PEGAEOPHYTON (Crucif.).
garhwalense *H.J. Chowdhery & S. Singh in Indian J. Forest.,* 8(4): 335 (1985 publ. 1986)—India.

PELARGONIUM (Geraniac.).
caucalifolium
subsp. convolvulifolium *(Schltr. ex Knuth) J.J.A. van der Walt in S. Afr. J. Bot.,* 52(5): 459 (1986): P. convolvulifolium.
desertorum *P. Vorster in S. Afr. J. Bot.,* 52(2): 184 (1986)—South Africa (Cape Province).
exhibens *P. Vorster in S. Afr. J. Bot.,* 52(5): 481 (1986)—South Africa (Cape Province).
longicaule
var. angustipetalum *D.A. Boucher in S. Afr. J. Bot.,* 52(5): 456 (1986)—South Africa (Cape Province).
loschgeanum *Mart., Hort. Acad. Erlangen:* 140 (1814); *cf. D.J. Mabberley in Taxon,* 33(3): 443 (1984)— Locality not stated.
luteum *(Andr.) Sm. in Rees, Cyclop.,* 26(2): Pelargonium n. 5 (1813); *cf. D.J. Mabberley in Taxon,* 31(1): 71 (1982): Basionym not stated.
molle *Hort. Vindob. ex Hornem., Hort. Hafn., Suppl.:* 77 (1819); *cf. D.J. Mabberley in Taxon,* 33(3): 437 (1984)— Locality not stated.
multicale
subsp. subherbaceum *(Knuth) J.J.A. van der Walt in S. Afr. J. Bot.,* 52(5): 449 (1986): P. subherbaceum.
suburbanum *Clifford ex D.A. Boucher in S. Afr. J. Bot.,* 52(5): 450 (1986), nom. nov.: Myrrhidium urbanum *Eckl. & Zeyh.*
suburbanum
subsp. bipinnatifidum *(Harv.) D.A. Boucher in S. Afr. J. Bot.,* 52(5): 453 (1986): P. urbanum var. bipinnatifidum.

×PELASTYLIS hort. in Orchid Rev., 94(1110): centre page pull-out p. 8 (1986). *ORCHIDACEAE.* (Pelatantheria × Rhynchostylis).

PELIOSANTHES (Convallariac.).
torulosa *Y. Wan in Bull. Bot. Res. North-East. Forest. Inst.,* 6(1): 147 (1986)—China.

PELTOPHORUM (Legumin.).
venezuelense *L. Cárdenas, H. Rodríguez-Rodríguez & A. Varela de Vega in Ernstia,* 41: 1 (1986)—Venezuela.

PENNISETUM (Gramin.).
felicianum *J. Asonganyi in Bull. Mus. Nation. Hist. Nat., B, Adansonia,* Ser. 4, 7(3): 251 (1985 publ. 1986)—Cameroun.
verticillatum *(L.) R. Br. ex Sweet, Hort. Suburb. Lond.:* 19 (1818); *cf. D.J. Mabberley in Taxon,* 31(1): 71 (1982): Basionym not stated.
viride *(L.) R. Br. ex Sweet, Hort. Suburb. Lond.:* 19 (1818); *cf. D.J. Mabberley in Taxon,* 31(1): 71 (1982): Basionym not stated.

PENSTEMON (Scrophulariac.).
colvillii *Huber, Cat. Graines 1868:* 5 (1868); *cf. D.J. Mabberley in Taxon,* 34(3): 451 (1985)— U.S.A. (California).
glaber
var. brandegei *(Porter) C.C. Freeman in Phytologia,* 60(2): 105 (1986): P. cyananthus var. brandegei.
penlandii *W.A. Weber in Phytologia,* 60(6): 459 (1986)—U.S.A. (Colorado).

PENTACOILANTHUS F. Rappa & V. Camarrone in Lav. Ist. Bot. Giard. Colon. Palermo, 14: 64 (1954). *AIZOACEAE.*
aitonis *(N.E. Br.) F. Rappa & V. Camarrone in Lav. Ist. Bot. Giard. Colon. Palermo,* 14: 65 (1954), without basionym ref.: Cryophytum aitonis.
crassicaulis *(Schw.) F. Rappa & V. Camarrone in Lav. Ist. Bot. Giard. Colon. Palermo,* 18: 21 (1962), without basionym ref.: Sceletium crassicaule.
crystallinus *(N.E. Br.) F. Rappa & V. Camarrone in Lav. Ist. Bot. Giard. Colon. Palermo,* 14: 65 (1954), without basionym ref.: Cryophytum crystallinum.
expansus *(L. Bol.) F. Rappa & V. Camarrone in Lav. Ist. Bot. Giard. Colon. Palermo,* 18: 21 (1962), without basionym ref.: Sceletium expansum.
granulicaulis *(Schwant.) F. Rappa & V. Camarrone in Lav. Ist. Bot. Giard. Colon. Palermo,* 14: 65 (1954), without basionym ref.: Psilocaulon granulicaule.
splendens *(Schwant.) F. Rappa & V. Camarrone in Lav. Ist. Bot. Giard. Colon. Palermo,* 14: 65 (1954), without basionym ref.: Aridaria splendens.

PENTACOILANTHUS :–
tortuosus *(N.E. Br.) F. Rappa & V. Camarrone in Lav. Ist. Bot. Giard. Colon. Palermo,* 18: 21 (1962), without basionym ref.: Sceletium tortuosum.

PENTALINON (Apocynac.).
andrieuxii *(Müller Argoviensis) B.F. Hansen & R.P. Wunderlin in Taxon,* 35(1): 168 (1986): Urechites andrieuxii.
luteum *(L.) B.F. Hansen & R.P. Wunderlin in Taxon,* 35(1): 167 (1986): Vinca lutea.

PENTASCHISTIS (Gramin.).
gracilis *S.M. Phillips in Kew Bull.,* 41(4): 1028 (1986)—Ethiopia.

PEPEROMIA (Piperac.).
boomii *J.A. Steyermark in Brittonia,* 38(3): 220 (1986)—Venezuela.
wolfgang-krahnii *W. Rauh in Kakt. And. Sukk.,* 37(7): 140 (1986)—Peru.

PERAPENTACOILANTHUS F. Rappa & V. Camarrone in Lav. Ist. Bot. Giard. Colon. Palermo, 15: 21 (1956). *AIZOACEAE.*

PERATETRACOILANTHUS F. Rappa & V. Camarrone in Lav. Ist. Bot. Giard. Colon. Palermo, 18: 22 (1962), without type. *AIZOACEAE.*
defoliatus *(Schw.) F. Rappa & V. Camarrone in Lav. Ist. Bot. Giard. Colon. Palermo,* 18: 23 (1962), without basionym ref.: Aridaria defoliata.
geniculiflorus *(N.E. Br.) F. Rappa & V. Camarrone in Lav. Ist. Bot. Giard. Colon. Palermo,* 18: 23 (1962), without basionym ref.: Aridaria geniculiflora.
haeckelianus *(N.E. Br.) F. Rappa & V. Camarrone in Lav. Ist. Bot. Giard. Colon. Palermo,* 18: 23 (1962), without basionym ref.: Platythyra haeckeliana.
junceus *(Schw.) F. Rappa & V. Camarrone in Lav. Ist. Bot. Giard. Colon. Palermo,* 18: 23 (1962), without basionym ref.: Psilocaulon junceum.
noctiflorus *(Schw.) F. Rappa & V. Camarrone in Lav. Ist. Bot. Giard. Colon. Palermo,* 18: 23 (1962), without basionym ref.: Aridaria noctiflora.
parviflorus *(Schw.) F. Rappa & V. Camarrone in Lav. Ist. Bot. Giard. Colon. Palermo,* 18: 23 (1962), without basionym ref.: Psilocaulon parviflorum.
spinuliferus *(N.E. Br.) F. Rappa & V. Camarrone in Lav. Ist. Bot. Giard. Colon. Palermo,* 18: 23 (1962), without basionym ref.: Aridaria spinulifera.
tetragonus *(L. Bol.) F. Rappa & V. Camarrone in Lav. Ist. Bot. Giard. Colon. Palermo,* 18: 23 (1962), without basionym ref.: Aridaria tetragona.

PERDICIUM (Compos.).
lyratum *(Willd.) Sweet, Hort. Brit., ed. 2:* 280 (1830); *cf. D.J. Mabberley in Taxon,* 31(1): 72 (1982): Basionym not stated.
tomentosum *C.P. Thunberg ex A. Murray in Linnaeus, Syst. Veg., ed. 14:* 769 (1784)—Japan. †

PERGULARIA (Asclepiadac.).
japonica *C.P. Thunberg ex A. Murray in Linnaeus, Syst. Veg., ed. 14:* 256 (1784)—Japan. †

PERISTROPHE (Acanthac.).
brassii *R.M. Barker in J. Adelaide Bot. Gard.,* 9: 192 (1986)—Australia (Queensland).

PERISTYLUS (Orchidac.).
nematocaulon *(J.D. Hook.) J.J. Wood in Kew Bull.,* 41(4): 811 (1986): Habenaria nematocaulon.
superanthus *J.J. Wood in Kew Bull.,* 41(4): 811 (1986)—Nepal, Bhutan, Sikkim, India.

PERSEA (Laurac.).
culilawan *(Bl.) Bischoff, Grund. Med. Bot.:* 342 (1831); *cf. D.J. Mabberley in Taxon,* 33(3): 443 (1984): Basionym not stated.
sintoc *(Bl.) Bischoff, Grund. Med. Bot.:* 342 (1831); *cf. D.J. Mabberley in Taxon,* 33(3): 443 (1984): Cinnamomum sintoc.

PETASITES (Compos.).
pyrenaica *(L.) G. López González in An. Jard. Bot. Madrid,* 42(2): 323 (1985 publ. 1986): Tussilago pyrenaica.

PETRORHAGIA (Caryophyllac.).
gasparrinii *(Guss.) J. Holub in Preslia,* 58(4): 302 (1986): Gypsophila gasparrinii.
zoharyana *A. Liston in Candollea,* 41(1): 179 (1986)—Israel, Jordan, Syria.

PEUCEDANUM (Umbellif.).
acaule *R.H. Shan & M.L. Sheh in Acta Phytotax. Sin.,* 24(4): 308 (1986)—China.
ampliatum *K.T. Fu in Fl. Tsinlingensis,* 1(3): 462, 427 (1981)—China.
elegans *(Spreng.) Sweet, Hort. Brit., ed. 2:* 250 (1830); *cf. D.J. Mabberley in Taxon,* 31(1): 72 (1982): Basionym not stated.
hirsutiusculum
var. subglabrum *R.H. Shan & M.L. Sheh in Acta Phytotax. Sin.,* 24(4): 310 (1986)—China.
japonicum *C.P. Thunberg ex A. Murray in Linnaeus, Syst. Veg., ed. 14:* 280 (1784)—Japan. †
ledebourielloides *K.T. Fu in Fl. Tsinlingensis,* 1(3): 463, 428 (1981)—China.
longshengense *R.H. Shan & M.L. Sheh in Acta Phytotax. Sin.,* 24(4): 306 (1986)—China.
mashanense *R.H. Shan & M.L. Sheh in Acta Phytotax. Sin.,* 24(4): 304 (1986)—China.

PEUCEDANUM :-
medicum
var. gracile *Dunn ex R.H. Shan & M.L. Sheh in Acta Phytotax. Sin.,* 24(4): 310 (1986)—China.
praeruptorum
var. grande *K.T. Fu in Fl. Tsinlingensis,* 1(3): 463, 428 (1981)—China.
quangxiense *R.H. Shan & M.L. Sheh in Acta Phytotax. Sin.,* 24(4): 308 (1986)—China.
refractum *Vill., Cat. Méth. Pl. Jard. Strasb.:* 235 (1807); *cf. D.J. Mabberley in Taxon,* 33(3): 443 (1984)— Locality not stated.
rubricaule *R.H. Shan & M.L. Sheh in Acta Phytotax. Sin.,* 24(4): 305 (1986)—China.
stridii *P. Hartvig in A. Strid (ed.), Mount. Fl. Greece,* 1: 720 (1986)—Greece.
vourinense *(Leute) P. Hartvig in A. Strid (ed.), Mount. Fl. Greece,* 1: 716 (1986): P. longifolium subsp. vourinense.
wulongense *R.H. Shan & M.L. Sheh in Acta Phytotax. Sin.,* 24(4): 309 (1986)—China.

PHACELIA (Hydrophyllac.).
aldea *R. Br. ex Sweet, Hort. Brit.:* 295 (1827); *cf. D.J. Mabberley in Taxon,* 30(1): 16 (1981)— Locality not stated.
demissa
var. minor *N.D. Atwood in Great Basin Nat.,* 46(2): 256 (1986)—U.S.A. (Utah).
douglasii
subsp. cryptantha *(Brand) R.M. Beauchamp in Phytologia,* 59(7): 440 (1986): P. douglasii var. cryptantha.
hintoniorum *B.L. Turner in Brittonia,* 38(2): 123 (1986)—Mexico.
magellanica *(Lam.) R. Br. ex Sweet, Hort. Brit.:* 295 (1827); *cf. D.J. Mabberley in Taxon,* 30(1): 16 (1981): Basionym not stated.
potosina *B.L. Turner in Brittonia,* 38(2): 124 (1986)—Mexico.
zaragozana *B.L. Turner in Brittonia,* 38(2): 127 (1986)—Mexico.

PHAGNALON (Compos.).
lavranosii *M. Qaiser & H.W. Lack in Willdenowia,* 15(2): 442 (1986)—Afars Issas Terr.

PHALARIS (Gramin.).
hispida *C.P. Thunberg ex A. Murray in Linnaeus, Syst. Veg., ed. 14:* 104 (1784)—Japan. †

PHANEROSTYLIS (Compos.).
hintoniorum *(B. Turner) R.M. King & H. Robinson in Phytologia,* 60(1): 80 (1986): Brickellia hintoniorum.

PHARBITIS (Convolvulac.).
coerulea *(Koen. ex Roxb.) Wall. ex O'Shaugnessy, Bengal Disp.:* 505 (1842); *cf. D.J. Mabberley in Taxon,* 32(1): 87 (1983): Basionym not stated.

PHASEOLUS (Legumin.).
deliciosus *Huber, Cat. Print. 1864:* 4 (1864); *cf. D.J. Mabberley in Taxon,* 34(3): 451 (1985)— Locality not stated.

PHILADELPHUS (Hydrangeac.).
laxus *C.C. Gmel., Hort. Mag. Duc. Bad. Carlsruh.:* 201 (1811); *cf. D.J. Mabberley in Taxon,* 33(3): 440 (1984)— Locality not stated.

PHILIPPIA (Ericac.).
drakensbergensis *E.G.H. Oliver in Bothalia,* 15(3–4): 550 (1985)—South Africa (Natal, Cape Province), Lesotho.

PHILODENDRON (Arac.).
sect. Philopsammos *G.S. Bunting in Phytologia,* 60(5): 306 (1986).
amplisinum *G.S. Bunting in Phytologia,* 60(5): 307 (1986)—Venezuela.
anaadu *G.S. Bunting in Phytologia,* 60(5): 309 (1986)—Venezuela.
appunii *G.S. Bunting in Phytologia,* 60(5): 310 (1986)—Venezuela.
borgesii *G.S. Bunting in Phytologia,* 60(5): 310 (1986)—Venezuela.
calatheifolium *G.S. Bunting in Phytologia,* 60(5): 311 (1986)—Venezuela.
canaimae *G.S. Bunting in Phytologia,* 60(5): 313 (1986)—Venezuela.
cataniapoense *G.S. Bunting in Phytologia,* 60(5): 314 (1986)—Venezuela.
consobrinum *G.S. Bunting in Phytologia,* 60(5): 314 (1986)—Venezuela.
davidsei *G.S. Bunting in Phytologia,* 60(5): 316 (1986)—Venezuela.
delascioi *G.S. Bunting in Phytologia,* 60(5): 317 (1986)—Venezuela.
dyscarpium
var. ventuarianum *G.S. Bunting in Phytologia,* 60(5): 317 (1986)—Venezuela.
exile *G.S. Bunting in Phytologia,* 60(5): 319 (1986)—Venezuela.
guaiquinimae *G.S. Bunting in Phytologia,* 60(5): 319 (1986)—Venezuela.
inaequilaterum
subsp. anthoblastum *G.S. Bunting in Phytologia,* 60(5): 319 (1986)—Venezuela.
liesneri *G.S. Bunting in Phytologia,* 60(5): 320 (1986)—Venezuela.
marahuacae *G.S. Bunting in Phytologia,* 60(5): 320 (1986)—Venezuela.
multinervum *G.S. Bunting in Phytologia,* 60(5): 322 (1986)—Venezuela.
orionis *G.S. Bunting in Phytologia,* 60(5): 322 (1986)—Venezuela.
peraiense *G.S. Bunting in Phytologia,* 60(5): 324 (1986)—Venezuela.
perplexum *G.S. Bunting in Phytologia,* 60(5): 325 (1986)—Venezuela.
rhodoaxis
var. angustifolium *G.S. Bunting in Phytologia,* 60(5): 325 (1986)—Venezuela.

PHILODENDRON :-
sabulosum *G.S. Bunting in Phytologia,* 60(5): 326 (1986)—Venezuela.
strictum *G.S. Bunting in Phytologia,* 60(5): 328 (1986)—Venezuela.
sucrense *G.S. Bunting in Phytologia,* 60(5): 329 (1986)—Venezuela.
tachirense *G.S. Bunting in Phytologia,* 60(5): 330 (1986)—Venezuela.
tatei
subsp. melanochlorum *(Bunt.) G.S. Bunting in Phytologia,* 60(5): 332 (1986): P. melanochlorum.
triangulare *G.S. Bunting in Phytologia,* 60(5): 332 (1986)—Venezuela.
victoriae *G.S. Bunting in Phytologia,* 60(5): 333 (1986)—Venezuela.
wurdackii *G.S. Bunting in Phytologia,* 60(5): 333 (1986)—Venezuela.
yavitense *G.S. Bunting in Phytologia,* 60(5): 335 (1986)—Venezuela.

PHILOXERUS (Amaranthac.).
brasiliensis *R. Br. ex Sweet, Hort. Suburb. Lond.:* 48 (1818); *cf. D.J. Mabberley in Taxon,* 30(1): 17 (1981)— Locality not stated.
vermiculatus *(L.) Sm. in Rees, Cyclop.,* 27(1): Philoxerus no. 3 (1814); *cf. D.J. Mabberley in Taxon,* 30(1): 17 (1981): Basionym not stated.

PHLOMIS (Labiat.).
×composita
nothovar. almijarensis *(Pau) I. Mateu in Acta Bot. Malacitana,* 11: 193 (1986): P. × almijarensis.
nothovar. malacitana *(Pau) I. Mateu in Acta Bot. Malacitana,* 11: 194 (1986): P. crinita var. malacitana.
nothovar. trullenquei *(Pau) I. Mateu in Acta Bot. Malacitana,* 11: 193 (1986): P. × trullenquei.
umbrosa
var. subaustralis *K.T. Fu & J.Q. Fu in Fl. Tsinlingensis,* 1(4): 398, 246 (1983)—China.
var. sylvatica *K.T. Fu & J.Q. Fu in Fl. Tsinlingensis,* 1(4): 398, 246 (1983)—China.

PHLOX (Polemoniac.).
austromontana
var. jonesii *(Wherry) S.L. Welsh in Great Basin Nat.,* 45(4): 792 (1985): P. jonesii.
var. lutescens *S.L. Welsh in Great Basin Nat.,* 45(4): 792 (1985)—U.S.A. (Utah).
azurea *G.L. Smith in Wasmann J. Biol.,* 43(1–2): 43 (1985 publ. 1986)—U.S.A. (California).

PHOEBE (Laurac.).
chavarriana *B.E. Hammel in J. Arnold Arbor.,* 67(1): 131 (1986)—Costa Rica.
puwenensis *W.C. Cheng in Sci. Silvae Sin.,* 8(1): 3 (1963)—China.

PHOTINIA (Rosac.).
austroguizhouensis *Y.K. Li in Bull. Bot. Res. North-East. Forest. Inst.*, 6(4): 107 (1986)—China.
subparvifolia *Y.K. Li & X.M. Wang in Bull. Bot. Res. North-East. Forest. Inst.*, 6(4): 108 (1986)—China.

PHRYNIUM (Marantac.).
rheedei *C.R. Suresh & D.H. Nicolson in Taxon*, 35(2): 355 (1986), nom. nov.: Pontederia ovata *L.*

PHTHIRUSA (Loranthac.).
geniculifera *C.T. Rizzini & A. de Mattos-Filho in Rev. Brasil. Biol.*, 46(2): 319 (1986)—Brazil.

PHYLLACTIS (Valerianac.).
rigida
var. tenuifolia *(Ruiz & Pav.) B.B. Larsen in Nordic J. Bot.*, 6(4): 437 (1986): Valeriana tenuifolia.

PHYLLANTHUS (Euphorbiac.).
barbarae *M.C. Johnston in Syst. Bot.*, 11(1): 35 (1986)—Mexico.
leptocladus
var. pubescens *P.T. Li & D.Y. Liu in Bull. Bot. Res. North-East. Forest. Inst.*, 6(1): 181 (1986)—China.

PHYLLOCEPHALUM (Compos.).
anthelminticum *(L.) S.R. Paul & S.L. Kapoor in J. Econ. Taxon. Bot.*, 6(3): 728 (1985): Conyza anthelmintica.
courtallense *(Wight) B.M. Narayana in Curr. Sci.*, 51(8): 438 (1982): Decaneurum courtallense.
mayurii *(C.E.C. Fischer) B.M. Narayana in Curr. Sci.*, 51(8): 438 (1982): Centratherum mayurii.
phyllolaenum *(DC.) B.M. Narayana in Curr. Sci.*, 51(8): 438 (1982): Decaneurum phyllolaenum.
rangacharii *(Gamble) B.M. Narayana in Curr. Sci.*, 51(8): 438 (1982): Centratherum rangacharii.
ritchiei *(Hook. f.) B.M. Narayana in Curr. Sci.*, 51(8): 438 (1982): Centratherum ritchiei.
sengaltherianum *(Narayana) B.M. Narayana in Curr. Sci.*, 51(8): 438 (1982): Centratherum sengaltherianum.
tenue *(Clarke) B.M. Narayana in Curr. Sci.*, 51(8): 439 (1982): Centratherum tenue.

PHYMASPERMUM (Compos.).
acerosum *(DC.) M. Källersjö in Nordic J. Bot.*, 5(6): 538 (1985 publ. 1986): Morysia acerosa.
athanasioides *(S. Moore) M. Källersjö in Nordic J. Bot.*, 5(6): 538 (1985 publ. 1986): Pentzia athanasioides.
bolusii *(Hutch.) M. Källersjö in Nordic J. Bot.*, 5(6): 538 (1985 publ. 1986): Brachymeris bolusii.

PHYMASPERMUM :–
erubescens *(Hutch.) M. Källersjö in Nordic J. Bot.*, 5(6): 538 (1985 publ. 1986): Brachymeris erubescens.
montanum *(Hutch.) M. Källersjö in Nordic J. Bot.*, 5(6): 538 (1985 publ. 1986): Brachymeris montana.
peglerae *(Hutch.) M. Källersjö in Nordic J. Bot.*, 5(6): 538 (1985 publ. 1986): Brachymeris peglerae.
pinnatifidum *(Oliv.) M. Källersjö in Nordic J. Bot.*, 5(6): 538 (1985 publ. 1986): Pentzia pinnatifida.
scoparium *(DC.) M. Källersjö in Nordic J. Bot.*, 5(6): 538 (1985 publ. 1986): Brachymeris scoparia.
villosum *(Hilliard) M. Källersjö in Nordic J. Bot.*, 5(6): 538 (1985 publ. 1986): Athanasia villosa.
woodii *(Thell.) M. Källersjö in Nordic J. Bot.*, 5(6): 538 (1985 publ. 1986): Pentzia woodii.

PHYSARIA (Crucif.).
acutifolia
var. stylosa *(Rollins) S.L. Welsh in Great Basin Nat.*, 46(2): 255 (1986): P. stylosa.
chambersii
var. sobolifera *S.L. Welsh in Great Basin Nat.*, 46(2): 255 (1986)—U.S.A. (Utah).

PHYTEUMA (Campanulac.).
barrelieri *Vill., Cat. Méth. Pl. Jard. Strasb.*: 183 (1807); *cf. D.J. Mabberley in Taxon*, 33(3): 443 (1984)— Locality not stated.

PHYTOLACCA (Phytolaccac.).
erubescens *Huber, Wochenschrift*, 13: 168 (1870); *cf. D.J. Mabberley in Taxon*, 34(3): 451 (1985)— Locality not stated.

PICEA (Pinac.).
meyeri
var. mongolica *H.Q. Wu in Bull. Bot. Res. North-East. Forest. Inst.*, 6(2): 153 (1986)—China.

PICRIS (Compos.).
japonica *C.P. Thunberg ex A. Murray in Linnaeus, Syst. Veg., ed. 14:* 711 (1784)—Japan. †
littoralis *P. Baltasar Merino, Algunas Pl. Rar. ... de La Guardia (Pontevedra):* 33 (1895)—Spain.

PILEA (Urticac.).
tabularis *C.C. Berg in Bull. Mus. Nation. Hist. Nat., B, Adansonia, Ser. 4*, 7(4): 429 (1985 publ. 1986)—French Guiana.

PILOGYNE (Cucurbitac.).
lucida *Naud. ex Huber, Cat. 1866:* 7 (1866); *cf. D.J. Mabberley in Taxon*, 34(3): 452 (1985)— Locality not stated.

PILOSOCEREUS (Cactac.).
cenepequei *C.T. Rizzini & A. de Mattos-Filho in Rev. Brasil. Biol.*, 46(2): 324 (1986)—Brazil.

PILOSOCEREUS :–
subsimilis *C.T. Rizzini & A. de Mattos-Filho in Rev. Brasil. Biol.*, 46(2): 327 (1986)—Brazil.

PIMPINELLA (Umbellif.).
valleculosa *K.T. Fu in Fl. Tsinlingensis,* 1(3): 457, 405 (1981)—China.
woodii *C.C. Townsend in Kew Bull.*, 41(1): 59 (1986)—Yemen.
xizangensis *R.H. Shan & F.T. Pu in Acta Phytotax. Sin.*, 24(4): 311 (1986)—Xizang.

PINDA P.K. Mukherjee & L. Constance in Kew Bull., 41(1): 224 (1986). *UMBELLIFERAE.*
concanensis *(Dalzell) P.K. Mukherjee & L. Constance in Kew Bull.*, 41(1): 226 (1986): Heracleum concanense.

PINELLIA (Arac.).
yaoluopingensis *X.H. Guo & X.L. Liu in Acta Bot. Yunnanica,* 8(2): 223 (1986)—China.

PINGUICULA (Lentibulariac.).
sect. Crassifolia *F. Speta & F. Fuchs in Stapfia,* 10: 113 (1982).
barbata *S.Z. Ruiz & J. Rzedowski in Phytologia,* 60(4): 256 (1986)—Mexico.
ehlersiae *F. Speta & F. Fuchs in Stapfia,* 10: 114 (1982), as 'ehlersae'—Mexico.
emarginata *S.Z. Ruiz & J. Rzedowski in Phytologia,* 60(4): 258 (1986)—Mexico.
takakii *S.Z. Ruiz & J. Rzedowski in Phytologia,* 60(4): 260 (1986)—Mexico.
zecheri *F. Speta & F. Fuchs in Stapfia,* 10: 111 (1982)—Mexico.

PINUS (Pinac.).
subsect. Rzedowskianae *S. Carvajal in Phytologia,* 59(2): 134 (1986).
lambertiana *Dougl. ex Taylor & Philips, Phil. Mag. n.s.,* 2: 449 (Dec. 1827); *cf. D.J. Mabberley in Taxon,* 30(1): 17 (1981)—Locality not stated.
lumholtzii
var. microphylla *S. Carvajal in Phytologia,* 59(2): 135 (1986)—Mexico.
macvaughii *S. Carvajal in Phytologia,* 59(2): 139 (1986)—Mexico.
montezumae
var. mezambrana *S. Carvajal in Phytologia,* 59(2): 138 (1986), as 'mezambranus'—Mexico.
novogaliciana *S. Carvajal in Phytologia,* 59(2): 131 (1986), without latin descr., as 'novo-galiciana'—Mexico.
pindrow *Royle ex D. Don in Brewster, Taylor & Phillips, Phil. Mag. & J. Sci.,* 8: 255 (March 1836); *cf. D.J. Mabberley in Taxon,* 31(1): 72 (1982)— Locality not stated.
pungens *Lamb. in Ann. Bot.,* 2: 198 (1805); *cf. D.J. Mabberley in Taxon,* 30(1): 17 (1981)—Locality not stated.
torreyana
subsp. insularis *J.R. Haller in Syst. Bot.,* 11(1): 45 (1986)—U.S.A. (California).

PIPER (Piperac.).
emeiense *Y.C. Tseng in Acta Phytotax. Sin.,* 24(5): 385 (1986)—China.
infossum *Y.C. Tseng in Acta Phytotax. Sin.,* 24(5): 383 (1986)—Xizang.
var. nudum *Y.C. Tseng in Acta Phytotax. Sin.,* 24(5): 385 (1986)—Xizang.
madidum *Y.C. Tseng in Acta Phytotax. Sin.,* 24(5): 382 (1986)—Xizang.

PIPTADENIA (Legumin.).
anolidurus *R.C. Barneby in Brittonia,* 38(3): 222 (1986)—Brazil, Ecuador, Peru.
buchtienii *R.C. Barneby in Brittonia,* 38(3): 224 (1986)—Bolivia.
cuzcoensis *R.C. Barneby in Brittonia,* 38(3): 226 (1986)—Peru.
imatacae *R.C. Barneby in Brittonia,* 38(3): 227 (1986)—Venezuela.
peruviana *(J.F. Macbr.) R.C. Barneby in Brittonia,* 38(3): 226 (1986): P. adiantoides var. peruviana.

PIPTANTHUS (Legumin.).
leiocarpus
var. sericopetalus *P.C. Li in Fl. Xizangica,* 2: 719 (1985)—Xizang.

PITCAIRNIA (Bromeliac.).
inaequalis *W. Weber in Feddes Repert.,* 97(3–4): 93 (1986)—Brazil.
juzepczukii *W. Weber in Feddes Repert.,* 97(3–4): 95 (1986)—Peru.
luschnathii *W. Weber in Feddes Repert.,* 97(3–4): 95 (1986)—Brazil.
rondonicola *L.B. Smith & R.W. Read in J. Bromeliad Soc.,* 36(1): 10 (1986)—Brazil.
woronowii *W. Weber in Feddes Repert.,* 97(3–4): 98 (1986)—Colombia.

PITHECELLOBIUM (Legumin.).
heterophyllum *(Roxb.) K. Haridasan & R.R. Rao, Forest Fl. Meghalaya:* 341 (1985): Mimosa heterophylla.
killipii *(Britton & Killip) C. Barbosa in Caldasia,* 15(71–75): 191 (1986): Punjuba killipii.
popayanense *C. Barbosa in Caldasia,* 15(71–75): 192 (1986), nom. nov.: Punjuba lehmannii *Britton & Killip.*

PITTOSPORUM (Pittosporac.).
angustilimbum *C.Y. Wu in Fl. Yunnanica,* 3: 328 (1983)—China.
buxifolium *K.M. Feng in Fl. Yunnanica,* 3: 318 (1983)—China.
calcicola *C.Y. Wu in Fl. Yunnanica,* 3: 326 (1983), with two types cited—China.
johnstonianum
var. glomerulatum *C.Y. Wu in Fl. Yunnanica,* 3: 325 (1983)—China.
ledoides *(Hand.-Mazz.) C.Y. Wu in Fl. Yunnanica,* 3: 328 (1983): P. heterophyllum var. ledoides.
longicarpum *S.K. Wu in Fl. Yunnanica,* 3: 316 (1983)—China.

PITTOSPORUM :–
membranifolium *S.C. Huang in Fl. Yunnanica,* 3: 314 (1983)—China.
monanthum *C.Y. Wu in Fl. Yunnanica,* 3: 318 (1983)—China.
podocarpifolium *C.Y. Wu in Fl. Yunnanica,* 3: 318 (1983)—China.
reflexisepalum *C.Y. Wu in Fl. Yunnanica,* 3: 326 (1983)—China.
ternstroemioides *C.Y. Wu in Fl. Yunnanica,* 3: 320 (1983)—China.

PLAKOTHIRA J. Florence in Bull. Mus. Nation. Hist. Nat., B, Adansonia, Ser. 4, 7(3): 239 (1985 publ. 1986). *LOASACEAE.*
frutescens *J. Florence in Bull. Mus. Nation. Hist. Nat., B, Adansonia,* Ser. 4, 7(3): 240 (1985 publ. 1986)—Marquesas Is.

PLANICHLOA B.K. Simon in Austrobaileya, 2(3): 212 (1986). *GRAMINEAE.*
nervilemma *B.K. Simon in Austrobaileya,* 2(3): 212 (1986)—Australia (Queensland).

PLANTAGO (Plantaginac.).
adspersus *Bernh., Sel. Sem. Hort. Erfurt.,* 1835; *Linnaea 11, Lit.*: 86 (1837); *cf. D.J. Mabberley in Taxon,* 32(1): 85 (1983)—Locality not stated.

PLATANTHERA (Orchidac.).
bhutanica *K. Inoue in J. Jap. Bot.,* 61(7): 193 (1986)—Bhutan.
handel-mazzettii *K. Inoue in J. Jap. Bot.,* 61(7): 195 (1986)—China.
praeclara *C.J. Sheviak & M.L. Bowles in Rhodora,* 88(854): 278 (1986)—U.S.A. (N. Dakota).

PLATYCAULOS (Restionac.).
acutus *E.E. Esterhuysen in Bothalia,* 15(3–4): 434 (1985)—South Africa (Cape Province).
anceps *(Mast.) H.P. Linder in Bothalia,* 15(3–4): 436 (1985): Hypolaena anceps.
callistachyus *(Kunth) H.P. Linder in Bothalia,* 15(3–4): 436 (1985): Restio callistachyus.
cascadensis *(Pillans) H.P. Linder in Bothalia,* 15(3–4): 436 (1985): Restio cascadensis.
depauperatus *(Kunth) H.P. Linder in Bothalia,* 15(3–4): 436 (1985): Restio depauperatus.
major *(Mast.) H.P. Linder in Bothalia,* 15(3–4): 436 (1985): Restio compressus var. major.
subcompressus *(Pillans) H.P. Linder in Bothalia,* 15(3–4): 437 (1985): Restio subcompressus.

PLATYCRATER (Hydrangeac.).
arguta
var. sinensis *H. Hara in J. Jap. Bot.,* 61(3): 70 (1986)—China.

PLATYSTELE (Orchidac.).
subgen. Teagueia *C.A. Luer, Ic. Pleurothallid. I-Syst. Pleurothallidinae (Monogr. Syst. Bot. Missouri Bot. Gard. 15):* 45 (1986).

PLATYSTELE :–
megaloglossa *C.A. Luer & R. Escobar R. in Orquideologia,* 16(3): 39 (1986)—Colombia.
phasmida *C.A. Luer & R. Escobar R. in Orquideologia,* 16(3): 42 (1986)—Colombia.

PLECTRANTHUS (Labiat.).
parviflorus *Henckel, Adumb. Pl. Hort. Hal.*: 17 (1806, pre 15 May); *cf. D.J. Mabberley in Taxon,* 32(1): 86 (1983)— Locality not stated.

PLECTRONIA (Rubiac.).
blepharopetala *K. Schum. ex A. Engler in Abh. Königl. Preuss. Akad. Wiss. Berlin,* 1894: 61 (1894)—Tropical East Africa. †

PLEIOSPILOS (Aizoac.).
subgen. Punctillaria *(N.E. Br.) H. Hartmann & S. Liede in Bot. Jahrb.,* 106(4): 475 (1986): gen. Punctillaria.
compactus
subsp. canus *(Haw.) H. Hartmann & S. Liede in Bot. Jahrb.,* 106(4): 476 (1986): Mesembryanthemum canum.
subsp. fergusoniae *(L. Bolus) H. Hartmann & S. Liede in Bot. Jahrb.,* 106(4): 478 (1986): P. fergusoniae.
subsp. minor *(L. Bolus) H. Hartmann & S. Liede in Bot. Jahrb.,* 106(4): 479 (1986): P. minor.
subsp. sororius *(N.E. Br.) H. Hartmann & S. Liede in Bot. Jahrb.,* 106(4): 478 (1986): Mesembryanthemum sororium.

PLEIOSTACHYA (Marantac.).
leiostachya *(Donn. Sm.) B.E. Hammel in Phytologia,* 60(1): 16 (1986): Ischnosiphon morlae var. leiostachya.

PLEROMA (Melastomatac.).
candidum *Bull, Cat.,* 358: 2 (1901); *cf. D.J. Mabberley in Taxon,* 34(3): 456 (1985)—Locality not stated.

PLEUROTHALLIS (Orchidac.).
balaeniceps *C.A. Luer & R.L. Dressler in Orquideologia,* 16(3): 47 (1986)—Panama.
holstii *G. Carnevali F.C. & I. Ramírez in Ernstia,* 39: 18 (1986)—Venezuela.
jalapensis *(Krzl.) L.A. Garay in Bot. Mus. Leafl. Harvard Univ.,* 30(3): (58) 192 (1985 publ. 1986): Masdevallia jalapensis.
nakatae *T. Hashimoto in Ann. Tsukuba Bot. Gard.,* 4: 3 (1986)—Peru.
vinealis *C.A. Luer & R. Escobar R. in Orquideologia,* 16(3): 30 (1986)—Colombia.

PLEUROTHYRIUM (Laurac.).
amplifolium *(Mez) J.G. Rohwer in Mitt. Inst. Allg. Bot. Hamburg,* 20: 43 (1986): Nectandra amplifolia.
intermedium *(Mez) J.G. Rohwer in Mitt. Inst. Allg. Bot. Hamburg,* 20: 43 (1986): Nectandra intermedia.

PLEUROTHYRIUM :–
palmanum *(Mez & Smith) J.G. Rohwer in Mitt. Inst. Allg. Bot. Hamburg,* 20: 41 (1986): Ocotea palmana.
trianae *(Mez) J.G. Rohwer in Mitt. Inst. Allg. Bot. Hamburg,* 20: 43 (1986): Nectandra trianae.
undulatum *(Meissn.) J.G. Rohwer in Mitt. Inst. Allg. Bot. Hamburg,* 20: 44 (1986): Oreodaphne undulata.

PLINIA (Myrtac.).
rivularis *(Camb.) A.D. Rotman in Bol. Soc. Argent. Bot.,* 24(1–2): 195 (1985): Eugenia rivularis.

PLUCHEA (Compos.).
arabica *(Boiss.) M. Qaiser & H.W. Lack in Willdenowia,* 15(2): 451 (1986): Varthemia arabica.

PLUMBAGO (Plumbaginac.).
americana *Weigel, Hort. Gryph.:* 12 (1782); *cf. D.J. Mabberley in Taxon,* 33(3): 443 (1984)— Locality not stated.

POA (Gramin.).
barbata *C.P. Thunberg ex A. Murray in Linnaeus, Syst. Veg., ed. 14:* 114 (1784)—Japan. †
bulbosa
var. hackelii *(Post) N. Feinbrun-Dothan, Fl. Palaestina,* 4: 246 (1986): P. hackelii.
chokensis *S.M. Phillips in Kew Bull.,* 41(4): 1027 (1986)—Ethiopia.
ferruginea *C.P. Thunberg ex A. Murray in Linnaeus, Syst. Veg., ed. 14:* 114 (1784)—Japan. †
glauca
subsp. frearitis *(Orph. ex Halácsy) H. Scholz in Willdenowia,* 15(2): 395 (1986): P. caesia var. frearitis.
hirta *C.P. Thunberg ex A. Murray in Linnaeus, Syst. Veg., ed. 14:* 113 (1784)—Japan. †
japonica *C.P. Thunberg ex A. Murray in Linnaeus, Syst. Veg., ed. 14:* 114 (1784)—Japan. †
ophiolithica *H. Scholz in Willdenowia,* 15(2): 396 (1986)—Greece.

POGONOPUS (Rubiac.).
erythroxylon *(Standl.) J.H. Kirkbride in Acta Amazonica,* 15(1–2): 54 (1985 publ. 1986): Capirona erythroxylon.

POGOSTEMON (Labiat.).
erectus
var. gracilis *(Dalz.) G. Panigrahi in J. Orissa Bot. Soc.,* 6(1): 52 (1984): Dysophylla gracilis.
kachinensis *(Mukherjee) G. Panigrahi in J. Orissa Bot. Soc.,* 6(1): 53 (1984): Dysophylla kachinensis.

POHLIDIUM G. Davidse, T.R. Soderstrom & R.P. Ellis in Syst. Bot., 11(1): 131 (1986). *GRAMINEAE.*

POHLIDIUM :–
petiolatum *G. Davidse, T.R. Soderstrom & R.P. Ellis in Syst. Bot.,* 11(1): 131 (1986)—Panama.

POINSETTIA (Euphorbiac.).
cyathophora *(J. Murr.) Bartl., Ind. Sem. Hort. Acad. Gott.,* 1839: 6 (1839); *cf. D.J. Mabberley in Taxon,* 32(1): 86 (1983): Euphorbia cyathophora.

POLEMANNIA (Umbellif.).
simplicior *O.M. Hilliard & B.L. Burtt in Notes Roy. Bot. Gard. Edinburgh,* 43(2): 225 (1986)—South Africa (Cape Province, Orange Free State, Natal), Lesotho.

POLHILLIA C.H. Stirton in S. Afr. J. Bot., 52(2): 170 (1986), nom. nov.: Lebeckia subgen. Plecolobium C.H. Stirton (Legumin.).
canescens *C.H. Stirton in S. Afr. J. Bot.,* 52(2): 177 (1986)—South Africa (Cape Province).
connatum *(Harv.) C.H. Stirton in S. Afr. J. Bot.,* 52(2): 174 (1986): Argyrolobium connatum.
pallens *C.H. Stirton in S. Afr. J. Bot.,* 52(2): 171 (1986)—South Africa (Cape Province).
waltersii *(C.H. Stirton) C.H. Stirton in S. Afr. J. Bot.,* 52(2): 173 (1986): Lebeckia waltersii.

POLYALTHIA (Annonac.).
australis *(Benth.) L.W. Jessup in Austrobaileya,* 2(3): 227 (1986): Popowia australis.

POLYCARPAEA (Caryophyllac.).
corymbosa
var. diffusa *(Wight & Arn.) S.R. Ramesh & B.A. Razi in J. Econ. Taxon. Bot.,* 7(3): 740 (1985 publ. 1986): P. diffusa.

POLYGALA (Polygalac.).
albicans
var. caracaensis *Glaz. ex M. do C.M. Marques in Rodriguésia,* 36(60): 7 (1984)—Brazil.
var. silvae *M. do C.M. Marques in Rodriguésia,* 36(60): 7 (1984)—Brazil.
angustifolia *(Chodat) R.N. Banerjee in Bull. Bot. Surv. India,* 26(1–2): 2 (1984 publ. 1985): P. arillata var. angustifolia.
var. scandens *(Mukerjee) G.S. Giri & R.N. Banerjee in Bull. Bot. Surv. India,* 26(1–2): 3 (1984 publ. 1985): P. arillata f. scandens.
arillata
var. chartacea *(Mukerjee) G.S. Giri in Bull. Bot. Surv. India,* 26(1–2): 5 (1984 publ. 1985): P. arillata f. chartacea.
var. revoluta *(Mukerjee) G.S. Giri in Bull. Bot. Surv. India,* 26(1–2): 5 (1984 publ. 1985): P. arillata f. revoluta.
caespitosa *T. Nees, Pl. Offic.,* sub t. 411 (?1828); *cf. D.J. Mabberley in Taxon,* 31(1): 72 (1982)— Locality not stated.

POLYGALA :–
globulifera
var. kachinensis *(Mukerjee) R.N. Banerjee in Bull. Bot. Surv. India,* 26(1–2): 6 (1984 publ. 1985): P. arillata f. kachinensis.
millspaughiana *J.A.R. Paiva in Fontqueria,* 7: 1 (1985), nom. nov.: P. uncinata *Wright ex Millsp.*

POLYGONATUM (Convallariac.).
cirrhifoliodes *D.M. Liu & W.Z. Zeng in Bull. Bot. Res. North-East. Forest. Inst.,* 6(2): 91 (1986)—China, Xizang.
strumulosum *D.M. Liu & W.Z. Zeng in Bull. Bot. Res. North-East. Forest. Inst.,* 6(2): 92 (1986)—China.

POLYGONUM (Polygonac.).
aboriginum *(E.L. Greene) F. Fedde in Just's Bot. Jahresber.,* 32(1): 417 (1905): Persicaria aboriginum.
abscissum *(E.L. Greene) F. Fedde in Just's Bot. Jahresber.,* 32(1): 417 (1905): Persicaria abscissa.
alismaefolium *(E.L. Greene) F. Fedde in Just's Bot. Jahresber.,* 32(1): 417 (1905): Persicaria alismaefolia.
ammophilum *(E.L. Greene) F. Fedde in Just's Bot. Jahresber.,* 32(1): 417 (1905): Persicaria ammophila.
asclepiadeum *(E.L. Greene) F. Fedde in Just's Bot. Jahresber.,* 32(1): 417 (1905): Persicaria asclepiadea.
calophyllum *(E.L. Greene) F. Fedde in Just's Bot. Jahresber.,* 32(1): 415 (1905): Bistorta calophylla.
canadense *(E.L. Greene) F. Fedde in Just's Bot. Jahresber.,* 32(1): 417 (1905): Persicaria canadensis.
caurianum
subsp. hudsonianum *S.J. Wolf & J. McNeill in Rhodora,* 88(856): 469 (1986)—Canada (Quebec, Labrador).
chelanicum *(E.L. Greene) F. Fedde in Just's Bot. Jahresber.,* 32(1): 417 (1905): Persicaria chelanica.
clarkei *C.W. Park in Brittonia,* 38(3): 217 (1986), nom. nov.: P. pedunculare var. assamicum *Hook. f.*
covillei *(E.L. Greene) F. Fedde in Just's Bot. Jahresber.,* 32(1): 417 (1905): Persicaria covillei.
cusickii *(E.L. Greene) F. Fedde in Just's Bot. Jahresber.,* 32(1): 417 (1905): Persicaria cusickii.
debilis
f. purpurea *N. Yonezawa in J. Phytogeogr. Taxon.,* 33(1): 56 (1985)—Japan.
filiforme *C.P. Thunberg ex A. Murray in Linnaeus, Syst. Veg., ed. 14:* 377 (1784)—Japan. †
fistulosum *(E.L. Greene) F. Fedde in Just's Bot. Jahresber.,* 32(1): 417 (1905): Persicaria fistulosa.

POLYGONUM :–
franciscanum *(E.L. Greene) F. Fedde in Just's Bot. Jahresber.,* 32(1): 417 (1905): Persicaria franciscana.
franktonii *S.J. Wolf & J. McNeill in Rhodora,* 88(856): 474 (1986)—Canada (Nova Scotia).
grandifolium *(E.L. Greene) F. Fedde in Just's Bot. Jahresber.,* 32(1): 417 (1905): Persicaria grandifolia.
hesperium *(E.L. Greene) F. Fedde in Just's Bot. Jahresber.,* 32(1): 417 (1905): Persicaria hesperia.
homalostachyum *(E.L. Greene) F. Fedde in Just's Bot. Jahresber.,* 32(1): 417 (1905): Persicaria homalostachya.
insigne *(E.L. Greene) F. Fedde in Just's Bot. Jahresber.,* 32(1): 417 (1905): Persicaria insignis.
laetevirens *(E.L. Greene) F. Fedde in Just's Bot. Jahresber.,* 32(1): 417 (1905): Persicaria laetevirens.
langloisii *(E.L. Greene) F. Fedde in Just's Bot. Jahresber.,* 32(1): 417 (1905): Persicaria langloisii.
laurinum *(E.L. Greene) F. Fedde in Just's Bot. Jahresber.,* 32(1): 417 (1905): Persicaria laurina.
leptophyllum *(E.L. Greene) F. Fedde in Just's Bot. Jahresber.,* 32(1): 415 (1905): Bistorta leptophylla.
lilacinum *(E.L. Greene) F. Fedde in Just's Bot. Jahresber.,* 32(1): 415 (1905): Bistorta lilacina.
littorale *(E.L. Greene) F. Fedde in Just's Bot. Jahresber.,* 32(1): 415 (1905): Bistorta littoralis.
lonchophyllum *(E.L. Greene) F. Fedde in Just's Bot. Jahresber.,* 32(1): 417 (1905): Persicaria lonchophylla.
mesochorum *(E.L. Greene) F. Fedde in Just's Bot. Jahresber.,* 32(1): 417 (1905): Persicaria mesochora.
multiflorum *C.P. Thunberg ex A. Murray in Linnaeus, Syst. Veg., ed. 14:* 379 (1784)—Japan. †
muriculatum *(E.L. Greene) F. Fedde in Just's Bot. Jahresber.,* 32(1): 417 (1905): Persicaria muriculata.
novae-angliae *(E.L. Greene) F. Fedde in Just's Bot. Jahresber.,* 32(1): 417 (1905): Persicaria novae-angliae.
ophiophilum *(E.L. Greene) F. Fedde in Just's Bot. Jahresber.,* 32(1): 417 (1905): Persicaria ophiophila.
oreganum *(E.L. Greene) F. Fedde in Just's Bot. Jahresber.,* 32(1): 417 (1905): Persicaria oregana.
otophyllum *(E.L. Greene) F. Fedde in Just's Bot. Jahresber.,* 32(1): 417 (1905): Persicaria otophylla.
plantagineum *(E.L. Greene) F. Fedde in Just's Bot. Jahresber.,* 32(1): 417 (1905): Persicaria plantaginea.
plattense *(E.L. Greene) F. Fedde in Just's Bot. Jahresber.,* 32(1): 417 (1905): Persicaria plattensis.

POLYGONUM :-
porteri *(E.L. Greene) F. Fedde in Just's Bot. Jahresber.,* 32(1): 417 (1905): Persicaria porteri.
pratincolum *(E.L. Greene) F. Fedde in Just's Bot. Jahresber.,* 32(1): 417 (1905): Persicaria pratincola.
propinquum *(E.L. Greene) F. Fedde in Just's Bot. Jahresber.,* 32(1): 417 (1905): Persicaria propinqua.
psychrophilum *(E.L. Greene) F. Fedde in Just's Bot. Jahresber.,* 32(1): 417 (1905): Persicaria psychrophila.
purpuratum *(E.L. Greene) F. Fedde in Just's Bot. Jahresber.,* 32(1): 417 (1905): Persicaria purpurata.
raii *Bab. in Brewster, Taylor & Phillips, Phil. Mag. & J. Sci.,* 8: 346 (Apr. 1836); *cf. D.J. Mabberley in Taxon,* 31(1): 72 (1982)— Locality not stated.
remotum *(E.L. Greene) F. Fedde in Just's Bot. Jahresber.,* 32(1): 417 (1905): Persicaria remota.
ribesioides *Huber, Cat. Végét.:* 14 (1875); *cf. D.J. Mabberley in Taxon,* 34(3): 451 (1985)— Locality not stated.
rigidulum *(E.L. Greene) F. Fedde in Just's Bot. Jahresber.,* 32(1): 417 (1905): Persicaria rigidula.
rothrockii *(E.L. Greene) F. Fedde in Just's Bot. Jahresber.,* 32(1): 417 (1905): Persicaria rothrockii.
scaberulum *(E.L. Greene) F. Fedde in Just's Bot. Jahresber.,* 32(1): 415 (1905): Bistorta scaberula.
scopulinum *(E.L. Greene) F. Fedde in Just's Bot. Jahresber.,* 32(1): 415 (1905): Bistorta scopulina.
senticosum
var. sagittifolium *(Léveillé & Vaniot) C.W. Park in Brittonia,* 38(3): 218 (1986): P. sagittifolium
spectabile *(E.L. Greene) F. Fedde in Just's Bot. Jahresber.,* 32(1): 417 (1905): Persicaria spectabilis.
subcoriaceum *(E.L. Greene) F. Fedde in Just's Bot. Jahresber.,* 32(1): 417 (1905): Persicaria subcoriacea.
subsagittatum *(De Wildeman) C.W. Park in Brittonia,* 38(3): 218 (1986): P. pedunculare var. subsagittatum.
vestitum *(E.L. Greene) F. Fedde in Just's Bot. Jahresber.,* 32(1): 417 (1905): Persicaria vestita.
villosulum *(E.L. Greene) F. Fedde in Just's Bot. Jahresber.,* 32(1): 417 (1905): Persicaria villosula.
wardii *(E.L. Greene) F. Fedde in Just's Bot. Jahresber.,* 32(1): 417 (1905): Persicaria wardii.

POLYPOGON (Gramin.).
monspeliensis
var. indicus *S. Bhattacharya & S.K. Jain in Bull. Bot. Surv. India,* 25(1–4): 208 (1983 publ. 1985)—India, Afghanistan.

PONTHIEVA (Orchidac.).
angustipetala *E.W. Greenwood in Orquídea (México),* 10(1): 9 (1986)—Mexico.

POPULUS (Salicac.).
lasiocarpa
var. longiamenta *P.Y. Mao & P.X. He in Bull. Bot. Res. North-East. Forest. Inst.,* 6(2): 79 (1986)—China.
wulianensis *S.B. Liang & X.W. Li in Bull. Bot. Res. North-East. Forest. Inst.,* 6(2): 135 (1986)—China.

PORANDRA (Commelinac.).
microphylla *Y. Wan in Bull. Bot. Res. North-East. Forest. Inst.,* 6(4): 153 (1986)—China.

PORFIRIA (Cactac.).
schwarzii *(Fric) Crkal, Lovec Kaktusu:* 395 (1983); *cf. Repert. Pl. Succ. (I.O.S.),* 34: 7 (1985): Haagea schwarzii.

PORPAX (Orchidac.).
chandrasekharanii *P. Bhargavan & C.N. Mohanan in Curr. Sci.,* 51(20): 990 (1982)—India.

PORTULACA (Portulacac.).
carrissoana *(Exell & Mend.) B.L. Nyananyo in Taxon,* 35(3): 592 (1986): Sedopsis carrissoana.

POTAMOGETON (Potamogetonac.).
austrosibiricus *L.I. Kashina in I.M. Krasnoborov & T.A. Safonova (eds.), Novoe Fl. Sibiri:* 243 (1986), as 'austro-sibiricus'—U.S.S.R. (Siberia).
fontigenus *Y.H. Guo, H.Q. Wang & X.Z. Sun in Acta Bot. Bor.-Occid. Sin.,* 5(4): 301 (1985)—China.
pectinatus
subsp. chakassiensis *L.I. Kashina in I.M. Krasnoborov & T.A. Safonova (eds.), Novoe Fl. Sibiri:* 245 (1986)—U.S.S.R. (Siberia).

POTENTILLA (Rosac.).
sect. Graciles *(Wolf) J. Soják in Čas. Nár. Muz. (Prague),* 152(1): 21 (1983): P. grex Graciles.
sect. Tridentatae *(Wolf) J. Soják in Čas. Nár. Muz. (Prague),* 152(1): 21 (1983): P. grex Tridentatae.
aphanes *J. Soják in Willdenowia,* 16(1): 131 (1986)—Mongolia, Xinjiang, China.
blanda *J. Soják in Čas. Nár. Muz. (Prague),* 151(4): 206 (1982 publ. 1983)—U.S.S.R. (Middle Asia).
×blanda
var. homotricha *J. Soják in Bot. Jahrb.,* 106(2): 208 (1986)—U.S.S.R. (Middle Asia).
×burjatica *J. Soják in Bot. Jahrb.,* 106(2): 176 (1986)—U.S.S.R. (Siberia). (P. arenosa × tergemina).

POTENTILLA :–

var. bigemina *J. Soják in Bot. Jahrb.*, 106(2): 207 (1986)—U.S.S.R. (Siberia).

var. laeta *J. Soják in Bot. Jahrb.*, 106(2): 206 (1986)—U.S.S.R. (Siberia).

burmanica *J. Soják in Čas. Nár. Muz. (Prague)*, 152(3): 160 (1983)—Burma.

×chamaeleo *J. Soják in Bot. Jahrb.*, 106(2): 184 (1986)—Mongolia. (P. crebridens × pamirica).

var. byssina *J. Soják in Bot. Jahrb.*, 106(2): 207 (1986)—Mongolia.

var. pamiricoides *J. Soják in Bot. Jahrb.*, 106(2): 208 (1986)—Mongolia.

var. quinata *J. Soják in Bot. Jahrb.*, 106(2): 207 (1986)—Mongolia.

var. tytthantha *J. Soják in Bot. Jahrb.*, 106(2): 207 (1986)—Mongolia.

×chionea

var. chadchalica *J. Soják in Bot. Jahrb.*, 106(2): 206 (1986)—Mongolia.

var. congesta *J. Soják in Bot. Jahrb.*, 106(2): 206 (1986)—Mongolia.

var. malyschevii *(Peschk.) J. Soják in Bot. Jahrb.*, 106(2): 209 (1986): P. malyschevii.

var. sericeoides *J. Soják in Bot. Jahrb.*, 106(2): 206 (1986)—Mongolia.

exuta *J. Soják in Willdenowia*, 16(1): 127 (1986)—Mongolia, U.S.S.R. (Middle Asia).

flavida *J. Soják in Čas. Nár. Muz. (Prague)*, 151(3): 149 (1982)—India.

geoides

subsp. halacsyana *(Degen) A. Strid in A. Strid (ed.), Mount. Fl. Greece*, 1: 407 (1986): P. halacsyana.

subsp. longisepala *A. Strid in A. Strid (ed.), Mount. Fl. Greece*, 1: 408 (1986)—Greece.

subsp. regis-borisii *(Stoj.) A. Strid in A. Strid (ed.), Mount. Fl. Greece*, 1: 408 (1986): P. regis-borisii.

hilbigii *J. Soják in Willdenowia*, 16(1): 135 (1986)—Mongolia.

inopinata *J. Soják in Willdenowia*, 16(1): 130 (1986)—Mongolia.

×insularis *J. Soják in Bot. Jahrb.*, 106(2): 203 (1986)—Svalbard, Greenland. (P. chamissonis × lyngei).

jacquemontii *(Franchet) J. Soják in Bot. Jahrb.*, 106(2): 191 (1986): P. saundersiana var. jacquemontii.

laevipes *J. Soják in Willdenowia*, 16(1): 125 (1986)—Mongolia.

×lenae *J. Soják in Bot. Jahrb.*, 106(2): 179 (1986)—U.S.S.R. (Siberia). (P. arenosa × conferta).

lineata *Trev., Ind. Sem. Vratislav.*, 1822; *cf. D.J. Mabberley in Taxon*, 33(3): 443 (1984)— Locality not stated.

neumanniana

var. campestris *(Wallr.) O. de Bolòs & J. Vigo, Fl. Països Catalans:* 392 (1984): P. verna var. campestris.

POTENTILLA :–

var. hirsuta *(DC.) O. de Bolòs & J. Vigo, Fl. Països Catalans:* 393 (1984): P. verna var. hirsuta.

oligandra *J. Soják in Čas. Nár. Muz. (Prague)*, 152(3): 160 (1983)—Xizang.

pedata *Willd. ex Hornem., Hort. Hafn.*: 477 (1815); *cf. D.J. Mabberley in Taxon*, 33(3): 437 (1984)— Locality not stated.

pensylvanica

subsp. oreodoxa *(Soják) O. de Bolòs & J. Vigo, Fl. Països Catalans:* 380 (1984): P. oreodoxa.

petrovskyi *J. Soják in Čas. Nár. Muz. (Prague)*, 153(2): 102 (1984)—U.S.S.R. (Soviet Far East).

pulchella *R. Br., Chloris melv.*: 19 (1823); *cf. D.J. Mabberley in Taxon*, 30(1): 17 (1981)— Locality not stated.

×rhipidophylla *J. Soják in Bot. Jahrb.*, 106(2): 196 (1986)—Mongolia. (P. crebridens × multifida).

×rubricaulis

var. furcata *(Pors.) J. Soják in Bot. Jahrb.*, 106(2): 209 (1986): P. furcata.

salwinensis *J. Soják in Čas. Nár. Muz. (Prague)*, 152(3): 160 (1983)—China.

saundersiana

var. segmentata *(Soják) J. Soják in Bot. Jahrb.*, 106(2): 209 (1986): P. forrestii var. segmentata.

serrata *J. Soják in Willdenowia*, 16(1): 133 (1986)—Mongolia.

×subquinata

var. niveoides *J. Soják in Bot. Jahrb.*, 106(2): 208 (1986)—Greenland.

tapetodes *J. Soják in Čas. Nár. Muz. (Prague)*, 152(3): 160 (1983)—Bhutan.

×thomsonii

var. caliginoides *J. Soják in Bot. Jahrb.*, 106(2): 208 (1986)—Xizang.

var. tibetica *(Ostenf.) J. Soják in Bot. Jahrb.*, 106(2): 209 (1986): P. hololeuca var. tibetica.

var. trijuga *J. Soják in Bot. Jahrb.*, 106(2): 208 (1986)—India.

POTHOS (Arac.).

umbraculifera *De Vriese, Hort. Spaarn-Berg.*: 41 (1839); *cf. D.J. Mabberley in Taxon*, 33(3): 443 (1984)— Locality not stated.

POUZOLZIA (Urticac.).

bracteosa *I. Friis in Bol. Soc. Brot.*, 58: 202 (1985)—Zambia.

PRENANTHES (Compos.).

chinensis *C.P. Thunberg ex A. Murray in Linnaeus, Syst. Veg., ed. 14:* 714 (1784)— Japan. †

debilis *C.P. Thunberg ex A. Murray in Linnaeus, Syst. Veg., ed. 14:* 714 (1784)— Japan. †

PRENANTHES :–
dentata *C.P. Thunberg ex A. Murray in Linnaeus, Syst. Veg., ed. 14:* 715 (1784)—Japan. †
hastata *C.P. Thunberg ex A. Murray in Linnaeus, Syst. Veg., ed. 14:* 715 (1784)—Japan. †
humilis *C.P. Thunberg ex A. Murray in Linnaeus, Syst. Veg., ed. 14:* 715 (1784)—Japan. †
integra *C.P. Thunberg ex A. Murray in Linnaeus, Syst. Veg., ed. 14:* 714 (1784)—Japan. †
lyrata *C.P. Thunberg ex A. Murray in Linnaeus, Syst. Veg., ed. 14:* 715 (1784)—Japan. †
multiflora *C.P. Thunberg ex A. Murray in Linnaeus, Syst. Veg., ed. 14:* 715 (1784)—Japan. †
squarrosa *C.P. Thunberg ex A. Murray in Linnaeus, Syst. Veg., ed. 14:* 715 (1784)—Japan. †

PRESTOEA (Palmae).
darienensis *A. Henderson in Brittonia,* 38(3): 266 (1986)—Panama.
simplicifolia *G. Galeano-Garcés in Brittonia,* 38(1): 62 (1986)—Colombia.

PRIMULA (Primulac.).
sect. Inayatii *(Smith & Fletcher) Y.J. Nasir in Willdenowia,* 15(2): 480 (1986): P. subsect. Inayatii.
sect. Sibirica *(Smith & Fletcher) Y.J. Nasir in Willdenowia,* 15(2): 480 (1986): P. subsect. Sibirica.
baokongensis *Chen & C.M. Hu ex J.X. Yang in Fl. Tsinlingensis,* 1(4): 394, 39 (1983)—China.
subansirica *G.D. Pal in J. Bombay Nat. Hist. Soc.,* 82(3): 617 (1985 publ. 1986)—India.

PRITZELAGO (Crucif.).
alpina
subsp. affinis *(Gren. ex F. Schultz) J. Soják in Čas. Nár. Muz. (Prague),* 152(1): 21 (1983): Hutchinsia affinis.

PRIVA (Verbenac.).
mitchellii *Endl., Cat. Horti Vindob.,* 2: 47 (1842); *cf. D.J. Mabberley in Taxon,* 33(3): 435 (1984)— Locality not stated.
roxburghii *Endl., Cat. Horti Vindob.,* 2: 47 (1842); *cf. D.J. Mabberley in Taxon,* 33(3): 435 (1984)— Locality not stated.

PROLONGOA (Compos.).
macrocarpa *(Cosson & Kralik) S.A. Alavi in Fl. Libya,* 107: 171 (1983): Pyrethrum macrocarpum.

PRUNUS (Rosac.).
aspera *C.P. Thunberg ex A. Murray in Linnaeus, Syst. Veg., ed. 14:* 463 (1784)—Japan. †

PRUNUS :–
cornuta
var. integrifolia *S.C. Ghora & G. Panigrahi in J. Econ. Taxon. Bot.,* 7(1): 228 (1985)—India.
elliptica *C.P. Thunberg ex A. Murray in Linnaeus, Syst. Veg., ed. 14:* 463 (1784)—Japan. †
glandulosa *C.P. Thunberg ex A. Murray in Linnaeus, Syst. Veg., ed. 14:* 463 (1784)—Japan. †
glauciphylla *S.C. Ghora & G. Panigrahi in J. Econ. Taxon. Bot.,* 7(1): 228 (1985)—India, Sikkim.
incisa *C.P. Thunberg ex A. Murray in Linnaeus, Syst. Veg., ed. 14:* 463 (1784)—Japan. †
japonica *C.P. Thunberg ex A. Murray in Linnaeus, Syst. Veg., ed. 14:* 463 (1784)—Japan. †
paniculata *C.P. Thunberg ex A. Murray in Linnaeus, Syst. Veg., ed. 14:* 463 (1784)—Japan. †
tomentosa *C.P. Thunberg ex A. Murray in Linnaeus, Syst. Veg., ed. 14:* 464 (1784)—Japan. †
wattii *S.C. Ghora & G. Panigrahi in J. Econ. Taxon. Bot.,* 7(1): 231 (1985)—India, Sikkim, Xizang.

PSAMMOPYRUM A. Löve in Veröff. Geobot. Inst. Rübel, 87: 50 (1986). *GRAMINEAE.*
athericum *(Link) A. Löve in Veröff. Geobot. Inst. Rübel,* 87: 50 (1986): Triticum athericum.
fontqueri *(Melderis) A. Löve in Veröff. Geobot. Inst. Rübel,* 87: 50 (1986): Elymus pungens subsp. fontqueri.
pungens *(Pers.) A. Löve in Veröff. Geobot. Inst. Rübel,* 87: 50 (1986): Triticum pungens.
subsp. campestre *(Gren. & Godron) A. Löve in Veröff. Geobot. Inst. Rübel,* 87: 50 (1986): Agropyron campestre.

PSEUDABUTILON (Malvac.).
cinereum *(Griseb.) A. Krapovickas in Bol. Soc. Argent. Bot.,* 24(1–2): 206 (1985): Abutilon cinereum.

PSEUDARRHENATHERUM (Gramin.).
setifolia *(Brot.) B.E. Smythies in Englera,* 3(3): 645 (1986): Avena setifolia.

PSEUDOACANTHOCEREUS (Cactac.).
boreominarum *C.T. Rizzini & A. de Mattos-Filho in Rev. Brasil. Biol.,* 46(2): 327 (1986)—Brazil.

PSEUDOFUMARIA (Papaverac.).
alba *(Miller) M. Lidén in Op. Bot.,* 88: 32 (1986): Fumaria alba.
subsp. acaulis *(Wulfen) M. Lidén in Op. Bot.,* 88: 32 (1986): Fumaria acaulis.
subsp. leiosperma *(Conrath) M. Lidén in Op. Bot.,* 88: 32 (1986): Corydalis leiosperma.

PSEUDOGOODYERA (Orchidac.).
gonzalezii *(L.O. Wms.) P. Burns-Balogh in Orquídea (México),* 10(1): 92 (1986): Spiranthes gonzalezii.

PSEUDOLYSIMACHION (Scrophulariac.).
spicatum
subsp. bashkiriense *(Tzvel.) J. Soják in Čas. Nár. Muz. (Prague),* 152(1): 21 (1983): Veronica spicata subsp. bashkiriensis.
subsp. klokovii *(Tzvel.) J. Soják in Čas. Nár. Muz. (Prague),* 152(1): 21 (1983): Veronica spicata subsp. klokovii.
subsp. maeoticum *(Klok.) J. Soják in Čas. Nár. Muz. (Prague),* 152(1): 21 (1983): Veronica maeotica.
subsp. petschoricum *(Tzvel.) J. Soják in Čas. Nár. Muz. (Prague),* 152(1): 21 (1983): Veronica spicata subsp. petschorica.
viscosulum *(Klok.) J. Soják in Čas. Nár. Muz. (Prague),* 152(1): 21 (1983): Veronica viscosula.

PSEUDOMERTENSIA (Boraginac.).
rosulata *(Ovchinnikov & Chukavina) P.N. Ovchinnikov & A.P. Chukavina in Izv. Akad. Nauk Tadzh. SSR, Biol Nauk,* 2(79): 104 (1980): Scapicephalus rosulatus.

PSEUDOXANDRA (Annonac.).
bahiensis *P.J.M. Maas in Proc. Kon. Nederl. Akad. Wetensch., C,* 89(3): 265 (1986)—Brazil.
cuspidata *P.J.M. Maas in Proc. Kon. Nederl. Akad. Wetensch., C,* 89(3): 267 (1986)—French Guiana, Brazil.
pacifica *P.J.M. Maas in Proc. Kon. Nederl. Akad. Wetensch., C,* 89(3): 270 (1986)—Colombia.
sclerocarpa *P.J.M. Maas in Proc. Kon. Nederl. Akad. Wetensch., C,* 89(3): 271 (1986)—Colombia.

PSEUDUVARIA (Annonac.).
froggattii *(F. Muell.) L.W. Jessup in Austrobaileya,* 2(3): 227 (1986): Mitrephora froggattii.

PSIDIUM (Myrtac.).
wrightii *D. Don ex W. Wright in Mitchell, Memoir W. Wright:* 278 (1828); *cf. D.J. Mabberley in Taxon,* 30(1): 12 (1981)—Locality not stated.

PSORALEA (Legumin.).
fruticosa *Spreng., Bot. Gart. Univ. Halle:* 75 (1800); *cf. D.J. Mabberley in Taxon,* 32(1): 86 (1983)— Locality not stated.
verrucosa *Willd. ex Spreng., Bot. Gart. Univ. Halle:* 75 (1800); *cf. D.J. Mabberley in Taxon,* 32(1): 86 (1983)— Locality not stated.

PSORALIDIUM (Legumin.).
lanceolatum
var. stenophyllum *(Rydb.) S.L. Welsh in Great Basin Nat.,* 46(2): 257 (1986): Psoralea stenophylla.
var. stenostachys *(Rydb.) S.L. Welsh in Great Basin Nat.,* 46(2): 257 (1986): Psoralea stenostachys.

PSYCHOTRIA (Rubiac.).
balakrishnii *D.B. Deb & M. Gangopadhyay in Bull. Bot. Surv. India,* 25(1–4): 211 (1983 publ. 1985)—Andaman Is.
bristolii *W.A. Whistler in J. Arnold Arbor.,* 67(3): 354 (1986)—Samoa.
burkillii *D.B. Deb & M. Gangopadhyay in Bull. Bot. Surv. India,* 25(1–4): 213 (1983 publ. 1985)—India.
carnea
subsp. oncocarpa *(Schumann) W.A. Whistler in J. Arnold Arbor.,* 67(3): 354 (1986): P. oncocarpa.
christophersenii *W.A. Whistler in J. Arnold Arbor.,* 67(3): 363 (1986)—Samoa.
grandistipulata *(Lauterbach) W.A. Whistler in J. Arnold Arbor.,* 67(3): 356 (1986): Randia grandistipulata.
keralensis *D.B. Deb & M. Gangopadhyay in Bull. Bot. Surv. India,* 25(1–4): 213 (1983 publ. 1985)—India.
nudiflora
var. latifolia *D.B. Deb & M. Gangopadhyay in Bull. Bot. Surv. India,* 25(1–4): 215 (1983 publ. 1985)—India.
sclerocarpa *W.A. Whistler in J. Arnold Arbor.,* 67(3): 351 (1986)—Samoa.
vaupelii *W.A. Whistler in J. Arnold Arbor.,* 67(3): 352 (1986)—Samoa.

PSYLLIUM (Plantaginac.).
majus *Ratier, Pharm. Franç.:* 118 (1827); *cf. D.J. Mabberley in Taxon,* 31(1): 72 (1982)—Locality not stated.

PTERICHIS (Orchidac.).
colombiana *G. Morales L. in Orquideologia,* 16(3): 72 (1986)—Colombia.
fernandezii *G. Morales L. in Orquideologia,* 16(3): 74 (1986)—Colombia.
galeata
var. quilinsayacoana *G. Morales L. in Orquideologia,* 16(3): 60 (1986)—Colombia.

PTERNOPETALUM (Umbellif.).
brevium *(Shan & Pu) K.T. Fu in Fl. Tsinlingensis,* 1(3): 400 (1981): P. longicaule var. brevium.
lamellosociliare *K.T. Fu in Fl. Tsinlingensis,* 1(3): 457, 401 (1981), as 'lamelloso-ciliare'—China.

PTEROCEPHALUS (Dipsacac.).
multisetus *(Vis.) Endl., Cat. Horti Vindob.,* 1: 303 (1842); *cf. D.J. Mabberley in Taxon,* 33(3): 435 (1984): Basionym not stated.

PTEROCEPHALUS :–
siamensis *(Craib) R. Verlaque in Rev. Cytol. Biol. Vég. Bot.*, 9(1): 30 (1986), without basionym ref.: Scabiosa siamensis.

PTEROPENTACOILANTHUS F. Rappa & V. Camarrone in Lav. Ist. Bot. Giard. Colon. Palermo, 18: 22 (1962), without type. *AIZOACEAE.*
hypertrophicus *(Dtr.) F. Rappa & V. Camarrone in Lav. Ist. Bot. Giard. Colon. Palermo,* 18: 22 (1962), without basionym ref.: Halenbergia hypertrophica.

PTEROSTYLIS (Orchidac.).
abrupta *D.L. Jones in Orchadian,* 8(6): 122 (1985)—Australia (New South Wales).

PTEROTETRACOILANTHUS F. Rappa & V. Camarrone in Lav. Ist. Bot. Giard. Colon. Palermo, 18: 22 (1962), without type. *AIZOACEAE.*

PTILOTRICHUM (Crucif.).
cyclocarpum
subsp. pindicum *P. Hartvig in A. Strid (ed.), Mount. Fl. Greece,* 1: 305 (1986)—Greece, Albania.

PULICARIA (Compos.).
inuloides *Vahl ex Hornem., Hort. Hafn.*: 823 (1815); *cf. D.J. Mabberley in Taxon,* 33(3): 437 (1984)— Locality not stated.
omanensis
subsp. milleri *E. Gamal-Eldin in Notes Roy. Bot. Gard. Edinburgh,* 43(3): 486 (1986)—Oman.
pulvinata *E. Gamal-Eldin in Notes Roy. Bot. Gard. Edinburgh,* 43(3): 481 (1986)—Oman.

PULMONARIA (Boraginac.).
tuberosa *Schmidt ex Mart. in Denkr. Kön. Akad. Wiss. Münch.*, 5e(7): 175 (1817); *cf. D.J. Mabberley in Taxon,* 33(3): 444 (1984)— Locality not stated.

PURSHIA (Rosac.).
ericifolia *(Torr. ex Gray) J. Henrickson in Phytologia,* 60(6): 468 (1986): Cowania ericifolia.
mexicana *(D. Don) J. Henrickson in Phytologia,* 60(6): 468 (1986): Cowania mexicana.
mexicana *(D. Don) S.L. Welsh in Great Basin Nat.*, 46(2): 260 (1986): Cowania mexicana.
var. stansburiana *(Torr.) S.L. Welsh in Great Basin Nat.,* 46(2): 260 (1986), as 'stansburyi': Cowania stansburiana.
plicata *(D. Don) J. Henrickson in Phytologia,* 60(6): 468 (1986): Cowania plicata.
stansburiana *(Torr.) J. Henrickson in Phytologia,* 60(6): 468 (1986): Cowania stansburiana.
subintegra *(Kearney) J. Henrickson in Phytologia,* 60(6): 468 (1986): Cowania subintegra.

PUYA (Bromeliac.).
textoragicolae *W. Weber in Feddes Repert.,* 97(3–4): 101 (1986), nom. nov.: Pitcairnia weberbaueri *Mez.*

PYCNOCYCLA (Umbellif.).
bashagardiana *V. Mozaffarian in Iranian J. Bot.*, 3(1): 84 (1985)—Iran.

PYCREUS (Cyperac.).
bolei *S.M. Almeida in J. Bombay Nat. Hist. Soc.*, 83(1): 180 (1986)—India.
lancelotii *S.M. Almeida in J. Bombay Nat. Hist. Soc.*, 83(1): 182 (1986)—India.

PYRETHRUM (Compos.).
argenteum *G.F. Hoffm., Hort. Mosq.*: 31 (1808); *cf. D.J. Mabberley in Taxon,* 31(1): 72 (1982)— Locality not stated.
grandiflorum *Hook. in Lyon, Repulse Bay:* 191 (1825); *cf. D.J. Mabberley in Taxon,* 30(1): 17 (1981)— Locality not stated.
lacustre *(Brot.) Hornem., Hort. Hafn.:* 828 (1815); *cf. D.J. Mabberley in Taxon,* 33(3): 437 (1984): Basionym not stated.

PYROLA (Ericac.).
rotundifolia
subsp. grandiflora *(Rad.) V.G. Sergienko, Fl. poluostrova Kanin:* 80 (1986): P. grandiflora.

PYROSTRIA (Rubiac.).
buxifolia *(Bak.) B.P.G. Hochreutiner in Annuaire Conserv. Jard. Bot. Genève,* 11–12: 99 (65) (1980): Plectronia buxifolia. †
subevenia *(K. Schum.) B.P.G. Hochreutiner in Annuaire Conserv. Jard. Bot. Genève,* 11–12: 99 (65) (1908): Plectronia subevenia.
syringaefolia *(Bak.) B.P.G. Hochreutiner in Annuaire Conserv. Jard. Bot. Genève,* 11–12: 99 (65) (1908): Plectronia syringaefolia.

PYRUS (Rosac.).
hostii *(Jacq.) Endl., Cat. Horti Vindob.*, 2. 440 (1842); *cf. D.J. Mabberley in Taxon,* 33(3): 436 (1984): Basionym not stated.
japonica *C.P. Thunberg ex A. Murray in Linnaeus, Syst. Veg., ed. 14:* 467 (1784)—Japan. †
latifolia *(Lam.) hort. in Hort. Soc. Lond., Cat. Fr.:* 183 (1826); *cf. D.J. Mabberley in Taxon,* 33(3): 443 (1984): Basionym not stated.

Q

QIONGZHUEA (Gramin.).
maculata *T.H. Wen in J. Bamboo Res.*, 5(2): 22 (1986)—China.

QUALEA (Vochysiac.).
brasiliana *Stafleu & L. Marcano Berti in Pittieria,* 13: 8 (1986)—Brazil.
hannekesaskiarum *L. Marcano Berti in Pittieria,* 13: 5 (1986)—Brazil.
wurdackii *L. Marcano Berti in Pittieria,* 13: 7 (1986)—Venezuela.

QUAMOCLIT (Convolvulac.).
oculatus *Naud. ex Huber, Cat. Gén. 1870 & 1871:* 6 (1870); *cf. D.J. Mabberley in Taxon,* 34(3): 452 (1985)— Locality not stated.
purpureus *Huber, Cat. Gén. 1870 & 1871:* 6 (1870); *cf. D.J. Mabberley in Taxon,* 34(3): 451 (1985)— Locality not stated.

QUELTIA (Amaryllidac.).
capax *Salisb. ex Haw., Suppl. Pl. Succ.:* 128 (1819); *cf. D.J. Mabberley in Taxon,* 31(1): 71 (1982)— Locality not stated.

QUERCUS (Fagac.).
acuta *C.P. Thunberg ex A. Murray in Linnaeus, Syst. Veg., ed. 14:* 858 (1784)—Japan. †
cuspidata *C.P. Thunberg ex A. Murray in Linnaeus, Syst. Veg., ed. 14:* 858 (1784)—Japan. †
dentata *C.P. Thunberg ex A. Murray in Linnaeus, Syst. Veg., ed. 14:* 858 (1784)—Japan. †
gambelii
var. bonina *S.L. Welsh in Great Basin Nat.,* 46(1): 108 (1986)—U.S.A. (Utah).
glabra *C.P. Thunberg ex A. Murray in Linnaeus, Syst. Veg., ed. 14:* 858 (1784)—Japan. †
glauca *C.P. Thunberg ex A. Murray in Linnaeus, Syst. Veg., ed. 14:* 858 (1784)—Japan. †
havardii
var. tuckeri *S.L. Welsh in Great Basin Nat.,* 46(1): 109 (1986)—U.S.A (Utah).
ningqiangensis *S.Z. Qu & W.H. Zhang in Acta Bot. Bor.-Occid. Sin.,* 6(1): 52 (1986)—China.
tsinglingensis *Liou ex S.Z. Qu & W.H. Zhang in Acta Bot. Bor.-Occid. Sin.,* 6(1): 54 (1986)—China.

R

RABDOSIA (Labiat.).
rubescens
f. lushanensis *Z.Y. Gao & Y.R. Li in Acta Phytotax. Sin.,* 24(1): 15 (1986)—China.
var. lushiensis *Z.Y. Gao & Y.R. Li in Acta Phytotax. Sin.,* 24(1): 16 (1986)—China.
var. taihangensis *Z.Y. Gao & Y.R. Li in Acta Phytotax. Sin.,* 24(1): 15 (1986)—China.

RACEMOBAMBOS (Gramin.).
microphylla *(Hsueh & Yi) Keng f. & T.H. Wen in J. Bamboo Res.,* 5(2): 13 (1986): Neomicrocalamus microphyllus.
prainii *(Gamble) Keng f. & T.H. Wen in J. Bamboo Res.,* 5(2): 13 (1986): Microcalamus prainii.
yunnanensis *T.H. Wen in J. Bamboo Res.,* 5(2): 11 (1986)—China.

RACOSPERMA C. Martius in Hortus Regius Monacensis Seminifer (1835); cf. L. Pedley in Bot. J. Linn. Soc., 92(3): 239 (1986), nom. nov.: Acacia sect. Phyllodineae A.P. de Candolle (Legumin.).
sect. Lycopodiifolia *(Pedley) L. Pedley in Bot. J. Linn. Soc.,* 92(3): 240 (1986): Acacia sect. Lycopodiifoliae.
sect. Plurinervia *(Bentham) L. Pedley in Bot. J. Linn. Soc.,* 92(3): 240 (1986): Acacia ser. Plurinerves.
sect. Pulchella *(Bentham) L. Pedley in Bot. J. Linn. Soc.,* 92(3): 240 (1986): Acacia ser. Pulchellae.
ammobium *(Maconochie) L. Pedley in Bot. J. Linn. Soc.,* 92(3): 247 (1986): Acacia ammobia.
auriculiforme *(Cunningham ex Bentham) L. Pedley in Bot. J. Linn. Soc.,* 92(3): 247 (1986): Acacia auriculiformis.
axillare *(Bentham) L. Pedley in Bot. J. Linn. Soc.,* 92(3): 248 (1986): Acacia axillaris.
beckleri *(Tindale) L. Pedley in Bot. J. Linn. Soc.,* 92(3): 248 (1986): Acacia beckleri.
concurrens *(Pedley) L. Pedley in Bot. J. Linn. Soc.,* 92(3): 248 (1986): Acacia concurrens.
confluens *(Maiden & Blakely) L. Pedley in Bot. J. Linn. Soc.,* 92(3): 248 (1986): Acacia confluens.
confusum *(Merrill) L. Pedley in Bot. J. Linn. Soc.,* 92(3): 248 (1986): Acacia confusa.
coriaceum *(DC.) L. Pedley in Bot. J. Linn. Soc.,* 92(3): 248 (1986): Acacia coriacea.
crombiei *(C. White) L. Pedley in Bot. J. Linn. Soc.,* 92(3): 247 (1986): Acacia crombiei.
deanei *(R.T. Baker) L. Pedley in Bot. J. Linn. Soc.,* 92(3): 248 (1986): Acacia decurrens var. deanei.
ensifolium *(Pedley) L. Pedley in Bot. J. Linn. Soc.,* 92(3): 248 (1986): Acacia ensifolia.
falcatum *(Willd.) L. Pedley in Bot. J. Linn. Soc.,* 92(3): 248 (1986): Acacia falcata.
georgense *(Tindale) L. Pedley in Bot. J. Linn. Soc.,* 92(3): 248 (1986): Acacia georgensis.
hylonomum *(Pedley) L. Pedley in Bot. J. Linn. Soc.,* 92(3): 248 (1986): Acacia hylonoma.
juliferum *(Bentham) L. Pedley in Bot. J. Linn. Soc.,* 92(3): 248 (1986): Acacia julifera.
kauaiense *(Hillebrand) L. Pedley in Bot. J. Linn. Soc.,* 92(3): 248 (1986): Acacia kauaiensis.
koa *(A. Gray) L. Pedley in Bot. J. Linn. Soc.,* 92(3): 248 (1986): Acacia koa.

RACOSPERMA :–
lycopodiifolium *(Cunn. ex Hooker) L. Pedley in Bot. J. Linn. Soc.*, 92(3): 240 (1986): Acacia lycopodiifolia.
macdonnellense *(Maconochie) L. Pedley in Bot. J. Linn. Soc.*, 92(3): 248 (1986): Acacia macdonnellensis.
maitlandii *(F. Mueller) L. Pedley in Bot. J. Linn. Soc.*, 92(3): 248 (1986): Acacia maitlandii.
mearnsii *(De Wildeman) L. Pedley in Bot. J. Linn. Soc.*, 92(3): 249 (1986): Acacia mearnsii.
melanoxylon *(R. Br.) L. Pedley in Bot. J. Linn. Soc.*, 92(3): 240 (1986): Acacia melanoxylon.
nelsonii *(Maslin) L. Pedley in Bot. J. Linn. Soc.*, 92(3): 249 (1986): Acacia nelsonii.
olganum *(Maconochie) L. Pedley in Bot. J. Linn. Soc.*, 92(3): 249 (1986): Acacia olgana.
orarium *(F. Mueller) L. Pedley in Bot. J. Linn. Soc.*, 93(2): 249 (1986): Acacia oraria.
penninerve *(Sieber ex DC.) L. Pedley in Bot. J. Linn. Soc.*, 92(3): 239 (1986): Acacia penninervis.
petraeum *(Pedley) L. Pedley in Bot. J. Linn. Soc.*, 92(3): 249 (1986): Acacia petraea.
peuce *(F. Mueller) L. Pedley in Bot. J. Linn. Soc.*, 92(3): 249 (1986): Acacia peuce.
pubifolium *(Pedley) L. Pedley in Bot. J. Linn. Soc.*, 92(3): 249 (1986): Acacia pubifolia.
pulchellum *(R. Br.) L. Pedley in Bot. J. Linn. Soc.*, 92(3): 240 (1986): Acacia pulchella.
pycnostachyum *(F. Mueller) L. Pedley in Bot. J. Linn. Soc.*, 92(3): 249 (1986): Acacia pycnostachya.
riceanum *(Henslow) L. Pedley in Bot. J. Linn. Soc.*, 92(3): 249 (1986): Acacia riceana.
simplex *(Sparrman) L. Pedley in Bot. J. Linn. Soc.*, 92(3): 249 (1986): Mimosa simplex.
validinervium *(Maiden & Blakely) L. Pedley in Bot. J. Linn. Soc.*, 92(3): 249 (1986): Acacia validinervia.
victoriae *(Bentham) L. Pedley in Bot. J. Linn. Soc.*, 92(3): 249 (1986): Acacia victoriae.
wetarense *(Pedley) L. Pedley in Bot. J. Linn. Soc.*, 92(3): 249 (1986): Acacia wetarensis.
xiphocladum *(Baker) L. Pedley in Bot. J. Linn. Soc.*, 92(3): 249 (1986): Acacia xiphoclada.

RADIUSIA (Legumin.).
alopecuroides *(L.) Endl., Cat. Horti Vindob.*, 2: 517 (1842); *cf. D.J. Mabberley in Taxon,* 33(3): 436 (1984): Sophora alopecuroides.
flavescens *(Ait.) Endl., Cat. Horti Vindob.*, 2: 517 (1842); *cf. D.J. Mabberley in Taxon,* 33(3): 436 (1984): Sophora flavescens.

RAJANIA (Dioscoreac.).
hexaphylla *C.P. Thunberg ex A. Murray in Linnaeus, Syst. Veg., ed. 14:* 888 (1784)—Japan. †
quinata *C.P. Thunberg ex A. Murray in Linnaeus, Syst. Veg., ed. 14:* 888 (1784)—Japan. †

RANDIA (Rubiac.).
subgen. Ceriscoides *(Hook. f.) D.D. Tirvengadum in Ceylon J. Sci., Biol. Sci.*, 14(1–2): 5 (1981): Gardenia sect. Ceriscoides.
subgen. Ceriscus *(Benth. & Hook. f.) D.D. Tirvengadum in Ceylon J. Sci., Biol. Sci.*, 14(1–2): 4 (1981), as '(Gaertn. f.)': R. sect. Ceriscus.
turgida *(Roxb.) D.D. Tirvengadum in Ceylon J. Sci., Biol. Sci.*, 14(1–2): 5 (1981): Gardenia turgida.

RANUNCULUS (Ranunculac.).
acris
var. aestivalis *(L. Benson) S.L. Welsh in Great Basin Nat.*, 46(2): 260 (1986): R. acriformis var. aestivalis.
aggregatus *Hort. Vind. ex Capelli, Cat. Stirp. Reg. Hort. Bot. Taur.:* 47 (1821); *cf. D.J. Mabberley in Taxon,* 31(1): 72 (1982)—Locality not stated.
andersonii
var. juniperinus *(Jones) S.L. Welsh in Great Basin Nat.*, 46(2): 260 (1986): R. juniperinus.
bonariensis
subsp. jaunarii *(Urban) J. Molero in Fontqueria,* 7: 19 (1985): R. bonariensis var. jaunarii.
subsp. phyteumifolius *(St.-Hil.) J. Molero in Fontqueria,* 7: 19 (1985): Casalea phyteumifolia.
subsp. trisepalus *(Gill. ex Hook. & Arnott) J. Molero in Fontqueria,* 7: 19 (1985): R. trisepalus.
collicolus *Y. Menadue in Brunonia,* 8(2): 373 (1986)—Tasmania.
jugosus *Y. Menadue in Brunonia,* 8(2): 377 (1986)—Tasmania.
minor *(L. Liou) W.T. Wang in Bull. Bot. Res. North-East. Forest. Inst.*, 6(1): 29 (1986): R. involucratus var. minor.
muscigenus *W.T. Wang in Bull. Bot. Res. North-East. Forest. Inst.*, 6(1): 30 (1986)—Xizang.
prasinus *Y. Menadue in Brunonia,* 8(2): 375 (1986)—Tasmania.
sinovaginatus *W.T. Wang in Bull. Bot. Res. North-East. Forest. Inst.*, 6(1): 34 (1986), nom. nov.: R. vaginatus *Hand.-Mazz.*
ternatus *C.P. Thunberg ex A. Murray in Linnaeus, Syst. Veg., ed. 14:* 516 (1784)—Japan. †
tuberculatus *Kit. ex Hornem., Hort. Hafn.*: 528 (1815); *cf. D.J. Mabberley in Taxon,* 33(3): 437 (1984)— Locality not stated.

RAPANEA (Myrsinac.).
clemensiae *H. Sleumer in Blumea,* 31(2): 254 (1986)—New Guinea (Papua New Guinea).
coriifolia *H. Sleumer in Blumea,* 31(2): 248 (1986)—New Guinea (Papua New Guinea).

RAPANEA :–
lamii *H. Sleumer in Blumea,* 31(2): 264 (1986)—New Guinea (Papua New Guinea, W. Irian).
minutifolia *J. Knoester, M. Wijn & H. Sleumer in Blumea,* 31(2): 252 (1986)—New Guinea (W. Irian, Papua New Guinea).
perakensis *(King & Gamble) B.C. Stone in Fed. Mus. J. (Kuala Lumpur),* 26(1): 121 (1981): Myrsine perakensis.
ralstoniae *P.S. Green in J. Arnold Arbor.,* 67(1): 113 (1986), nom. nov.: Myrsine crassifolia *Endl.*
velutina *J. Knoester, M. Wijn & H. Sleumer in Blumea,* 31(2): 251 (1986), nom. nov.: R. nummularia *van Royen.*
womersleyi *H. Sleumer in Blumea,* 31(2): 266 (1986)—New Guinea (Papua New Guinea, W. Irian).

RAPHANISTRUM (Crucif.).
segetale *Henckel, Enum. Pl. Regiom. Boruss.:* 172 (1817); *cf. D.J. Mabberley in Taxon,* 33(3): 444 (1984)— Locality not stated.

RAPHIONACME (Asclepiadac.).
namibiana *H.J.T. Venter & R.L. Verhoeven in S. Afr. J. Bot.,* 52(4): 332 (1986)—Namibia.
palustris *H.J.T. Venter & R.L. Verhoeven in S. Afr. J. Bot.,* 52(2): 149 (1986)—South Africa (Natal).

REBUTIA (Cactac.).
sumayana *W. Rausch in Succulenta,* 65(4): 74 (1986)—Bolivia.

REHDERODENDRON (Styracac.).
conostyle *C.Y. Wu in Fl. Yunnanica,* 3: 410 (1983)—China.
membranifolium *C.Y. Wu in Fl. Yunnanica,* 3: 412 (1983)—China.
microcarpum *K.M. Feng in Fl. Yunnanica,* 3: 410 (1983)—China.

REINWARDTIODENDRON (Meliac.).
dubium *(Merr.) X.M. Chen in J. Wuhan Bot. Res.,* 4(2): 183 (1986): Lansium dubium.

REMIJIA (Rubiac.).
chelomaphylla *G.A. Sullivan in Syst. Bot.,* 11(2): 298 (1986)—Peru.

RENANTHERA (Orchidac.).
auyongii *E.A. Christenson in Orchid Rev.,* 94(1114): 264 (1986)—Borneo (Sarawak).
auyongii *E.A. Christenson in Orchid Dig.,* 50(5): 169 (1986)—Borneo (Sarawak).

RESEDA (Resedac.).
abyssinica *Fres., Sem. Hort. Bot. Franc.,* 1835; *Linnaea 11, Lit.:* 86 (1837); *cf. D.J. Mabberley in Taxon,* 32(1): 86 (1983)—Locality not stated.

RESTIO (Restionac.).
colliculospermus *H.P. Linder in Bothalia,* 15(3–4): 443 (1985)—South Africa (Cape Province).

RESTIO :–
corneolus *E.E. Esterhuysen in Bothalia,* 15(3–4): 443 (1985)—South Africa (Cape Province).
decipiens *(N.E. Br.) H.P. Linder in Bothalia,* 15(3–4): 444 (1985): Hypolaena decipiens.
fragilis *E.E. Esterhuysen in Bothalia,* 15(3–4): 447 (1985)—South Africa (Cape Province).
glaber *(Rodway) L.A.S. Johnson & B.G. Briggs in Telopea,* 2(6): 738 (1986): R. oligocephalus var. glaber.
implicatus *E.E. Esterhuysen in Bothalia,* 15(3–4): 447 (1985)—South Africa (Cape Province).
inconspicuus *E.E. Esterhuysen in Bothalia,* 15(3–4): 449 (1985)—South Africa (Cape Province).
ingens *E.E. Esterhuysen in Bothalia,* 15(3–4): 449 (1985)—South Africa (Cape Province).
inveteratus *E.E. Esterhuysen in Bothalia,* 15(3–4): 450 (1985)—South Africa (Cape Province).
mahonii
subsp. humbertii *(Cherm.) H.P. Linder in Kew Bull.,* 41(1): 103 (1986): R. madagascariensis var. humbertii.
montanus *E.E. Esterhuysen in Bothalia,* 15(3–4): 453 (1985)—South Africa (Cape Province).
nuwebergensis *E.E. Esterhuysen in Bothalia,* 15(3–4): 453 (1985)—South Africa (Cape Province).
peculiaris *E.E. Esterhuysen in Bothalia,* 15(3–4): 455 (1985)—South Africa (Cape Province).
perseverans *E.E. Esterhuysen in Bothalia,* 15(3–4): 457 (1985)—South Africa (Cape Province).
pillansii *H.P. Linder in Bothalia,* 15(3–4): 457 (1985), nom. nov.: Leptocarpus stokoei *Pillans.*
pulvinatus *E.E. Esterhuysen in Bothalia,* 15(3–4): 457 (1985)—South Africa (Cape Province).
pumilus *E.E. Esterhuysen in Bothalia,* 15(3–4): 458 (1985)—South Africa (Cape Province).
quartziticola *H.P. Linder in Kew Bull.,* 41(1): 103 (1986)—Zimbabwe, Mozambique.
rarus *E.E. Esterhuysen in Bothalia,* 15(3–4): 459 (1985)—South Africa (Cape Province).
rupicola *E.E. Esterhuysen in Bothalia,* 15(3–4): 459 (1985)—South Africa (Cape Province).
secundus *(Pillans) H.P. Linder in Bothalia,* 15(3–4): 460 (1985): Leptocarpus secundus.
singularis *E.E. Esterhuysen in Bothalia,* 15(3–4): 460 (1985)—South Africa (Cape Province).
vallis-simius *H.P. Linder in Bothalia,* 15(3–4): 462 (1985)—South Africa (Cape Province).
verrucosus *E.E. Esterhuysen in Bothalia,* 15(3–4): 462 (1985)—South Africa (Cape Province).
versatilis *H.P. Linder in Bothalia,* 15(3–4): 463 (1985), nom. nov.: Hypolaena diffusa *Mast.*
zuluensis *H.P. Linder in Bothalia,* 15(3–4): 463 (1985)—South Africa (Natal).

RESTREPIOPSIS (Orchidac.).
subgen. Endresia *C.A. Luer, Ic. Pleurothallid. I-Syst. Pleurothallidinae (Monogr. Syst. Bot. Missouri Bot. Gard. 15):* 55 (1986).

RETAMA (Legumin.).
raetam
subsp. gussonei *(Webb) W. Greuter in Willdenowia,* 15(2): 429 (1986): R. gussonei.

REYNOUTRIA (Polygonac.).
×bohemica *J. Chrtek & A. Chrtková in Čas. Nár. Muz. (Prague),* 152(2): 120 (1983)—Czechoslovakia. (R. japonica × sachaliensis).

RHAMNONEURON (Thymelaeac.).
rubriflorum *C.Y. Wu ex S.C. Huang in Acta Bot. Yunnanica,* 7(3): 278 (1985)—China.

RHAMNUS (Rhamnac.).
calcicolus *Q.H. Chen in Acta Bot. Yunnanica,* 7(4): 413 (1985)—China.
chiennanensis *Z.R. Xu in Acta Sci. Nat. Univ. Sunyatseni,* 1986(2): 100 (1986), without type—China.
erythroxyloides *Hoffmanns., Preiss-Verz.:* 38 (1833); *cf. D.J. Mabberley in Taxon,* 33(3): 433 (1984)— Locality not stated.
kiusiana *S. Hatusima in J. Phytogeogr. Taxon.,* 33(2): 71 (1985)—Japan.
qianweiensis *Z.Y. Zhu in Acta Bot. Bor.-Occid. Sin.,* 6(2): 135 (1986)—China.
saxatilis
subsp. prunifolius *(Sibth. & Sm.) B. Aldén in A. Strid (ed.), Mount. Fl. Greece,* 1: 586 (1986): R. prunifolius.
subsp. rhodopeaus *(Velen.) B. Aldén in A. Strid (ed.), Mount. Fl. Greece,* 1: 587 (1986): R. rhodopeaus.

RHAPHIDOSPORA (Acanthac.).
bonneyana *(F. Muell.) R.M. Barker in J. Adelaide Bot. Gard.,* 9: 233 (1986): Chloanthes bonneyana.
cavernarum *(F. Muell.) R.M. Barker in J. Adelaide Bot. Gard.,* 9: 232 (1986): Justicia cavernarum.

RHAPONTICUM (Compos.).
behen *(L.) Ratier, Pharm. Franç.:* 121 (1827); *cf. D.J. Mabberley in Taxon,* 31(1): 72 (1982): Centaurea behen.

RHEUM (Polygonac.).
×svetlanae *L.S. Krasovskaya in L.S. Krasovskaya & I.G. Levichev, Fl. Chatkal'skogo Zapovednika:* 170 (1986)—U.S.S.R. (Middle Asia). (R. cordatum × maximowiczii).

RHIPIDOGLOSSUM (Orchidac.).
adoxum *(Rasm.) K. Senghas in Schlechter Orchideen,* 1(16–18): 1110 (1986): Diaphananthe adoxa.

RHIPIDOGLOSSUM :-
candidum *(Cribb) K. Senghas in Schlechter Orchideen,* 1(16–18): 1110 (1986): Diaphananthe candida.
laticalcar *(Hall) K. Senghas in Schlechter Orchideen,* 1(16–18): 1110 (1986): Diaphananthe laticalcar.
melianthum *(Cribb) K. Senghas in Schlechter Orchideen,* 1(16–18): 1110 (1986): Diaphananthe meliantha.
montanum *(Piers) K. Senghas in Schlechter Orchideen,* 1(16–18): 1111 (1986): Angraecum montanum.
oxycentron *(Cribb) K. Senghas in Schlechter Orchideen,* 1(16–18): 1111 (1986): Diaphananthe oxycentron.
pachyrhizum *(Cribb) K. Senghas in Schlechter Orchideen,* 1(16–18): 1111 (1986): Diaphananthe pachyrhiza.
tanneri *(Cribb) K. Senghas in Schlechter Orchideen,* 1(16–18): 1111 (1986): Diaphananthe tanneri.

×RHIZANTHERA (Orchidac.).
martysiensis *M. Balayer in Bull. Soc. Bot. France, Lett. Bot.,* 133(3): 283 (1986)—France. (Dactylorhiza fuchsii × Platanthera chlorantha).
thilensis *(G. Keller & Jeanjean) L.V. Aver'yanov in Bot. Zhurn.,* 71(1): 93 (1986): ×Orchiplatanthera thilensis.

RHODAMNIA (Myrtac.).
dumicola *G.P. Guymer & L.W. Jessup in Austrobaileya,* 2(3): 228 (1986)—Australia (Queensland).
glabrescens *G.P. Guymer & L.W. Jessup in Austrobaileya,* 2(3): 231 (1986)—Australia (Queensland).
whiteana *G.P. Guymer & L.W. Jessup in Austrobaileya,* 2(3): 229 (1986)—Australia (Queensland, New South Wales).

RHODAX (Cistac.).
allionii *(Tineo) J. Holub in Preslia,* 58(4): 302 (1986): Helianthemum allionii.
frigidulus *(Cuatrec.) J. Holub in Preslia,* 58(4): 302 (1986): Helianthemum frigidulum.
nebrodensis *(Heldr.) J. Holub in Preslia,* 58(4): 302 (1986): Helianthemum nebrodense.
piloselloides *(Lapeyr.) J. Holub in Preslia,* 58(4): 302 (1986): Cistus piloselloides.
pourretii *(Timb.-Lagr.) J. Holub in Preslia,* 58(4): 302 (1986): Helianthemum pourretii.
rotundifolius *(Dunal) J. Holub in Preslia,* 58(4): 302 (1986): Helianthemum rotundifolium.
serrae *(Camb.) J. Holub in Preslia,* 58(4): 302 (1986): Helianthemum serrae.
viscarioides *(Hervier) J. Holub in Preslia,* 58(4): 302 (1986): Helianthemum viscarioides.

RHODIOLA (Crassulac.).
imbricata
var. lobulata *N.B. Singh & U.C. Bhattacharyya in Bull. Bot. Surv. India,* 25(1–4): 246 (1983 publ. 1985)—India.
lobulata *(Singh & Bhattacharyya) H. Ohba in J. Jap. Bot.*, 61(7): 205 (1986): R. imbricata var. lobulata.
phariensis *(H. Ohba) S.H. Fu in Fl. Xizangica,* 2: 427 (1985): Sedum phariense.
purpureoviridis
subsp. phariensis *(H. Ohba) H. Ohba in J. Jap. Bot.*, 61(7): 206 (1986): Sedum phariense.
sacra
var. tsuiana *(S.H. Fu) S.H. Fu in Fl. Xizangica,* 2: 426 (1985): R. tsuiana.
staminea *(O. Pauls.) S.H. Fu in Fl. Xizangica,* 2: 428 (1985): Sedum stamineum.
tangutica *(Maxim.) S.H. Fu in Bull. Bot. Res. North-East. Forest. Inst.*, 6(4): 158 (1986): Sedum algidum var. tanguticum.

RHODOCOCCUM (Ericac.).
vitis-idaea
subsp. minus *(Lodd.) V.G. Sergienko, Fl. poluostrova Kanin:* 83 (1986), without basionym ref.: Basionym not stated.

RHODOCOMA (Restionac.).
fruticosa *(Thunb.) H.P. Linder in Bothalia,* 15(3–4): 479 (1985): Restio fruticosus.
gigantea *(Kunth) H.P. Linder in Bothalia,* 15(3–4): 479 (1985): Thamnochortus giganteus.

RHODODENDRON (Ericac.).
clementinae
subsp. aureodorsale *W.P. Fang ex J.Q. Fu in Fl. Tsinlingensis,* 1(4): 393, 20 (1983)—China.
danbaense *L.C. Hu in Bull. Bot. Res. North-East. Forest. Inst.*, 6(1): 155 (1986)—China.
degronianum
f. amagianum *(Yamazaki) H. Hara in J. Jap. Bot.*, 61(8): 246 (1986): R. metternichii f. amagianum.
subsp. heptamerum *(Maxim.) H. Hara in J. Jap. Bot.*, 61(8): 246 (1986): R. metternichii α. heptamerum.
var. hondoense *(Nakai) H. Hara in J. Jap. Bot.*, 61(8): 246 (1986): R. metternichii var. hondoense.
var. intermedium *(Sugimoto) H. Hara in J. Jap. Bot.*, 61(8): 247 (1986): R. metternichii var. intermedium.
var. kyomaruense *(Yamazaki) H. Hara in J. Jap. Bot.*, 61(8): 246 (1986): R. metternichii var. kyomaruense.
subsp. yakushimanum *(Nakai) H. Hara in J. Jap. Bot.*, 61(8): 246 (1986): R. yakushimanum.
tsinlingense *W.P. Fang ex J.Q. Fu in Fl. Tsinlingensis,* 1(4): 393, 10 (1983)—China.

RHODODENDRON :–
wasonii
var. wenchuanense *L.C. Hu in Bull. Bot. Res. North-East. Forest. Inst.*, 6(1): 156 (1986), as 'uenchuanense'—China.
xizangense *W.P. Fang & W.K. Hu ex Q.Z. Yu in Bull. Bot. Res. North-East. Forest. Inst.*, 6(1): 167 (1986)—Xizang.

RHODOSPATHA (Arac.).
falconensis *G.S. Bunting in Phytologia,* 60(5): 336 (1986)—Venezuela.
guasareensis *G.S. Bunting in Phytologia,* 60(5): 337 (1986)—Venezuela.
steyermarkii *G.S. Bunting in Phytologia,* 60(5): 337 (1986)—Venezuela.

RHODOSTEMONODAPHNE (Laurac.).
anomala *(Mez) J.G. Rohwer in Mitt. Inst. Allg. Bot. Hamburg,* 20: 83 (1986): Nectandra anomala.
celiana *(C.K. Allen) J.G. Rohwer in Mitt. Inst. Allg. Bot. Hamburg,* 20: 85 (1986): Ocotea celiana.
dioica *(Mez) J.G. Rohwer in Mitt. Inst. Allg. Bot. Hamburg,* 20: 83 (1986): Nectandra dioica.
grandis *(Mez) J.G. Rohwer in Mitt. Inst. Allg. Bot. Hamburg,* 20: 84 (1986): Endlicheria grandis.
kunthiana *(Nees) J.G. Rohwer in Mitt. Inst. Allg. Bot. Hamburg,* 20: 84 (1986): Acrodiclidium kunthianum.
laxa *(Meissn.) J.G. Rohwer in Mitt. Inst. Allg. Bot. Hamburg,* 20: 83 (1986): Synandrodaphne laxa.
miranda *(Sandw.) J.G. Rohwer in Mitt. Inst. Allg. Bot. Hamburg,* 20: 85 (1986): Nectandra miranda.
mirecolorata *(C.K. Allen) J.G. Rohwer in Mitt. Inst. Allg. Bot. Hamburg,* 20: 86 (1986): Ocotea mirecolorata.

RHOICISSUS (Vitac.).
tridentata
subsp. cuneifolia *(Eckl. & Zeyh.) N.R. Urton in S. Afr. J. Bot.*, 52(5): 393 (1986): Cissus cuneifolia.

RHUS (Anacardiac.).
aromatica
var. simplicifolia *(Greene) A. Cronquist in Great Basin Nat.*, 46(2): 254 (1986): R. canadensis var. simplicifolia.
glutinosa
subsp. abyssinica *(Oliv.) M.G. Gilbert in Nordic J. Bot.*, 6(5): 572 (1986): R. abyssinica.
subsp. neoglutinosa *(M.G. Gilbert) M.G. Gilbert in Nordic J. Bot.*, 6(5): 572 (1986): R. neoglutinosa.
heterophylla *C.C. Gmel., Hort. Mag. Duc. Bad. Carlsruh.*: 225 (1811); *cf. D.J. Mabberley in Taxon,* 33(3): 440 (1984)— Locality not stated.
neoglutinosa *M.G. Gilbert in Nordic J. Bot.*, 6(2): 139 (1986)—Ethiopia.

RHYNCHOSPORA (Cyperac.).
durangensis *R. Kral & W.W. Thomas in Brittonia,* 38(3): 214 (1986)—Mexico.
oaxacana *R. Kral & W.W. Thomas in Brittonia,* 38(3): 210 (1986)—Mexico.

RIBES (Grossulariac.).
palmatum *Hort. Par. ex Fée, Cat. Méth. Pl. Jard. Bot. Strasb.:* 92 (1836); *cf. D.J. Mabberley in Taxon,* 32(1): 86 (1983)—Locality not stated.

RICINUS (Euphorbiac.).
compactus *Huber, Cat. Print. 1865:* 8 (1865); *cf. D.J. Mabberley in Taxon,* 34(3): 451 (1985)— Locality not stated.

RINZIA (Myrtac.).
affinis *M.E. Trudgen in Nuytsia,* 5(3): 431 (1986)—Australia (W. Australia).
carnosa *(S. Moore) M.E. Trudgen in Nuytsia,* 5(3): 426 (1986): Baeckea carnosa.
communis *M.E. Trudgen in Nuytsia,* 5(3): 435 (1986)—Australia (W. Australia).
dimorphandra *(F. Muell. ex Benth.) M.E. Trudgen in Nuytsia,* 5(3): 430 (1986): Baeckea dimorphandra.
morrisonii *M.E. Trudgen in Nuytsia,* 5(3): 423 (1986)—Australia (W. Australia).
rubra *M.E. Trudgen in Nuytsia,* 5(3): 427 (1986)—Australia (W. Australia).
schollerifolia *(Lehm.) M.E. Trudgen in Nuytsia,* 5(3): 422 (1986): Baeckea schollerifolia.
sessilis *M.E. Trudgen in Nuytsia,* 5(3): 436 (1986)—Australia (W. Australia).

RIOCREUXIA (Asclepiadac.).
bubalina *W. Elliot, Fl. Andhirica,* 59 (1859); *cf. D.J. Mabberley in Taxon,* 32(1): 87 (1983)— Locality not stated.

RIPOGONUM (Liliac.).
brevifolium *J.G. Conran & H.T. Clifford in Fl. Australia,* 46: 231, 189 (1986)—Australia (Queensland, New South Wales).

RIVINA (Phytolaccac.).
herbacea *Huber, Cat. Gén. 1871 & 1872:* 4 (1871); *cf. D.J. Mabberley in Taxon,* 34(3): 451 (1985)— Locality not stated.

ROEZLIA (Amaryllidac.).
regia *Laurentius' Gärtn. Leipzig in Bonplandia,* 9: 274 (15 Sept. 1861); *cf. D.J. Mabberley in Taxon,* 34(3): 454 (1985)—Locality not stated.

ROMNALDA (Xanthorrhoeac.).
strobilacea *R.J.F. Henderson & P.R. Sharpe in Fl. Australia,* 46: 224, 92 (1986)—Australia (Queensland).

RONDELETIA (Rubiac.).
acunae *A. Borhidi & M. Fernandez in Acta Bot. Hung.,* 31(1–4): 147 (1985)—Cuba.

RONDELETIA :–
combsioides *M. Fernandez & A. Borhidi in Acta Bot. Hung.,* 31(1–4): 149 (1985)—Cuba.
convoluta *M. Fernandez & A. Borhidi in Acta Bot. Hung.,* 31(1–4): 151 (1985)—Cuba.
galanensis *M. Fernandez & A. Borhidi in Acta Bot. Hung.,* 31(1–4): 152 (1985)—Cuba.
×incerta *A. Borhidi & M. Fernandez in Acta Bot. Hung.,* 31(1–4): 155 (1985)—Cuba. (R. miraflorensis × subcanescens).
intermixta
subsp. turquinensis *M. Fernandez & A. Borhidi in Acta Bot. Hung.,* 31(1–4): 157 (1985)—Cuba.
lucida *M. Fernandez & A. Borhidi in Acta Bot. Hung.,* 31(1–4): 158 (1985)—Cuba.
miraflorensis *M. Fernandez & A. Borhidi in Acta Bot. Hung.,* 31(1–4): 154 (1985)—Cuba.
×obscura *A. Borhidi & M. Fernandez in Acta Bot. Hung.,* 31(1–4): 161 (1985)—Cuba. (R. paucinervis × plicatula).
pachyphylla
subsp. myrtilloides *M. Fernandez & A. Borhidi in Acta Bot. Hung.,* 31(1–4): 163 (1985)—Cuba.
papayoensis *M. Fernandez & A. Borhidi in Acta Bot. Hung.,* 31(1–4): 165 (1985)—Cuba.
peninsularis *M. Fernandez & A. Borhidi in Acta Bot. Hung.,* 31(1–4): 166 (1985)—Cuba.
subcanescens *M. Fernandez & A. Borhidi in Acta Bot. Hung.,* 31(1–4): 168 (1985)—Cuba.
toaensis *M. Fernandez & A. Borhidi in Acta Bot. Hung.,* 31(1–4): 170 (1985)—Cuba.

ROSA (Rosac.).
sect. Kokanicae *I.O. Buzunova in Bot. Zhurn.,* 71(4): 486 (1986).
multiflora *C.P. Thunberg ex A. Murray in Linnaeus, Syst. Veg., ed. 14.* 474 (1784)—Japan †
rugosa *C.P. Thunberg ex A. Murray in Linnaeus, Syst. Veg., ed. 14:* 473 (1784)—Japan. †

ROSTELLULARIA (Acanthac.).
adscendens *(R. Br.) R.M. Barker in J. Adelaide Bot. Gard.,* 9: 250 (1986): Justicia adscendens.
adscendens
subsp. clementii *(Domin) R.M. Barker in J. Adelaide Bot. Gard.,* 9: 267 (1986): Justicia clementii.
var. clementii *(Domin) R.M. Barker in J. Adelaide Bot. Gard.,* 9: 269 (1986): Justicia clementii.
subsp. dallachyi *R.M. Barker in J. Adelaide Bot. Gard.,* 9: 274 (1986)—Australia (Queensland).
subsp. glaucoviolacea *(Domin) R.M. Barker in J. Adelaide Bot. Gard.,* 9: 272 (1986): Justicia glaucoviolacea.

ROSTELLULARIA :–
var. hispida *(Domin) R.M. Barker in J. Adelaide Bot. Gard.,* 9: 261 (1986): Justicia procumbens var. hispida.
var. juncea *(R. Br.) R.M. Barker in J. Adelaide Bot. Gard.,* 9: 258 (1986): Justicia juncea.
var. largiflorens *R.M. Barker in J. Adelaide Bot. Gard.,* 9: 270 (1986)—Australia (N. Terr., Queensland, W. Australia).
var. latifolia *(Domin) R.M. Barker in J. Adelaide Bot. Gard.,* 9: 266 (1986): Justicia procumbens var. latifolia.
var. pogonanthera *(F. Muell.) R.M. Barker in J. Adelaide Bot. Gard.,* 9: 264 (1986): R. pogonanthera.
andamanica *M.K. Vasudeva Rao in J. Econ. Taxon. Bot.,* 6(3): 719 (1985)—Andaman Is.

ROTTBOELLIA (Gramin.).
paradoxa *R. de Koning & M.S.M. Sosef in Blumea,* 31(2): 306 (1986)—Philippine Is.
salina *Kitaib. ex Spreng., Nachtr. Bot. Gart. Univ. Halle:* 34 (1801); *cf. D.J. Mabberley in Taxon,* 32(1): 86 (1983)— Locality not stated.

ROUSSELIA (Urticac.).
subgen. Bissea *I.A. Grudzinskaya in Bot. Zhurn.,* 71(7): 939 (1986).
cubensis *I.A. Grudzinskaya in Bot. Zhurn.,* 71(7): 939 (1986)—Cuba.
impariflora *I.A. Grudzinskaya in Bot. Zhurn.,* 71(7): 940 (1986)—Cuba.

RUBIA (Rubiac.).
cordifolia
var. coriacea *Z.Y. Zhang in Fl. Tsinlingensis,* 1(5): 420, 15 (1985)—China.
indica *Ten., Cat. Ort. Bot. Nap.:* 102 (1845); *cf. D.J. Mabberley in Taxon,* 32(1): 86 (1983)— Locality not stated.
membranacea
var. caudata *Z.Y. Zhang in Fl. Tsinlingensis,* 1(5): 421, 17 (1985)—China.
var. incurvata *Z.Y. Zhang in Fl. Tsinlingensis,* 1(5): 421, 17 (1985)—China.
munjith *Roxb. ex Playfair, Taleef Shareef:* 150 (1833); *cf. D.J. Mabberley in Taxon,* 32(1): 87 (1983)— Locality not stated.
ovatifolia *Z.Y. Zhang in Fl. Tsinlingensis,* 1(5): 420, 15 (1985), with two types cited—China.

RUBUS (Rosac.).
alexeterius
var. acaenocalyx *(Hara) T.T. Yü & L.T. Lu in Fl. Xizangica,* 2: 613 (1985), without basionym ref.: R. acaenocalyx.
aurora *A. van de Beek, R.J. Bijlsma & F.M. Muller in Gorteria,* 13(2): 38 (1986)—Netherlands.
canescens
subsp. lloydianus *(Genev.) O. de Bolòs & J. Vigo, Fl. Països Catalans:* 351 (1984): R. lloydianus.

RUBUS :–
var. lloydianus *(Genev.) O. de Bolòs & J. Vigo, Fl. Països Catalans:* 351 (1984): R. lloydianus.
var. subparilis *(Sudre) O. de Bolòs & J. Vigo, Fl. Països Catalans:* 351 (1984): R. lloydianus f. subparilis.
glaucifolius
subsp. ganderi *(Bailey) R.M. Beauchamp in Phytologia,* 59(7): 440 (1986): R. ganderi.
incisus *C.P. Thunberg ex A. Murray in Linnaeus, Syst. Veg., ed. 14:* 476 (1784)—Japan. †
palmatus *C.P. Thunberg ex A. Murray in Linnaeus, Syst. Veg., ed. 14:* 475 (1784)—Japan. †
rosaefolius
var. wuyishanensis *Z.X. Yu in Bull. Bot. Res. North-East. Forest. Inst.,* 6(1): 152 (1986)—China.
stans
var. soulieanus *(Card.) T.T. Yü & L.T. Lu in Fl. Xizangica,* 2: 618 (1985), without basionym ref.: R. soulieanus.
trifidus *C.P. Thunberg ex A. Murray in Linnaeus, Syst. Veg., ed. 14:* 476 (1784)—Japan. †
triphyllus *C.P. Thunberg ex A. Murray in Linnaeus, Syst. Veg., ed. 14:* 475 (1784)—Japan. †
villosus *C.P. Thunberg ex A. Murray in Linnaeus, Syst. Veg., ed. 14:* 475 (1784)—Japan. †

RUELLIA (Acanthac.).
japonica *C.P. Thunberg ex A. Murray in Linnaeus, Syst. Veg., ed. 14:* 576 (1784)—Japan. †
obovata *Roxb. ex Hornem., Hort. Hafn., Suppl.:* 144 (1819); *cf. D.J. Mabberley in Taxon,* 33(3): 437 (1984)— Locality not stated.
sindica *A. Ghafoor & H. Heine in Willdenowia,* 16(1): 122 (1986), nom. nov.: Dipteracanthus longifolius *Stocks.*

RUILOPEZIA (Compos.).
emmanuelis *J. Cuatrecasas in Phytologia,* 61(1): 56 (1986)—Venezuela.
usubillagae *J. Cuatrecasas in Phytologia,* 61(1): 53 (1986)—Venezuela.
vergarae *J. Cuatrecasas & López-Figueiras in Phytologia,* 61(1): 58 (1986)—Venezuela.

RULINGIA (Sterculiac.).
hermaniifolia *(DC.) Endl., Cat. Horti Vindob.,* 2: 358 (1842); *cf. D.J. Mabberley in Taxon,* 33(3): 436 (1984): Basionym not stated.

RUMEX (Polygonac.).
altissimus
subsp. ellipticus *(Greene) Á. Löve in Taxon,* 35(3): 612 (1986): R. ellipticus.
aquaticus
subsp. nematopodus *(K.H. Rech.) Á. Löve in Taxon,* 35(3): 612 (1986): R. nematopodus.

RUMEX :–
subsp. tomentellus *(K.H. Rech.) Á. Löve in Taxon,* 35(3): 612 (1986): R. tomentellus.
densiflorus
subsp. orthoneurus *(K.H. Rech.) Á. Löve in Taxon,* 35(3): 612 (1986): R. orthoneurus.
subsp. pycnanthus *(K.H. Rech.) Á. Löve in Taxon,* 35(3): 612 (1986): R. pycnanthus.
pallidus
subsp. subarcticus *(Lepage) Á. Löve in Taxon,* 35(3): 613 (1986): R. subarcticus.
undulatus *Hort. Par. ex Hornem., Hort. Hafn.:* 351 (1813); *cf. D.J. Mabberley in Taxon,* 33(3): 437 (1984)— Locality not stated.
verticillatus
subsp. fascicularis *(Small) Á. Löve in Taxon,* 35(3): 613 (1986): R. fasçicularis.
subsp. floridanus *(Meisn.) Á. Löve in Taxon,* 35(3): 613 (1986): R. floridanus.

RUPICAPNOS (Papaverac.).
subsect. Sarcocapnoides *(Pugsley) M. Lidén in Op. Bot.,* 88: 92 (1986): R. sect. Sarcocapnoides.
calcarata *M. Lidén in Op. Bot.,* 88: 104 (1986), without type—Algeria.
longipes
subsp. aurasiaca *Maire ex M. Lidén in Op. Bot.,* 88: 101 (1986)—Algeria.
subsp. reboudiana *(Pomel) M. Lidén in Op. Bot.,* 88: 102 (1986): R. reboudiana.

S

SACCIOLEPIS (Gramin.).
indica
var. intermedia *S.M. Almeida in J. Bombay Nat. Hist. Soc.,* 83(1): 184 (1986)—India.

SACHOKIELLA A.A. Kolakovskiĭ in Soobshch. Akad. Nauk Gruz. SSR, 118(3): 595 (1985). *CAMPANULACEAE.*
macrochlamys *(Boiss. & Huet.) A.A. Kolakovskiĭ in Soobshch. Akad. Nauk Gruz. SSR,* 118(3): 595 (1985): Campanula macrochlamys.

SAINTMORYSIA (Compos.).
dentata *(L.) Endl., Cat. Horti Vindob.,* 1: 367 (1842); *cf. D.J. Mabberley in Taxon,* 33(3): 436 (1984): Athanasia dentata.

SALACIA (Celastrac.).
lovettii *N. Hallé & B. Mathew in Kew Bull.,* 41(2): 399 (1986)—Tanzania.

SALIX (Salicac.).
babylonica
var. glandulipilosa *P.Y. Mao & W.Z. Li in Bull. Bot. Res. North-East. Forest. Inst.,* 6(2): 79 (1986)—China.

SALIX :–
cupularis
var. acutifolia *S.Q. Zhou in Acta Bot. Bor.-Occid. Sin.,* 4(1): 2 (1984)—China.
fengxianica *N. Chao in Acta Bot. Bor.-Occid. Sin.,* 5(2): 115 (1985)—China.
forbesii *Sweet, Hort. Brit., ed. 2:* 469 (1830); *cf. D.J. Mabberley in Taxon,* 32(1): 87 (1983)—Locality not stated.
fruticulosa *S. de Lacroix in Bull. Soc. Bot. France,* 6: 566 (1860)—France.
integra *C.P. Thunberg ex A. Murray in Linnaeus, Syst. Veg., ed. 14:* 880 (1784)—Japan. †
japonica *C.P. Thunberg ex A. Murray in Linnaeus, Syst. Veg., ed. 14:* 879 (1784)—Japan. †
kungmuensis *P.Y. Mao & W.Z. Li in Bull. Bot. Res. North-East. Forest. Inst.,* 6(2): 80 (1986)—China.
longissimipedicellaris *N. Chao ex P.Y. Mao in Bull. Bot. Res. North-East. Forest. Inst.,* 6(2): 81 (1986)—China.
lucida
subsp. caudata *(Nutt.) G.W. Argus in Canad. J. Bot.,* 64(3): 545 (1986): S. pentandra var. caudata.
subsp. lasiandra *(Benth.) G.W. Argus in Canad. J. Bot.,* 64(3): 545 (1986): S. lasiandra.
qinlingica *C. Wang & N. Chao in Acta Bot. Bor.-Occid. Sin.,* 5(2): 116 (1985)—China.
suchowensis *W.C. Cheng in Sci. Silvae Sin.,* 8(1): 4 (1963), with two types cited—China.
sungkianica
f. brevistachys *Y.L. Chou & Tung in Bull. Bot. Res. North-East. Forest. Inst.,* 6(3): 145 (1986)—China.

SALOMONIA (Polygalac.).
cantoniensis
var. edentula *(DC.) C.Y. Wu in Fl. Yunnanica,* 3: 291 (1983): S. edentula.

SALPISTELE (Orchidac.).
dielsii *(Mansf.) C.A. Luer, Ic. Pleurothallid. I-Syst. Pleurothallidinae (Monogr. Syst. Bot. Missouri Bot. Gard. 15):* 57 (1986): Lepanthes dielsii.
pensilis *(Schltr.) C.A. Luer, Ic. Pleurothallid. I-Syst. Pleurothallidinae (Monogr. Syst. Bot. Missouri Bot. Gard. 15):* 57 (1986): Lepanthes pensilis.

SALSOLA (Chenopodiac.).
sect. Irania *V.P. Bochantsev in Bot. Zhurn.,* 71(10): 1400 (1986).
dentata *(Willd.) Germann, Verz. Pfl. Bot. Gart. Kais. Univ. Dorpat,* 1807: 116 (1807); *cf. D.J. Mabberley in Taxon,* 32(1): 86 (1983): Basionym not stated.

SALVIA (Labiat.).
blancoana
subsp. hegelmaieri *(Porta & Rigo) R. Figuerola in An. Jard. Bot. Madrid,* 42(2): 538 (1985 publ. 1986): S. hegelmaieri.

SALVIA :-
subsp. mariolensis *R. Figuerola in An. Jard. Bot. Madrid,* 42(2): 538 (1985 publ. 1986)—Spain.
camphorata *Huber, Cat. Gén. 1871 & 1872:* 5 (1871); *cf. D.J. Mabberley in Taxon,* 34(3): 453 (1985)— Locality not stated.
cana *Wall. ex Benth. in Lindl., Bot. Reg.,* sub t. 1292 (1830); *cf. D.J. Mabberley in Taxon,* 31(1): 72 (1982)— Locality not stated.
inconspicua *Bertol., Syll. Pl. Hort. Bot. Bonon.,* 1827: 7 (1827); *cf. D.J. Mabberley in Taxon,* 32(1): 82 (1983)— Locality not stated.
japonica *C.P. Thunberg ex A. Murray in Linnaeus, Syst. Veg., ed. 14:* 72 (1784)—Japan. †
lavandulifolia
subsp. amethystea *(Emberger & Maire) J.L. Rosua & G. Blanca in Acta Bot. Malacitana,* 11: 252 (1986): S. aucheri var. amethystea.
var. aurasiaca *(Maire) J.L. Rosua & G. Blanca in Acta Bot. Malacitana,* 11: 259 (1986): S. aucheri var. aurasiaca.
subsp. blancoana *(Webb & Heldr.) J.L. Rosua & G. Blanca in Acta Bot. Malacitana,* 11: 256 (1986): S. blancoana.
var. blancoana *(Webb & Heldr.) J.L. Rosua & G. Blanca in Acta Bot. Malacitana,* 11: 258 (1986): S. blancoana.
subsp. maurorum *(Ball) J.L. Rosua & G. Blanca in Acta Bot. Malacitana,* 11: 254 (1986): S. candelabrum subsp. maurorum.
subsp. mesatlantica *(Maire) J.L. Rosua & G. Blanca in Acta Bot. Malacitana,* 11: 249 (1986): S. aucheri var. mesatlantica.
var. vellerea *(Cuatr.) J.L. Rosua & G. Blanca in Acta Bot. Malacitana,* 11: 261 (1986): S. officinalis var. vellerea.
nigrescens *Bull, Retail List,* 121: 170 (1876); *cf. D.J. Mabberley in Taxon,* 34(3): 455 (1985)— Locality not stated.

SAMBUCUS (Caprifoliac.).
henriana *M.L. Samutina in Bot. Zhurn.,* 71(8): 1121 (1986)—China.
japonica *C.P. Thunberg ex A. Murray in Linnaeus, Syst. Veg., ed. 14:* 295 (1784)—Japan. †

SANANGO (Buddlejac.).
racemosum *(Ruiz & Pavon) K. Barringer in Phytologia,* 59(5): 363 (1986): Gomara racemosa.

SANICULA (Umbellif.).
orthacantha
var. costata *(Wolff) K.T. Fu in Fl. Tsinlingensis,* 1(3): 375 (1981): S. costata.

SANSEVIERIA (Dracaenac.).
fischeri *(Baker) W. Marais in Kew Bull.,* 41(1): 58 (1986): Buphane fischeri.

SAPIUM (Euphorbiac.).
discolor
var. wenhsienensis *S.B. Ho in Fl. Tsinlingensis,* 1(3): 451, 180 (1981)—China.

SARCOCAPNOS (Papaverac.).
crassifolia
subsp. atlantis *(Emb. & Maire) M. Lidén in Op. Bot.,* 88: 34 (1986): S. baetica var. atlantis.

SARCOCOCCA (Buxac.).
wallichii
f. membranacea *J.R. Sealy in Bot. J. Linn. Soc.,* 92(2): 146 (1986)—India.
zeylanica
var. brevifolia *(Muell. Arg.) J.R. Sealy in Bot. J. Linn. Soc.,* 92(2): 138 (1986): S. saligna var. brevifolia.

SARCOGLYPHIS (Orchidac.).
potamophila *(Schltr.) L.A. Garay & W. Kittredge in Bot. Mus. Leafl. Harvard Univ.,* 30(3): (58) 192 (1985 publ. 1986): Sarcanthus potamophilus.

SARCOMELICOPE (Rutac.).
follicularis *T.G. Hartley in Bull. Mus. Nation. Hist. Nat., B, Adansonia,* Ser. 4, 8(2): 183 (1986)—New Caledonia.
megistophylla *T.G. Hartley in Bull. Mus. Nation. Hist. Nat., B, Adansonia,* Ser. 4, 8(2): 185 (1986)—New Caledonia.
pembaiensis *T.G. Hartley in Bull. Mus. Nation. Hist. Nat., B, Adansonia,* Ser. 4, 8(2): 185 (1986)—New Caledonia.

SARCOPHYTE (Balanophorac.).
sanguinea
subsp. piriei *(Hutchinson) B. Hansen in Bot. Jahrb.,* 106(3): 366 (1986): S. piriei.

SARCOTOECHIA (Sapindac.).
heterophylla *S.T. Reynolds in Fl. Australia,* 25: 201, 84 (1985)—Australia (Queensland).
lanceolata *(C. White) S.T. Reynolds in Fl. Australia,* 25: 201 (1985): Toechima lanceolatum.
serrata *S.T. Reynolds in Fl. Australia,* 25: 201, 83 (1985)—Australia (Queensland).
villosa *S.T. Reynolds in Fl. Australia,* 25: 201, 84 (1985)—Australia (Queensland).

SARGENTODOXA (Sargentodoxac.).
simplicifolia *S.Z. Qu & C.L. Min in Bull. Bot. Res. North-East. Forest. Inst.,* 6(2): 87 (1986)—China.

SAROJUSTICIA (Acanthac.).
kempeana
subsp. muelleri *R.M. Barker in J. Adelaide Bot. Gard.,* 9: 243 (1986)—Australia (W. Australia).

SATUREJA (Labiat.).
linearifolia *(Brullo & Furnari) W. Greuter in Willdenowia,* 15(2): 422 (1986): Euhesperida linearifolia.
sardoa *(Ascherson & Levier) W. Greuter & H.M. Burdet in W. Greuter, H.M. Burdet & G. Long (eds.), Med-checklist,* 3: 325 (1986): Calamintha alpina var. sardoa.
zuvandica *D.A. Kapanadze in Soobshch. Akad. Nauk Gruz. SSR,* 119(2): 375 (1985), nom. nov.: S. densiflora *Zeinalova.*

SAURAUIA (Actinidiac.).
hirsuta *C.F. Liang in Guihaia,* 6(3): 175 (1986)—Xizang.

SAUSSUREA (Compos.).
carduiformis
var. megaphylla *X.Y. Wu in Fl. Tsinlingensis,* 1(5): 423, 367 (1985)—China.
dutaillyana
var. macrocephala *(Ling) X.Y. Wu in Fl. Tsinlingensis,* 1(5): 360 (1985): S. rufostrigillosa var. macrocephala.
huashanensis *(Ling) X.Y. Wu in Fl. Tsinlingensis,* 1(5): 365 (1985): S. eriolepis var. huashanensis.
iodostegia
var. glandulifera *X.Y. Wu in Fl. Tsinlingensis,* 1(5): 423, 347 (1985)—China.
oligantha
var. oligolepis *(Ling) X.Y. Wu in Fl. Tsinlingensis,* 1(5): 361 (1985): S. oligolepis.

SAUVAGESIA (Ochnac.).
falcisepala *C. Sastre in Phytologia,* 59(5): 313 (1986)—Venezuela.
falcisepala *C. Sastre in Bull. Mus. Nation. Hist. Nat., B, Adansonia, Ser. 4,* 8(1): 13 (1986)—Venezuela.

SAXIFRAGA (Saxifragac.)
auriculata
var. conaensis *J.T. Pan in Fl. Xizangica,* 2: 471 (1985)—Xizang.
carnosa *Sweet, Hort. Suburb. Lond.:* 98 (1818); *cf. D.J. Mabberley in Taxon,* 31(1): 72 (1982)— Locality not stated.
ciliatopetala *(Engl. & Irmsch.) J.T. Pan in Fl. Xizangica,* 2: 481 (1985): S. hirculus f. ciliatopetala.
densifoliata
var. nedongensis *J.T. Pan in Fl. Xizangica,* 2: 473 (1985)—Xizang.
diffusicallosa *C.Y. Wu in Fl. Xizangica,* 2: 472 (1985), nom. nov.: S. taylorii *H. Sm.*
egregia
var. eciliata *J.T. Pan in Fl. Xizangica,* 2: 476 (1985)—Xizang.
flaccida *J.T. Pan in Fl. Xizangica,* 2: 506 (1985)—Xizang.
×guadarramica *J. Fernández Casas in Fontqueria,* 2: 25 (1982): Spain.

SAXIFRAGA :–
harry-smithii *B.M. Wadhwa in Kew Bull.,* 41(1): 61 (1986)—Bhutan.
haworthii *Sweet, Hort. Suburb. Lond.:* 98 (1818); *cf. D.J. Mabberley in Taxon,* 31(1): 72 (1982)— Locality not stated.
heterocladoides *J.T. Pan in Fl. Xizangica,* 2: 474 (1985)—Xizang.
hirculus
var. major *(Engl. & Irmsch.) J.T. Pan in Fl. Xizangica,* 2: 480 (1985): S. hirculus f. major.
montanella
var. retusa *J.T. Pan in Fl. Xizangica,* 2: 482 (1985)—Xizang.
nakaoides *J.T. Pan in Fl. Xizangica,* 2: 485 (1985)—Xizang.
nanella
var. glabrisepala *J.T. Pan in Fl. Xizangica,* 2: 503 (1985)—Xizang.
nayarii *B.M. Wadhwa in Bull. Bot. Surv. India,* 26(1–2): 102 (1984 publ. 1985)—Xizang.
parnassifolia
var. obscuricallosa *J.T. Pan in Fl. Xizangica,* 2: 478 (1985)—Xizang.
propinqua *R. Br. in Ross, Voy. Expl. Baffin's Bay:* cxlii (April 1819); *cf. D.J. Mabberley in Taxon,* 30(1): 17 (1981)— Locality not stated.
punctulata
var. minuta *J.T. Pan in Fl. Xizangica,* 2: 492 (1985)—Xizang.
punctulatoides *J.T. Pan in Fl. Xizangica,* 2: 493 (1985)—Xizang.
rotundipetala *J.T. Pan in Fl. Xizangica,* 2: 513 (1985)—Xizang.
serrata *(Haw.) Lehm., Sem. Hort. Bot. Hamburg,* 1824: 16 (1824); *cf. D.J. Mabberley in Taxon,* 32(1): 86 (1983): Basionym not stated.
subpeltata *Fröbel, Revue horticole,* 44: 147 (1872); *cf. D.J. Mabberley in Taxon,* 34(3): 455 (1985)— Locality not stated.
substrigosa *J.T. Pan in Fl. Xizangica,* 2: 463 (1985)—Xizang.
umbellulata
var. muricola *(Marq. & Shaw) J.T. Pan in Fl. Xizangica,* 2: 494 (1985): S. muricola.
wardii
var. glabripedicellata *J.T. Pan in Fl. Xizangica,* 2: 468 (1985)—Xizang.

SCABIOSA (Dipsacac.).
longepedunculata *Hornem., Hort. Hafn.:* 985 (1815); *cf. D.J. Mabberley in Taxon,* 33(3): 437 (1984)— Locality not stated.
scopolii *Hort. Vindob. ex Hornem., Hort. Hafn., Suppl.:* 16 (1819); *cf. D.J. Mabberley in Taxon,* 33(3): 437 (1984)— Locality not stated.

SCAGEA G. McPherson in Bull. Mus. Nation. Hist. Nat., B, Adansonia, Ser. 4, 7(3): 247 (1985 publ. 1986). *EUPHORBIACEAE.*

SCAGEA :–
depauperata *(Baillon) G. McPherson in Bull. Mus. Nation. Hist. Nat., B, Adansonia,* Ser. 4, 7(3): 248 (1985 publ. 1986): Longetia depauperata.
oligostemon *(Guillaumin) G. McPherson in Bull. Mus. Nation. Hist. Nat., B, Adansonia,* Ser. 4, 7(3): 248 (1985 publ. 1986): Baloghia oligostemon.

SCALESIA (Compos.).
gordilloi *O. Hamann & S. Wium-Andersen in Nordic J. Bot.,* 6(1): 35 (1986)—Galapagos Is.

SCANDIX (Umbellif.).
nemorosa *(Bieb.) Hornem., Hort. Hafn., Suppl.:* 34 (1819); *cf. D.J. Mabberley in Taxon,* 33(3): 437 (1984): Basionym not stated.

SCHIEDEELLA (Orchidac.).
chartacea *(L.O. Wms.) P. Burns-Balogh in Orquídea (México),* 10(1): 92 (1986): Spiranthes chartacea.
dendroneura *(Sheviak & Bye) P. Burns-Balogh in Orquídea (México),* 10(1): 92 (1986): Spiranthes dendroneura.
diaphana *(Lindl.) P. Burns-Balogh & Greenwood in Orquídea (México),* 10(1): 93 (1986): Spiranthes diaphana.
durangensis *(A. & S.) P. Burns-Balogh in Orquídea (México),* 10(1): 93 (1986): Spiranthes durangensis.

SCHISTOCARPHA (Compos.).
longiligula
var. seleri *(Rydb.) B.L. Turner in Phytologia,* 59(4): 280 (1986): S. seleri.

SCHIZACHYRIUM (Gramin.).
villosum *(Poir.) J.F. Veldkamp in Blumea,* 31(2): 306 (1986): Rottboellia villosa.

SCHIZOSTACHYUM (Gramin.).
annulatum *C.J. Hsueh & W.P. Zhang in J. Bamboo Res.,* 5(1): 77 (1986)—China.

SCHOENOIDES O. Seberg in Willdenowia, 16(1): 181 (1986). *CYPERACEAE.*
oligocephalus *(W.M. Curtis) O. Seberg in Willdenowia,* 16(1): 182 (1986): Oreobolus oligocephalus.

SCHOENOXIPHIUM (Cyperac.).
altum *I. Kukkonen in Notes Roy. Bot. Gard. Edinburgh,* 43(3): 365 (1986)—South Africa (Cape Province).
bracteosum *I. Kukkonen in Notes Roy. Bot. Gard. Edinburgh,* 43(3): 365 (1986)—South Africa (Cape Province, Natal), Lesotho.
burttii *I. Kukkonen in Notes Roy. Bot. Gard. Edinburgh,* 43(3): 365 (1986)—South Africa (Natal).
molle *I. Kukkonen in Notes Roy. Bot. Gard. Edinburgh,* 43(3): 366 (1986)—South Africa (Natal).

SCHOENOXIPHIUM :–
strictum *I. Kukkonen in Notes Roy. Bot. Gard. Edinburgh,* 43(3): 366 (1986)—South Africa (Natal).

SCHRANKIA (Legumin.).
microphylla
var. floridana *(Chapman) D. Isely in Castanea,* 51(3): 204 (1986): S. floridana.
nuttallii
var. hystricina *(Small ex Britt. & Rose) D. Isely in Castanea,* 51(3): 205 (1986): Leptoglottis hystricina.

SCILLA (Hyacinthac.).
clusiana *Endl., Cat. Horti Vindob.,* 1: 129 (1842); *cf. D.J. Mabberley in Taxon,* 33(3): 436 (1984)— Locality not stated.
japonica *C.P. Thunberg ex A. Murray in Linnaeus, Syst. Veg., ed. 14:* 329 (1784)—Japan. †
scilloides
var. mounsei *(Lévl.) D.R. McKean in Notes Roy. Bot. Gard. Edinburgh,* 44(1): 195 (1986): S. chinensis var. mounsei.

SCIRPOIDES (Cyperac.).
holoschoenus
subsp. panormitanus *(Parlat.) J. Soják in Čas. Nár. Muz. (Prague),* 152(1): 21 (1983): Scirpus panormitanus.

SCLAREA (Labiat.).
amplexicaulis *(Lam.) J. Soják in Čas. Nár. Muz. (Prague),* 152(1): 21 (1983): Salvia amplexicaulis.
austriaca *(Jacq.) J. Soják in Čas. Nár. Muz. (Prague),* 152(1): 21 (1983): Salvia austriaca.
bicolor *(Lam.) J. Soják in Čas. Nár. Muz. (Prague),* 152(1): 21 (1983): Salvia bicolor.
candidissima *(Vahl) J. Soják in Čas. Nár. Muz. (Prague),* 152(1): 21 (1983): Salvia candidissima.
dumetorum *(Andrzejowski ex Besser) J. Soják in Čas. Nár. Muz. (Prague),* 152(1): 22 (1983): Salvia dumetorum.
nutans *(L.) J. Soják in Čas. Nár. Muz. (Prague),* 152(1): 22 (1983): Salvia nutans.
rhodantha *(Zef.) J. Soják in Čas. Nár. Muz. (Prague),* 152(1): 22 (1983): Salvia rhodantha.
sibthorpii *(Smith) J. Soják in Čas. Nár. Muz. (Prague),* 152(1): 22 (1983): Salvia sibthorpii.
sonclarii *(Pant.) J. Soják in Čas. Nár. Muz. (Prague),* 152(1): 22 (1983): Salvia sonclari.
transsilvanica *(Schur) J. Soják in Čas. Nár. Muz. (Prague),* 152(1): 22 (1983): Salvia transsilvanica.
verbenacea *(L.) J. Soják in Čas. Nár. Muz. (Prague),* 152(1): 22 (1983): Salvia verbenacea.
virgata *(Jacq.) J. Soják in Čas. Nár. Muz. (Prague),* 152(1): 22 (1983): Salvia virgata.
viridis *(L.) J. Soják in Čas. Nár. Muz. (Prague),* 152(1): 22 (1983): Salvia viridis.

SCLERIA (Cyperac.).
cerradicola *T. Koyama in Acta Amazonica,* 14(1–2 Suppl.): 112 (1984 publ. 1986)—Brazil, Venezuela, Guyana.
latifolia
subsp. foliosa *(Nees) T. Koyama in Acta Amazonica,* 14(1–2 Suppl.): 108 (1984 publ. 1986): Schizolepis foliosa.
mitis
subsp. eggersiana *(Boeckeler) T. Koyama in Acta Amazonica,* 14(1–2 Suppl.): 112 (1984 publ. 1986): S. eggersiana.

SCLEROCACTUS (Cactac.).
pubispinus
var. spinosior *(Engelm.) S.L. Welsh in Great Basin Nat.,* 44(1): 67 (1984): Echinocactus whipplei var. spinosior.
whipplei
var. glaucus *(K. Schum.) S.L. Welsh in Great Basin Nat.,* 44(1): 68 (1984): Echinocactus glaucus.

SCLERORHACHIS (Compos.).
platyrachis *(Boiss.) Podlech ex K.H. Rechinger in Fl. Iran.,* 158: 47 (1986): Pyrethrum platyrachis.

SCORDIUM (Labiat.).
procumbens *Lag., Elenchus:* 14 (1816); *cf. D.J. Mabberley in Taxon,* 31(1): 72 (1982)—Locality not stated.

SCURRULA (Loranthac.).
chingii *(Cheng) H.S. Kiu in Fl. Yunnanica,* 3: 363 (1983): Loranthus chingii.
var. yunnanensis *H.S. Kiu in Fl. Yunnanica,* 3: 364 (1983)—China.
parasitica
var. graciliflora *(Wall. ex DC.) H.S. Kiu in Fl. Yunnanica,* 3: 363 (1983): Loranthus graciliflorus.

SCUTELLARIA (Labiat.).
albida
subsp. velenovskyi *(Rech. f.) W. Greuter & H.M. Burdet in W. Greuter, H.M. Burdet & G. Long (eds.), Med-checklist,* 3: 342 (1986): S. velenovskyi.
hyssopifolia
var. hispida *(Benth.) J.L. Reveal in Bot. J. Linn. Soc.,* 92(2): 175 (1986): S. integrifolia var. hispida.

SEBAEA (Gentianac.).
madagascariensis *J. Klackenberg in Taxon,* 35(3): 595 (1986), nom. nov.: Belmontia stricta *Schinz.*

SEDUM (Crassulac.).
anhuiense *S.H. Fu & X.W. Wang in Bull. Bot. Res. North-East. Forest. Inst.,* 6(4): 137 (1986)—China.

SEDUM :–
creticum
var. hierapetrae *(Rech. f.) H.'t. Hart & I. Hagemann in A. Strid (ed.), Mount. Fl. Greece,* 1: 353 (1986): S. hierapetrae.
grisebachii
subsp. flexuosum *(Wettst.) W. Greuter & H.M. Burdet in W. Greuter, H.M. Burdet & G. Long (eds.), Med-checklist,* 3: 20 (1986): S. flexuosum.
subsp. kostovii *(Stefanov) W. Greuter & H.M. Burdet in W. Greuter, H.M. Burdet & G. Long (eds.), Med-checklist,* 3: 20 (1986): S. kostovii.
gypsophilum *B.L. Turner in Phytologia,* 59(5): 321 (1986)—Mexico.
hengduanense *K.T. Fu in Acta Bot. Bor.-Occid. Sin.,* 6(2): 105 (1986)—China, Xizang.
lineare *C.P. Thunberg ex A. Murray in Linnaeus, Syst. Veg., ed. 14:* 430 (1784)—Japan. †
longyanense *K.T. Fu in Acta Bot. Bor.-Occid. Sin.,* 6(2): 108 (1986)—Xizang.
magellense
subsp. olympicum *(Boiss.) W. Greuter & H.M. Burdet in W. Greuter, H.M. Burdet & G. Long (eds.), Med-checklist,* 3: 24 (1986): S. olympicum.
multicaule
subsp. rugosum *K.T. Fu in Acta Bot. Bor.-Occid. Sin.,* 6(2): 105 (1986)—Xizang.
muyaicum *K.T. Fu in Acta Bot. Bor.-Occid. Sin.,* 6(2): 107 (1986)—China.
obtusipetalum
subsp. dandyanum *H. Ohba in J. Jap. Bot.,* 61(9): 274 (1986)—Nepal.
subtile
subsp. chinense *H. Ohba in J. Jap. Bot.,* 61(8): 227 (1986)—China.
taiwanianum *S.S. Ying in Quart. J. Chinese Forest.,* 19(2): 101 (1986)—Taiwan.
uniflorum
subsp. rugosum *(Fröd.) K.T. Fu in Acta Bot. Bor.-Occid. Sin.,* 6(2): 110 (1986): S. japonicum f. rugosum.

SEMILIQUIDAMBAR (Hamamelidac.).
chingii
var. longipes *Y.K. Li & X.M. Wang in Acta Bot. Yunnanica,* 8(3): 275 (1986)—China.

SEMPERVIVUM (Crassulac.).
×alidae *B.J.M. Zonneveld in Brit. Cact. Succ. J.,* 4(3): 65 (1986)—Cultivation. (S. grandiflorum × wulfenii).
malacophyllum *(Pallas) Salm-Dyck, Ind. Pl. Succ.:* 53 (1822); *cf. D.J. Mabberley in Taxon,* 31(1): 72 (1982): Basionym not stated.

SENECIO (Compos.).
ser. Arborei *(O. Hoffm.) C. Jeffrey in Kew Bull.,* 41(4): 887 (1986): S. sect. Arborei.
ser. Lanati *(Muschl.) C. Jeffrey in Kew Bull.,* 41(4): 898 (1986): S. sect. Lanati.

SENECIO :–
sect. Peltati *(Muschl.) C. Jeffrey in Kew Bull.*, 41(4): 897 (1986): S. subsect. Peltati.
ser. Plantaginei *(Harv.) C. Jeffrey in Kew Bull.*, 41(4): 892 (1986): S. sect. Plantaginei.
sect. Rowleyani *C. Jeffrey in Kew Bull.*, 41(4): 934 (1986).
ser. Sinuosi *(Harv.) C. Jeffrey in Kew Bull.*, 41(4): 901 (1986): S. sect. Sinuosi.
amplificatus *C. Jeffrey in Kew Bull.*, 41(4): 900 (1986)—Kenya.
barkleyi *B.L. Turner in Phytologia*, 59(2): 89 (1986)—Mexico.
carbonellii *S. Díaz-Piedrahíta in Caldasia*, 15(71–75): 35 (1986)—Colombia.
crispatipilosus *C. Jeffrey in Kew Bull.*, 41(4): 893 (1986)—Uganda, Kenya, Sudan.
dentato-alatus *Mildbr. ex C. Jeffrey in Kew Bull.*, 41(4): 893 (1986)—Tanzania.
discokaraguensis *C. Jeffrey in Kew Bull.*, 41(4): 895 (1986)—Tanzania.
doryphoroides *C. Jeffrey in Kew Bull.*, 41(4): 895 (1986)—Tanzania.
exarachnoideus *C. Jeffrey in Kew Bull.*, 41(4): 896 (1986)—Tanzania.
fragrans *Lehm., Sem. Hort. Bot. Hamb.*, 1823: 14 (1824); *Linnaea 3, Lit.*: 8 (1828); *cf. D.J. Mabberley in Taxon*, 32(1): 86 (1983)—Locality not stated.
garlandii *F. Muell. ex R.O. Belcher in Muelleria*, 6(3–4): 173 (1986)—Australia (New South Wales).
gramineticola *C. Jeffrey in Kew Bull.*, 41(4): 895 (1986)—Tanzania, Malawi.
greenwayi *C. Jeffrey in Kew Bull.*, 41(4): 903 (1986)—Tanzania, Malawi.
gregatus *O.M. Hilliard in Notes Roy. Bot. Gard. Edinburgh*, 43(3): 352 (1986), nom. nov.: S. serratuloides var. gracilis *Harvey*.
hedbergii *C. Jeffrey in Kew Bull.*, 41(4): 899 (1986)—Tanzania, Kenya.
immixtus *C. Jeffrey in Kew Bull.*, 41(4): 896 (1986)—Tanzania.
jacobaea
subsp. dunensis *(Dumort.) J.W. Kadereit & P.D. Sell in Watsonia*, 16(1): 23 (1986): S. dunensis.
japonicus *C.P. Thunberg ex A. Murray in Linnaeus, Syst. Veg., ed. 14:* 756 (1784)—Japan. †
johnstonii
subsp. adnivalis *(Stapf) C. Jeffrey in Kew Bull.*, 41(4): 889 (1986): S. adnivalis.
var. adnivalis *(Stapf) C. Jeffrey in Kew Bull.*, 41(4): 890 (1986): S. adnivalis.
var. alticola *(T.C.E. Fr.) C. Jeffrey in Kew Bull.*, 41(4): 890 (1986): S. alticola.
var. battiscombei *(R.E. & T.C.E. Fr.) C. Jeffrey in Kew Bull.*, 41(4): 891 (1986): S. battiscombei.
var. cheranganiensis *(Cotton & Blakel.) C. Jeffrey in Kew Bull.*, 41(4): 891 (1986): S. cheranganiensis.
var. cottonii *(Hutch. & Tayl.) C. Jeffrey in Kew Bull.*, 41(4): 889 (1986): S. cottonii.

SENECIO :–
var. dalei *(Cotton & Blakel.) C. Jeffrey in Kew Bull.*, 41(4): 891 (1986): S. dalei.
var. elgonensis *(T.C.E. Fr.) C. Jeffrey in Kew Bull.*, 41(4): 889 (1986): S. elgonensis.
var. erici-rosenii *(R.E. & T.C.E. Fr.) C. Jeffrey in Kew Bull.*, 41(4): 890 (1986): S. erici-rosenii.
var. kilimanjari *(Mildbr.) C. Jeffrey in Kew Bull.*, 41(4): 888 (1986): S. kilimanjari.
var. ligulatus *(Cotton & Blakel.) C. Jeffrey in Kew Bull.*, 41(4): 889 (1986): S. gardneri var. ligulatus.
var. meruensis *(Cotton & Blakel.) C. Jeffrey in Kew Bull.*, 41(4): 888 (1986): S. meruensis.
keniensis
subsp. brassiciformis *(R.E. & T.C.E. Fr.) C. Jeffrey in Kew Bull.*, 41(4): 892 (1986)—S. brassiciformis.
linifoliaster *G. López González in An. Jard. Bot. Madrid*, 42(2): 323 (1985 publ. 1986)—Spain.
mabberleyi *C. Jeffrey in Kew Bull.*, 41(4): 893 (1986)—Tanzania.
margaritae *C. Jeffrey in Kew Bull.*, 41(4): 898 (1986)—Kenya.
meyeri-johannis
subsp. olomotiensis *C. Jeffrey in Kew Bull.*, 41(4): 900 (1986)—Tanzania.
microalatus *C. Jeffrey in Kew Bull.*, 41(4): 894 (1986)—Tanzania.
mooreioides *C. Jeffrey in Kew Bull.*, 41(4): 899 (1986)—Tanzania.
morotonensis *C. Jeffrey in Kew Bull.*, 41(4): 898 (1986)—Uganda.
navugabensis *C. Jeffrey in Kew Bull.*, 41(4): 901 (1986)—Uganda.
plantagineoides *C. Jeffrey in Kew Bull.*, 41(4): 898 (1986)—Kenya.
pseudosubsessilis *C. Jeffrey in Kew Bull.*, 41(4): 894 (1986)—Kenya.
richardsonii *B.L. Turner in Phytologia*, 60(3): 157 (1986)—Mexico.
sororius *C. Jeffrey in Kew Bull.*, 41(4): 895 (1986)—Tanzania.
subfractiflexus *C. Jeffrey in Kew Bull.*, 41(4): 894 (1986)—Tanzania.
transmarinus
var. major *C. Jeffrey in Kew Bull.*, 41(4): 902 (1986)—Uganda.
var. virungae *C. Jeffrey in Kew Bull.*, 41(4): 903 (1986)—Uganda, Zaïre.
volcanicola *C. Jeffrey in Kew Bull.*, 41(4): 893 (1986)—Tanzania.

SENEGALIA (Legumin.).
sect. Filicinae *(Bentham) L. Pedley in Bot. J. Linn. Soc.*, 92(3): 238 (1986): Acacia ser. Filicinae.
albizioides *(Pedley) L. Pedley in Bot. J. Linn. Soc.*, 92(3): 250 (1985): Acacia albizioides.
angustissima *(Miller) L. Pedley in Bot. J. Linn. Soc.*, 92(3): 238 (1986): Mimosa angustissima.

SENEGALIA :–
ferruginea *(DC.) L. Pedley in Bot. J. Linn. Soc.*, 92(3): 250 (1986): Acacia ferruginea.

SENNA (Legumin.).
aphylla
subsp. divaricata *(Hieronymus) L.D. Bravo in Brittonia,* 38(3): 270 (1986): Cassia aphylla var. divaricata.
holosericea *(Fresen.) W. Greuter in Willdenowia,* 15(2): 429 (1986): Cassia holosericea.
rigida
var. inermis *(Bravo) L.D. Bravo in Brittonia,* 38(3): 270 (1986): Cassia rigida var. inermis.

SERAPIAS (Orchidac.).
erecta *C.P. Thunberg ex A. Murray in Linnaeus, Syst. Veg., ed. 14:* 816 (1784)—Japan. †
falcata *C.P. Thunberg ex A. Murray in Linnaeus, Syst. Veg., ed. 14:* 816 (1784)—Japan. †
lingua
var. flavescens *M. Balayer in Bull. Soc. Bot. France, Lett. Bot.*, 133(3): 280 (1986)—France.
var. rubescens *M. Balayer in Bull. Soc. Bot. France, Lett. Bot.*, 133(3): 280 (1986)—France.
subsp. veneris *M. Balayer in Bull. Soc. Bot. France, Lett. Bot.*, 133(3): 279 (1986)—France.
vomeracea
var. cordigeroides *(Nelson) N.R. & A.K. Campbell in Willdenowia,* 16(1): 54 (1986): S. orientalis var. cordigeroides.
subsp. flava *M. Balayer in Bull. Soc. Bot. France, Lett. Bot.*, 133(3): 280 (1986)—France.

SERIPHIDIUM (Compos.).
barrelieri *(Bess.) J. Soják in Čas. Nár. Muz. (Prague),* 152(1): 22 (1983): Artemisia barrelieri.
caerulescens *(L.) J. Soják in Čas. Nár. Muz. (Prague),* 152(1): 22 (1983): Artemisia caerulescens.
subsp. gallicum *(Willd.) J. Soják in Čas. Nár. Muz. (Prague),* 152(1): 22 (1983): Artemisia gallica.
dzevanovskyi *(Leonova) J. Soják in Čas. Nár. Muz. (Prague),* 152(1): 22 (1983): Artemisia dzevanovskyi.
herba-alba *(Asso) J. Soják in Čas. Nár. Muz. (Prague),* 152(1): 22 (1983): Artemisia herba-alba.
nutans *(Willd.) J. Soják in Čas. Nár. Muz. (Prague),* 152(1): 22 (1983): Artemisia nutans.
santonicum *(L.) J. Soják in Čas. Nár. Muz. (Prague),* 152(1): 22 (1983): Artemisia santonica.

SERIPHIDIUM :–
subsp. patens *(Neilr.) J. Soják in Čas. Nár. Muz. (Prague),* 152(1): 22 (1983): Artemisia maritima β var. patens.
vallesiacum *(All.) J. Soják in Čas. Nár. Muz. (Prague),* 152(1): 22 (1983): Artemisia vallesiaca.

SERJANIA (Sapindac.).
bahiana *M.S. Ferrucci in Bol. Soc. Argent. Bot.*, 24(1–2): 107 (1985)—Brazil.

SERRATULA (Compos.).
japonica *C.P. Thunberg ex A. Murray in Linnaeus, Syst. Veg., ed. 14:* 723 (1784)—Japan. †
pulchella *Fisch. ex Hornem., Hort. Hafn.:* 775 (1815); *cf. D.J. Mabberley in Taxon,* 33(3): 437 (1984)— Locality not stated.

SESAMOIDES (Resedac.).
purpurascens *(L.) G. López González in An. Jard. Bot. Madrid,* 42(2): 321 (1985 publ. 1986): Reseda purpurascens.

SESELI (Umbellif.).
incisodentatum *K.T. Fu in Fl. Tsinlingensis,* 1(3): 459, 412 (1981), as 'inciso-dentatum'—China.
peucedanifolium *Trev., Ind. Sem. Wratislav., App. II:* 3 (1820); *cf. D.J. Mabberley in Taxon,* 31(1): 72 (1982)— Locality not stated.
sclerophyllum *E.P. Korovin in Izv. Akad. Nauk Tadzh. SSR, Biol. Nauk,* 3(60): 7 (1975)—U.S.S.R. (Middle Asia).
subaphyllum *E.P. Korovin in Izv. Akad. Nauk Tadzh. SSR, Biol. Nauk,* 3(60): 6 (1975)—U.S.S.R. (Middle Asia).

SETARIA (Gramin.).
italica
subsp. colchica *I.I. Maisaya & A.D. Gorgidze in Soobshch. Akad. Nauk Gruz. SSR,* 119(3): 589 (1985), without type—?U.S.S.R. (Caucasus).

SHERARDIA (Rubiac.).
arvensis
subsp. maritima *(Griseb.) J. Soják in Čas. Nár. Muz. (Prague),* 152(1): 22 (1983): S. arvensis var. maritima.

SIDA (Malvac.).
massaica *K. Vollesen in Kew Bull.*, 41(1): 97 (1986)—Kenya, Tanzania.
shinyangensis *K. Vollesen in Kew Bull.*, 41(1): 97 (1986)—Tanzania.
tanaensis *K. Vollesen in Kew Bull.*, 41(1): 95 (1986)—Kenya.
tenuicarpa *K. Vollesen in Kew Bull.*, 41(1): 92 (1986)—Ethiopia, Zaïre, Uganda, Kenya, Tanzania.

SIDERITIS (Labiat.).
ciliata *C.P. Thunberg ex A. Murray in Linnaeus, Syst. Veg., ed. 14:* 532 (1784)—Japan. †

SIDERITIS :–
gomeraea
subsp. perezii *M.L. Negrin-Sosa in Vieraea,* 16(1–2): 273 (1986)—Canary Is.

×**SIDRANARA** hort. in Orchid Rev., 94(1107): centre page pull-out p. 8 (1986). *ORCHIDACEAE.* (Ascocentrum × Phalaenopsis × Renanthera).

SIEVEKINGIA (Orchidac.).
herklotziana *R. Jenny in Orchidee,* 37(2): 75 (1986)—Colombia.
herrenhusana *R. Jenny in Orchidee,* 37(2): 78 (1986)—Ecuador.

SIEVERSIA (Rosac.).
paradoxa *D. Don ex Tilloch & Taylor, Phil. Mag.,* 64: 374 (Nov. 1824); *cf. D.J. Mabberley in Taxon,* 30(1): 17 (1981)— Locality not stated.

SILENE (Caryophyllac.).
colpophylla *F. Wrigley in Ann. Bot. Fenn.,* 23(1): 77 (1986)—France.
decumbens *Hornem., Hort. Hafn.:* 415 (1813); *cf. D.J. Mabberley in Taxon,* 33(3): 437 (1984)— Locality not stated.
firma
var. pubescens *(Makino) S.Y. He in Fl. Beijing,* 1: 209 (1984), without basionym ref.: S. firma f. pubescens.
ibosii
subsp. grosiana *(Pau & Font Quer) J. Fernández Casas in Fontqueria,* 1: 9 (1982): S. grosiana.
multifurcata *C.L. Tang in Acta Phytotax. Sin.,* 24(5): 391 (1986)—Xizang.
otites
subsp. hungarica *F. Wrigley in Ann. Bot. Fenn.,* 23(1): 74 (1986)—Hungary, Austria, Czechoslovakia.
seoulensis
var. angustata *C.L. Tang in Acta Phytotax. Sin.,* 24(5): 391 (1986)—China.
sericata *C.L. Tang in Acta Phytotax. Sin.,* 24(5): 389 (1986)—China.
suavis *Schrad., Ind. Sem. Hort. Acad. Gott.,* 1835: 6 (1835); *Linnaea 11, Lit.:* 90 (1837); *cf. D.J. Mabberley in Taxon,* 32(1): 86 (1983)— Locality not stated.
tenorii *Schrad., Ind. Sem. Hort. Acad. Gott.,* 1821: 4 (1821); *cf. D.J. Mabberley in Taxon,* 32(1): 86 (1983)— Locality not stated.
tubiformis *C.L. Tang in Acta Phytotax. Sin.,* 24(5): 387 (1986)—China.
uniflora
subsp. cratericola *(Franco) J. do Amaral Franco in Ann. Bot. Fenn.,* 23(1): 91 (1986): S. vulgaris subsp. cratericola.
velebitica *(Degen) F. Wrigley in Ann. Bot. Fenn.,* 23(1): 70 (1986): S. otites subsp. velebitica.

SINAPIS (Crucif.).
cernua *C.P. Thunberg ex A. Murray in Linnaeus, Syst. Veg., ed. 14:* 602 (1784)— Japan. †
japonica *C.P. Thunberg ex A. Murray in Linnaeus, Syst. Veg., ed. 14:* 602 (1784)— Japan. †
viminea *(L.) Lag., Elenchus:* 15 (1816); *cf. D.J. Mabberley in Taxon,* 31(1): 72 (1982): Basionym not stated.

SINOJOHNSTONIA (Boraginac.).
chekiangensis *(Migo) W.T. Wang ex Z.Y. Zhang in Fl. Tsinlingensis,* 1(4): 186 (1983): Omphalodes chekiangensis.
moupinensis *(Franch.) W.T. Wang ex Z.Y. Zhang in Fl. Tsinlingensis,* 1(4): 185 (1983): Omphalodes moupinensis.

SINOSASSAFRAS H.W. Li in Acta Bot. Yunnanica, 7(2): 134 (1985). *LAURACEAE.*
flavinervia *(Allen) H.W. Li in Acta Bot. Yunnanica,* 7(2): 134 (1985): Lindera flavinervia.

SIPHOCAMPYLUS (Campanulac.).
cavanillesii *(Mart.) J.W. Loud., Lad. Fl. Gard. Orn. Greenh. Pl.:* 184 (1848); *cf. D.J. Mabberley in Taxon,* 32(1): 87 (1983): Basionym not stated.

SISYMBRIUM (Crucif.).
praetermissum *T.K. Mardaleĭshvili in Soobshch. Akad. Nauk Gruz. SSR,* 98(2): 397 (1980)—U.S.S.R. (Caucasus).

SITODIUM (Morac.).
macrocarpon *Thunb. in Phil. Trans. Roy. Soc. London,* 69: 467 (1779); *cf. D.J. Mabberley in Taxon,* 30(1): 12 (1981)— Locality not stated.

SIUM (Umbellif.).
decumbens *C.P. Thunberg ex A. Murray in Linnaeus, Syst. Veg., ed. 14:* 285 (1784)— Japan. †
japonicum *C.P. Thunberg ex A. Murray in Linnaeus, Syst. Veg., ed. 14:* 284 (1784)— Japan. †

SLOANEA (Elaeocarpac.).
bathiei *C. Tirel in Fl. Madagascar,* Fam. 125: 48 (1985)—Madagascar.
longisepala *C. Tirel in Fl. Madagascar,* Fam. 125: 47 (1985)—Madagascar.

SMILACINA (Convallariac.).
bifolia *(L.) Ker in J. Sci. Arts,* 1: 182 (1816); *cf. D.J. Mabberley in Taxon,* 31(1): 72 (1982): Basionym not stated.

SMILAX (Smilacac.).
aculeatissima *J.G. Conran in Fl. Australia,* 46: 231, 187 (1986)—Australia (Queensland).
rigida *Soland. in Russell, Aleppo, ed. 2,* 2: 271 (1794); *cf. D.J. Mabberley in Taxon,* 31(1): 72 (1982)— Locality not stated.

SMILAX :–
umbrosa *J.M. Xu in Bull. Bot. Res. North-East. Forest. Inst.*, 6(2): 67 (1986)—China.

SMYRNIUM (Umbellif.).
perfoliatum
subsp. rotundifolium *(Miller) P. Hartvig in A. Strid (ed.), Mount. Fl. Greece*, 1: 672 (1986): S. rotundifolium.

×**SOBENNIGRAECUM** hort. in Orchid Rev., 94(1107): centre page pull-out p. 8 (1986). *ORCHIDACEAE*. (Angraecum × Sobennikoffia).

SOCRATEA (Palmae).
hecatonandra *(Dugand) R. Bernal-Gonzalez in Kew Bull.*, 41(1): 152 (1986): Metasocratea hecatonandra.
montana *R. Bernal-González & A. Henderson in Brittonia*, 38(1): 55 (1986)—Colombia.

SOLANECIO (Compos.).
angulatus *(Vahl) C. Jeffrey in Kew Bull.*, 41(4): 922 (1986): Cacalia angulata.
biafrae *(Oliv. & Hiern) C. Jeffrey in Kew Bull.*, 41(4): 922 (1986): Senecio biafrae.
buchwaldii *(O. Hoffm.) C. Jeffrey in Kew Bull.*, 41(4): 921 (1986): Senecio buchwaldii.
cydoniifolius *(O. Hoffm.) C. Jeffrey in Kew Bull.*, 41(4): 921 (1986): Senecio cydoniifolius.
epidendricus *(Mattf.) C. Jeffrey in Kew Bull.*, 41(4): 921 (1986): Senecio epidendricus.
gigas *(Vatke) C. Jeffrey in Kew Bull.*, 41(4): 923 (1986): Senecio gigas.
goetzei *(O. Hoffm.) C. Jeffrey in Kew Bull.*, 41(4): 921 (1986): Senecio goetzei.
gymnocarpus *C. Jeffrey in Kew Bull.*, 41(4): 920 (1986)—Tanzania.
gynuroides *C. Jeffrey in Kew Bull.*, 41(4): 923 (1986)—Uganda.
kanzibiensis *(Humbert & Staner) C. Jeffrey in Kew Bull.*, 41(4): 921 (1986): Senecio kanzibiensis.
mannii *(Hook. f.) C. Jeffrey in Kew Bull.*, 41(4): 922 (1986): Senecio mannii.
mirabilis *(Muschl.) C. Jeffrey in Kew Bull.*, 41(4): 921 (1986): Senecio mirabilis.
nandensis *(S. Moore) C. Jeffrey in Kew Bull.*, 41(4): 921 (1986): Senecio nandensis.
tuberosus *(Sch. Bip. ex A. Rich.) C. Jeffrey in Kew Bull.*, 41(4): 921 (1986): Senecio tuberosus.

SOLANUM (Solanac.).
sect. Glaucophyllum *A. Child in Feddes Repert.*, 97(3–4): 144 (1986).
abitaguense *S. Knapp in Brittonia*, 38(3): 290 (1986)—Ecuador, Peru.
amnicola *S. Knapp in Brittonia*, 38(3): 295 (1986)—Peru.
bellum *S. Knapp in Brittonia*, 38(3): 294 (1986)—Ecuador.

SOLANUM :–
callicarpifolium *Kunth & Bouché, Ind. Sem. Hort. Bot. Berol.*, 1845: 10 (1845); *Ann. Sci. Nat. Sér. III*, 5: 355 (1846); *cf. D.J. Mabberley in Taxon*, 32(1): 86 (1983)—Locality not stated.
corniculatum *Huber, Cat. Print. 1865:* 9 (1865); *cf. D.J. Mabberley in Taxon*, 34(3): 451 (1985)— Locality not stated.
cucullatum *S. Knapp in Brittonia*, 38(3): 292 (1986)—Ecuador.
donachui *C. Ochoa in Phytologia*, 59(7): 461 (1986)—Colombia.
erosomarginatum *S. Knapp in Brittonia*, 38(3): 273 (1986)—Venezuela.
foetens *Pittier ex S. Knapp in Brittonia*, 38(3): 275 (1986)—Venezuela.
guzmanguense *M.D. Whalen & A. Sagástegui in Brittonia*, 38(1): 9 (1986)—Peru.
haematocarpum *Huber, Cat. Gén. 1871 & 1872:* 5 (1871); *cf. D.J. Mabberley in Taxon*, 34(3): 451 (1985)— Locality not stated.
incahuasinum *C. Ochoa in Kurtziana*, 12–13: 183 (1979)—Peru.
lasiocladum *S. Knapp in Brittonia*, 38(3): 278 (1986)—Venezuela.
linnaeanum *F.N. Hepper & P.-M.L. Jaeger in Kew Bull.*, 41(2): 435 (1986)—South Africa (Cape Province).
lucens *S. Knapp in Brittonia*, 38(3): 280 (1986)—Venezuela, Colombia.
lyratum *C.P. Thunberg ex A. Murray in Linnaeus, Syst. Veg., ed. 14:* 224 (1784)—Japan. †
mahoriensis *W.G. D'Arcy & A. Rakotozafy in Ann. Missouri Bot. Gard.*, 73(2): 498 (1986)—Madagascar.
malacothrix *S. Knapp in Brittonia*, 38(1): 89 (1986)—Mexico.
mapiricum *S. Knapp in Brittonia*, 38(3): 299 (1986)—Bolivia.
melongena
subsp. agreste *S.P. Dikiĭ in Trudӯ Prikl. Bot. Genet. Selek.*, 88: 105 (1984), as 'agrestis'—India.
var. angustum *S.P. Dikiĭ in Trudӯ Prikl. Bot. Genet. Selek.*, 88: 105 (1984), as 'angusta'—India.
var. cylindricum *S.P. Dikiĭ in Trudӯ Prikl. Bot. Genet. Selek.*, 88: 105 (1984), as 'cylindrica'—India.
var. giganteum *(Alef.) S.P. Dikiĭ in Trudӯ Prikl. Bot. Genet. Selek.*, 88: 105 (1984), without basionym ref. but name same as avowed basionym: S. melongena gigantea.
var. globosi *S.P. Dikiĭ in Trudӯ Prikl. Bot. Genet. Selek.*, 88: 106 (1984)—Vietnam.
var. leucoum *(Alef.) S.P. Dikiĭ in Trudӯ Prikl. Bot. Genet. Selek.*, 88: 105 (1984), name same as avowed basionym: S. melongena leucoum.
var. racemiflorum *S.P. Dikiĭ in Trudӯ Prikl. Bot. Genet. Selek.*, 88: 105 (1984), as 'racemiflora'—India.

SOLANUM :–
var. racemosum *S.P. Dikiĭ in Trudȳ Prikl. Bot. Genet. Selek.*, 88: 105 (1984), as 'racemosa'—India.
var. stenoleucum *(Alef.) S.P. Dikiĭ in Trudȳ Prikl. Bot. Genet. Selek.*, 88: 104 (1984), name same as avowed basionym: S. melongena stenoleuca.
var. variegatum *(Alef.) S.P. Dikiĭ in Trudȳ Prikl. Bot. Genet. Selek.*, 88: 105 (1984), name same as avowed basionym: S. melongena variegata.
var. violaceum *(Alef.) S.P. Dikiĭ in Trudȳ Prikl. Bot. Genet. Selek.*, 88: 105 (1984), name same as avowed basionym: S. melongena violacea.
var. viride *S.P. Dikiĭ in Trudȳ Prikl. Bot. Genet. Selek.*, 88: 104 (1984), as 'viridis'—Malaya.
mortonii *A.T. Hunziker in Kurtziana,* 12–13: 133 (1979)—Argentina.
ombrophilum *Pittier ex S. Knapp in Brittonia,* 38(3): 282 (1986)—Venezuela.
physalifolium
var. nitidibaccatum *(Bitter) J.M. Edmonds in Bot. J. Linn. Soc.*, 92(1): 27 (1986): S. nitidibaccatum.
tanysepalum *S. Knapp in Brittonia,* 38(3): 284 (1986)—Venezuela.
tenuiflagellatum *Bitter ex S. Knapp in Brittonia,* 38(3): 286 (1986)—Venezuela.
turgidum *S. Knapp in Brittonia,* 38(3): 288 (1986)—Venezuela.
verrucosum *Schlecht., Ind. Sem. Hort. Acad. Hal.*, 1839; *Linnaea 14, Lit.*: 129 (?1841); *cf. D.J. Mabberley in Taxon,* 32(1): 86 (1983)—Locality not stated.
×viirsooi *K.A. Okada & A.M. Clausen in Euphytica,* 34(1): 227 (1985)—Argentina. (S. acaule subsp. acaule × infundibuliforme).
warzcewiczoides *Huber, Cat. Graines 1870:* 3 (1870); *Wochenschrift,* 13: 181 (1870); *cf. D.J. Mabberley in Taxon,* 34(3): 451 (1985)— Locality not stated.
yanamonense *S. Knapp in Brittonia,* 38(3): 298 (1986)—Peru.

SOLDANELLA (Primulac.).
ser. Pindicae *F.K. Meyer in Haussknechtia, Mitt. Thüring. Bot. Ges.*, 2: 22 (1985).
ser. Pusillae *F.K. Meyer in Haussknechtia, Mitt. Thüring. Bot. Ges.*, 2: 32 (1985).
alpicola *F.K. Meyer in Haussknechtia, Mitt. Thüring. Bot. Ges.*, 2: 34 (1985)—Austria, Switzerland, Italy, Jugoslavia.
cyanaster *O. Schwarz ex F.K. Meyer in Haussknechtia, Mitt. Thüring. Bot. Ges.*, 2: 8 (1985)—Bulgaria.
macedonica *F.K. Meyer in Haussknechtia, Mitt. Thüring. Bot. Ges.*, 2: 23 (1985)—Jugoslavia.
pirinica *F.K. Meyer in Haussknechtia, Mitt. Thüring. Bot. Ges.*, 2: 36 (1985)—Bulgaria.

SOLDANELLA :–
pseudomontana *F.K. Meyer in Haussknechtia, Mitt. Thüring. Bot. Ges.*, 2: 15 (1985)—Romania, Czechoslovakia.
rhodopaea *F.K. Meyer in Haussknechtia, Mitt. Thüring. Bot. Ges.*, 2: 25 (1985)—Bulgaria.
stiriaca *F.K. Meyer in Haussknechtia, Mitt. Thüring. Bot. Ges.*, 2: 20 (1985)—Austria.

SOLIDAGO (Compos.).
ouachitensis *C.E.S. & R.J. Taylor in Sida,* 11(3): 334 (1986)—U.S.A. (Oklahoma, Arkansas).

SOLIVA (Compos.).
anthemifolia *(A. Juss.) Sweet, Hort. Brit.*: 243 (1827); *cf. D.J. Mabberley in Taxon,* 30(1): 17 (1981): Basionym not stated.
stolonifera *(Pers.) R. Br. ex Sweet, Hort. Brit.*: 243 (1827); *cf. D.J. Mabberley in Taxon,* 30(1): 17 (1981): Basionym not stated.

SORBUS (Rosac.).
sect. Duales *T.I. Zaikonnikova in Bot. Zhurn.*, 71(6): 813 (1986).
subsect. Lucidae *T.I. Zaikonnikova in Bot. Zhurn.*, 71(6): 814 (1986).
aria
subsp. mougeotii *(Soyer-Will. & Godr.) O. de Bolòs & J. Vigo, Fl. Països Catalans:* 413 (1984): S. mougeotii.
scepusiensis *M. Kovanda in Willdenowia,* 16(1): 119 (1986)—Czechoslovakia.
yuana *S.A. Spongberg in J. Arnold Arbor.*, 67(2): 257 (1986)—China.

SORGHUM (Gramin.).
compactum *(Lam.) Lag., Elenchus:* 15 (1816); *cf. D.J. Mabberley in Taxon,* 31(1): 72 (1982): Basionym not stated.
melanocarpum *Huber, Cat. Graines 1868:* 7 (1868); *Wochenschrift,* 11: 182 (1868); *cf. D.J. Mabberley in Taxon,* 34(3): 451 (1985)— Ethiopia.
nankinense *Huber, Cat. Graines 1867:* 6 (1867); *cf. D.J. Mabberley in Taxon,* 34(3): 451 (1985)— Locality not stated.
tataricum *Huber, Cat. Graines 1867:* 6 (1867); *cf. D.J. Mabberley in Taxon,* 34(3): 451 (1985)— Locality not stated.

SOTEROSANTHUS Lehmann ex R. Jenny in Orchidee, 37(2): 73 (1986). *ORCHIDACEAE.*
shepheardii *(Rolfe) R. Jenny in Orchidee,* 37(2): 74 (1986): Sievekingia shepheardii.

SOYMIDA A. Jussieu in Mirb. & Cass. apud Guill. in Bull. Sci. Nat. Geol., 23: 238 (1830); cf. D.J. Mabberley in Taxon, 31(1): 66 (1982). *MELIACEAE.*
febrifuga *(Roxb.) A. Juss. in Mém. Mus. Paris,* 19: 251 (1832); *cf. D.J. Mabberley in Taxon,* 31(1): 66 (1982): Swietenia febrifuga.

SPARREA A.T. Hunziker & N. Dottori in Kurtziana, 11: 35 (1978). *ULMACEAE.*

SPARREA :–
schippii *(Standley) A.T. Hunziker & N. Dottori in Kurtziana,* 11: 37 (1978): Celtis schippii.

SPATHIPAPPUS (Compos.).
chitralensis *D. Podlech in Fl. Iran.,* 158: 151 (1986)—Pakistan.

SPATHOLIRION (Commelinac.).
elegans *(Cherfils) C.Y. Wu in Fl. Yunnanica,* 3: 685 (1983): Streptolirion elegans.

SPERGULA (Caryophyllac.).
rubra *(L.) F.G. Bartling in Linnaea,* 7: 625 (1832): Arenaria rubra.

SPERGULARIA (Caryophyllac.).
melanocaulos *P. Baltasar Merino, Algunas Pl. Rar. ... de La Guardia (Pontevedra):* 26 (1895)—Spain.

SPERMACOCE (Rubiac.).
intricans *(Hepper) H.M. Burkill in Kew Bull.,* 41(4): 1006 (1986): Borreria intricans.
macrantha *(Hepper) H.M. Burkill in Kew Bull.,* 41(4): 1006 (1986): Borreria macrantha.
ramanii *V.V. Sivarajan & R.V. Nair in Taxon,* 35(2): 367 (1986)—India.
saxicola *(K. Schum.) H.M. Burkill in Kew Bull.,* 41(4): 1006 (1986): Borreria saxicola.

SPHAERADENIA (Cyclanthac.).
pendula *B.E. Hammel in Phytologia,* 60(1): 11 (1986)—Costa Rica, Panama.

SPHAEROCORYNE (Annonac.).
gracilis *(Engl. & Diels) B. Verdcourt in Kew Bull.,* 41(2): 295 (1986): Popowia gracilis.
subsp. engleriana *(Exell & Mendonça) B. Verdcourt in Kew Bull.,* 41(2): 295 (1986): Popowia engleriana.

SPHENOSTEMON (Sphenostemonac.).
sect. Apetalae *(Steen.) C.G.G.J. van Steenis in Fl. Males.,* 10(2): 147 (1986): S. ser. Apetalae.

SPIRADICLIS (Rubiac.).
coccinea *H.S. Lo in Bull. Bot. Res. North-East. Forest. Inst.,* 6(4): 38 (1986)—China.
howii *H.S. Lo in Bull. Bot. Res. North-East. Forest. Inst.,* 6(4): 41 (1986)—China.
purpureocaerulea *H.S. Lo in Bull. Bot. Res. North-East. Forest. Inst.,* 6(4): 39 (1986)—China.
umbelliformis *H.S. Lo in Bull. Bot. Res. North-East. Forest. Inst.,* 6(4): 36 (1986)—China.
xizangensis *H.S. Lo in Bull. Bot. Res. North-East. Forest. Inst.,* 6(4): 43 (1986)—Xizang.

SPIRAEA (Rosac.).
arunachalensis *G. Panigrahi & K.M. Purohit in Bull. Bot. Surv. India,* 26(1–2): 83 (1984 publ. 1985)—India.
callosa *C.P. Thunberg ex A. Murray in Linnaeus, Syst. Veg., ed. 14:* 471 (1784)—Japan. †

SPIRAEA :–
chambaensis *K.M. Purohit & G. Panigrahi in Bull. Bot. Surv. India,* 25(1–4): 230 (1983 publ. 1985)—India.
darjeelingensis *G. Panigrahi & K.M. Purohit in Bull. Bot. Surv. India,* 26(1–2): 83 (1984 publ. 1985)—India, Sikkim.
emarginata *K.M. Purohit & G. Panigrahi in Bull. Bot. Surv. India,* 26(1–2): 88 (1984 publ. 1985)—India.
expansa *Wall. ex C. Koch in A. Br., Klotzsch & Bouché, App. Spec. Nov. Hort. Reg. Bot. Berol.,* 1853: 12 (1853); *cf. D.J. Mabberley in Taxon,* 33(3): 443 (1984)— Locality not stated.
incisa *C.P. Thunberg ex A. Murray in Linnaeus, Syst. Veg., ed. 14:* 472 (1784)—Japan. †
nayarii *K.M. Purohit & G. Panigrahi in Bull. Bot. Surv. India,* 26(1–2): 78 (1984 publ. 1985)—Sikkim.
palmata *C.P. Thunberg ex A. Murray in Linnaeus, Syst. Veg., ed. 14:* 472 (1784)—Japan. †
panchananii *G. Panigrahi & K.M. Purohit in Bull. Bot. Surv. India,* 26(1–2): 80 (1984 publ. 1985)—India.
raizadae *G. Panigrahi & K.M. Purohit in M.R. Sharma & B.K. Gupta, Glimpses Pl. Sci.:* 39 (1986)—India.
rhamniphylla *G. Panigrahi & K.M. Purohit in Bull. Bot. Surv. India,* 26(1–2): 76 (1984 publ. 1985)—India.
subdioica *K.M. Purohit & G. Panigrahi in Bull. Bot. Surv. India,* 26(1–2): 86 (1984 publ. 1985)—India.
subrotundifolia *G. Panigrahi & K.M. Purohit in Bull. Bot. Surv. India,* 26(1–2): 88 (1984 publ. 1985)—Sikkim.
tanguensis *K.M. Purohit & G. Panigrahi in Bull. Bot. Surv. India,* 26(1–2): 80 (1984 publ. 1985)—Sikkim.

SPODIOPOGON (Gramin.).
jainii *V.J. Nair, A.N. Singh & N.C. Nair in Curr. Sci.,* 50(16): 730 (1981)—India.

SPOROBOLUS (Gramin.).
fertilis
var. purpureasuffusus *(J. Ohwi) P.C. Keng & X.S. Shen in Acta Bot. Bor.-Occid. Sin.,* 5(2): 161 (1985): S. elongatus var. purpurea-suffusus.

SPRYGINIA (Crucif.).
winkleri
subsp. undulata *(Botsch.) I.T. Vasil'chenko & L.I. Vasil'eva in I.T. Vasil'chenko (ed.), Rast. Sred. Azii:* 70 (1986): S. undulata.

STABEROHA (Restionac.).
ornata *E.E. Esterhuysen in Bothalia,* 15(3–4): 396 (1985)—South Africa (Cape Province).

STACHYS (Labiat.).
comosa *L.E. Codd in Bothalia,* 16(1): 51 (1986)—South Africa (Natal).

STACHYURUS (Stachyurac.).
calcareus *H.T. Chang & Z.R. Xu in Acta Sci. Nat. Univ. Sunyatseni,* 1986(2): 98 (1986), without type—China.
callosus *C.Y. Wu in Fl. Yunnanica,* 3: 344 (1983)—China.
chinensis
var. brachystachyus *C.Y. Wu & S.K. Chen in Fl. Yunnanica,* 3: 348 (1983)—China, Xizang.
himalaicus
var. alatipes *C.Y. Wu in Fl. Yunnanica,* 3: 343 (1983)—China.
var. dasyrachis *C.Y. Wu in Fl. Yunnanica,* 3: 344 (1983)—China.
var. microphyllus *C.Y. Wu in Fl. Yunnanica,* 3: 344 (1983)—China.

STATICE (Plumbaginac.).
baetica *(Boiss.) Font Quer & Rothmaler, Sched. Fl. Iber. Select., Cent. II-III,* n. 165 (1935); *cf. C. Navarro, M. Gutiérrez-Bustillo & A.G. Bueno in Fontqueria,* 7: 13 (1985): Armeria baetica.
capitella *(Pau) Font Quer & Rothmaler, Sched. Fl. Iber. Select., Cent. II-III,* n. 168 (1935); *cf. C. Navarro, M. Gutiérrez-Bustillo & A.G. Bueno in Fontqueria,* 7: 13 (1985): Armeria capitella.
dertosensis *Font Quer & Rothmaler, Sched. Fl. Iber. Select., Cent. II-III,* n. 171 (1935); *cf. C. Navarro, M. Gutiérrez-Bustillo & A.G. Bueno in Fontqueria,* 7: 13 (1985), nom. nov.: Armeria fontqueri *Pau ex Font Quer.*
floribunda *Lem. ex Huber, Wochenschrift,* 9: 81 (1866); *cf. D.J. Mabberley in Taxon,* 34(3): 451 (1985)— Locality not stated.
isernii *(Vicioso & Beltrán) Font Quer & Rothmaler, Sched. Fl. Iber. Select., Cent. II-III,* n. 174 (1935); *cf. C. Navarro, M. Gutiérrez-Bustillo & A.G. Bueno in Fontqueria,* 7: 13 (1985): Armeria caespitosa var. isernii.
longearistata *(Boiss. & Reuter) Font Quer & Rothmaler, Sched. Fl. Iber. Select., Cent. II-III,* n. 169 (1935); *cf. C. Navarro, M. Gutiérrez-Bustillo & A.G. Bueno in Fontqueria,* 7: 13 (1985): Armeria longearistata.
macrophylla *(Boiss. & Reuter) Font Quer & Rothmaler, Sched. Fl. Iber. Select., Cent. II-III,* n. 167 (1935); *cf. C. Navarro, M. Gutiérrez-Bustillo & A.G. Bueno in Fontqueria,* 7: 13 (1985): Armeria macrophylla.

STAUROGYNE (Acanthac.).
leptocaulis
subsp. decumbens *R.M. Barker in J. Adelaide Bot. Gard.,* 9: 62 (1986)—Australia (W. Australia, N. Terr., Queensland).

STELIS (Orchidac.).
sect. Nexipous *(Garay) C.A. Luer, Ic. Pleurothallid. I-Syst. Pleurothallidinae (Monogr. Syst. Bot. Missouri Bot. Gard. 15):* 62 (1986): S. subgen. Nexipous.

STELLARIA (Caryophyllac.).
bistylata *W.Z. Di & Y. Ren in Acta Bot. Bor.-Occid. Sin.,* 5(3): 231 (1985)—China.
longipes
var. monantha *(Hulten) S.L. Welsh in Great Basin Nat.,* 46(2): 255 (1986): S. monantha.
palustris
subsp. fennica *(Murb.) V.G. Sergienko, Fl. poluostrova Kanin:* 56 (1986): S. palustris var. fennica.
undulata *C.P. Thunberg ex A. Murray in Linnaeus, Syst. Veg., ed. 14:* 422 (1784)—Japan. †

STELLERA (Thymelaeac.).
chamaejasme
f. chrysantha *S.C. Huang in Acta Bot. Yunnanica,* 7(3): 291 (1985)—China, Xizang.

STEMONA (Stemonac.).
angusta *I.R.H. Telford in Fl. Australia,* 46: 230, 180 (1986)—Australia (Queensland).
prostrata *I.R.H. Telford in Fl. Australia,* 46: 230, 178 (1986)—Australia (N. Terr.).

STENORRHYNCHOS (Orchidac.).
aphyllus *(Hook.) Sweet, Hort. Brit., ed. 2:* 485 (1830); *cf. D.J. Mabberley in Taxon,* 30(1): 17 (1981): Basionym not stated.
petensis *(L.O. Wms.) P. Burns-Balogh & Greenwood in Orquídea (México),* 10(1): 93 (1986): Spiranthes petensis.
seminudum *(Schltr.) P. Burns-Balogh in Orquídea (México),* 10(1): 93 (1986): Spiranthes seminuda.

STENOSPERMATION (Arac.).
ammiticum
subsp. neblinae *G.S. Bunting in Phytologia,* 60(5): 339 (1986)—Venezuela.

STEPHANIA (Menispermac.).
brevipedunculata *C.Y. Wu & D.D. Tao in Fl. Xizangica,* 2: 159 (1985)—Xizang.
dicentrinifera *H.S. Lo & M. Yang in Fl. Yunnanica,* 3: 251 (1983)—China.
intermedia *H.S. Lo in Fl. Yunnanica,* 3: 247 (1983)—China.
longipes *H.S. Lo in Fl. Yunnanica,* 3: 254 (1983)—China.
viridiflavens *H.S. Lo & M. Yang in Fl. Yunnanica,* 3: 248 (1983)—China.
yunnanensis *H.S. Lo in Fl. Yunnanica,* 3: 250 (1983)—China.

STEVIA (Compos.).
mollis *Schrad., Ind. Sem. Hort. Acad. Gott.,* 1831: 5 (1831); *Linnaea 8, Lit.:* 25 (1834); *cf. D.J. Mabberley in Taxon,* 32(1): 86 (1983)— Locality not stated.

STIEFIA (Labiat.).
bucharica *(M. Pop.) J. Soják in Čas. Nár. Muz. (Prague),* 152(1): 22 (1983): Salvia bucharica.
compressa *(Vent.) J. Soják in Čas. Nár. Muz. (Prague),* 152(1): 22 (1983): Salvia compressa.
hydrangea *(DC. ex Benth.) J. Soják in Čas. Nár. Muz. (Prague),* 152(1): 22 (1983): Salvia hydrangea.
korolkovii *(Reg. & Schmalh.) J. Soják in Čas. Nár. Muz. (Prague),* 152(1): 22 (1983): Salvia korolkovi.
molucellae *(Benth.) J. Soják in Čas. Nár. Muz. (Prague),* 152(1): 22 (1983): Salvia molucellae.
multicaulis *(Vahl) J. Soják in Čas. Nár. Muz. (Prague),* 152(1): 22 (1983): Salvia multicaulis.

STIGMAPHYLLON (Malpighiac.).
aberrans *C. Anderson in Syst. Bot.,* 11(1): 121 (1986)—Peru.
adenophorum *C. Anderson in Syst. Bot.,* 11(1): 120 (1986)—Costa Rica.
floribundum *(DC.) C. Anderson in Syst. Bot.,* 11(1): 128 (1986): Banisteria floribunda.
florosum *C. Anderson in Syst. Bot.,* 11(1): 123 (1986)—Peru, Ecuador, Brazil, Bolivia.
singulare *C. Anderson in Syst. Bot.,* 11(1): 126 (1986)—Venezuela, Colombia.

STIPA (Gramin.).
hirticulmis *S.L. Hatch, J. Valdés R. & C.W. Morden in Syst. Bot.,* 11(1): 186 (1986)—Mexico.
majalis
var. setulosissima *(Klok.) N.N. Tsvelev in Byull. Mosk. Obshch. Ispȳt. Prir., Biol.,* 91(1): 121 (1986): S. setulosissima.
pulcherrima
var. karadagensis *N.N. Tsvelev in Byull. Mosk. Obshch. Ispȳt. Prir., Biol.,* 91(1): 121 (1986), nom. nov.: S. heterophylla *Klok.*
sicula *B. Moraldo, V. La Valva, M. Ricciardi & G. Caputo in Delpinoa,* 23–24: 139 (1981–1982 publ. 1985)—Sicily.
zalesskii
var. glabrata *(P. Smirn.) N.N. Tsvelev in Byull. Mosk. Obshch. Ispȳt. Prir., Biol.,* 91(1): 120 (1986): S. rubens proles glabrata.
var. maeotica *(Klok.) N.N. Tsvelev in Byull. Mosk. Obshch. Ispȳt. Prir., Biol.,* 91(1): 121 (1986): S. maeotica.

STREPTOCARPUS (Gesneriac.).
arcuatus *O.M. Hilliard & B.L. Burtt in Notes Roy. Bot. Gard. Edinburgh,* 43(2): 229 (1986)—Malawi.
dolichanthus *O.M. Hilliard & B.L. Burtt in Notes Roy. Bot. Gard. Edinburgh,* 43(2): 230 (1986)—Malawi.
eylesii
subsp. silvicola *O.M. Hilliard & B.L. Burtt in Notes Roy. Bot. Gard. Edinburgh,* 43(2): 231 (1986)—Malawi.

STREPTOLIRION (Commelinac.).
volubile
subsp. subalpinum *C.Y. Wu in Fl. Yunnanica,* 3: 683 (1983)—China.

STROBILANTHES (Acanthac.).
guangxiensis *S.Z. Huang in Guihaia,* 6(3): 179 (1986)—China.

STROMBOCACTUS (Cactac.).
disciformis
var. seidelii *(Fric) Crkal, Lovec Kaktusu:* 401 (1983); *cf. Repert. Pl. Succ. (I.O.S.),* 34: 8 (1985): S. turbiniformis var. seidelii.

STRUTHIOLA (Thymelaeac.).
angustiloba *Peterson & O.M. Hilliard in Notes Roy. Bot. Gard. Edinburgh,* 43(2): 219 (1986)—South Africa (Natal).

STYLOCHITON (Arac.).
bogneri *S.J. Mayo in Fl. Trop. E. Afr., Arac.:* 53 (1985)—Tanzania, Kenya.
milneanus *S.J. Mayo in Fl. Trop. E. Afr., Arac.:* 47 (1985)—Tanzania.

STYPHELIA (Epacridac.).
violaceospicata *(Guillaumin) G. McPherson in Bull. Mus. Nation. Hist. Nat., B, Adansonia, Ser. 4,* 7(4): 453 (1985 publ. 1986): Leucopogon violaceo-spicatus.

STYRAX (Styracac.).
japonica
var. longipedunculata *Z.Y. Zhang in Fl. Tsinlingensis,* 1(4): 395, 63 (1983)—China.
var. nervillosa *Z.Y. Zhang in Fl. Tsinlingensis,* 1(4): 395, 63 (1983)—China.
macrosperma *C.Y. Wu in Fl. Yunnanica,* 3: 427 (1983)—China.
mallotifolia *C.Y. Wu in Fl. Yunnanica,* 3: 420 (1983)—China.

SUCCISA (Dipsacac.).
lacerifolia *(Hayata) R. Verlaque in Rev. Cytol. Biol. Vég. Bot.,* 9(1): 13 (1986), without basionym ref.: Scabiosa lacerifolia.
pratensis
subsp. hirsuta *(Opiz) J. Chrtek in Čas. Nár. Muz. (Prague),* 152(2): 100 (1983): S. hirsuta.

SUCCISA :-
subsp. scotiaca *(Baksay) J. Chrtek in Čas. Nár. Muz. (Prague),* 152(2): 100 (1983): S. pratensis var. scotiaca.

SULCOREBUTIA (Cactac.).
kruegeri
var. hoffmanniana *(Backbg.) J.D. Donald in Succulenta,* 65(10): 208 (1986): Lobivia hoffmanniana.
menesesii
var. kamiensis *A.J. Brederoo & J.D. Donald in Succulenta,* 65(8): 155 (1986)—Bolivia.
var. muschii *(Vasq.) J.D. Donald in Succulenta,* 65(10): 208 (1986): S. muschii.
steinbachii
var. australis *W. Rausch in Succulenta,* 65(11): 240 (1986)—Bolivia.
vizcarrae
var. laui *A.J. Brederoo & J.D. Donald in Succulenta,* 65(3): 52 (1986)—Bolivia.

SUTERA (Scrophulariac.).
beverlyana *O.M. Hilliard & B.L. Burtt in Notes Roy. Bot. Gard. Edinburgh,* 43(2): 216 (1986)—Lesotho.
silenioides *O.M. Hilliard in Notes Roy. Bot. Gard. Edinburgh,* 43(2): 216 (1986)—South Africa (Natal).

SWERTIA (Gentianac.).
dichotoma
var. punctata *T.N. He & J.X. Yang in Fl. Tsinlingensis,* 1(4): 397, 126 (1983), as 'punctatum'—China.
elongata *T.N. He & S.W. Liu ex J.X. Yang in Fl. Tsinlingensis,* 1(4): 397, 124 (1983)—China.

SWIETENIA (Meliac.).
febrifuga *Roxb., Bot. Descr. Swietenia:* 1 (1793); *cf. D.J. Mabberley in Taxon,* 31(1): 66 (1982)— India.

SYMPHYTUM (Boraginac.).
davisii
subsp. cycladense *(Pawl.) W.T. Stearn in Ann. Mus. Goulandris,* 7: 196 (1986): S. cycladense.
subsp. icaricum *(Pawl.) W.T. Stearn in Ann. Mus. Goulandris,* 7: 198 (1986): S. icaricum.
subsp. naxicola *(Pawl.) W.T. Stearn in Ann. Mus. Goulandris,* 7: 194 (1986): S. naxicola.

SYMPLOCARPUS (Arac.).
nipponicus
f. viridispathus *J. Ōhara in J. Phytogeogr. Taxon.,* 33(2): 81 (1985)—Japan.

SYMPLOCOS (Symplocac.).
sect. Glomeratae *Y.F. Wu in Acta Phytotax. Sin.,* 24(3): 193 (1986).
sect. Singuliflorae *Y.F. Wu in Acta Phytotax. Sin.,* 24(3): 193 (1986).

SYMPLOCOS :-
adenopus
var. vestita *Huang & Y.F. Wu in Acta Phytotax. Sin.,* 24(3): 202 (1986)—China.
azuensis *D.H. Mai in Feddes Repert.,* 97(1–2): 13 (1986)—Hispaniola (Dominican Republic).
cochinchinensis
var. puberula *Huang & Y.F. Wu in Acta Phytotax. Sin.,* 24(3): 202 (1986)—China.
costatifructa *H.P. Nooteboom in Blumea,* 31(2): 277 (1986)—Borneo (Sarawak, Brunei, Sabah).
dolichostylosa *Y.F. Wu in Acta Phytotax. Sin.,* 24(3): 195 (1986)—China.
hookeri
var. tomentosa *Y.F. Wu in Acta Phytotax. Sin.,* 24(3): 202 (1986)—China.
iliaspaiensis *H.P. Nooteboom in Blumea,* 31(2): 279 (1986)—Borneo (Sarawak, ?Kalimantan).
martinicensis
subsp. jamaicensis *(Kr. & Urb.) D.H. Mai in Feddes Repert.,* 97(1–2): 15 (1986): S. jamaicensis.
subsp. strigillosa *(Kr. & Urb.) D.H. Mai in Feddes Repert.,* 97(1–2): 15 (1986): S. strigillosa.
oblanceolata *Y.F. Wu in Acta Phytotax. Sin.,* 24(3): 198 (1986)—China.
ovatibracteata *Y.F. Wu in Acta Phytotax. Sin.,* 24(3): 195 (1986)—China.
persistens *Huang & Y.F. Wu in Acta Phytotax. Sin.,* 24(3): 199 (1986)—China.
rachitricha *Y.F. Wu in Acta Phytotax. Sin.,* 24(3): 196 (1986)—China.
tetragona *Chen ex Y.F. Wu in Acta Phytotax. Sin.,* 24(3): 194 (1986)—China.
wenshanensis *Huang & Y.F. Wu in Acta Phytotax. Sin.,* 24(3): 199 (1986)—China.
wuliangshanensis *Huang & Y.F. Wu in Acta Phytotax. Sin.,* 24(3): 199 (1986)—China.
yizhangensis *Y.F. Wu in Acta Phytotax. Sin.,* 24(3): 200 (1986)—China.

SYNIMA (Sapindac.).
macrophylla *S.T. Reynolds in Fl. Australia,* 25: 202, 82 (1985)—Australia (Queensland).

SYNOUM A. Jussieu in Mirb. & Cass. apud Guill. in Bull. Sci. Nat. Geol., 23: 237 (1830); cf. D.J. Mabberley in Taxon, 31(1): 66 (1982). *MELIACEAE.*

SYRINGA (Oleac.).
suspensa *C.P. Thunberg ex A. Murray in Linnaeus, Syst. Veg., ed. 14:* 57 (1784)—Japan. †

SYZYGIUM (Myrtac.).
jainii *K. Haridasan & R.R. Rao, Forest Fl. Meghalaya:* 398 (1985)—India.
nervosum
var. obovatum *(Kurz) A. Kumar in J. Econ. Taxon. Bot.,* 7(3): 664 (1985 publ. 1986): Eugenia operculata var. obovata.

SYZYGIUM :-
praecox *(Roxb.) K. Haridasan & R.R. Rao, Forest Fl. Meghalaya:* 401 (1985): Eugenia praecox.

T

TABEBUIA (Bignoniac.).
gemmiflora *C.T. Rizzini & A. de Mattos-Filho in Rev. Brasil. Biol.,* 46(2): 320 (1986)—Brazil.

TABERNAEMONTANA (Apocynac.).
celastroides *(Kerr) P.T. Li in Guihaia,* 6(3): 173 (1986): Ervatamia celastroides.
ceratocarpa *(Kerr) P.T. Li in Guihaia,* 6(3): 173 (1986): Ervatamia ceratocarpa.
chengkiangensis *(Tsiang) P.T. Li in Guihaia,* 6(3): 169 (1986): Ervatamia chengkiangensis.
continentalis *(Tsiang) P.T. Li in Guihaia,* 6(3): 170 (1986): Ervatamia continentalis.
var. pubiflora *(Tsiang) P.T. Li in Guihaia,* 6(3): 170 (1986): Ervatamia continentalis var. pubiflora.
elliptica *C.P. Thunberg ex A. Murray in Linnaeus, Syst. Veg., ed. 14:* 255 (1784)—Japan. †
flabelliformis *(Tsiang) P.T. Li in Guihaia,* 6(3): 170 (1986): Ervatamia flabelliformis.
guangdongensis *P.T. Li in Guihaia,* 6(3): 169 (1986), nom. nov.: Ervatamia puberula *Tsiang & P.T. Li.*
hainanensis *(Tsiang) P.T. Li in Guihaia,* 6(3): 169 (1986): Ervatamia hainanensis.
kwangsiensis *(Tsiang) P.T. Li in Guihaia,* 6(3): 171 (1986): Ervatamia kwangsiensis.
kweichowensis *(Tsiang) P.T. Li in Guihaia,* 6(3): 171 (1986): Ervatamia kweichowensis.
latifolia *(Michx.) Parmentier, Cat. Pl. Enghien:* 77 (1818); *cf. D.J. Mabberley in Taxon,* 33(3): 443 (1984): Basionym not stated.
officinalis *(Tsiang) P.T. Li in Guihaia,* 6(3): 172 (1986): Ervatamia officinalis.
peninsularis *(Kerr) P.T. Li in Guihaia,* 6(3): 173 (1986): Pagiantha peninsularis.
rotensis *(Kanehira) P.T. Li in Guihaia,* 6(3): 172 (1986): Ervatamia rotensis.
thailandensis *P.T. Li in Guihaia,* 6(3): 172 (1986), nom. nov.: Ervatamia calcicola *Kerr.*
tsiangiana *P.T. Li in Guihaia,* 6(3): 171 (1986), nom. nov.: Ervatamia tenuiflora *Tsiang.*
yunnanensis *(Tsiang) P.T. Li in Guihaia,* 6(3): 171 (1986): Ervatamia yunnanensis.
var. heterosepala *(Tsiang) P.T. Li in Guihaia,* 6(3): 171 (1986): Ervatamia yunnanensis var. heterosepala.

TAINIA (Orchidac.).
ovalifolia *(Ames & Schweinf.) L.A. Garay & W. Kittredge in Bot. Mus. Leafl. Harvard Univ.,* 30(3): (59) 193 (1985 publ. 1986): Eulophia ovalifolia.

TANACETUM (Compos.).
sect. Afghanica *D. Podlech in Fl. Iran.,* 158: 124 (1986).
sect. Ajania *(Poljak.) D. Podlech in Fl. Iran.,* 158: 125 (1986): gen. Ajania.
sect. Hemipappus *(C. Koch) D. Podlech in Fl. Iran.,* 158: 135 (1986): gen. Hemipappus.
sect. Hippolytia *(Poljak.) D. Podlech in Fl. Iran.,* 158: 132 (1986): gen. Hippolytia.
sect. Lepidolopsis *(Poljak.) D. Podlech in Fl. Iran.,* 158: 135 (1986): gen. Lepidolopsis.
sect. Paleacea *D. Podlech in Fl. Iran.,* 158: 142 (1986).
sect. Paradoxa *D. Podlech in Fl. Iran.,* 158: 143 (1986).
sect. Parthenium *(Briq.) D. Podlech in Fl. Iran.,* 158: 98 (1986): T. subsect. Parthenium.
sect. Pyrethrellum *(Tzvel.) D. Podlech in Fl. Iran.,* 158: 100 (1986): Pyrethrum sect. Pyrethrellum.
sect. Richteria *(Kar. & Kir.) D. Podlech in Fl. Iran.,* 158: 106 (1986): gen. Richteria.
sect. Tanacetopsis *(Tzvel.) D. Podlech in Fl. Iran.,* 158: 127 (1986): Cancrinia sect. Tanacetopsis.
sect. Xylanthemum *(Tzvel.) D. Podlech in Fl. Iran.,* 158: 137 (1986): gen. Xylanthemum.
sect. Xylopyrethrum *(Tzvel.) D. Podlech in Fl. Iran.,* 158: 102 (1986): Pyrethrum sect. Xylopyrethrum.
afghanicum *(Gilli) D. Podlech in Fl. Iran.,* 158: 127 (1986): Chrysanthemum afghanicum.
archibaldii *D. Podlech in Fl. Iran.,* 158: 105 (1986)—Iran.
bamianicum *D. Podlech in Fl. Iran.,* 158: 123 (1986)—Afghanistan.
crassicollum *(Rech. f.) D. Podlech in Fl. Iran.,* 158: 133 (1986): Chrysanthemum crassicollum.
djilgense *(Franch.) D. Podlech in Fl. Iran.,* 158: 107 (1986): Chrysanthemum djilgense.
doabense *D. Podlech in Fl. Iran.,* 158: 131 (1986)—Afghanistan.
eriobasis
subsp. tricholepis *D. Podlech in Fl. Iran.,* 158: 130 (1986)—Afghanistan.
freitagii *D. Podlech in Fl. Iran.,* 158: 132 (1986)—Afghanistan.
ghoratense *D. Podlech in Fl. Iran.,* 158: 125 (1986)—Afghanistan.
gillettii *D. Podlech in Fl. Iran.,* 158: 140 (1986)—Iraq.
hedgei *D. Podlech in Fl. Iran.,* 158: 130 (1986)—Afghanistan.
hololeucum *(Bornm.) D. Podlech in Fl. Iran.,* 158: 105 (1986): Pyrethrum hololeucum.
maymanense *D. Podlech in Fl. Iran.,* 158: 124 (1986)—Afghanistan.

TANACETUM :–
nuristanicum *D. Podlech in Fl. Iran.*, 158: 146 (1986)—Afghanistan.
paghmanense *D. Podlech in Fl. Iran.*, 158: 141 (1986)—Afghanistan.
pakistanicum *D. Podlech in Fl. Iran.*, 158: 101 (1986)—Pakistan.
paleaceum *D. Podlech in Fl. Iran.*, 158: 143 (1986)—Afghanistan.
polycephalum
subsp. argyrophyllum *(C. Koch) D. Podlech in Fl. Iran.*, 158: 117 (1986): Gymnocline argyrophylla.
subsp. azerbaidjanicum *D. Podlech in Fl. Iran.*, 158: 119 (1986)—Iran, Iraq.
subsp. duderanum *(Boiss.) D. Podlech in Fl. Iran.*, 158: 120 (1986): Pyrethrum duderanum.
subsp. farsicum *D. Podlech in Fl. Iran.*, 158: 118 (1986)—Iran.
subsp. heterophyllum *(Boiss.) D. Podlech in Fl. Iran.*, 158: 117 (1986): T. heterophyllum.
subsp. junesarense *(Bornm.) D. Podlech in Fl. Iran.*, 158: 119 (1986): Chrysanthemum junesarense.
salsugineum *D. Podlech in Fl. Iran.*, 158: 147 (1986)—Iran.
silvicola *D. Podlech in Fl. Iran.*, 158: 144 (1986)—Afghanistan.
stapfianum *(Rech. f.) D. Podlech in Fl. Iran.*, 158: 112 (1986): Chrysanthemum stapfianum.
tenuisectum *(Boiss.) D. Podlech in Fl. Iran.*, 158: 103 (1986): Pyrethrum tenuisectum.
tirinense *D. Podlech in Fl. Iran.*, 158: 145 (1986)—Afghanistan.
trifoliolatum *D. Podlech in Fl. Iran.*, 158: 147 (1986)—Iran.

TANQUANA H. Hartmann & S. Liede in Bot. Jahrb., 106(4): 479 (1986). *AIZOACEAE.*
archeri *(L. Bolus) H. Hartmann & S. Liede in Bot. Jahrb.*, 106(4): 480 (1986): Pleiospilos archeri.
hilmarii *(L. Bolus) H. Hartmann & S. Liede in Bot. Jahrb.*, 106(4): 481 (1986): Pleiospilos hilmarii.
prismatica *(Schwantes) H. Hartmann & S. Liede in Bot. Jahrb.*, 106(4): 481 (1986), basionym nom. nud.: Mesembryanthemum prismaticum.

TAPIRIRA (Anacardiac.).
guianensis
subsp. subandina *A. Barfod & L.B. Holm-Nielsen in Nordic J. Bot.*, 6(4): 425 (1986)—Colombia, Ecuador.
velutinifolia *(Cowan) L. Marcano Berti in Pittieria,* 13: 23 (1986): Bursera velutinifolia.

TAPURA (Dichapetalac.).
carinata *F.J. Breteler in Agric. Univ. Wageningen Pap.*, 86(3): 53 (1986)—Gabon, Congo - Brazzaville.

TARAXACUM (Compos.).
bohemicum *J. Kirschner & J. Štěpánek in Preslia,* 58(2): 99 (1986)—Czechoslovakia.
cognatum *J. Kirschner & J. Štěpánek in Preslia,* 58(2): 106 (1986)—Czechoslovakia, Austria, Hungary.
discretum *H. øllgaard in Nordic J. Bot.*, 6(1): 21 (1986)—Denmark, Sweden, Great Britain, Netherlands, France, Germany, Poland, Austria, U.S.A. (Maine, Vermont, Connecticut).
domabile *J. Kirschner & J. Štěpánek in Preslia,* 58(2): 102 (1986)—Czechoslovakia, Hungary.
multilepis *J. Kirschner & J. Štěpánek in Preslia,* 58(2): 114 (1986)—France, Germany, Switzerland.
subolivaceum *C.E. Sonck in Ann. Bot. Fenn.*, 23(2): 165 (1986)—Greece.
telmatophilum *J. Kirschner & J. Štěpánek in Preslia,* 58(2): 104 (1986)—Czechoslovakia, Hungary.
vexatum *C.E. Sonck in Ann. Bot. Fenn.*, 23(2): 167 (1986), nom. nov.: T. lacistophylloides *Sonck.*

TAVERNIERA (Legumin.).
albida *M. Thulin in Symb. Bot. Upsal.*, 25(1): 89 (1985)—South Yemen.
brevialata *M. Thulin in Symb. Bot. Upsal.*, 25(1): 82 (1985)—Oman.
longisetosa *M. Thulin in Symb. Bot. Upsal.*, 25(1): 85 (1985)—Somali Republic.
multinoda *M. Thulin in Symb. Bot. Upsal.*, 25(1): 83 (1985)—Somali Republic, Oman, ?South Yemen.
oligantha *(Franch.) M. Thulin in Symb. Bot. Upsal.*, 25(1): 75 (1985): T. schimperi var. oligantha.

TAXILLUS (Loranthac.).
caloreas
var. fargesii *(Lecte.) H.S. Kiu in Fl. Yunnanica,* 3: 368 (1983): Loranthus caloreas var. fargesii.
limprichtii *(Grüning) H.S. Kiu in Fl. Yunnanica,* 3: 368 (1983): Loranthus limprichtii.
var. longiflorus *(Lecte.) H.S. Kiu in Fl. Yunnanica,* 3: 368 (1983): Loranthus estipitatus var. longiflorus.
sutchuenensis
var. duclouxii *(Lecte.) H.S. Kiu in Fl. Yunnanica,* 3: 369 (1983): Loranthus duclouxii.

TAXUS (Taxac.).
macrophylla *C.P. Thunberg ex A. Murray in Linnaeus, Syst. Veg., ed. 14:* 895 (1784)—Japan. †
verticillata *C.P. Thunberg ex A. Murray in Linnaeus, Syst. Veg., ed. 14:* 895 (1784)—Japan. †

TECLEA (Rutac.).
borenensis *M.G. Gilbert in Nordic J. Bot.*, 6(2): 141 (1986)—Kenya, Ethiopia.

TECOMA (Bignoniac.).
stans
var. mollis *(H.B.K.) M.A. Siddiqi in Fl. Libya,* 103: 6 (1983): T. mollis.

TELANTHERA (Amaranthac.).
brasiliensis *(L. mut. Jacq.) Endl., Cat. Horti Vindob.*, 1: 265 (1842); *cf. D.J. Mabberley in Taxon,* 33(3): 436 (1984): Basionym not stated.
porrigens *(Jacq.) Endl., Cat. Horti Vindob.*, 1: 265 (1842); *cf. D.J. Mabberley in Taxon,* 33(3): 436 (1984): Basionym not stated.

TEPHROSIA (Legumin.).
albifoliolis *A. Nongonierma & T. Sarr in Bull. Inst. Fond. Afr. Noire, A,* 44(3–4): 249 (1982 publ. 1986), without type—Niger, Senegal, Sudan.
pondoensis *(Codd) B.D. Schrire in Bothalia,* 15(3–4): 552 (1985): Mundulea pondoensis.
tricolor *(Hook.) Sweet, Hort. Brit., ed. 2:* 142 (1830); *cf. D.J. Mabberley in Taxon,* 31(1): 72 (1982): Basionym not stated.

TETRACHYRON (Compos.).
oaxacana *B.L. Turner in Phytologia,* 60(4): 251 (1986)—Mexico.

TETRACOILANTHUS F. Rappa & V. Camarrone in Lav. Ist. Bot. Giard. Colon. Palermo, 14: 66 (1954). *AIZOACEAE.*
cordifolius *(?) F. Rappa & V. Camarrone in Lav. Ist. Bot. Giard. Colon. Palermo,* 14: 64, table opp. p. 62 (1954), without basionym ref.: Aptenia cordifolia.

×**TETRADIACRIUM** hort. in Orchid Rev., 94(1117): centre page pull-out p. 8 (1986). *ORCHIDACEAE.* (Diacrium × Tetramicra).

TETRAGONIA (Aizoac.).
japonica *C.P. Thunberg ex A. Murray in Linnaeus, Syst. Veg., ed. 14:* 467 (1784)—Japan. †

TETRAMERIUM (Acanthac.).
sect. Siphonanthus *T.F. Daniel, Syst. Tetramerium (Acanthac.) (Syst. Bot. Monog., 12):* 119 (1986).
sect. Torreyella *Happ ex T.F. Daniel, Syst. Tetramerium (Acanthac.) (Syst. Bot. Monog., 12):* 109 (1986).
abditum *(T.S. Brandegee) T.F. Daniel, Syst. Tetramerium (Acanthac.) (Syst. Bot. Monog., 12):* 113 (1986): Anisacanthus abditus.
angustius *(Nees) T.F. Daniel, Syst. Tetramerium (Acanthac.) (Syst. Bot. Monog., 12):* 83 (1986): Blechum angustius.
butterwickianum *T.F. Daniel, Syst. Tetramerium (Acanthac.) (Syst. Bot. Monog., 12):* 89 (1986)—Mexico.

TETRAMERIUM :-
crenatum *T.F. Daniel, Syst. Tetramerium (Acanthac.) (Syst. Bot. Monog., 12):* 85 (1986)—Mexico.
denudatum *T.F. Daniel, Syst. Tetramerium (Acanthac.) (Syst. Bot. Monog., 12):* 42 (1986)—Peru.
emilyanum *T.F. Daniel, Syst. Tetramerium (Acanthac.) (Syst. Bot. Monog., 12):* 74 (1986)—Mexico.
guerrerense *T.F. Daniel, Syst. Tetramerium (Acanthac.) (Syst. Bot. Monog., 12):* 101 (1986)—Mexico.
macvaughii *T.F. Daniel, Syst. Tetramerium (Acanthac.) (Syst. Bot. Monog., 12):* 92 (1986), as 'mcvaughii'—Mexico.
oaxacanum *T.F. Daniel, Syst. Tetramerium (Acanthac.) (Syst. Bot. Monog., 12):* 77 (1986)—Mexico.
obovatum *T.F. Daniel, Syst. Tetramerium (Acanthac.) (Syst. Bot. Monog., 12):* 78 (1986)—Mexico.
ochoterenae *(Miranda) T.F. Daniel, Syst. Tetramerium (Acanthac.) (Syst. Bot. Monog., 12):* 109 (1986): Anisacanthus ochoterenae.
peruvianum *(Lindau) T.F. Daniel, Syst. Tetramerium (Acanthac.) (Syst. Bot. Monog., 12):* 119 (1986): Siphonoglossa peruviana.
rzedowskii *T.F. Daniel, Syst. Tetramerium (Acanthac.) (Syst. Bot. Monog., 12):* 104 (1986)—Mexico.
sagasteguianum *T.F. Daniel, Syst. Tetramerium (Acanthac.) (Syst. Bot. Monog., 12):* 110 (1986)—Peru.
surcubambense *T.F. Daniel, Syst. Tetramerium (Acanthac.) (Syst. Bot. Monog., 12):* 44 (1986)—Peru.
tetramerioides *(Lindau) T.F. Daniel, Syst. Tetramerium (Acanthac.) (Syst. Bot. Monog., 12):* 82 (1986): Cardiacanthus tetramerioides.
wasshausenii *T.F. Daniel, Syst. Tetramerium (Acanthac.) (Syst. Bot. Monog., 12):* 71 (1986)—Peru, Bolivia.
yaquianum *T.F. Daniel, Syst. Tetramerium (Acanthac.) (Syst. Bot. Monog., 12):* 102 (1986)—Mexico.
zeta *T.F. Daniel, Syst. Tetramerium (Acanthac.) (Syst. Bot. Monog., 12):* 66 (1986)—Peru.

TETRASTIGMA (Vitac.).
obtectum
subsp. dichotomum *W.T. Wang in Bull. Bot. Res. North-East. Forest. Inst.*, 6(4): 22 (1986)—China.
pycnanthum *(Collett & Hemsley) P. Singh & B.V. Shetty in Taxon,* 35(3): 597 (1986): Vitis pycnantha.
yunnanense
var. glabrum *W.T. Wang in Bull. Bot. Res. North-East. Forest. Inst.*, 6(4): 23 (1986)—China.

TEUCRIUM (Labiat.).
cylindraceum *W. Greuter & Burdet in Willdenowia,* 15(2): 423 (1986), nom. nov.: T. polium var. cylindricum *Maire ex Batt.*
luteum
subsp. gabesianum *(Puech) Greuter & Burdet ex W. Greuter in Willdenowia,* 15(2): 423 (1986): T. aureum subsp. gabesianum.
rotundifolium
subsp. transatlanticum *Emberger ex W. Greuter, H.M. Burdet & G. Long (eds.), Med-checklist,* 3: xxix (1986): Morocco.

THALICTRUM (Ranunculac.).
flavum
subsp. simplex *(L.) O. de Bolòs & J. Vigo, Fl. Països Catalans:* 243 (1984): T. simplex.
glaucum *Schrad., Hort. Gott.:* 14, t. 8 (1809); *cf. D.J. Mabberley in Taxon,* 33(3): 443 (1984)— Locality not stated.
minus
subsp. olympicum *(Boiss. & Heldr.) A. Strid in A. Strid (ed.), Mount. Fl. Greece,* 1: 229 (1986): T. olympicum.

THAMNOCHORTUS (Restionac.).
arenarius *E.E. Esterhuysen in Bothalia,* 15(3–4): 472 (1985)—South Africa (Cape Province).
cinereus *H.P. Linder in Bothalia,* 15(3–4): 473 (1985)—South Africa (Cape Province).
lucens *(Poir.) H.P. Linder in Bothalia,* 15(3–4): 475 (1985): Restio lucens.
rigidus *E.E. Esterhuysen in Bothalia,* 15(3–4): 477 (1985)—South Africa (Cape Province).

THELOCACTUS (Cactac.).
lausseri *J. Riha & J. Busek in Kakt. And. Sukk.,* 37(8): 164 (1986)—Mexico.

THELYPODIOPSIS (Crucif.).
sagittata
var. ovalifolia *(Rydb.) S.L. Welsh in Great Basin Nat.,* 46(2): 256 (1986): Thelypodium ovalifolium.

THEMISTOCLESIA (Ericac.).
pariensis *J.L. Luteyn in Bull. Torrey Bot. Club,* 112(4): 449 (1985)—Venezuela.

THERMOPSIS (Legumin.).
barbata
f. chrysantha *P.C. Li in Fl. Xizangica,* 2: 725 (1985)—Xizang.
macrophylla
subsp. semota *(Jepson) R.M. Beauchamp in Phytologia,* 59(7): 439 (1986): T. macrophylla var. semota.
mollis *M.A. Curtis ex A. Gray, Thermopsis mollis:* (Feb. 1846) *et in Mem. Acad. Arts & Sci. n.s.,* 3: 47, t. 9 (after May 1846); *cf. D.J. Mabberley in Taxon,* 30(1): 17 (1981)—Locality not stated.

THEROCISTUS J. Holub in Preslia, 58(4): 300 (1986). *CISTACEAE.*
acuminatus *(Viv.) J. Holub in Preslia,* 58(4): 303 (1986): Cistus acuminatus.
brevipes *(Boiss. & Reut.) J. Holub in Preslia,* 58(4): 303 (1986): Helianthemum brevipes.
bupleurifolius *(Lam.) J. Holub in Preslia,* 58(4): 303 (1986): Cistus bupleurifolius.
echioides *(Lam.) J. Holub in Preslia,* 58(4): 303 (1986): Cistus echioides.
glomeratus *(Willk.) J. Holub in Preslia,* 58(4): 303 (1986): Tuberaria glomerata.
guttatus *(L.) J. Holub in Preslia,* 58(4): 303 (1986): Cistus guttatus.
subsp. littoralis *(Rouy & Fouc.) J. Holub in Preslia,* 58(4): 303 (1986): Helianthemum guttatum f. littorale.
inconspicuus *(Thibaud ex Pers.) J. Holub in Preslia,* 58(4): 303 (1986): Helianthemum inconspicuum.
lipopetalus *(Murb.) J. Holub in Preslia,* 58(4): 303 (1986): Helianthemum guttatum subsp. lipopetalum.
macrosepalus *(Cosson) J. Holub in Preslia,* 58(4): 303 (1986): Helianthemum guttatum var. macrosepalum.
praecox *(W. Grosser) J. Holub in Preslia,* 58(4): 303 (1986): Tuberaria praecox.
villosissimus *(Pomel) J. Holub in Preslia,* 58(4): 303 (1986): Helianthemum villosissimum.

THIBAUDIA (Ericac.).
setigera *(D. Don ex G. Don f.) Endl., Cat. Horti Vindob.,* 2: 135 (1842); *cf. D.J. Mabberley in Taxon,* 33(3): 436 (1984): Agapetes setigera.

THISMIA (Burmanniac.).
sect. Myostoma *(Miers) P.J.M. Maas & H. Maas-van de Kamer in Fl. Neotrop.,* 42: 158 (1986): gen. Myostoma.
subgen. Ophiomeris *(Miers) P.J.M. Maas & H. Maas-van de Kamer in Fl. Neotrop.,* 42: 144 (1986): gen. Ophiomeris.
sect. Ophiomeris *(Miers) P.J.M. Maas & H. Maas-van de Kamer in Fl. Neotropica,* 42: 145 (1986): gen. Ophiomeris.
sect. Pyramidalis *P.J.M. Maas & H. Maas-van de Kamer in Fl. Neotrop.,* 42: 161 (1986).
caudata *P.J.M. Maas & H. Maas-van de Kamer in Fl. Neotrop.,* 42: 162 (1986), nom. nov.: Glaziocharis macahensis *Taubert ex Warming.*
fungiformis *(Taubert ex Warming) P.J.M. Maas & H. Maas-van de Kamer in Fl. Neotrop.,* 42: 165 (1986): Triscyphus fungiformis.
singeri *(de la Sota) P.J.M. Maas & H. Maas-van de Kamer in Fl. Neotrop.,* 42: 166 (1986): Mamorea singeri.

THLASPI (Crucif.).
cariense *A. Carlström in Willdenowia,* 16(1): 73 (1986)—Turkey.

THLASPI :–
zaffranii *(F.K. Meyer) W. Greuter & Burdet in Willdenowia,* 15(2): 420 (1986): Noccaea zaffranii.

THORNCROFTIA (Labiat.).
media *L.E. Codd in Bothalia,* 16(1): 52 (1986)—South Africa (Transvaal).

THOTTEA (Aristolochiac.).
duchartrei *V.V. Sivarajan, A. Babu & Indu Balachandran in Indian J. Forest.,* 8(4): 267 (1985 publ. 1986)—India.

THRINAX (Palmae).
sect. Macrocarpae *(León ex Muñiz) A. Borhidi & O. Muñiz in Acta Bot. Hung.,* 31(1–4): 226 (1985): gen. Macrocarpae.
compacta *(Griseb. & Wendl. ex Griseb.) A. Borhidi & O. Muñiz in Acta Bot. Hung.,* 31(1–4): 226 (1985): Trithrinax compacta.
ekmaniana *(Burret) A. Borhidi & O. Muñiz in Acta Bot. Hung.,* 31(1–4): 227 (1985): Hemithrinax ekmaniana.
rivularis *(León) A. Borhidi & O. Muñiz in Acta Bot. Hung.,* 31(1–4): 226 (1985): Hemithrinax rivularis.
var. savannarum *(León) A. Borhidi & O. Muñiz in Acta Bot. Hung.,* 31(1–4): 226 (1985): Hemithrinax savannarum.

THYMUS (Labiat.).
subsect. Anomali *(Rouy) R. Morales Valverde in Ruizia,* 3: 146 (1986): T. sect. Anomali.
subsect. Pseudothymbra *(Bentham) R. Morales Valverde in Ruizia,* 3: 146 (1986): T. sect. Pseudothymbra.
subsect. Thymastra *(Nyman ex Velen.) R. Morales Valverde in Ruizia,* 3: 146 (1986): T. sect. Thymastra.
praecox
subsp. caucasicus *(Ronniger) J. Jalas in Willdenowia,* 15(2): 423 (1986): T. caucasicus.
var. grossheimii *(Ronniger) J. Jalas in Willdenowia,* 15(2): 424 (1986): T. grossheimii.
subsp. jankae *(Čelak.) J. Jalas in Willdenowia,* 15(2): 423 (1986): T. jankae.

TILIA (Tiliac.).
argentea
f. subglobularis *Wagn. ex L.Zs. Vöröss in Acta Bot. Hung.,* 31(1–4): 173 (1985)—Hungary.
caucasica
f. seiuntobracteata *L.Zs. Vöröss in Acta Bot. Hung.,* 31(1–4): 179 (1985)—Hungary.
euchlora
f. pseudomultibracteata *L.Zs. Vöröss in Acta Bot. Hung.,* 31(1–4): 179 (1985), as 'pseudo-multibracteata'—Hungary.
grandifolia
f. praestabilis *Wagn. ex L.Zs. Vöröss in Acta Bot. Hung.,* 31(1–4): 175 (1985)—Cultivation.

TILIA :–
platyphyllos
f. grandibracteata *L.Zs. Vöröss in Acta Bot. Hung.,* 31(1–4): 175 (1985)—Romania.
f. pseudotrichogyna *L.Zs. Vöröss in Acta Bot. Hung.,* 31(1–4): 175 (1985), as 'pseudo-trichogyna'—Hungary.
f. sublaevigata *Wagn. ex L.Zs. Vöröss in Acta Bot. Hung.,* 31(1–4): 175 (1985)—Hungary.
×praelustris *Wagn. ex L.Zs. Vöröss in Acta Bot. Hung.,* 31(1–4): 179 (1985)—Hungary. (T.× jakabiana × petiolaris).
×pseudolongirostris *Wagn. ex L.Zs. Vöröss in Acta Bot. Hung.,* 31(1–4): 173 (1985)—Hungary. (T. argentea × petiolaris).
×pseudopulchra *Wagn. ex L.Zs. Vöröss in Acta Bot. Hung.,* 31(1–4): 175 (1985)—Cultivation. (T. argentea var. calvescens × petiolaris).

TILLANDSIA (Bromeliac.).
imperialis *Morren ex Roezl, Deut. Gärtn.-Zeit.,* 1881; *Belgique horticole,* 33: 133 (1883); *cf. D.J. Mabberley in Taxon,* 34(3): 455 (1985)— Locality not stated.
jaliscopinicola *L. Hromadnik & P. Schneider in Haussknechtia, Mitt. Thüring. Bot. Ges.,* 2: 43 (1985)—Mexico.
krukoffiana
var. piepenbringii *W. Rauh in J. Bromeliad Soc.,* 36(6): 259 (1986)—Peru.
maxima *Strangew., Sketch Mosq. Shore:* 138 (1822); *cf. D.J. Mabberley in Taxon,* 33(3): 444 (1984)— Locality not stated.
muhrae *W. Weber in Feddes Repert.,* 97(3–4): 101 (1986)—Argentina.
pamelae *W. Rauh in J. Bromeliad Soc.,* 36(3): 117 (1986)—Mexico.
pseudocardenasii *W. Weber in Feddes Repert.,* 97(3–4): 103 (1986)—Bolivia.
spiralipetala *E.J. Gouda in J. Bromeliad Soc.,* 36(4): 165 (1986)—Bolivia.

TINOSPORA (Menispermac.).
guangxiensis *H.S. Lo in Guihaia,* 6(1–2): 52 (1986)—China.
hainanensis *H.S. Lo & Z.X. Li in Guihaia,* 6(1–2): 51 (1986)—Hainan.
sagittata
var. yunnanensis *(S.Y. Hu) H.S. Lo in Fl. Yunnanica,* 3: 232 (1983): T. yunnanensis.

TITHYMALUS (Euphorbiac.).
duvalii *(Lecoq & Lamotte) J. Soják in Čas. Nár. Muz. (Prague),* 152(1): 22 (1983): Euphorbia duvali.
ericetorum *(Zumagl.) J. Soják in Čas. Nár. Muz. (Prague),* 149(3–4): 210 (1980 publ. 1981): Euphorbia ericetorum.
esula
subsp. riparia *(Schur) J. Chrtek & B. Křísa in Čas. Nár. Muz. (Prague),* 152(4): 226 (1983 publ. 1984): Euphorbia esula var. d riparia.

TITHYMALUS :–
subsp. tristis *(Bess. ex Bieb.) J. Chrtek & B. Křísa in Čas. Nár. Muz. (Prague),* 152(4): 226 (1983 publ. 1984): Euphorbia tristis.
flavicomus
subsp. costeanus *(Rouy) J. Soják in Čas. Nár. Muz. (Prague),* 152(1): 22 (1983): Euphorbia flavicoma Race E. costeana.
gulestanicus *(Podlech) J. Soják in Čas. Nár. Muz. (Prague),* 152(1): 22 (1983): Euphorbia gulestanica.
nicaeensis
subsp. cadrilateri *(Prod.) J. Soják in Čas. Nár. Muz. (Prague),* 152(1): 22 (1983): Euphorbia cadrilateri.
subsp. glareosus *(Pall. ex Marsch.-Bieb.) J. Soják in Čas. Nár. Muz. (Prague),* 152(1): 22 (1983): Euphorbia glareosa.
subsp. goldei *(Prokh.) J. Soják in Čas. Nár. Muz. (Prague),* 152(1): 22 (1983): Euphorbia goldei.
subsp. pannonicus *(Host) J. Soják in Čas. Nár. Muz. (Prague),* 152(1): 22 (1983): Euphorbia pannonica.
subsp. stepposus *(Zoz) J. Soják in Čas. Nár. Muz. (Prague),* 152(1): 23 (1983): Euphorbia stepposa.
seguieranus
subsp. loiseleuri *(Rouy) J. Soják in Čas. Nár. Muz. (Prague),* 152(1): 23 (1983): Euphorbia gerardiana Race E. loiseleuri.

TOECHIMA (Sapindac.).
monticola *S.T. Reynolds in Fl. Australia,* 25: 202, 79 (1985)—Australia (Queensland).
pterocarpum *S.T. Reynolds in Fl. Australia,* 25: 202, 81 (1985)—Australia (Queensland).

TOLLATIA (Compos.).
chrysanthemoides *(DC.) Endl., Cat. Horti Vindob.,* 1: 353 (1842); *cf. D.J. Mabberley in Taxon,* 33(3): 436 (1984): Basionym not stated.

TOLUMNIA (Orchidac.).
subgen. Variegata *G.J. Braem in Orchidee,* 37(2): 59 (1986).
apiculata *(Moir) G.J. Braem in Orchidee,* 37(2): 58 (1986): Oncidium apiculatum.
bahamensis *(Nash ex Britt. & Millsp.) G.J. Braem in Orchidee,* 37(2): 59 (1986): Oncidium bahamense.
berenyce *(Rchb. f.) G.J. Braem in Orchidee,* 37(2): 58 (1986): Oncidium berenyce.
calochila *(Cogn.) G.J. Braem in Orchidee,* 37(2): 58 (1986): Oncidium calochilum.
caribensis *(Moir) G.J. Braem in Orchidee,* 37(2): 58 (1986): Oncidium caribense.
caymanensis *(Moir) G.J. Braem in Orchidee,* 37(2): 58 (1986): Oncidium caymanense.
compressicaulis *(Withner) G.J. Braem in Orchidee,* 37(2): 58 (1986): Oncidium compressicaule.
concava *(Moir) G.J. Braem in Orchidee,* 37(2): 58 (1986): Oncidium concavum.

TOLUMNIA :–
cuneilabia *(Moir) G.J. Braem in Orchidee,* 37(2): 58 (1986): Oncidium cuneilabium.
guianensis *(Aubl.) G.J. Braem in Orchidee,* 37(2): 58 (1986): Ophrys guianensis.
guibertiana *(A. Rich.) G.J. Braem in Orchidee,* 37(2): 58 (1986): Oncidium guibertianum.
haitensis *(Leonard & Ames ex Ames) G.J. Braem in Orchidee,* 37(2): 58 (1986): Oncidium haitense.
hawkesiana *(Moir) G.J. Braem in Orchidee,* 37(2): 59 (1986): Oncidium hawkesianum.
jimenezii *(Moir) G.J. Braem in Orchidee,* 37(2): 59 (1986): Oncidium jimenezii.
leiboldii *(Rchb. f.) G.J. Braem in Orchidee,* 37(2): 58 (1986): Oncidium leiboldii.
lemoniana *(Lindley) G.J. Braem in Orchidee,* 37(2): 58 (1986): Oncidium lemonianum.
lucayana *(Nash ex Britt. & Millsp.) G.J. Braem in Orchidee,* 37(2): 58 (1986): Oncidium lucayanum.
lyrata *(Withner) G.J. Braem in Orchidee,* 37(2): 58 (1986): Oncidium lyratum.
moiriana *(Osment) G.J. Braem in Orchidee,* 37(2): 58 (1986): Oncidium moirianum.
×osmentii *(Withner) G.J. Braem in Orchidee,* 37(2): 59 (1986): Oncidium osmentii.
prionochila *(Krzl.) G.J. Braem in Orchidee,* 37(2): 58 (1986): Oncidium prionochilum.
quadrilobia *(C. Schweinf.) G.J. Braem in Orchidee,* 37(2): 58 (1986): Oncidium quadrilobium.
sasseri *(Moir) G.J. Braem in Orchidee,* 37(2): 59 (1986), without exact basionym page: Oncidium sasseri.
scandens *(Moir) G.J. Braem in Orchidee,* 37(2): 59 (1986): Oncidium scandens.
sylvestris *(Lindley) G.J. Braem in Orchidee,* 37(2): 59 (1986): Oncidium sylvestre.
tetrapetala *(Jacq.) G.J. Braem in Orchidee,* 37(2): 58 (1986): Epidendrum tetrapetalum.
tuerckheimii *(Cogn.) G.J. Braem in Orchidee,* 37(2): 59 (1986): Oncidium tuerckheimii.
urophylla *(Lodd. ex Lindley) G.J. Braem in Orchidee,* 37(2): 59 (1986): Oncidium urophyllum.
usneoides *(Lindley) G.J. Braem in Orchidee,* 37(2): 59 (1986): Oncidium usneoides.
variegata *(Sw.) G.J. Braem in Orchidee,* 37(2): 59 (1986): Epidendrum variegatum.
velutina *(Lindley ex Paxton) G.J. Braem in Orchidee,* 37(2): 59 (1986): Oncidium velutinum.

TONGOLOA (Umbellif.).
cnidiifolia *K.T. Fu in Fl. Tsinlingensis,* 1(3): 456, 385 (1981)—China.
filicaudicis *K.T. Fu in Fl. Tsinlingensis,* 1(3): 456, 385 (1981)—China.

TOONA (Meliac.).
sinensis
var. schensiana *(C.DC.) X.M. Chen in J. Wuhan Bot. Res.,* 4(2): 187 (1986): Cedrela sinensis var. schensiana.

TORULARIA (Crucif.).
shuanghuica *K.C. Kuan & An in Fl. Xizangica,* 2: 404 (1985)—Xizang.

TOXICODENDRON (Anacardiac.).
quinquefoliolatum *Q.H. Chen in Acta Bot. Yunnanica,* 7(4): 415 (1985)—China.
vernicifluum
var. shaanxiense *J.Z. Zhang & Z.Y. Shang in Acta Bot. Bor.-Occid. Sin.,* 5(4): 314 (1985)—China.

TRACHYDIUM (Umbellif.).
involucellatum *R.H. Shan & F.T. Pu in Acta Phytotax. Sin.,* 24(4): 313 (1986)—Xizang.

TRACHYMENE (Umbellif.).
humilis
subsp. breviscapa *(Domin) P.S. Short in Muelleria,* 6(3–4): 166 (1986): Didiscus humilis f. breviscapus.

TRACHYSPERMUM (Umbellif.).
aethusifolium
var. verruculosum *C.C. Townsend in Kew Bull.,* 41(2): 456 (1986), as 'verruculosa'—Kenya.

TRADESCANTIA (Commelinac.).
sect. Campelia *(L.C. Rich.) D.R. Hunt in Kew Bull.,* 41(2): 404 (1986): gen. Campelia.
sect. Corinna *D.R. Hunt in Kew Bull.,* 41(2): 405 (1986).
sect. Rhoeo *(Hance) D.R. Hunt in Kew Bull.,* 41(2): 401 (1986): gen. Rhoeo.
sect. Zebrina *(Schnizlein) D.R. Hunt in Kew Bull.,* 41(2): 404 (1986): gen. Zebrina.
exaltata *D.R. Hunt in Kew Bull.,* 41(2): 406 (1986)—Mexico.
zebrina
var. flocculosa *(G. Brueckner) D.R. Hunt in Kew Bull.,* 41(2): 406 (1986): Zebrina flocculosa.
var. mollipila *D.R. Hunt in Kew Bull.,* 41(2): 406 (1986)—Mexico.

TRAGACANTHA (Legumin.).
albanica *(Sirj.) J. Soják in Čas. Nár. Muz. (Prague),* 152(1): 23 (1983): Astragalus albanicus.
alexandri *(Sirj.) J. Soják in Čas. Nár. Muz. (Prague),* 152(1): 23 (1983): Astragalus alexandri.
chodzha-bakirganica *M.R. Rasulova in Izv. Akad. Nauk Tadzh. SSR, Biol. Nauk,* 4(81): 93 (1980), as '(B. Kom.)' but basionym invalid—U.S.S.R. (Middle Asia).
cyllenea *(Boiss. & Heldr.) J. Soják in Čas. Nár. Muz. (Prague),* 152(1): 23 (1983): Astragalus cylleneus.
monachorum *(Sirj.) J. Soják in Čas. Nár. Muz. (Prague),* 152(1): 23 (1983): Astragalus monachorum.

TRAGOPOGON (Compos.).
parviflorus *Trev., Ind. Sem. Wratislav. App. II:* 3 (1820); *cf. D.J. Mabberley in Taxon,* 31(1): 72 (1982)— Locality not stated.
rhodanthus *Sweet, Hort. Brit., ed. 2:* 279 (1830); *cf. D.J. Mabberley in Taxon,* 31(1): 72 (1982)— Locality not stated.

TRAPA (Trapac.).
natans
var. pumila *Nakano ex B. Verdcourt in Kew Bull.,* 41(2): 448 (1986)—Vietnam, Thailand, Cambodia, Laos, Japan, ?India, Cameroun, South Africa (Natal).

TRIBULUS (Zygophyllac.).
bimucronatus
subsp. inermis *(Kralik) H. Hosni ex M.N. El Hadidi in Fl. Trop. E. Afr., Zygophyllac.:* 3 (1985): T. inermis.
pentandrus
var. micropteris *(Kralik) H. Hosni ex M.N. El Hadidi in Fl. Trop. E. Afr., Zygophyllac.:* 3 (1985): T. alatus var. micropteris.
subramanyamii *P. Singh, G.S. Giri & V. Singh in Bull. Bot. Surv. India,* 25(1–4): 197 (1983 publ. 1985)—India.
zeyheri
subsp. macranthus *(Hassk.) M.N. El Hadidi in Fl. Trop. E. Afr., Zygophyllac.:* 7 (1985): T. macranthus.

TRICALYSIA (Rubiac.).
bridsoniana *E. Robbrecht in Bull. Jard. Bot. Nation. Belg.,* 56(1–2): 146 (1986)—Kenya.
var. pandensis *E. Robbrecht in Bull. Jard. Bot. Nation. Belg.,* 56(1–2): 147 (1986)—Tanzania.
somaliensis *E. Robbrecht in Bull. Jard. Bot. Nation. Belg.,* 56(1–2): 149 (1986)—Somali Republic.

TRICHANTHEMIS (Compos.).
afghanica *D. Podlech in Fl. Iran.,* 158: 155 (1986)—Afghanistan.

TRICHOCAULON (Asclepiadac.).
felinum *D.T. Cole in Aloe,* 22(1): 6 (1985)—South Africa (Cape Province).

TRICHOPYRUM A. Löve in Veröff. Geobot. Inst. Rübel, 87: 49 (1986). *GRAMINEAE.*
intermedium *(Host) A. Löve in Veröff. Geobot. Inst. Rübel,* 87: 49 (1986): Triticum intermedium.
subsp. afghanicum *(Melderis) A. Löve in Veröff. Geobot. Inst. Rübel,* 87: 49 (1986), without basionym ref. see Art. 33.3 ICBN 1983: Basionym not stated.
subsp. barbulatum *(Schur) A. Löve in Taxon,* 35(1): 198 (1986): Agropyron barbulatum.

TRICHOPYRUM :-
subsp. barbulatum *(Schur) A. Löve in Veröff. Geobot. Inst. Rübel,* 87: 49 (1986), without basionym ref. see Art. 33.3 ICBN 1983: Basionym not stated.
subsp. epiroticum *(Melderis) A. Löve in Veröff. Geobot. Inst. Rübel,* 87: 49 (1986), without basionym ref. see Art. 33.3 ICBN 1983: Basionym not stated.
subsp. gentryi *(Melderis) A. Löve in Veröff. Geobot. Inst. Rübel,* 87: 49 (1986), without basionym ref. see Art. 33.3 ICBN 1983: Basionym not stated.
subsp. graecum *(Melderis) A. Löve in Veröff. Geobot. Inst. Rübel,* 87: 49 (1986), without basionym ref. see Art. 33.3 ICBN 1983: Basionym not stated.
subsp. podperae *(Nabelek) A. Löve in Veröff. Geobot. Inst. Rübel,* 87: 49 (1986), without basionym ref. see Art. 33.3 ICBN 1983: Basionym not stated.
subsp. pouzolzii *(Godron) Á. Löve in Taxon,* 35(1): 198 (1986): Triticum pouzolzii.
subsp. pouzolzii *(Godron) A. Löve in Veröff. Geobot. Inst. Rübel,* 87: 49 (1986), without basionym ref. see Art. 33.3. ICBN 1983: Basionym not stated.
subsp. pulcherrimum *(Grossh.) Á. Löve in Taxon,* 35(1): 198 (1986): Agropyron pulcherrimum.
subsp. pulcherrimum *(Grossh.) A. Löve in Veröff. Geobot. Inst. Rübel,* 87: 49 (1986), without basionym ref. see Art. 33.3 ICBN 1983: Basionym not stated.
varnense *(Velen.) A. Löve in Veröff. Geobot. Inst. Rübel,* 87: 49 (1986): Triticum varnense.

TRICHOSALPINX (Orchidac.).
sect. Apodae *C.A. Luer, Ic. Pleurothallid. I-Syst. Pleurothallidinae (Monogr. Syst. Bot. Missouri Bot. Gard. 15):* 66 (1986).
sect. Pseudolepanthes *C.A. Luer, Ic. Pleurothallid. I-Syst. Pleurothallidinae (Monogr. Syst. Bot. Missouri Bot. Gard. 15):* 68 (1986).
subgen. Tubella *C.A. Luer, Ic. Pleurothallid. I-Syst. Pleurothallidinae (Monogr. Syst. Bot. Missouri Bot. Gard. 15):* 66 (1986).
sect. Tubellae *C.A. Luer, Ic. Pleurothallid. I-Syst. Pleurothallidinae (Monogr. Syst. Bot. Missouri Bot. Gard. 15):* 68 (1986).
farrago *C.A. Luer & A.C. Hirtz in Orchidee,* 37(3): 141 (1986)—Ecuador.
microlepanthes *(Griseb.) C.A. Luer, Ic. Pleurothallid. I-Syst. Pleurothallidinae (Monogr. Syst. Bot. Missouri Bot. Gard. 15):* 68 (1986): Pleurothallis microlepanthes.

TRICORYNE (Anthericac.).
corynothecoides *G.J. Keighery in Willdenowia,* 15(2): 473 (1986)—Australia (W. Australia).

TRIFOLIUM (Legumin.).
campestre *C.C. Gmel., Hort. Mag. Duc. Bad. Carlsruh.:* 270 (1811); *cf. D.J. Mabberley in Taxon,* 33(3): 440 (1984)— Locality not stated.
creticum *(L.) G.F. Hoffm., Hort. Mosq.:* 39 (1808); *cf. D.J. Mabberley in Taxon,* 31(1): 72 (1982): Basionym not stated.
italicum *(L.) G.F. Hoffm., Hort. Mosq.:* 39 (1808); *cf. D.J. Mabberley in Taxon,* 31(1): 72 (1982): Basionym not stated.

TRIGONELLA (Legumin.).
brachycarpa *(Fisch.) Moris, Enum. Sem. Hort. Reg. Bot. Taur.,* 1831; *Flora 15 Lit.,* 2: 80 (1832); *cf. D.J. Mabberley in Taxon,* 33(3): 443 (1984): Basionym not stated.

TRIGONIA (Trigoniac.).
brasiliensis *Jacob-Makoy, Cat. Prix Cour. 1837:* 10 (1837); *cf. D.J. Mabberley in Taxon,* 34(3): 456 (1985)— Locality not stated.

TRIGONOCAPNOS (Papaverac.).
lichtensteinii *(Cham. & Schlecht.) M. Lidén in Op. Bot.,* 88: 105 (1986): Fumaria lichtensteinii.

TRIGONOTIS (Boraginac.).
chengkouensis *W.T. Wang in Bull. Bot. Res. North-East. Forest. Inst.,* 6(3): 81 (1986)—China.
corispermoides
var. sessilis *W.T. Wang in Bull. Bot. Res. North-East. Forest. Inst.,* 6(3): 81 (1986)—China.
leucantha *W.T. Wang in Bull. Bot. Res. North-East. Forest. Inst.,* 6(3): 87 (1986)—China.
longipes *W.T. Wang in Bull. Bot. Res. North-East. Forest. Inst.,* 6(3): 85 (1986)—China.
longiramosa *W.T. Wang in Bull. Bot. Res. North-East. Forest. Inst.,* 6(3): 83 (1986)—China.
muliensis *W.T. Wang in Bull. Bot. Res. North-East. Forest. Inst.,* 6(3): 84 (1986)—China.
peduncularis
var. amblyosepala *(Nakai & Kitag.) W.T. Wang in Bull. Bot. Res. North-East. Forest. Inst.,* 6(3): 90 (1986): T. amblyosepala.
var. macrantha *W.T. Wang in Bull. Bot. Res. North-East. Forest. Inst.,* 6(3): 89 (1986)—China, Xizang.

TRIMENIA (Trimeniac.).
macrura *(Gilg & Schltr.) W.R. Philipson in Fl. Males.,* 10(2): 332 (1986): Piptocalyx macrurus.
moorei *(Oliv.) W.R. Philipson in Fl. Males.,* 10(2): 330 (1986): Piptocalyx moorei.

TRINIA (Umbellif.).
glauca
subsp. pindica *P. Hartvig in A. Strid (ed.), Mount. Fl. Greece,* 1: 694 (1986)—Greece, Albania.

TRIPLARIS (Polygonac.).
melaenodendron
subsp. colombiana *(Meisner) J. Brandbyge in Nordic J. Bot.,* 6(5): 566 (1986): T. colombiana.
setosa
var. woytkowskii *J. Brandbyge in Nordic J. Bot.,* 6(5): 561 (1986)—Peru.

TRIPLEUROSPERMUM (Compos.).
caucasicum
var. melanolepis *(Boiss. & Buhse) D. Podlech in Fl. Iran.,* 158: 75 (1986): Chamaemelum melanolepis.
philaenorum *(Maire & Weiller) S.A. Alavi in Fl. Libya,* 107: 148 (1983): Heteromera philaenorum.

TRIPODION (Legumin.).
graecum *(Boiss.) P. Lassen in Willdenowia,* 16(1): 114 (1986): Hammatolobium graecum.
kremerianum *(Cosson) P. Lassen in Willdenowia,* 16(1): 114 (1986): Ludovicia kremeriana.

TRIPOGON (Gramin.).
anantaswamianus *P.V. Sreekumar, V.J. Nair & N.C. Nair in Bull. Bot. Surv. India,* 25(1–4): 185 (1983 publ. 1985)—India.

TRIPTEROSPERMUM (Gentianac.).
distylum *J. Murata & T. Yahara in Acta Phytotax. Geobot.,* 36(4–6). 162 (1985)—Japan.

TRISETELLA (Orchidac.).
andreettae *C.A. Luer in Lindleyana,* 1(3): 186 (1986)—Ecuador.
escobarii *C.A. Luer in Lindleyana,* 1(3): 188 (1986)—Colombia.
hirtzii *C.A. Luer in Lindleyana,* 1(3): 190 (1986)—Ecuador.
hoeijeri *C.A. Luer & Hirtz in Lindleyana,* 1(3): 192 (1986)—Ecuador.

TRISETUM (Gramin.).
longifolium *Nees, Ind. Sem. Hort. Bot. Vratislav.,* 1835; *Linnaea 11, Lit.:* 131 (1837); *cf. D.J. Mabberley in Taxon,* 32(1): 86 (1983)— Locality not stated.

TRISTEMMA (Melastomatac.).
vestitum *H. Jacques-Félix in Bull. Mus. Nation. Hist. Nat., B, Adansonia,* Ser. 4, 8(2): 191 (1986)—Gabon.

TRITICUM (Gramin.).
durum
var. sphaeroaffine *P. Popov & I. Stankov in Dokl. Bolg. Akad. Nauk,* 34(12): 1699 (1981), without type—Cultivation.
var. sphaeroafricanum *P. Popov & I. Stankov in Dokl. Bolg. Akad. Nauk,* 34(12): 1670 (1981), without type—Cultivation.
var. sphaeroleucurum *P. Popov & I. Stankov in Dokl. Bolg. Akad. Nauk,* 34(12): 1699 (1981), without type—Cultivation.
var. sphaeromelanopus *P. Popov & I. Stankov in Dokl. Bolg. Akad. Nauk,* 34(12): 1699 (1981), without type—Cultivation.
ramosum *Weigel, Hort. Gryph.:* 10 (1782); *cf. D.J. Mabberley in Taxon,* 33(3): 443 (1984)— Locality not stated.

TRIUMFETTA (Tiliac.).
malebarica *Koen. ex Rottb., Pl. Hort. Univ. Rar. Prog.:* 16 (1773); *cf. D.J. Mabberley in Taxon,* 33(3): 439 (1984)— Locality not stated.
trilocularis *Roxb. ex Hornem., Hort. Hafn., Suppl.:* 140 (1819); *cf. D.J. Mabberley in Taxon,* 33(3): 438 (1984)— Locality not stated.

TRIUNIA (Proteac.).
montana *(C. White) D.B. Foreman in Muelleria,* 6(3–4): 195 (1986): Helicia youngiana var. montana.
robusta *(C. White) D.B. Foreman in Muelleria,* 6(3–4): 196 (1986): Helicia youngiana var. robusta.

TRIURIS (Triuridac.).
hexophthalma *P.J.M. Maas in Fl. Neotrop.,* 40: 50 (1986)—Guyana.

TRUDELIA L.A. Garay in Orchid Dig., 50(2): 73 (1986). *ORCHIDACEAE.*
alpina *(Lindl.) L.A. Garay in Orchid Dig.,* 50(2): 76 (1986): Luisia alpina.
chlorosantha *L.A. Garay in Orchid Dig.,* 50(2): 76 (1986)—Bhutan.
griffithii *(Lindl.) L.A. Garay in Orchid Dig.,* 50(2): 76 (1986): Vanda griffithii.

×**TRUDELIANDA** L.A. Garay in Orchid Dig., 50(2): 76 (1986). *ORCHIDACEAE.* (Trudelia × Vanda).

TSIANGIA P.P.H. But, H.H. Hsue & P.T. Li in Blumea, 31(2): 311 (1986). *RUBIACEAE.*
hongkongensis *(Seem.) P.P.H. But, H.H. Hsue & P.T. Li in Blumea,* 31(2): 311 (1986): Gaertnera hongkongensis.

TUBERARIA (Cistac.).
glomerata *Willk., Ic. Pl. Hisp.,* 2: 80, t. 117 (1859)—Algeria. †

TULIPA (Liliac.).
agenensis
subsp. sharonensis *(Dinsmore) N. Feinbrun-Dothan, Fl. Palaestina,* 4: 41 (1986): T. sharonensis.
persica *(Lindl.) Sweet, Hort. Brit., ed. 2:* 536 (1830); *cf. D.J. Mabberley in Taxon,* 31(1): 72 (1982): T. oculus-solis var. persica.

TURANIPHYTUM (Compos.).
codringtonii *(Rech. f.) D. Podlech in Fl. Iran.,* 158: 225 (1986): Artemisia codringtonii.

TURBINICARPUS (Cactac.).
gielsdorfianus *(Werderm.) John & Riha in Kaktusy,* 19(1): 22 (1983); *cf. Repert. Pl. Succ. (I.O.S.),* 34: 8 (1985): Echinocactus gielsdorfianus.
horripilus *(Lem.) John & Riha in Kaktusy,* 19(1): 22 (1983); *cf. Repert. Pl. Succ. (I.O.S.),* 34: 8 (1985): Mammillaria horripila.
knuthianus *(Böd.) John & Riha in Kaktusy,* 19(1): 22 (1983); *cf. Repert. Pl. Succ. (I.O.S.),* 34: 8 (1985): Echinocactus knuthianus.
saueri *(Böd.) John & Riha in Kaktusy,* 19(1): 22 (1983); *cf. Repert. Pl. Succ. (I.O.S.),* 34: 8 (1985): Echinocactus saueri.
viereckii *(Werderm.) John & Riha in Kaktusy,* 19(1): 22 (1983); *cf. Repert. Pl. Succ. (I.O.S.),* 34: 8 (1985): Echinocactus viereckii.
var. major *(Glass & Foster) John & Riha in Kaktusy,* 19(1): 22 (1983); *cf. Repert. Pl. Succ. (I.O.S.),* 34: 8 (1985): Gymnocactus viereckii var. major.
ysabelae *(K. Schlange) John & Riha in Kaktusy,* 19(1): 22 (1983); *cf. Repert. Pl. Succ. (I.O.S.),* 34: 8 (1985): Thelocactus ysabelae.
var. brevispinus *(K. Schlange) John & Riha in Kaktusy,* 19(1): 22 (1983); *cf. Repert. Pl. Succ. (I.O.S.),* 34: 8 (1985): Thelocactus ysabelae var. brevispinus.

TURNERA (Turnerac.).
concinna *M.M. Arbo in Candollea,* 41(1): 209 (1986)—Paraguay, Bolivia, Brazil.

TYLANTHUS (Rhamnac.).
acerosus *(Willd.) Endl., Cat. Horti Vindob.,* 2: 386 (1842); *cf. D.J. Mabberley in Taxon,* 33(3): 436 (1984): Basionym not stated.
ericoides *(L.) Endl., Cat. Horti Vindob.,* 2: 386 (1842); *cf. D.J. Mabberley in Taxon,* 33(3): 436 (1984): Phylica ericoides.

TYLERIA (Ochnac.).
apiculata *C. Sastre in Phytologia,* 59(5): 313 (1986)—Venezuela.
apiculata *C. Sastre in Bull. Mus. Nation. Hist. Nat., B, Adansonia, Ser. 4,* 8(1): 14 (1986)—Venezuela.

TZVELEVIA E.B. Alekseev in Byull. Mosk. Obshch. Ispȳt. Prir., Biol., 90(5): 103 (1985). *GRAMINEAE.*
kerguelensis *(Hook. f.) E.B. Alekseev in Byull. Mosk. Obshch. Ispȳt. Prir., Biol.,* 90(5): 103 (1985): Triodia kerguelensis.

U

ULEX (Legumin.).
parviflorus
subsp. africanus *(Webb) W. Greuter in Willdenowia,* 15(2): 429 (1986): U. africanus.

UNONOPSIS (Annonac.).
aviceps *P.J.M. Maas in Proc. Kon. Nederl. Akad. Wetensch., C,* 89(3): 272 (1986)—Colombia.
velutina *P.J.M. Maas in Proc. Kon. Nederl. Akad. Wetensch., C,* 89(3): 275 (1986)—Venezuela.

UROMYRTUS (Myrtac.).
metrosideros *(F.M. Bail.) A.J. Scott in Kew Bull.,* 41(2): 286 (1986): Myrtus metrosideros.

URTICA (Urticac.).
frutescens *C.P. Thunberg ex A. Murray in Linnaeus, Syst. Veg., ed. 14:* 851 (1784)—Japan. †
glabrata *Clem. in Atti 3 Riun. Sci. Ital.:* 1518 (1841); *cf. D.J. Mabberley in Taxon,* 31(1): 73 (1982)— Locality not stated.
japonica *C.P. Thunberg ex A. Murray in Linnaeus, Syst. Veg., ed. 14:* 851 (1784)—Japan. †
macrophylla *C.P. Thunberg ex A. Murray in Linnaeus, Syst. Veg., ed. 14:* 850 (1784)—Japan. †
spicata *C.P. Thunberg ex A. Murray in Linnaeus, Syst. Veg., ed. 14:* 850 (1784)—Japan. †
villosa *C.P. Thunberg ex A. Murray in Linnaeus, Syst. Veg., ed. 14:* 851 (1784)—Japan. †
whitlowii *Whitlow, Bot. Char. Desc. Urtica whitlowi:* 1 (before ?1814); *cf. D.J. Mabberley in Taxon,* 33(3): 444 (1984)—Locality not stated.

URVILLEA (Sapindac.).
andersonii *M.S. Ferrucci in Bol. Soc. Argent. Bot.,* 24(1–2): 113 (1985)—Brazil.
stipularis *M.S. Ferrucci in Bol. Soc. Argent. Bot.,* 24(1–2): 110 (1985)—Brazil.

UTRICULARIA (Lentibulariac.).
sect. Aranella *(Barnh.) P. Taylor in Kew Bull.*, 41(1): 7 (1986): gen. Aranella.
sect. Benjaminia *P. Taylor in Kew Bull.*, 41(1): 11 (1986).
sect. Candollea *P. Taylor in Kew Bull.*, 41(1): 6 (1986).
sect. Chelidon *P. Taylor in Kew Bull.*, 41(1): 10 (1986).
sect. Enskide *(Raf.) P. Taylor in Kew Bull.*, 41(1): 9 (1986): gen. Enskide.
sect. Iperua *P. Taylor in Kew Bull.*, 41(1): 10 (1986).
sect. Kamienskia *P. Taylor in Kew Bull.*, 41(1): 11 (1986).
sect. Lloydia *P. Taylor in Kew Bull.*, 41(1): 6 (1986).
sect. Martinia *P. Taylor in Kew Bull.*, 41(1): 7 (1986).
sect. Meionula *(Raf.) P. Taylor in Kew Bull.*, 41(1): 5 (1986): gen. Meionula.
sect. Nelipus *(Raf.) P. Taylor in Kew Bull.*, 41(1): 15 (1986): gen. Nelipus.
sect. Oliveria *P. Taylor in Kew Bull.*, 41(1): 13 (1986).
subgen. Polypompholyx *(Lehm.) P. Taylor in Kew Bull.*, 41(1): 1 (1986): gen. Polypompholyx.
sect. Polypompholyx *(Lehm.) P. Taylor in Kew Bull.*, 41(1): 2 (1986): gen. Polypompholyx.
sect. Psyllosperma *P. Taylor in Kew Bull.*, 41(1): 8 (1986).
sect. Setiscapella *(Barnh.) P. Taylor in Kew Bull.*, 41(1): 15 (1986): gen. Setiscapella.
sect. Sprucea *P. Taylor in Kew Bull.*, 41(1): 13 (1986).
sect. Tridentaria *P. Taylor in Kew Bull.*, 41(1): 2 (1986).
sect. Vesiculina *(Raf.) P. Taylor in Kew Bull.*, 41(1): 17 (1986): gen. Vesiculina.
antennifera *P. Taylor in Kew Bull.*, 41(1): 2 (1986)—Australia (W. Australia).
arnhemica *P. Taylor in Kew Bull.*, 41(1): 3 (1986)—Australia (N. Terr.).
benthamii *P. Taylor in Kew Bull.*, 41(1): 3 (1986)—Australia (W. Australia).
biovularioides *(Kuhlmann) P. Taylor in Kew Bull.*, 41(1): 16 (1986): Saccolaria biovularioides.
cheiranthos *P. Taylor in Kew Bull.*, 41(1): 3 (1986)—Australia (N. Terr.).
choristotheca *P. Taylor in Kew Bull.*, 41(1): 14 (1986)—Surinam.
christopheri *P. Taylor in Kew Bull.*, 41(1): 12 (1986)—Nepal.
circumvoluta *P. Taylor in Kew Bull.*, 41(1): 9 (1986)—Australia (N. Terr., Queensland).
corynephora *P. Taylor in Kew Bull.*, 41(1): 12 (1986)—Burma, Thailand.
costata *P. Taylor in Kew Bull.*, 41(1): 7 (1986)—Venezuela, Brazil.
determannii *P. Taylor in Kew Bull.*, 41(1): 14 (1986)—Surinam.
dunlopii *P. Taylor in Kew Bull.*, 41(1): 3 (1986)—Australia (N. Terr., W. Australia).

UTRICULARIA :-
fistulosa *P. Taylor in Kew Bull.*, 41(1): 3 (1986)—Australia (W. Australia).
forrestii *P. Taylor in Kew Bull.*, 41(1): 13 (1986)—China, Burma.
garrettii *P. Taylor in Kew Bull.*, 41(1): 13 (1986)—Thailand.
georgei *P. Taylor in Kew Bull.*, 41(1): 3 (1986)—Australia (W. Australia).
grandiflora *Mertens in Bull. Acad. Belg.*, 8: 66 (1841); *cf. D.J. Mabberley in Taxon*, 33(3): 444 (1984)— Locality not stated.
helix *P. Taylor in Kew Bull.*, 41(1): 4 (1986)—Australia (W. Australia).
hintonii *P. Taylor in Kew Bull.*, 41(1): 8 (1986)—Mexico.
huntii *P. Taylor in Kew Bull.*, 41(1): 8 (1986)—Brazil.
kenneallyi *P. Taylor in Kew Bull.*, 41(1): 4 (1986)—Australia (W. Australia).
khasiana *J. Joseph & J. Mani in Bull. Bot. Surv. India,* 25(1–4): 192 (1983 publ. 1985)—India.
macrocheilos *(P. Taylor) P. Taylor in Kew Bull.*, 41(1): 9 (1986): U. micropetala var. macrocheilos.
mirabilis *P. Taylor in Kew Bull.*, 41(1): 14 (1986)—Venezuela.
moniliformis *P. Taylor in Kew Bull.*, 41(1): 13 (1986)—Sri Lanka.
panamensis *Steyermark ex P. Taylor in Kew Bull.*, 41(1): 8 (1986)—Panama.
parthenopipes *P. Taylor in Kew Bull.*, 41(1): 7 (1986)—Brazil.
peranomala *P. Taylor in Kew Bull.*, 41(1): 12 (1986)—China.
perversa *P. Taylor in Kew Bull.*, 41(1): 16 (1986)—Mexico.
petersoniae *P. Taylor in Kew Bull.*, 41(1): 8 (1986)—Mexico.
physoceras *P. Taylor in Kew Bull.*, 41(1): 15 (1986)—Brazil.
quinquedentata *F. Mueller ex P. Taylor in Kew Bull.*, 41(1): 4 (1986)—Australia (N. Terr., Queensland, W. Australia).
raynalii *P. Taylor in Kew Bull.*, 41(1): 17 (1986)—Cameroun, Senegal, Upper Volta, Sudan, Rwanda.
recta *P. Taylor in Kew Bull.*, 41(1): 10 (1986)—Bhutan, India, China.
rhododactylos *P. Taylor in Kew Bull.*, 41(1): 5 (1986)—Australia (N. Terr.).
steenisii *P. Taylor in Kew Bull.*, 41(1): 13 (1986)—Sumatra.
terrae-reginae *P. Taylor in Kew Bull.*, 41(1): 5 (1986)—Australia (Queensland).
tridactyla *P. Taylor in Kew Bull.*, 41(1): 5 (1986)—Australia (W. Australia).
triflora *P. Taylor in Kew Bull.*, 41(1): 5 (1986)—Australia (N. Terr.).
westonii *P. Taylor in Kew Bull.*, 41(1): 2 (1986)—Australia (W. Australia).
wightiana *P. Taylor in Kew Bull.*, 41(1): 10 (1986), nom. nov.: U. squamosa *Wight.*

UVARIA (Annonac.).
grandiflora *Roxb. ex Hornem., Hort. Hafn., Suppl.:* 141 (1819); *cf. D.J. Mabberley in Taxon,* 33(3): 438 (1984)— Locality not stated.
tanzaniae *B. Verdcourt in Kew Bull.,* 41(2): 287 (1986)—Tanzania.

UVARIODENDRON (Annonac.).
oligocarpum *B. Verdcourt in Kew Bull.,* 41(2): 289 (1986)—Tanzania.

UVARIOPSIS (Annonac.).
bisexualis *B. Verdcourt in Kew Bull.,* 41(2): 289 (1986)—Tanzania.

UVULARIA (Convallariac.).
cirrhosa *C.P. Thunberg ex A. Murray in Linnaeus, Syst. Veg., ed. 14:* 325 (1784)—Japan. †
hirta *C.P. Thunberg ex A. Murray in Linnaeus, Syst. Veg., ed. 14:* 325 (1784)—Japan. †

V

VACCARIA (Caryophyllac.).
hispanica *(Miller) S. Rauschert in Wiss. Zeitschr. Univ. Halle,* 14: 496 (1965): Saponaria hispanica. †
hispanica *(Miller) S. Rauschert in Wiss. Zeitschr. Martin-Luther-Univ. Halle-Wittemberg, Math.-Nat.,* 14: 496 (1965): Saponaria hispanica. †

VACCINIUM (Ericac.).
bracteatum *C.P. Thunberg ex A. Murray in Linnaeus, Syst. Veg., ed. 14:* 363 (1784)—Japan. †
ciliatum *C.P. Thunberg ex A. Murray in Linnaeus, Syst. Veg., ed. 14:* 363 (1784)—Japan. †
hirtum *C.P. Thunberg ex A. Murray in Linnaeus, Syst. Veg., ed. 14:* 363 (1784)—Japan. †
pipolyi *J.L. Luteyn in Brittonia,* 38(2): 101 (1986)—Brazil.
pterocalyx *J.L. Luteyn in Bull. Torrey Bot. Club,* 112(4): 451 (1985)—Venezuela.
steyermarkii *J.L. Luteyn in Bull. Torrey Bot. Club,* 112(4): 451 (1985)—Venezuela.

VALERIANA (Valerianac.).
sect. Bracteata *B.B. Larsen in Nordic J. Bot.,* 6(4): 438 (1986).
capensis
var. nana *B.L. Burtt in Notes Roy. Bot. Gard. Edinburgh,* 43(3): 404 (1986)—South Africa (Natal, Cape Province), Lesotho.

VALERIANA :–
convallarioides *(Schmale) B.B. Larsen in Nordic J. Bot.,* 6(4): 442 (1986): Phyllactis convallarioides.
crinii
subsp. epirotica *(Phitos) R. Franzén in Willdenowia,* 15(2): 353 (1986): V. epirotica.
spiroflora *B.B. Larsen in Nordic J. Bot.,* 6(4): 443 (1986)—Colombia.
venusta *L.Q. Qiu in Acta Bot. Yunnanica,* 8(1): 45 (1986)—China.
villosa *C.P. Thunberg ex A. Murray in Linnaeus, Syst. Veg., ed. 14:* 81 (1784)—Japan. †
xiaheensis *L.Q. Qiu in Acta Bot. Yunnanica,* 8(1): 46 (1986)—China.

VALERIANELLA (Valerianac.).
sect. Tuberculatae *B.A. Sharipova in Dokl. Akad. Nauk Tadzh. SSR,* 27(2): 106 (1984).
gibosa *(Guss.) Tin., Cat. Pl. Panorm.:* 264 (1827); *cf. D.J. Mabberley in Taxon,* 31(1): 73 (1982): Basionym not stated.

VALLOTA (Amaryllidac.).
×hybrida *Bull, Cat.,* 324: 2 (tab.), 6 (1898); *cf. D.J. Mabberley in Taxon,* 34(3): 456 (1985)— Locality not stated. (Cyrtanthus sanguineus × V. speciosa).

VANDA (Orchidac.).
kwangtungensis *S.J. Cheng & Z.Z. Tang in Acta Bot. Yunnanica,* 8(2): 218 (1986)—China.

VELLA (Crucif.).
charpinii *J. Fernández Casas in Fontqueria,* 1: 9 (1982)—Morocco.
mairei
var. macrantha *J. Fernández Casas in Fontqueria,* 2: 25 (1982)—Morocco.
pseudocytisus
subsp. glabrata *W. Greuter in W. Greuter, H.M. Burdet & G. Long (eds.), Med-checklist,* 3: 172 (1986), nom. nov.: V. glabrescens *Cosson.*

VELLOZIA (Velloziac.).
capiticola *L.B. Smith in Bradea,* 4(30): 211 (1986)—Brazil.
goiasensis *L.B. Smith in Bradea,* 4(30): 212 (1986)—Brazil.

VEPRIS (Rutac.).
mandangoana *S. Lisowski in Fragm. Flor. Geobot.,* 29(2): 214 (1983 publ. 1985)—Zaïre.

VERBASCUM (Scrophulariac.).
erosum *Lag., Elenchus:* 17 (1816); *cf. D.J. Mabberley in Taxon,* 31(1): 73 (1982)—Locality not stated.
ovatifolium *DC. ex Hornem., Hort. Hafn.:* 211 (1813); *cf. D.J. Mabberley in Taxon,* 33(3): 438 (1984)— Locality not stated.

VERBENA (Verbenac.).
pilosiuscula *(Kunth) Endl., Cat. Horti Vindob.,* 2: 46 (1842); *cf. D.J. Mabberley in Taxon,* 33(3): 436 (1984): Basionym not stated.
salisburii *Endl., Cat. Horti Vindob.,* 2: 46 (1842); *cf. D.J. Mabberley in Taxon,* 33(3): 436 (1984)— Locality not stated.

VERBESINA (Compos.).
cuneifolia *C.C. Gmel., Hort. Mag. Duc. Bad. Carlsruh.:* 278 (1811); *cf. D.J. Mabberley in Taxon,* 33(3): 440 (1984)— Australia.
durangensis *B.L. Turner in Phytologia,* 60(4): 254 (1986)—Mexico.

VEREIA (Crassulac.).
rhizophylla *Bertol., Elenchus Hort. Bot. Bonon.:* 8 (1820); *cf. D.J. Mabberley in Taxon,* 32(1): 82 (1983)— Locality not stated.

VERNONIA (Compos.).
bractifimbriata *(Mendonça) G.V. Pope in Kew Bull.,* 41(1): 42 (1986): V. poskeana var. bractifimbriata.
caboverdeana *W. Lobin in Courier Forschungsinst. Senckenberg,* 81: 117 (1986), nom. nov.: Gymnanthemum bolleanum *Steetz.*
filisquama *M.G. Gilbert in Kew Bull.,* 41(1): 33 (1986)—Tanzania.
galamensis
subsp. afromontana *(R.E. Fries) M.G. Gilbert in Kew Bull.,* 41(1): 32 (1986): V. afromontana.
var. australis *M.G. Gilbert in Kew Bull.,* 41(1): 26 (1986)—Zimbabwe, Tanzania, Mozambique, Malawi.
var. ethiopica *M.G. Gilbert in Kew Bull.,* 41(1): 26 (1986)—Ethiopia.
subsp. gibbosa *M.G. Gilbert in Kew Bull.,* 41(1): 32 (1986)—Kenya.
subsp. lushotoensis *M.G. Gilbert in Kew Bull.,* 41(1): 30 (1986)—Tanzania, Uganda.
subsp. mutomoensis *M.G. Gilbert in Kew Bull.,* 41(1): 31 (1986)—Kenya.
subsp. nairobensis *M.G. Gilbert in Kew Bull.,* 41(1): 28 (1986)—Kenya, Tanzania.
var. petitiana *(A. Rich.) M.G. Gilbert in Kew Bull.,* 41(1): 25 (1986): V. petitiana.
jelfiae
var. albida *G.V. Pope in Kew Bull.,* 41(2): 394 (1986)—Zambia, Zaïre, Tanzania, Malawi, Zimbabwe.
mbalensis *G.V. Pope in Kew Bull.,* 41(2): 395 (1986)—Zambia, Tanzania.
nudiflora
f. albiflora *N.I. Matzenbacher in Comun. Mus. Ciênc. PUCRGS Sér. Bot.,* 30–39: 117 (1985)—Brazil.

VERNONIA :–
poskeana
subsp. botswanica *G.V. Pope in Kew Bull.,* 41(1): 39 (1986)—Botswana, Zimbabwe, Angola, Namibia, South Africa (Transvaal).
samfyana *G.V. Pope in Kew Bull.,* 41(1): 42 (1986)—Zambia, Angola.
sylvicola *G.V. Pope in Kew Bull.,* 41(2): 395 (1986)—Zambia, Zaïre, Tanzania, Mozambique, Malawi, Zimbabwe, Angola.

VERONICA (Scrophulariac.).
antalyensis *M.A. Fischer, S. Erik & H. Sümbül in Notes Roy. Bot. Gard. Edinburgh,* 44(1): 153 (1986)—Turkey.
miniana *P. Baltasar Merino, Algunas Pl. Rar. ... de La Guardia (Pontevedra):* 24 (1895)—Spain. †

VIBURNUM (Caprifoliac.).
cotinifolium
var. lacei *T.R. Dudley in Feddes Repert.,* 97(3–4): 129 (1986)—Pakistan, India, Himalayas.
var. wallichii *T.R. Dudley in Feddes Repert.,* 97(3–4): 131 (1986)—Nepal.
cuspidatum *C.P. Thunberg ex A. Murray in Linnaeus, Syst. Veg., ed. 14:* 295 (1784)—Japan. †
dilatatum *C.P. Thunberg ex A. Murray in Linnaeus, Syst. Veg., ed. 14:* 295 (1784)—Japan. †
erosum *C.P. Thunberg ex A. Murray in Linnaeus, Syst. Veg., ed. 14:* 295 (1784)—Japan. †
hirtum *C.P. Thunberg ex A. Murray in Linnaeus, Syst. Veg., ed. 14:* 295 (1784)—Japan. †
macrophyllum *C.P. Thunberg ex A. Murray in Linnaeus, Syst. Veg., ed. 14:* 295 (1784)—Japan. †
serratum *C.P. Thunberg ex A. Murray in Linnaeus, Syst. Veg., ed. 14:* 294 (1784) Japan. †
tomentosum *C.P. Thunberg ex A. Murray in Linnaeus, Syst. Veg., ed. 14:* 295 (1784)—Japan. †
virens *C.P. Thunberg ex A. Murray in Linnaeus, Syst. Veg., ed. 14:* 294 (1784)—Japan. †

VICATIA (Umbellif.).
bipinnata *R.H. Shan & F.T. Pu in Acta Phytotax. Sin.,* 24(4): 313 (1986)—China.

VICIA (Legumin.).
argaea *(P. Davis) W. Greuter in Willdenowia,* 16(1): 114 (1986): V. canescens subsp. argaea.
canescens
subsp. argentea *(Lapeyr.) O. de Bolòs & J. Vigo, Fl. Països Catalans:* 505 (1984): V. argentea.

VICIA :–
cracca
subsp. stenophylla *(Velen.) C.D. Preston in A. Strid (ed.), Mount. Fl. Greece,* 1: 485 (1986): V. tenuifolia subsp. stenophylla.
latibracteolata *K.T. Fu in Fl. Tsinlingensis,* 1(3): 449, 92 (1981)—China.
var. acerosa *K.T. Fu in Fl. Tsinlingensis,* 1(3): 449, 93 (1981)—China.
perelegans *K.T. Fu in Fl. Tsinlingensis,* 1(3): 449, 93 (1981)—China.
taipaica *K.T. Fu in Fl. Tsinlingensis,* 1(3): 448, 92 (1981)—China.
tenuifolia
subsp. dalmatica *(A. Kerner) W. Greuter in Willdenowia,* 16(1): 114 (1986): V. dalmatica.
subsp. villosa *(Battand.) W. Greuter in Willdenowia,* 16(1): 114 (1986): V. tenuifolia var. villosa.
venosa
var. glabristyla *Y. Endo & H. Ohashi in Sci. Rep. Tohoku Univ.,* Ser. 4, 39(2): 135 (1986)—Japan.
var. stolonifera *Y. Endo & H. Ohashi in Sci. Rep. Tohoku Univ.,* Ser. 4, 39(2): 135 (1986)—Japan.
var. yamanakae *Y. Endo & H. Ohashi in Sci. Rep. Tohoku Univ.,* Ser. 4, 39(2): 135 (1986)—Japan.

VIGNA (Legumin.).
sublobata *(Roxb.) G.C. Bairiganjan, P.C. Panda, B.P. Choudhury & S.N. Patnaik in J. Econ. Taxon. Bot.,* 7(2): 274 (1985 publ. 1986): Phaseolus sublobatus.
subramaniana *(Babu ex Raizada) M. Sharma in J. Econ. Taxon. Bot.,* 6(3): 736 (1985): Phaseolus subramanianus.
trilobata
var. pusilla *V.N. Naik & D.S. Pokle in J. Econ. Taxon. Bot.,* 7(3): 670 (1985 publ. 1986)—India.

VINCETOXICUM (Asclepiadac.).
fuscatum *(Hornem.) Endl., Cat. Horti Vindob.,* 1: 474 (1842); *cf. D.J. Mabberley in Taxon,* 33(3): 436 (1984): Basionym not stated.

VIOLA (Violac.).
arvensis
subsp. megalantha *J.D. Nauenburg, Untersuch. Variab. Ökol. & Syst. Viola tricolor-Gruppe Mitteleuropa:* 81 (1986)—Switzerland, Austria, Italy.
×brauniae *Grover ex T.S. Cooperrider in Mich. Bot.,* 25(3): 108 (1986)—U.S.A. (Ohio). (V. rostrata × striata).
chassanica *R.I. Korkishko in Bot. Zhurn.,* 71(1): 87 (1986)—U.S.S.R. (Soviet Far East), Korea.
guestphalica *J.D. Nauenburg, Untersuch. Variab. Ökol. & Syst. Viola tricolor-Gruppe Mitteleuropa:* 98 (1986)—Germany.

VIOLA :–
lutea
subsp. calaminaria *(Gingins) J.D. Nauenburg, Untersuch. Variab. Ökol. & Syst. Viola tricolor-Gruppe Mitteleuropa:* 92 (1986): V. sudetica var. (β) calaminaria.
pyrenaica
subsp. montserratii *M.A. Fernández Casado & H.S. Nava Fernández in Candollea,* 41(1): 102 (1986)—Spain.

VIRGILIA (Legumin.).
oroboides
subsp. ferruginea *B.-E. van Wyk in S. Afr. J. Bot.,* 52(4): 351 (1986)—South Africa (Cape Province).

VITEX (Verbenac.).
ovata *C.P. Thunberg ex A. Murray in Linnaeus, Syst. Veg., ed. 14:* 578 (1784)—Japan. †

VITIS (Vitac.).
amurensis
f. hermaphrodita *J.Y. Li in Bull. Bot. Res. North-East. Forest. Inst.,* 6(2): 151 (1986)—Cultivation.
heterophylla *C.P. Thunberg ex A. Murray in Linnaeus, Syst. Veg., ed. 14:* 244 (1784)—Japan. †
japonica *C.P. Thunberg ex A. Murray in Linnaeus, Syst. Veg., ed. 14:* 244 (1784)—Japan. †
pentaphylla *C.P. Thunberg ex A. Murray in Linnaeus, Syst. Veg., ed. 14:* 244 (1784)—Japan. †

VOCHYSIA (Vochysiac.).
steyermarkiana *L. Marcano Berti in Pittieria,* 13: 10 (1986)—Venezuela.

VOLKAMERIA (Verbenac.).
japonica *C.P. Thunberg ex A. Murray in Linnaeus, Syst. Veg., ed. 14:* 578 (1784)—Japan. †

VOYRIA (Gentianac.).
corymbosa
subsp. alba *(Standley) P. Ruyters & P.J.M. Maas in Fl. Neotrop.,* 41: 57 (1986): Leiphaimos albus.

VRIESEA (Bromeliac.).
bi-beatricis *G. Morillo in Ernstia,* 39: 2 (1986)—Venezuela.
graciliscapa *W. Weber in Feddes Repert.,* 97(3–4): 104 (1986)—Brazil.
ouroensis *W. Weber in Feddes Repert.,* 97(3–4): 106 (1986)—Brazil.
roberto-seidelii *W. Weber in J. Bromeliad Soc.,* 36(1): 12 (1986)—Brazil.
seideliana *W. Weber in Feddes Repert.,* 97(3–4): 108 (1986)—Brazil.

W

WAHLENBERGIA (Campanulac.).
appressifolia *O.M. Hilliard & B.L. Burtt in Notes Roy. Bot. Gard. Edinburgh,* 43(3): 350 (1986)—South Africa (Natal), Lesotho.
doleritica *O.M. Hilliard & B.L. Burtt in Notes Roy. Bot. Gard. Edinburgh,* 43(2): 196 (1986)—South Africa (Natal).
hispidula *(Thunb.) Schrad., Ind. Sem. Hort. Acad. Gott.,* 1821: 4 (1821); *cf. D.J. Mabberley in Taxon,* 32(1): 86 (1983): Basionym not stated.
lurida *(Lindl.) Manetti, Cat. Pl. Caes. Reg. Hort. Modic.:* 106 (1842); *cf. D.J. Mabberley in Taxon,* 33(3): 443 (1984): Codonopsis lurida.
pallidiflora *O.M. Hilliard & B.L. Burtt in Notes Roy. Bot. Gard. Edinburgh,* 43(3): 351 (1986)—South Africa (Natal).
polytrichifolia
subsp. dracomontana *O.M. Hilliard & B.L. Burtt in Notes Roy. Bot. Gard. Edinburgh,* 43(2): 197 (1986)—South Africa (Natal), Lesotho.

WAITZIA (Iridac.).
crispa *(L.f.) Kreysig, Verz. Warm Kalt Hauspfl. Dresden:* 52 (1829); *cf. D.J. Mabberley in Taxon,* 33(3): 434 (1984): Basionym not stated.
deusta *(Ait.) Kreysig, Verz. Warm Kalt Hauspfl. Dresden:* 52 (1829); *cf. D.J. Mabberley in Taxon,* 33(3): 434 (1984): Basionym not stated.
fenestrata *(Jacq.) Kreysig, Verz. Warm Kalt Hauspfl. Dresden:* 52 (1829); *cf. D.J. Mabberley in Taxon,* 33(3): 434 (1984): Basionym not stated.
flava *(Ait.) Kreysig, Verz. Warm Kalt Hauspfl. Dresden:* 52 (1829); *cf. D.J. Mabberley in Taxon,* 33(3): 434 (1984): Basionym not stated.
lineata *(Salisb.) Kreysig, Verz. Warm Kalt Hauspfl. Dresden:* 52 (1829); *cf. D.J. Mabberley in Taxon,* 33(3): 434 (1984): Basionym not stated.
longiflora *(Berg) Kreysig, Verz. Warm Kalt Hauspfl. Dresden:* 52 (1829); *cf. D.J. Mabberley in Taxon,* 33(3): 434 (1984): Basionym not stated.
miniata *(Jacq.) Kreysig, Verz. Warm Kalt Hauspfl. Dresden:* 52 (1829); *cf. D.J. Mabberley in Taxon,* 33(3): 434 (1984): Basionym not stated.
squalida *(Ait.) Kreysig, Verz. Warm Kalt Hauspfl. Dresden:* 52 (1829); *cf. D.J. Mabberley in Taxon,* 33(3): 434 (1984): Basionym not stated.
viridis *(Ait.) Kreysig, Verz. Warm Kalt Hauspfl. Dresden:* 52 (1829); *cf. D.J. Mabberley in Taxon,* 33(3): 434 (1984): Basionym not stated.

WALLENIA (Myrsinac.).
formonensis *W.S. Judd in Sida,* 11(3): 329 (1986)—Hispaniola (Haiti).

WEBEROCEREUS (Cactac.).
glaber
var. mirandae *(Bravo) U. Eliasson in Kaktus,* 21(2): 44 (1986): Selenicereus mirandae.
rosei *(Kimnach) U. Eliasson in Kaktus,* 21(2): 44 (1986): Eccremocactus rosei.

WEDELIA (Compos.).
ayerscottiana *B.L. Turner in Phytologia,* 60(2): 125 (1986)—Mexico.
montana
var. pilosa *H. Koyama in Acta Phytotax. Geobot.,* 36(4–6): 170 (1985)—Thailand, Nepal, India, Assam, Burma, China.
var. wallichii *(Less.) H. Koyama in Acta Phytotax. Geobot.,* 36(4–6): 168 (1985): W. wallichii.
thailandica *H. Koyama in Acta Phytotax. Geobot.,* 36(4–6): 168 (1985)—Thailand.

WEINGARTIA (Cactac.).
albaoides *F. Brandt in Letzeb. Cacteefren,* 4(8): 2–8 (1983); *cf. Repert. Pl. Succ. (I.O.S.),* 34: 8 (1985)— Locality not stated.
subsp. subfusca *F. Brandt in Letzeb. Cacteefren,* 4(11): 1–7 (1983); *cf. Repert. Pl. Succ. (I.O.S.),* 34: 8 (1985)— Locality not stated.
hoffmanniana *F. Brandt in Letzeb. Cacteefrenn,* 5(11): 231–237 (1984), nom. nov.; *cf. Repert. Pl. Succ. (I.O.S.),* 35: 5 (1985): Lobivia hoffmanniana *Backeb.*
miranda *F.H. Brandt in Kakt. Orchid. Rundschau,* 10(4–5): 42 (1985)—Bolivia.
tarabucina *F. Brandt in Letzeb. Cacteefren,* 6(4): 57 (1985); *cf. Repert. Pl. Succ. (I.O.S.),* 36: 10 (1985 publ. 1986)—Locality not stated.

WETTINIA (Palmae).
longipetala *A.H. Gentry in Ann. Missouri Bot. Gard.,* 73(1): 160 (1986)—Peru.

WIKSTROEMIA (Thymelaeac.).
ser. Brachystachyae *S.C. Huang in Acta Bot. Yunnanica,* 7(3): 282 (1985).
ser. Haoianae *S.C. Huang in Acta Bot. Yunnanica,* 7(3): 290 (1985).
ser. Lichiangense *S.C. Huang in Acta Bot. Yunnanica,* 7(3): 284 (1985).
ser. Micranthae *S.C. Huang in Acta Bot. Yunnanica,* 7(3): 282 (1985).
ser. Parviflorae *S.C. Huang in Acta Bot. Yunnanica,* 7(3): 288 (1985).
ser. Scytophyllae *S.C. Huang in Acta Bot. Yunnanica,* 7(3): 288 (1985).
baimashanensis *S.C. Huang in Acta Bot. Yunnanica,* 7(3): 286 (1985)—China.
capitatoracemosa *S.C. Huang in Acta Bot. Yunnanica,* 7(3): 287 (1985), as 'capitato-racemosa'—China, Xizang.

WIKSTROEMIA :–
cochlearifolia *S.C. Huang in Acta Bot. Yunnanica,* 7(3): 290 (1985)—China.
glabra
f. purpurea *(Cheng) S.C. Huang in Acta Bot. Yunnanica,* 7(3): 282 (1985): W. glabra var. purpurea.
huidongensis *C.Y. Chang in Bull. Bot. Res. North-East. Forest. Inst.,* 6(4): 145 (1986)—China.
longipaniculata *S.C. Huang in Acta Bot. Yunnanica,* 7(3): 283 (1985)—China.
lungtzeensis *S.C. Huang in Acta Bot. Yunnanica,* 7(3): 285 (1985)—Xizang.
micrantha
var. paniculata *(Li) S.C. Huang in Acta Bot. Yunnanica,* 7(3): 283 (1985): W. paniculata.
parviflora *S.C. Huang in Acta Bot. Yunnanica,* 7(3): 289 (1985)—China.
pilosa
var. kulingensis *(Domke) S.C. Huang in Acta Bot. Yunnanica,* 7(3): 291 (1985): W. kulingensis.
stenophylla
var. ziyangensis *C.Y. Yu in Fl. Tsinlingensis,* 1(3): 455, 330 (1981)—China.
techinensis *S.C. Huang in Acta Bot. Yunnanica,* 7(3): 284 (1985)—China.

WILLDENOWIA (Restionac.).
glomerata *(Thunb.) H.P. Linder in Bothalia,* 15(3–4): 494 (1985): Restio glomeratus.
incurvata *(Thunb.) H.P. Linder in Bothalia,* 15(3–4): 494 (1985): Restio incurvatus.
rugosa *E.E. Esterhuysen in Bothalia,* 15(3–4): 495 (1985)—South Africa (Cape Province).

WINIFREDIA L.A.S. Johnson & B.G. Briggs in Telopea, 2(6): 737 (1986). *RESTIONACEAE.*
sola *L.A.S. Johnson & B.G. Briggs in Telopea,* 2(6): 738 (1986)—Tasmania.

WURMBEA (Colchicac.).
dolichantha *B. Nordenstam in Op. Bot.,* 87: 14 (1986)—South Africa (Cape Province).
graniticola *T.D. Macfarlane in Nuytsia,* 5(3): 407 (1986)—Australia (W. Australia).
hiemalis *B. Nordenstam in Op. Bot.,* 87: 17 (1986)—South Africa (Cape Province).
inusta *(Baker) B. Nordenstam in Op. Bot.,* 87: 18 (1986): W. capensis var. inusta.
marginata *(Desr.) B. Nordenstam in Op. Bot.,* 87: 20 (1986): Melanthium marginatum.
monopetala *(L.f.) B. Nordenstam in Op. Bot.,* 87: 22 (1986): Melanthium monopetalum.
murchisoniana *T.D. Macfarlane in Nuytsia,* 5(3): 410 (1986)—Australia (W. Australia).
recurva *B. Nordenstam in Op. Bot.,* 87: 23 (1986)—South Africa (Cape Province).
robusta *B. Nordenstam in Op. Bot.,* 87: 25 (1986)—South Africa (Cape Province).
spicata
var. ustulata *(B. Nord.) B. Nordenstam in Op. Bot.,* 87: 28 (1986): W. ustulata.

WURMBEA :–
variabilis *B. Nordenstam in Op. Bot.,* 87: 29 (1986)—South Africa (Cape Province).

X

XANTHOPHYTUM (Rubiac.).
attopevense *(Pierre ex Pitard) H.S. Lo in Bull. Bot. Res. North-East. Forest. Inst.,* 6(4): 32 (1986): Paedicalyx attopevensis.
balansae *(Pitard) H.S. Lo in Bull. Bot. Res. North-East. Forest. Inst.,* 6(4): 31 (1986): Xanthophytopsis balansae.
kwangtungense *(Chun & How) H.S. Lo in Bull. Bot. Res. North-East. Forest. Inst.,* 6(4): 32 (1986): Xanthophytopsis kwangtungensis.

XANTHORRHOEA (Xanthorrhoeac.).
acaulis *(A. Lee) D.J. Bedford in Fl. Australia,* 46: 225 (1986): X. australis subsp. acaulis.
arenaria *D.J. Bedford in Fl. Australia,* 46: 225, 167 (1986)—Tasmania.
brunonis
subsp. semibarbata *D.J. Bedford in Fl. Australia,* 46: 225, 157 (1986)—Australia (W. Australia).
caespitosa *D.J. Bedford in Fl. Australia,* 46: 226, 167 (1986)—Australia (S. Australia).
concava *(A. Lee) D.J. Bedford in Fl. Australia,* 46: 226 (1986): X. resinosa subsp. concava.
fulva *(A. Lee) D.J. Bedford in Fl. Australia,* 46: 226 (1986): X. resinosa subsp. fulva.
glauca *D.J. Bedford in Fl. Australia,* 46: 226, 164 (1986)—Australia (Queensland, New South Wales).
subsp. angustifolia *D.J. Bedford in Fl. Australia,* 46: 226, 165 (1986)—Australia (New South Wales).
latifolia *(A. Lee) D.J. Bedford in Fl. Australia,* 46: 227 (1986): X. media subsp. latifolia.
subsp. maxima *D.J. Bedford in Fl. Australia,* 46: 227, 163 (1986)—Australia (New South Wales).
malacophylla *D.J. Bedford in Fl. Australia,* 46: 227, 164 (1986)—Australia (New South Wales).
minor
subsp. lutea *D.J. Bedford in Fl. Australia,* 46: 228, 166 (1986)—Australia (Victoria, S. Australia).
platyphylla *D.J. Bedford in Fl. Australia,* 46: 228, 157 (1986)—Australia (W. Australia).
semiplana
subsp. tateana *(F. Muell.) D.J. Bedford in Fl. Australia,* 46: 229 (1986): X. tateana.

XANTHOSOMA (Arac.).
bolivaranum *G.S. Bunting in Phytologia,* 60(5): 339 (1986)—Venezuela.

XANTHOSOMA :–
caladioides *M.H. Grayum in Ann. Missouri Bot. Gard.*, 73(2): 471 (1986)—Panama.
maroae *G.S. Bunting in Phytologia*, 60(5): 341 (1986)—Venezuela.
plowmanii *J. Bogner in Aroideana*, 8(4): 112 (1985 publ. 1986)—Brazil.

XEROLIRION A.S. George in Fl. Australia, 46: 229, 98 (1986). *XANTHORRHOEACEAE.*
divaricata *A.S. George in Fl. Australia*, 46: 230, 98 (1986)—Australia (W. Australia).

XEROMPHIS (Rubiac.).
uliginosa *(Retz.) J.K. Maheshwari in Bull. Bot. Soc. Univ. Saugar*, 10(1–2): 39 (1958 publ. 1961); *et in Bull. Bot. Surv. Ind.*, 3: 92 (1962): Gardenia uliginosa. †

XEROPHYLLUM (Melanthiac.).
sabadilla *(Retz.) D. Don ex G. Don f. in Loud., Hort. Brit., ed. 2:* 602 (1832); *cf. D.J. Mabberley in Taxon*, 30(1): 17 (1981): Veratrum sabadilla.

XEROPHYTA (Velloziac.).
longicaulis *O.M. Hilliard in Notes Roy. Bot. Gard. Edinburgh*, 43(3): 405 (1986)—South Africa (Natal).

XEROSPIRAEA J. Henrickson in Aliso, 11(2): 206 (1985 publ. 1986). *ROSACEAE.*
hartwegiana *(Rydb.) J. Henrickson in Aliso*, 11(2): 206 (1985 publ. 1986): Spiraea hartwegiana.

XEROTHAMNELLA (Acanthac.).
herbacea *R.M. Barker in J. Adelaide Bot. Gard.*, 9: 169 (1986)—Australia (Queensland).

XIZANGIA D.Y. Hong in Acta Phytotax. Sin., 24(2): 139 (1986). *SCROPHULARIACEAE.*
serrata *D.Y. Hong in Acta Phytotax. Sin.*, 24(2): 141 (1986)—Xizang.

XYLOBIUM (Orchidac.).
racemosum *(Hook.) Sweet, Hort. Brit., ed. 2:* 489 (1830); *cf. D.J. Mabberley in Taxon*, 30(1): 17 (1981): Basionym not stated.

XYLOPHYLLA (Euphorbiac.).
epiphyllanthus *(L.) Hornem., Hort. Hafn.:* 961 (1815); *cf. D.J. Mabberley in Taxon*, 33(3): 438 (1984): Phyllanthus epiphyllanthus.

XYLOSMA (Flacourtiac.).
japonicum
var. pubescens *(Rehd. & Wils.) C.Y. Chang in Fl. Tsinlingensis*, 1(3): 324 (1981): X. racemosum var. pubescens.

Y

YUCCA (Agavac.).
boscii *Hort. Par. ex Hornem., Hort. Hafn.:* 342 (1813); *cf. D.J. Mabberley in Taxon*, 33(3): 438 (1984)— Locality not stated.
harrimaniae
var. sterilis *E. Neese & S.L. Welsh in Great Basin Nat.*, 45(4): 789 (1985)—U.S.A. (Utah).

YUSHANIA (Gramin.).
sect. Confusae *T.P. Yi in J. Bamboo Res.*, 5(1): 8 (1986).
andropogonoides *(Hand.-Mazz.) T.P. Yi in J. Bamboo Res.*, 5(1): 66 (1986): Indocalamus andropogonoides.
bojieiana *T.P. Yi in J. Bamboo Res.*, 5(1): 8 (1986)—China.
brevipaniculata *(Hand.-Mazz.) T.P. Yi in J. Bamboo Res.*, 5(1): 44 (1986): Arundinaria brevipaniculata.
brevis *T.P. Yi in J. Bamboo Res.*, 5(1): 11 (1986)—China.
chingii *T.P. Yi in J. Bamboo Res.*, 5(1): 45 (1986)—China.
collina *T.P. Yi in J. Bamboo Res.*, 5(1): 13 (1986)—China.
complanata *T.P. Yi in J. Bamboo Res.*, 5(1): 15 (1986)—China.
elevata *T.P. Yi in J. Bamboo Res.*, 5(1): 17 (1986)—China.
exilis *T.P. Yi in J. Bamboo Res.*, 5(1): 20 (1986)—China.
falcatiaurita *Hsueh & T.P. Yi in J. Bamboo Res.*, 5(1): 22 (1986)—China.
farcticaulis *T.P. Yi in J. Bamboo Res.*, 5(1): 24 (1986)—China.
levigata *T.P. Yi in J. Bamboo Res.*, 5(1): 27 (1986)—China.
longiuscula *T.P. Yi in J. Bamboo Res.*, 5(1): 30 (1986)—China.
mabianensis *T.P. Yi in J. Bamboo Res.*, 5(1): 47 (1986)—China.
maculata *T.P. Yi in J. Bamboo Res.*, 5(1): 33 (1986)—China.
monophylla *T.P. Yi & B.M. Yang in J. Bamboo Res.*, 5(1): 50 (1986)—China.
oblonga *T.P. Yi in J. Bamboo Res.*, 5(1): 52 (1986)—China.
pachyclada *T.P. Yi in J. Bamboo Res.*, 5(1): 54 (1986)—China.
polytricha *Hsueh & T.P. Yi in J. Bamboo Res.*, 5(1): 58 (1986)—China.
punctulata *T.P. Yi in J. Bamboo Res.*, 5(1): 59 (1986)—China.
qiaojiaensis *Hsueh & T.P. Yi in J. Bamboo Res.*, 5(1): 35 (1986)—China.
rugosa *T.P. Yi in J. Bamboo Res.*, 5(1): 61 (1986)—China.
uniramosa *Hsueh & T.P. Yi in J. Bamboo Res.*, 5(1): 64 (1986)—China.
varians *T.P. Yi in J. Bamboo Res.*, 5(1): 38 (1986)—China.

YUSHANIA :–
vigens *T.P. Yi in J. Bamboo Res.*, 5(1): 40 (1986)—China.
violascens *(Keng) T.P. Yi in J. Bamboo Res.*, 5(1): 45 (1986): Arundinaria violascens.
weixiensis *T.P. Yi in J. Bamboo Res.*, 5(1): 42 (1986)—China.

Z

ZAMIA (Zamiac.).
mexicana *G. Don f. in Loud., Encycl. Pl., new ed.:* 1526 (1855); *cf. D.J. Mabberley in Taxon,* 30(1): 17 (1981)— Locality not stated.

ZANTHOXYLUM (Rutac.).
molokaiense *B.C. Stone in Proc. Acad. Nat. Sci. Philadelphia,* 137(2): 213 (1985)—Hawaiian Is.
rolandii *C. Beurton in Feddes Repert.*, 97(1–2): 35 (1986)—Cuba.

ZEA (Gramin.).
odontosperma *Ten., Cat. Ort. Bot. Nap.:* 99 (1845); *cf. D.J. Mabberley in Taxon,* 30(1): 17 (1981)— Locality not stated.

ZEHNERIA (Cucurbitac.).
suavis *(Schrad.) Endl., Cat. Horti Vindob.*, 2: 279 (1842); *cf. D.J. Mabberley in Taxon,* 33(3): 436 (1984): Basionym not stated.

ZEPHYRANTHES (Amaryllidac.).
×flaggii *L.B. Spencer in Phytologia,* 59(2): 87 (1986)—Panama. (Z. albiella × rosea).
×floryi *L.B. Spencer in Phytologia,* 59(2): 88 (1986)—Cultivation. (Z. × ajax × grandiflora).
guatemalensis *L.B. Spencer in Phytologia,* 59(2): 85 (1986)—Guatemala.

ZEPHYRANTHES :–
katheriniae *L.B. Spencer in Phytologia,* 59(2): 86 (1986)—Mexico.
latissimafolia *L.B. Spencer in Phytologia,* 59(2): 86 (1986)—Mexico.
subflava *L.B. Spencer in Phytologia,* 59(2): 87 (1986)—Mexico.

ZEUGITES (Gramin.).
americana
var. mexicana *(Kunth) R. McVaugh, Fl. Novo-Galiciana,* 14: 413 (1983): Despretzia mexicana.
var. pringlei *(Scribn.) R. McVaugh, Fl. Novo-Galiciana,* 14: 413 (1983): Z. pringlei.

ZEUXINE (Orchidac.).
ovata *(Gaud.) L.A. Garay & W. Kittredge in Bot. Mus. Leafl. Harvard Univ.*, 30(3): (59) 193 (1985 publ. 1986): Nervilia ovata.

ZINNIA (Compos.).
sulphurea *Hort. ex J.W. Loud., Lad. Fl. Gard. Orn. Ann.:* 204 (1840); *cf. D.J. Mabberley in Taxon,* 32(1): 87 (1983)— Locality not stated.

ZOOTROPHION (Orchidac.).
schenckii *(Cogn.) C.A. Luer, Ic. Pleurothallid. I-Syst. Pleurothallidinae (Monogr. Syst. Bot. Missouri Bot. Gard. 15):* 69 (1986): Cryptophoranthus schenckii.

ZUCCAGNIA (Liliac.).
serotina *(L.) Dennst., Hort. Belved.:* 104 (1820); *cf. D.J. Mabberley in Taxon,* 31(1): 73 (1982): Basionym not stated.

ZUCKIA (Chenopodiac.).
brandegei *(Gray) S.L. Welsh & Stutz in Great Basin Nat.*, 44(2): 208 (1984): Grayia brandegei.
var. arizonica *(Standl.) S.L. Welsh in Great Basin Nat.*, 44(2): 209 (1984): Z. arizonica.

APPENDIX — FERNS & FERN ALLIES

NAMES AT RANK OF FAMILY

CYSTODIACEAE
fam. **Cystodiaceae** J.R. Croft in Kew Bull., 41(4): 797 (1986).

NAMES AT GENERIC RANK AND BELOW

A

ACROSTICHUM
hastatum *C.P. Thunberg ex A. Murray in Linnaeus, Syst. Veg., ed. 14:* 928 (1784)— Japan. †
incisum *(Link) Endl., Cat. Horti Vindob.*, 1: 1 (1842); *cf. D.J. Mabberley in Taxon,* 33(3): 434 (1984): Basionym not stated.
lingua *C.P. Thunberg ex A. Murray in Linnaeus, Syst. Veg., ed. 14:* 928 (1784)— Japan. †
olfersia *Endl., Cat. Horti Vindob.*, 1: 2 (1842); *cf. D.J. Mabberley in Taxon,* 33(3): 434 (1984)— Locality not stated.

ADIANTUM
cardiochlaena *Kunze, Ind. Fil. Hort. Bot. Lips.*, 1837-1843: 2 (1843); *cf. D.J. Mabberley in Taxon,* 32(1): 84 (1983)— Locality not stated.

ALEURITOPTERIS
doniana *S.K. Wu ex R.C. Ching in Indian Fern J.*, 1(1–2): 49 (1984 publ. 1985), nom. nov.: Cheilanthes dealbata *Don*

ALLOSORUS
falcatus *(R. Br.) Kunze, Ind. Fil. Hort. Bot. Lips.*, 1837-1843: 2 (1843); *cf. D.J. Mabberley in Taxon,* 32(1): 84 (1983): Basionym not stated.

ALSOPHILA
acanthophora *(Holtt.) R. Tryon in Contrib. Gray Herb.*, 200: 32 (1970): Cyathea acanthophora Holtt.
acuminata *(Copel.) R. Tryon in Contrib. Gray Herb.*, 200: 32 (1970): Cyathea acuminata Copel.
acutula *R. Tryon in Contrib. Gray Herb.*, 200: 29 (1970), nom. nov.: Cyathea tsaratananensis *Tard.*
albida *(Tard.) R. Tryon in Contrib. Gray Herb.*, 200: 29 (1970): Cyathea albida Tard.
alderwereltii *(Copel.) R. Tryon in Contrib. Gray Herb.*, 200: 32 (1970): Cyathea alderwereltii Copel.
alleniae *(Holtt.) R. Tryon in Contrib. Gray Herb.*, 200: 32 (1970): Cyathea alleniae Holtt.
alta *(Copel.) R. Tryon in Contrib. Gray Herb.*, 200: 36 (1970): Cyathea alta Copel.
alticola *(Tard.) R. Tryon in Contrib. Gray Herb.*, 200: 29 (1970): Gymnosphaera alticola Tard.
aneitensis *(Hook.) R. Tryon in Contrib. Gray Herb.*, 200: 36 (1970): Cyathea aneitensis Hook.
apoensis *(Copel.) R. Tryon in Contrib. Gray Herb.*, 200: 32 (1970): Cyathea apoensis Copel.
appendiculata *(Baker) R. Tryon in Contrib. Gray Herb.*, 200: 29 (1970): Cyathea appendiculata Baker
approximata *(Bonap.) R. Tryon in Contrib. Gray Herb.*, 200: 29 (1970): Cyathea approximata Bonap.
archboldii *(C. Chr.) R. Tryon in Contrib. Gray Herb.*, 200: 32 (1970): Cyathea archboldii C. Chr.
auriculata *(Tard.) R. Tryon in Contrib. Gray Herb.*, 200: 29 (1970): Cyathea auriculata Tard.

ALSOPHILA :–

ballardii *(Tard.) R. Tryon in Contrib. Gray Herb.*, 200: 29 (1970): Cyathea ballardii Tard.

bellisquamata *(Bonap.) R. Tryon in Contrib. Gray Herb.*, 200: 29 (1970): Cyathea bellisquamata Bonap.

borbonica *(Desv.) R. Tryon in Contrib. Gray Herb.*, 200: 30 (1970): Cyathea borbonica Desv.

borneensis *(Copel.) R. Tryon in Contrib. Gray Herb.*, 200: 31 (1970): Cyathea borneensis Copel.

brausei *R. Tryon in Contrib. Gray Herb.*, 200: 33 (1970), nom. nov.: Cyathea hunsteiniana *Brause*

brevipinna *(Benth.) R. Tryon in Contrib. Gray Herb.*, 200: 36 (1970): Cyathea brevipinna Benth.

buennemeijeri *(vAvR.) R. Tryon in Contrib. Gray Herb.*, 200: 33 (1970): Cyathea buennemeijeri vAvR.

callosa *(Christ) R. Tryon in Contrib. Gray Herb.*, 200: 33 (1970): Cyathea callosa Christ

camerooniana *(Hook.) R. Tryon in Contrib. Gray Herb.*, 200: 30 (1970): Cyathea camerooniana Hook.

campanulata *R. Tryon in Contrib. Gray Herb.*, 200: 30 (1970), nom. nov.: Cyathea holstii *Hieron.*

catillifera *(Holtt.) R. Tryon in Contrib. Gray Herb.*, 200: 33 (1970): Cyathea catillifera Holtt.

celsa *R. Tryon in Contrib. Gray Herb.*, 200: 30 (1970), nom. nov.: Cyathea excelsa *Sw.*

cicatricosa *(Holtt.) R. Tryon in Contrib. Gray Herb.*, 200: 36 (1970): Cyathea cicatricosa Holtt.

cincinnata *(Brause) R. Tryon in Contrib. Gray Herb.*, 200: 33 (1970): Cyathea cincinnata Brause

cinerea *(Copel.) R. Tryon in Contrib. Gray Herb.*, 200: 33 (1970): Cyathea cinerea Copel.

coactilis *(Holtt.) R. Tryon in Contrib. Gray Herb.*, 200: 33 (1970): Cyathea coactilis Holtt.

costalisora *(Copel.) R. Tryon in Contrib. Gray Herb.*, 200: 33 (1970): Cyathea costalisora Copel.

crassicaula *R. Tryon in Contrib. Gray Herb.*, 200: 33 (1970), nom. nov.: Cyathea ledermannii *Brause*

cucullifera *(Holtt.) R. Tryon in Contrib. Gray Herb.*, 200: 33 (1970): Cyathea cucullifera Holtt.

cunninghamii *(Hook. f.) R. Tryon in Contrib. Gray Herb.*, 200: 36 (1970): Cyathea cunninghamii Hook. f.

deckenii *(Kuhn) R. Tryon in Contrib. Gray Herb.*, 200: 30 (1970): Cyathea deckenii Kuhn

decrescens *(Kuhn) R. Tryon in Contrib. Gray Herb.*, 200: 30 (1970): Cyathea decrescens Kuhn

ALSOPHILA :–

dicksonioides *(Holtt.) R. Tryon in Contrib. Gray Herb.*, 200: 33 (1970): Cyathea dicksonioides Holtt.

doctersii *(vAvR.) R. Tryon in Contrib. Gray Herb.*, 200: 33 (1970): Cyathea doctersii vAvR.

dregei *(Kze.) R. Tryon in Contrib. Gray Herb.*, 200: 30 (1970): Cyathea dregei Kze.

edanoi *(Copel.) R. Tryon in Contrib. Gray Herb.*, 200: 33 (1970): Cyathea edanoi Copel.

eriophora *(Holtt.) R. Tryon in Contrib. Gray Herb.*, 200: 33 (1970): Cyathea eriophora Holtt.

everta *(Copel.) R. Tryon in Contrib. Gray Herb.*, 200: 33 (1970): Cyathea everta Copel.

excavata *(Holtt.) R. Tryon in Contrib. Gray Herb.*, 200: 33 (1970): Cyathea excavata Holtt.

ferdinandii *R. Tryon in Contrib. Gray Herb.*, 200: 37 (1970), nom. nov.: Hemitelia macarthurii *F.v.Muell.*

ferruginea *(Christ) R. Tryon in Contrib. Gray Herb.*, 200: 33 (1970): Cyathea ferruginea Christ

foersteri *(Rosenst.) R. Tryon in Contrib. Gray Herb.*, 200: 33 (1970): Cyathea foersteri Rosenst.

geluensis *(Rosenst.) R. Tryon in Contrib. Gray Herb.*, 200: 33 (1970): Cyathea geluensis Rosenst.

glaberrima *(Holtt.) R. Tryon in Contrib. Gray Herb.*, 200: 33 (1970): Cyathea glaberrima Holtt.

glaucifolia *R. Tryon in Contrib. Gray Herb.*, 200: 30 (1970), nom. nov.: Cyathea glauca *Bory*

gleichenioides *(C. Chr.) R. Tryon in Contrib. Gray Herb.*, 200: 33 (1970): Cyathea gleichenioides C. Chr.

halconensis *(Christ) R. Tryon in Contrib. Gray Herb.*, 200: 34 (1970): Cyathea halconensis Christ

havilandii *(Baker) R. Tryon in Contrib. Gray Herb.*, 200: 34 (1970): Cyathea havilandii Baker

hermannii *R. Tryon in Contrib. Gray Herb.*, 200: 34 (1970), nom. nov.: Cyathea christii *Copel.*

heterochlamydea *(Copel.) R. Tryon in Contrib. Gray Herb.*, 200: 34 (1970): Cyathea heterochlamydea Copel.

hildebrandtii *(Kuhn) R. Tryon in Contrib. Gray Herb.*, 200: 30 (1970): Cyathea hildebrandtii Kuhn

hooglandii *(Holtt.) R. Tryon in Contrib. Gray Herb.*, 200: 34 (1970): Cyathea hooglandii Holtt.

hookeri *(Thwaites) R. Tryon in Contrib. Gray Herb.*, 200: 32 (1970): Cyathea hookeri Thwaites

horridula *(Copel.) R. Tryon in Contrib. Gray Herb.*, 200: 34 (1970): Cyathea horridula Copel.

ALSOPHILA :–

humbertiana *(C. Chr.) R. Tryon in Contrib. Gray Herb.,* 200: 30 (1970): Hemitelia humbertiana C. Chr.

hyacinthei *R. Tryon in Contrib. Gray Herb.,* 200: 30 (1970), nom. nov.: Cyathea boivinii *Kuhn*

hymenodes *(Mett.) R. Tryon in Contrib. Gray Herb.,* 200: 34 (1970): Cyathea hymenodes Mett.

imbricata *(vAvR.) R. Tryon in Contrib. Gray Herb.,* 200: 34 (1970): Cyathea imbricata vAvR.

inquinans *(Christ) R. Tryon in Contrib. Gray Herb.,* 200: 34 (1970): Cyathea inquinans Christ

isaloensis *(C. Chr.) R. Tryon in Contrib. Gray Herb.,* 200: 30 (1970): Cyathea isaloensis C. Chr.

javanica *(Bl.) R. Tryon in Contrib. Gray Herb.,* 200: 34 (1970): Cyathea javanica Bl.

kermadecensis *(Oliver) R. Tryon in Contrib. Gray Herb.,* 200: 37 (1970): Cyathea kermadecensis Oliver

kirkii *(Hook.) R. Tryon in Contrib. Gray Herb.,* 200: 30 (1970): Cyathea kirkii Hook.

klossii *(Ridley) R. Tryon in Contrib. Gray Herb.,* 200: 34 (1970): Cyathea klossii Ridley

lastii *(Baker) R. Tryon in Contrib. Gray Herb.,* 200: 30 (1970): Cyathea lastii Baker

latipinnula *(Copel.) R. Tryon in Contrib. Gray Herb.,* 200: 34 (1970): Cyathea latipinnula Copel.

leptochlamys *(Baker) R. Tryon in Contrib. Gray Herb.,* 200: 30 (1970): Cyathea leptochlamys Baker

ligulata *(Baker) R. Tryon in Contrib. Gray Herb.,* 200: 30 (1970): Cyathea ligulata Baker

lilianiae *R. Tryon in Contrib. Gray Herb.,* 200: 34 (1970), nom. nov.: Cyathea arfakensis *Gepp*

loerzingii *(Holtt.) R. Tryon in Contrib. Gray Herb.,* 200: 34 (1970): Cyathea loerzingii Holtt.

loheri *(Christ) R. Tryon in Contrib. Gray Herb.,* 200: 32 (1970): Cyathea loheri Christ

longipes *(Copel.) R. Tryon in Contrib. Gray Herb.,* 200: 34 (1970): Cyathea longipes Copel.

longipinnata *(Bonap.) R. Tryon in Contrib. Gray Herb.,* 200: 30 (1970): Cyathea longipinnata Bonap.

macgregorii *(F.v.Muell.) R. Tryon in Contrib. Gray Herb.,* 200: 34 (1970): Cyathea macgregorii F.v.Muell.

macropoda *(Domin) R. Tryon in Contrib. Gray Herb.,* 200: 34 (1970): Cyathea macropoda Domin

magnifolia *(vAvR.) R. Tryon in Contrib. Gray Herb.,* 200: 34 (1970): Cyathea magnifolia vAvR.

manniana *(Hook.) R. Tryon in Contrib. Gray Herb.,* 200: 30 (1970): Cyathea manniana Hook.

ALSOPHILA :–

marattioides *(Kaulf.) R. Tryon in Contrib. Gray Herb.,* 200: 30 (1970): Cyathea marattioides Kaulf.

marcescens *(N.A. Wakef.) R. Tryon in Contrib. Gray Herb.,* 200: 37 (1970): Cyathea marcescens N.A. Wakef.

masapilidensis *(Copel.) R. Tryon in Contrib. Gray Herb.,* 200: 34 (1970): Cyathea masapilidensis Copel.

matitanensis *R. Tryon in Contrib. Gray Herb.,* 200: 30 (1970), nom. nov.: Cyathea madagascarica *Bonap.*

media *(Wagn. & Greth.) R. Tryon in Contrib. Gray Herb.,* 200: 35 (1970): Cyathea media Wagn. & Greth.

melanocaula *(Desv.) R. Tryon in Contrib. Gray Herb.,* 200: 30 (1970): Cyathea melanocaula Desv.

melleri *(Baker) R. Tryon in Contrib. Gray Herb.,* 200: 31 (1970): Hemitelia melleri Baker

mesosora *(Holtt.) R. Tryon in Contrib. Gray Herb.,* 200: 35 (1970): Cyathea mesosora Holtt.

micra *R. Tryon in Contrib. Gray Herb.,* 200: 35 (1970), nom. nov.: Cyathea parva *Copel.*

microchlamys *(Holtt.) R. Tryon in Contrib. Gray Herb.,* 200: 35 (1970): Cyathea microchlamys Holtt.

microphylloides *(Rosenst.) R. Tryon in Contrib. Gray Herb.,* 200: 35 (1970): Cyathea microphylloides Rosenst.

milnei *(Hook. f.) R. Tryon in Contrib. Gray Herb.,* 200: 37 (1970): Cyathea milnei Hook. f.

montana *(vAvR.) R. Tryon in Contrib. Gray Herb.,* 200: 35 (1970): Hemitelia montana vAvR.

mossambicensis *(Baker) R. Tryon in Contrib. Gray Herb.,* 200: 31 (1970): Cyathea mossambicensis Baker

muelleri *(Baker) R. Tryon in Contrib. Gray Herb.,* 200: 35 (1970): Cyathea muelleri Baker

negrosiana *(Christ) R. Tryon in Contrib. Gray Herb.,* 200: 35 (1970): Cyathea negrosiana Christ

nicklesii *(Tard. & Ballard) R. Tryon in Contrib. Gray Herb.,* 200: 31 (1970): Gymnosphaera nicklesii Tard. & Ballard

nigrolineata *(Holtt.) R. Tryon in Contrib. Gray Herb.,* 200: 35 (1970): Cyathea nigrolineata Holtt.

nigropaleata *(Holtt.) R. Tryon in Contrib. Gray Herb.,* 200: 35 (1970): Cyathea nigropaleata Holtt.

nilgirensis *(Holtt.) R. Tryon in Contrib. Gray Herb.,* 200: 32 (1970): Cyathea nilgirensis Holtt.

oinops *(Hassk.) R. Tryon in Contrib. Gray Herb.,* 200: 35 (1970): Cyathea oinops Hassk.

oosora *(Holtt.) R. Tryon in Contrib. Gray Herb.,* 200: 35 (1970): Cyathea oosora Holtt.

ALSOPHILA :–

orientalis *(Kze.) R. Tryon in Contrib. Gray Herb.*, 200: 35 (1970): Disphenia orientalis Kze.

orthogonalis *(Bonap.) R. Tryon in Contrib. Gray Herb.*, 200: 31 (1970): Cyathea orthogonalis Bonap.

pachyrrhachis *(Copel.) R. Tryon in Contrib. Gray Herb.*, 200: 35 (1970): Cyathea pachyrrhachis Copel.

pallidipaleata *(Holtt.) R. Tryon in Contrib. Gray Herb.*, 200: 35 (1970): Cyathea pallidipaleata Holtt.

patellifera *(vAvR.) R. Tryon in Contrib. Gray Herb.*, 200: 35 (1970): Cyathea patellifera vAvR.

percrassa *(C. Chr.) R. Tryon in Contrib. Gray Herb.*, 200: 35 (1970): Cyathea percrassa C. Chr.

perpelvigera *(vAvR.) R. Tryon in Contrib. Gray Herb.*, 200: 35 (1970): Cyathea perpelvigera vAvR.

perpunctulata *(vAvR.) R. Tryon in Contrib. Gray Herb.*, 200: 35 (1970): Hemitelia perpunctulata vAvR.

perrieriana *(C. Chr.) R. Tryon in Contrib. Gray Herb.*, 200: 31 (1970): Cyathea perrieriana C. Chr.

physolepidota *(Alston) R. Tryon in Contrib. Gray Herb.*, 200: 35 (1970): Cyathea physolepidota Alston

pilosula *(Tard.) R. Tryon in Contrib. Gray Herb.*, 200: 31 (1970): Cyathea pilosula Tard.

plagiostegia *(Copel.) R. Tryon in Contrib. Gray Herb.*, 200: 37 (1970): Cyathea plagiostegia Copel.

polycarpa *(Jungh.) R. Tryon in Contrib. Gray Herb.*, 200: 35 (1970): Cyathea polycarpa Jungh.

pruinosa *(Rosenst.) R. Tryon in Contrib. Gray Herb.*, 200: 35 (1970): Cyathea pruinosa Rosenst.

pseudomuelleri *(Holtt.) R. Tryon in Contrib. Gray Herb.*, 200: 35 (1970): Cyathea pseudomuelleri Holtt.

pycnoneura *(Holtt.) R. Tryon in Contrib. Gray Herb.*, 200: 35 (1970): Cyathea pycnoneura Holtt.

quadrata *(Baker) R. Tryon in Contrib. Gray Herb.*, 200: 31 (1970): Cyathea quadrata Baker

rigens *(Rosenst.) R. Tryon in Contrib. Gray Herb.*, 200: 36 (1970): Cyathea rigens Rosenst.

rolandii *R. Tryon in Contrib. Gray Herb.*, 200: 31 (1970), nom. nov.: Cyathea costularis *Roland Bonap.*

rubella *(Holtt.) R. Tryon in Contrib. Gray Herb.*, 200: 36 (1970): Cyathea rubella Holtt.

rufopannosa *(Christ) R. Tryon in Contrib. Gray Herb.*, 200: 36 (1970): Cyathea rufopannosa Christ

ALSOPHILA :–

saccata *(Christ) R. Tryon in Contrib. Gray Herb.*, 200: 36 (1970): Cyathea saccata Christ

salletii *(Tard. & C. Chr.) R. Tryon in Contrib. Gray Herb.*, 200: 32 (1970): Cyathea salletii Tard. & C. Chr.

sechellarum *(Mett.) R. Tryon in Contrib. Gray Herb.*, 200: 31 (1970): Cyathea sechellarum Mett.

semiamplectens *(Holtt.) R. Tryon in Contrib. Gray Herb.*, 200: 36 (1970): Cyathea semiamplectens Holtt.

serratifolia *(Baker) R. Tryon in Contrib. Gray Herb.*, 200: 31 (1970): Cyathea serratifolia Baker

setulosa *(Copel.) R. Tryon in Contrib. Gray Herb.*, 200: 36 (1970): Cyathea setulosa Copel.

similis *(C. Chr.) R. Tryon in Contrib. Gray Herb.*, 200: 31 (1970): Cyathea similis C. Chr.

sinuata *(Hook. & Grev.) R. Tryon in Contrib. Gray Herb.*, 200: 32 (1970): Cyathea sinuata Hook. & Grev.

smithii *(Hook. f.) R. Tryon in Contrib. Gray Herb.*, 200: 37 (1970): Cyathea smithii Hook. f.

solomonensis *(Holtt.) R. Tryon in Contrib. Gray Herb.*, 200: 37 (1970): Cyathea solomonensis Holtt.

spinulosa *(Hook.) R. Tryon in Contrib. Gray Herb.*, 200: 32 (1970): Cyathea spinulosa Hook.

stelligera *(Holtt.) R. Tryon in Contrib. Gray Herb.*, 200: 37 (1970): Cyathea stelligera Holtt.

stokesii *(E. Brown) R. Tryon in Contrib. Gray Herb.*, 200: 37 (1970): Cyathea stokesii E. Brown

stuhlmannii *(Hieron.) R. Tryon in Contrib. Gray Herb.*, 200: 31 (1970): Cyathea stuhlmannii Hieron.

subtripinnata *(Holtt.) R. Tryon in Contrib. Gray Herb.*, 200: 36 (1970): Cyathea subtripinnata Holtt.

sumatrana *(Baker) R. Tryon in Contrib. Gray Herb.*, 200: 36 (1970): Cyathea sumatrana Baker

tanzaniana *R. Tryon in Contrib. Gray Herb.*, 200: 31 (1970), nom. nov.: Cyathea schliebenii *Reim.*

ternatea *(vAvR.) R. Tryon in Contrib. Gray Herb.*, 200: 36 (1970): Cyathea ternatea vAvR.

thomsonii *(Baker) R. Tryon in Contrib. Gray Herb.*, 200: 31 (1970): Cyathea thomsonii Baker

tricolor *(Colenso) R. Tryon in Contrib. Gray Herb.*, 200: 37 (1970): Cyathea tricolor Colenso

tsaratananensis *(C. Chr.) R. Tryon in Contrib. Gray Herb.*, 200: 31 (1970): Cyathea tsaratananensis C. Chr.

ALSOPHILA :-

tsilotsilensis *(Tard.) R. Tryon in Contrib. Gray Herb.*, 200: 31 (1970): Cyathea tsilotsilensis Tard.

vandeusenii *(Holtt.) R. Tryon in Contrib. Gray Herb.*, 200: 36 (1970): Cyathea vandeusenii Holtt.

vieillardii *(Mett.) R. Tryon in Contrib. Gray Herb.*, 200: 37 (1970): Cyathea vieillardii Mett.

viguieri *(Tard.) R. Tryon in Contrib. Gray Herb.*, 200: 31 (1970): Cyathea viguieri Tard.

welwitschii *(Hook.) R. Tryon in Contrib. Gray Herb.*, 200: 31 (1970): Cyathea welwitschii Hook.

zakamenensis *(Tard.) R. Tryon in Contrib. Gray Herb.*, 200: 31 (1970): Cyathea zakamenensis Tard.

AMAUROPELTA

knysnaensis *(N.C. Anthony & Schelpe) B.S. Parris in Kew Bull.*, 41(1): 70 (1986): Thelypteris knysnaensis N.C. Anthony & Schelpe

AMPHINEURON

queenslandicum *R.E. Holttum in Kew Bull.*, 41(3): 518 (1986)—Australia (Queensland).

ANEMIA

blechnoides *Sm. in Rees, Cyclop.*, 39(1): Anemia n. 3 (1818); *cf. D.J. Mabberley in Taxon*, 32(1): 81 (1983)— Brazil.

ARACHNIODES

abrupta *R.C. Ching in Bull. Bot. Res. North-East. Forest. Inst.*, 6(3): 35 (1986)—China.

ailaoshanensis *R.C. Ching in Bull. Bot. Res. North-East. Forest. Inst.*, 6(3): 60 (1986)—China.

anshunensis *R.C. Ching & Y.T. Hsieh in Bull. Bot. Res. North-East. Forest. Inst.*, 6(3): 67 (1986) China.

aristatissima *R.C. Ching in Bull. Bot. Res. North-East. Forest. Inst.*, 6(3): 1 (1986)—China.

attenuata *R.C. Ching in Bull. Bot. Res. North-East. Forest. Inst.*, 6(3): 2 (1986)—China.

austroyunnanensis *R.C. Ching in Bull. Bot. Res. North-East. Forest. Inst.*, 6(3): 3 (1986), as 'austro-yunnanensis'—China.

baiseensis *R.C. Ching in Bull. Bot. Res. North-East. Forest. Inst.*, 6(3): 25 (1986)—China.

borealis *S. Serizawa in J. Jap. Bot.*, 61(2): 48 (1986)—Japan, Korea.

f. ciliata *(Nakaike) S. Serizawa in J. Jap. Bot.*, 61(2): 52 (1986): Leptorumohra miqueliana f. ciliata Nakaike

calcarata *R.C. Ching in Bull. Bot. Res. North-East. Forest. Inst.*, 6(3): 30 (1986)—China.

chingii *Y.T. Hsieh in Bull. Bot. Res. North-East. Forest. Inst.*, 6(4): 4 (1986)—China.

cornopteris *R.C. Ching in Bull. Bot. Res. North-East. Forest. Inst.*, 6(3): 4 (1986)—China.

costulisora *R.C. Ching in Bull. Bot. Res. North-East. Forest. Inst.*, 6(3): 62 (1986)—China.

ARACHNIODES :-

cyrtomifolia *R.C. Ching in Bull. Bot. Res. North-East. Forest. Inst.*, 6(3): 31 (1986)—China.

damiaoshanensis *Y.T. Hsieh in Bull. Bot. Res. North-East. Forest. Inst.*, 6(4): 6 (1986)—China.

decomposita *R.C. Ching in Bull. Bot. Res. North-East. Forest. Inst.*, 6(3): 49 (1986)—China.

elevatas *R.C. Ching in Bull. Bot. Res. North-East. Forest. Inst.*, 6(3): 40 (1986)—China.

emeiensis *R.C. Ching in Bull. Bot. Res. North-East. Forest. Inst.*, 6(3): 5 (1986)—China.

erythrosora *R.C. Ching in Bull. Bot. Res. North-East. Forest. Inst.*, 6(3): 42 (1986)—China.

falcata *R.C. Ching in Bull. Bot. Res. North-East. Forest. Inst.*, 6(3): 7 (1986)—China.

fengii *R.C. Ching in Bull. Bot. Res. North-East. Forest. Inst.*, 6(3): 8 (1986)—China.

foeniculacea *R.C. Ching in Bull. Bot. Res. North-East. Forest. Inst.*, 6(3): 45 (1986)—China.

fujiangensis *R.C. Ching in Bull. Bot. Res. North-East. Forest. Inst.*, 6(3): 29 (1986)—China.

futeshanensis *Y.T. Hsieh in Bull. Bot. Res. North-East. Forest. Inst.*, 6(4): 5 (1986)—China.

gigantea *R.C. Ching in Bull. Bot. Res. North-East. Forest. Inst.*, 6(3): 66 (1986)—China.

gijiangensis *R.C. Ching in Bull. Bot. Res. North-East. Forest. Inst.*, 6(3): 33 (1986)—China.

gizushanensis *R.C. Ching in Bull. Bot. Res. North-East. Forest. Inst.*, 6(3): 41 (1986)—China.

gongshanensis *R.C. Ching & Y.T. Hsieh in Bull. Bot. Res. North-East. Forest. Inst.*, 6(3): 68 (1986)—China.

gradata *R.C. Ching in Bull. Bot. Res. North-East. Forest. Inst.*, 6(3): 39 (1986)—China.

guangtongensis *R.C. Ching in Bull. Bot. Res. North-East. Forest. Inst.*, 6(3): 58 (1986)—China.

guangxiensis *R.C. Ching in Bull. Bot. Res. North-East. Forest. Inst.*, 6(3): 27 (1986)—China.

guanxianensis *R.C. Ching in Bull. Bot. Res. North-East. Forest. Inst.*, 6(3): 50 (1986)—China.

hekouensis *R.C. Ching in Bull. Bot. Res. North-East. Forest. Inst.*, 6(3): 57 (1986)—China.

heyuanensis *R.C. Ching in Bull. Bot. Res. North-East. Forest. Inst.*, 6(3): 9 (1986)—China.

huapingensis *R.C. Ching & Chiu in Bull. Bot. Res. North-East. Forest. Inst.*, 6(3): 53 (1986)—China.

hunanensis *R.C. Ching in Bull. Bot. Res. North-East. Forest. Inst.*, 6(3): 10 (1986)—China.

×ikeminensis *S. Serizawa in J. Jap. Bot.*, 61(2): 55 (1986)—Japan. (A. miqueliana × quadripinnata subsp. fimbriata).

jiangxiensis *R.C. Ching in Bull. Bot. Res. North-East. Forest. Inst.*, 6(3): 43 (1986)—China.

ARACHNIODES :-

jinfushanensis *R.C. Ching in Bull. Bot. Res. North-East. Forest. Inst.*, 6(3): 11 (1986)—China.

jingdungensis *R.C. Ching in Bull. Bot. Res. North-East. Forest. Inst.*, 6(3): 64 (1986)—China.

leuconeura *R.C. Ching in Bull. Bot. Res. North-East. Forest. Inst.*, 6(3): 12 (1986)—China.

longipinna *R.C. Ching in Bull. Bot. Res. North-East. Forest. Inst.*, 6(3): 13 (1986)—China.

lushanensis *R.C. Ching in Bull. Bot. Res. North-East. Forest. Inst.*, 6(3): 61 (1986)—China.

maguanensis *Ching & Y.T. Hsieh in Bull. Bot. Res. North-East. Forest. Inst.*, 6(4): 2 (1986)—China.

maoshanensis *R.C. Ching in Bull. Bot. Res. North-East. Forest. Inst.*, 6(3): 54 (1986)—China.

mengziensis *R.C. Ching in Bull. Bot. Res. North-East. Forest. Inst.*, 6(3): 14 (1986)—China.

multifida *R.C. Ching in Bull. Bot. Res. North-East. Forest. Inst.*, 6(3): 15 (1986)—China.

nanqingensis *R.C. Ching in Bull. Bot. Res. North-East. Forest. Inst.*, 6(3): 38 (1986)—China.

neoaristata *R.C. Ching in Bull. Bot. Res. North-East. Forest. Inst.*, 6(3): 34 (1986)—China.

nibashanensis *Y.T. Hsieh in Bull. Bot. Res. North-East. Forest. Inst.*, 6(4): 7 (1986)—China.

nitidula *R.C. Ching in Bull. Bot. Res. North-East. Forest. Inst.*, 6(3): 59 (1986)—China.

pianmaensis *R.C. Ching in Bull. Bot. Res. North-East. Forest. Inst.*, 6(3): 65 (1986)—China.

pseudoassamica *R.C. Ching in Bull. Bot. Res. North-East. Forest. Inst.*, 6(3): 16 (1986), as 'pseudo-assamica'—China.

pseudolongipinna *R.C. Ching in Bull. Bot. Res. North-East. Forest. Inst.*, 6(3): 17 (1986), as 'pseudo-longipinna'—China.

pseudosimplicior *R.C. Ching in Bull. Bot. Res. North-East. Forest. Inst.*, 6(3): 47 (1986), as 'pseudo-simplicior'—China.

quadripinnata *(Hayata) S. Serizawa in J. Jap. Bot.*, 61(2): 53 (1986): Microlepia quadripinnata Hayata

quadripinnata
subsp. fimbriata *(Koidz.) S. Serizawa in J. Jap. Bot.*, 61(2): 53 (1986): Dryopteris miqueliana var. fimbriata Koidz.

semifertilis *R.C. Ching in Bull. Bot. Res. North-East. Forest. Inst.*, 6(3): 18 (1986)—China.

setifera *R.C. Ching in Bull. Bot. Res. North-East. Forest. Inst.*, 6(3): 52 (1986)—China.

shuangbaiensis *R.C. Ching in Bull. Bot. Res. North-East. Forest. Inst.*, 6(3): 21 (1986)—China.

sichuanensis *R.C. Ching in Bull. Bot. Res. North-East. Forest. Inst.*, 6(3): 36 (1986)—China.

similis *R.C. Ching in Bull. Bot. Res. North-East. Forest. Inst.*, 6(3): 19 (1986)—China.

ARACHNIODES :-

sinoaristata *R.C. Ching in Bull. Bot. Res. North-East. Forest. Inst.*, 6(3): 20 (1986), as 'sino-aristata'—China.

sinorhomboidea *R.C. Ching in Bull. Bot. Res. North-East. Forest. Inst.*, 6(3): 55 (1986), as 'sino-rhombiodea'—China.

sparsa *R.C. Ching in Bull. Bot. Res. North-East. Forest. Inst.*, 6(3): 22 (1986)—China.

spinoserrulata *R.C. Ching in Bull. Bot. Res. North-East. Forest. Inst.*, 6(3): 46 (1986), as 'spino-serrulata'—China.

subamoena *R.C. Ching in Bull. Bot. Res. North-East. Forest. Inst.*, 6(3): 51 (1986)—China.

suijiangensis *Ching & Y.T. Hsieh in Bull. Bot. Res. North-East. Forest. Inst.*, 6(4): 1 (1986)—China.

×takayamensis *S. Serizawa in J. Jap. Bot.*, 61(2): 54 (1986)—Japan. (A. borealis × miqueliana).

triangularis *R.C. Ching in Bull. Bot. Res. North-East. Forest. Inst.*, 6(3): 26 (1986)—China.

xinpingensis *R.C. Ching in Bull. Bot. Res. North-East. Forest. Inst.*, 6(3): 23 (1986)—China.

yaomashanensis *R.C. Ching in Bull. Bot. Res. North-East. Forest. Inst.*, 6(3): 32 (1986)—China.

yinjiangensis *R.C. Ching in Bull. Bot. Res. North-East. Forest. Inst.*, 6(3): 44 (1986)—China.

yunnanensis *R.C. Ching in Bull. Bot. Res. North-East. Forest. Inst.*, 6(3): 24 (1986)—China.

yunqiensis *Y.T. Hsieh in Bull. Bot. Res. North-East. Forest. Inst.*, 6(4): 3 (1986)—China.

ASPLENIUM

adulterinum
subsp. presolanense *F. Mokry, H. Rasbach & T. Reichstein in Bot. Helvetica*, 96(1): 8 (1986)—Italy, Switzerland.

×barrancense *(W. Bennert & D.E. Meyer) J.J. Pericàs & J.A. Rosselló in Acta Bot. Malacitana*, 11: 297 (1986): × Asplenoceterach barrancense W. Bennert & D.E. Meyer

japonicum *C.P. Thunberg ex A. Murray in Linnaeus, Syst. Veg., ed. 14*: 934 (1784)—Japan. †

japonicum *(Thunb.) Kunze, Ind. Fil. Hort. Bot. Lips.*, 1837-1843: 2 (1843); *cf. D.J. Mabberley in Taxon*, 32(1): 84 (1983): Basionym not stated.

lanceum *C.P. Thunberg ex A. Murray in Linnaeus, Syst. Veg., ed. 14*: 933 (1784)—Japan. †

macrurum *Mickel & R.G. Stolze in Fl. Ecuador*, 23(14:6): 41 (1986), nom. nov.: Asplenium longicaudatum *Mickel & Stolze*

monanthes
var. castaneum *(Schlecht. & Cham.) R.G. Stolze in Fl. Ecuador*, 23(14:6): 45 (1986): Asplenium castaneum Schlecht. & Cham.

ASPLENIUM :–

var. wagneri *(Mett. ex Kuhn) R.G. Stolze in Fl. Ecuador,* 23(14:6): 46 (1986): Asplenium wagneri Mett. ex Kuhn

×recoderi *I. Aizpuru Oiarbide & P. Catalán Rodríguez in An. Jard. Bot. Madrid,* 42(2): 531 (1985 publ. 1986)—Spain. (A. fontanum subsp. fontanum × ruta-muraria subsp. ruta-muraria).

×trichomaniforme

nothosubsp. calcicolum *H. Rasbach & T. Reichstein in Bot. Helvetica,* 96(1): 14 (1986)—Italy. (A. adulterinum subsp. presolanense × trichomanes subsp. quadrivalens).

unilaterale

var. saxicola *M. Kato & K. Iwatsuki in J. Fac. Sci. Univ. Tokyo, Bot.,* 14(1): 46 (1986)—Moluccas.

ATHYRIUM

ser. Biserrulata *Ching & Y.T. Hsieh in Bull. Bot. Res. North-East. Forest. Inst.,* 6(4): 132 (1986).

subgen. Echinoathyrium *Ching & Y.T. Hsieh in Bull. Bot. Res. North-East. Forest. Inst.,* 6(4): 132 (1986).

sect. Echinoathyrium *Ching & Y.T. Hsieh in Bull. Bot. Res. North-East. Forest. Inst.,* 6(4): 132 (1986).

ser. Epiraches *Ching & Y.T. Hsieh in Bull. Bot. Res. North-East. Forest. Inst.,* 6(4): 133 (1986).

ser. Exindusiata *Ching & Y.T. Hsieh in Bull. Bot. Res. North-East. Forest. Inst.,* 6(4): 131 (1986).

sect. Filix-femina *Ching & Y.T. Hsieh in Bull. Bot. Res. North-East. Forest. Inst.,* 6(4): 131 (1986).

ser. Filix-femina *Ching & Y.T. Hsieh in Bull. Bot. Res. North-East. Forest. Inst.,* 6(4): 131 (1986).

ser. Fimbriata *Ching & Y.T. Hsieh in Bull. Bot. Res. North-East. Forest. Inst.,* 6(4): 130 (1986).

ser. Mackinnoiana *Ching & Y.T. Hsieh in Bull. Bot. Res. North-East. Forest. Inst.,* 6(4): 132 (1986).

ser. Macrocarpa *Ching & Y.T. Hsieh in Bull. Bot. Res. North-East. Forest. Inst.,* 6(4): 130 (1986).

sect. Niponica *Ching & Y.T. Hsieh in Bull. Bot. Res. North-East. Forest. Inst.,* 6(4): 131 (1986).

ser. Niponica *Ching & Y.T. Hsieh in Bull. Bot. Res. North-East. Forest. Inst.,* 6(4): 131 (1986).

ser. Pectinata *Ching & Y.T. Hsieh in Bull. Bot. Res. North-East. Forest. Inst.,* 6(4): 133 (1986).

sect. Polystichoides *Ching & Y.T. Hsieh in Bull. Bot. Res. North-East. Forest. Inst.,* 6(4): 130 (1986).

ser. Strigillosa *Ching & Y.T. Hsieh in Bull. Bot. Res. North-East. Forest. Inst.,* 6(4): 133 (1986).

ATHYRIUM :–

sect. Strigoathyrium *Ching & Y.T. Hsieh in Bull. Bot. Res. North-East. Forest. Inst.,* 6(4): 133 (1986).

aculeatum *R.C. Ching in Acta Bot. Bor.-Occid. Sin.,* 6(1): 15 (1986)—China.

acutum *R.C. Ching in Acta Bot. Bor.-Occid. Sin.,* 6(2): 101 (1986)—China.

adpressum *R.C. Ching & W.M. Chu in Acta Bot. Bor.-Occid. Sin.,* 6(1): 13 (1986)—China.

adscendens *R.C. Ching in Acta Bot. Bor.-Occid. Sin.,* 6(2): 100 (1986)—China.

aequilaterale *R.C. Ching in Acta Bot. Bor.-Occid. Sin.,* 6(1): 20 (1986)—China.

amabile *R.C. Ching in Acta Bot. Bor.-Occid. Sin.,* 6(1): 16 (1986)—China.

araiostegioides *R.C. Ching in Acta Bot. Bor.-Occid. Sin.,* 6(2): 103 (1986)—China.

aridum *R.C. Ching in Acta Bot. Bor.-Occid. Sin.,* 6(1): 10 (1986)—China.

bambusicola *R.C. Ching in Acta Bot. Bor.-Occid. Sin.,* 6(1): 18 (1986)—China.

baoxingense *R.C. Ching in Acta Bot. Bor.-Occid. Sin.,* 6(1): 14 (1986)—China.

×bicolor

nothosubsp. shiibaense *S. Serizawa in J. Phytogeogr. Taxon.,* 33(2): 77 (1985)—Japan. (A. neglectum subsp. australe × vidalii).

brevistipes *R.C. Ching in Acta Bot. Bor.-Occid. Sin.,* 6(1): 12 (1986)—China.

caudatum *R.C. Ching in Acta Bot. Bor.-Occid. Sin.,* 6(1): 21 (1986)—China.

caudiforme *R.C. Ching in Acta Bot. Bor.-Occid. Sin.,* 6(1): 21 (1986)—China.

costulalisorum *R.C. Ching in Acta Bot. Bor.-Occid. Sin.,* 6(1): 15 (1986)—China.

crassipes *R.C. Ching in Acta Bot. Bor.-Occid. Sin.,* 6(2): 103 (1986)—China.

dailingense *R.C. Ching in Acta Bot. Bor.-Occid. Sin.,* 6(1): 18 (1986)—China.

dajinense *R.C. Ching in Acta Bot. Bor.-Occid. Sin.,* 6(1): 17 (1986)—China.

daxianglingense *R.C. Ching & H.S. Kung in Acta Bot. Bor.-Occid. Sin.,* 6(1): 16 (1986) China.

decorum *R.C. Ching in Acta Bot. Bor.-Occid. Sin.,* 6(1): 13 (1986)—China.

dissectifolium *R.C. Ching in Acta Bot. Bor.-Occid. Sin.,* 6(2): 104 (1986)—China.

dissitifolium

var. kulhaitense *(W.S. Atkinson ex Clarke) R.C. Ching in Acta Bot. Bor.-Occid. Sin.,* 6(1): 22 (1986): Asplenium oxyphyllum var. kulhaitense W.S. Atkinson ex Clarke

elongatum *R.C. Ching in Acta Bot. Bor.-Occid. Sin.,* 6(2): 101 (1986)—China.

emeicola *R.C. Ching in Acta Bot. Bor.-Occid. Sin.,* 6(1): 13 (1986)—China.

ensiferum *R.C. Ching & H.S. Kung in Acta Bot. Bor.-Occid. Sin.,* 6(2): 102 (1986)—China.

excelsius *R.C. Ching in Acta Bot. Bor.-Occid. Sin.,* 6(1): 20 (1986)—China.

exindusiatum *R.C. Ching in Acta Bot. Bor.-Occid. Sin.,* 6(2): 102 (1986)—China.

ATHYRIUM :-

×flavosorum *S. Serizawa in J. Phytogeogr. Taxon.,* 33(2): 77 (1985)—Japan. (A. arisanense × delavayi).

glandulosum *R.C. Ching in Acta Bot. Bor.-Occid. Sin.,* 6(1): 19 (1986)—China.

guangnanense *R.C. Ching in Acta Bot. Bor.-Occid. Sin.,* 6(2): 103 (1986), nom. nov.: Athyrium yunnanicum *Ching*

guizhouense *R.C. Ching in Acta Bot. Bor.-Occid. Sin.,* 6(1): 11 (1986)—China.

hainanense *R.C. Ching in Acta Bot. Bor.-Occid. Sin.,* 6(2): 101 (1986)—Hainan.

×hohuanshanense *N. Yoshikawa in J. Phytogeogr. Taxon.,* 33(2): 80 (1985)—Taiwan. (A. arisanense × vidalii).

infrapuberulum *R.C. Ching in Acta Bot. Bor.-Occid. Sin.,* 6(2): 100 (1986)—China.

interjectum *R.C. Ching in Acta Bot. Bor.-Occid. Sin.,* 6(1): 15 (1986)—China.

intermixtum *R.C. Ching & P.S. Chui in Acta Bot. Bor.-Occid. Sin.,* 6(1): 21 (1986)—China.

jinshajiangense *R.C. Ching & Shing in Acta Bot. Bor.-Occid. Sin.,* 6(1): 14 (1986)—China.

kumaonicum *N. Punetha in Indian Fern J.,* 2(1–2): 29 (1985)—India.

lancipinnulum *R.C. Ching in Acta Bot. Bor.-Occid. Sin.,* 6(1): 18 (1986)—China.

latibasis *R.C. Ching in Acta Bot. Bor.-Occid. Sin.,* 6(2): 104 (1986)—China.

liangwangshanicum *R.C. Ching in Acta Bot. Bor.-Occid. Sin.,* 6(1): 21 (1986)—China.

lineare *R.C. Ching in Acta Bot. Bor.-Occid. Sin.,* 6(1): 11 (1986)—China.

longipinnum *R.C. Ching in Acta Bot. Bor.-Occid. Sin.,* 6(1): 15 (1986)—China.

machangense *R.C. Ching in Acta Bot. Bor.-Occid. Sin.,* 6(2): 100 (1986)—China.

muliense *R.C. Ching in Acta Bot. Bor.-Occid. Sin.,* 6(1): 16 (1986)—China.

neglectum

subsp. australe *S. Serizawa in J. Phytogeogr. Taxon.,* 33(2): 76 (1985)—Japan.

neodelavayi *R.C. Ching & H.S. Kung in Acta Bot. Bor.-Occid. Sin.,* 6(1): 22 (1986)—China.

nitidum *R.C. Ching in Acta Bot. Bor.-Occid. Sin.,* 6(1): 19 (1986)—China.

oblongum *R.C. Ching in Acta Bot. Bor.-Occid. Sin.,* 6(1): 17 (1986)—China.

obtusilimbum *R.C. Ching in Acta Bot. Bor.-Occid. Sin.,* 6(1): 11 (1986)—China.

pachyphyllum *R.C. Ching in Acta Bot. Bor.-Occid. Sin.,* 6(2): 102 (1986)—China.

praticola *R.C. Ching in Acta Bot. Bor.-Occid. Sin.,* 6(1): 18 (1986)—China.

sericellum *R.C. Ching in Acta Bot. Bor.-Occid. Sin.,* 6(1): 17 (1986)—China.

serratodentatum *R.C. Ching in Acta Bot. Bor.-Occid. Sin.,* 6(2): 99 (1986), as 'serrato-dentatum'—China.

squamipes *R.C. Ching in Acta Bot. Bor.-Occid. Sin.,* 6(1): 17 (1986)—China.

ATHYRIUM :-

×subcrassipes *S. Serizawa in J. Phytogeogr. Taxon.,* 33(2): 77 (1985)—Japan. (A. masamunei × nigripes).

sublineare *R.C. Ching in Acta Bot. Bor.-Occid. Sin.,* 6(2): 99 (1986)—China.

suprapuberulum *R.C. Ching in Acta Bot. Bor.-Occid. Sin.,* 6(1): 14 (1986)—China.

suprapubescense *R.C. Ching in Acta Bot. Bor.-Occid. Sin.,* 6(1): 10 (1986)—China.

tarulakaense *R.C. Ching in Acta Bot. Bor.-Occid. Sin.,* 6(1): 12 (1986)—China.

tenuifrons *(Wall. ex Blanford) N. Punetha in Indian Fern J.,* 2(1–2): 24 (1985), error (Wall. ex Hope): Asplenium tenuifrons Wall. ex Blanford

tsaii *R.C. Ching in Acta Bot. Bor.-Occid. Sin.,* 6(1): 19 (1986)—China.

uniforme *R.C. Ching in Acta Bot. Bor.-Occid. Sin.,* 6(1): 11 (1986)—China.

uniseriatum *R.C. Ching in Acta Bot. Bor.-Occid. Sin.,* 6(2): 99 (1986)—China.

venulosum *R.C. Ching in Acta Bot. Bor.-Occid. Sin.,* 6(1): 13 (1986)—China.

wumonshanicum *R.C. Ching in Acta Bot. Bor.-Occid. Sin.,* 6(1): 20 (1986)—China.

wuyishanense *R.C. Ching in Acta Bot. Bor.-Occid. Sin.,* 6(2): 104 (1986)—China.

yaanense *R.C. Ching in Acta Bot. Bor.-Occid. Sin.,* 6(1): 19 (1986)—China.

B

BLECHNUM

auriculatum *(Desv.) Endl., Cat. Horti Vindob.,* 1: 9 (1842); *cf. D.J. Mabberley in Taxon,* 33(3): 434 (1984): Basionym not stated.

australe

var. aberrans *N.C. Anthony & E.A. Schelpe in Bothalia,* 15(3–4): 555 (1985)—South Africa (Cape Province).

BOTRYCHIUM

ascendens *W.H. Wagner in Amer. Fern J.,* 76(2): 36 (1986)—U.S.A. (Oregon, Montana, Nevada), Canada (Alberta, British Columbia, Ontario, Yukon).

campestre *W.H. Wagner & D.R. Farrar in Amer. Fern J.,* 76(2): 39 (1986)—U.S.A. (Iowa, Minnesota, Nebraska, N. Dakota), Canada (Alberta, Saskatchewan).

fumarianum *Sm. in Rees, Cyclop.,* 39(2): Botrychium n. 4 (1819); *cf. D.J. Mabberley in Taxon,* 32(1): 82 (1983)— Locality not stated.

matricarianum *Sm. in Rees, Cyclop.,* 39(2): Botrychium n. 3 (1819); *cf. D.J. Mabberley in Taxon,* 32(1): 82 (1983)— Locality not stated.

pedunculosum *W.H. Wagner in Amer. Fern J.,* 76(2): 43 (1986)—U.S.A. (Oregon), Canada (British Columbia, Saskatchewan).

C

CALYMMODON
curtus *B.S. Parris in Kew Bull.*, 41(3): 506 (1986)—Malaya.

CHEILANTHES
brevifrons *(Khullar) S.P. Khullar in Indian Fern J.*, 1(1–2): 90 (1984 publ. 1985): Cheilanthes anceps var. brevifrons Khullar
linkiana *Kunze, Ind. Fil. Hort. Bot. Lips.*, 1837-1843: 2 (1843); *cf. D.J. Mabberley in Taxon,* 32(1): 84 (1983)— Locality not stated.

CHINGIA
australis *R.E. Holttum in Kew Bull.*, 41(3): 518 (1986)—Australia (Queensland).

×**CHRISMATOPTERIS** N. Quansah & D.S. Edwards in Kew Bull., 41(4): 805 (1986). (Christella × Pneumatopteris).
holttumii *N. Quansah & D.S. Edwards in Kew Bull.*, 41(4): 805 (1986)—Ghana, Sierra Leone, Liberia, Nigeria, Cameroun. (Christella dentata × Pneumatopteris afra).

CHRISTELLA
dentata
var. caespitosa *R.E. Holttum in Kew Bull.*, 41(3): 518 (1986)—Australia (Queensland).

CORNOPTERIS
seramensis *M. Kato in J. Jap. Bot.*, 61(8): 233 (1986)—Moluccas.

CRYPSINUS
ebenipes
var. subebenipes *(Ching) K. Iwatsuki, S.K. Wu, S. Mitsuta & X. Cheng in J. Fac. Sci. Univ. Tokyo, Bot.*, 14(1): 33 (1986): Phymatopsis subebenipes Ching
induratus *(Baker) B.S. Parris in Kew Bull.*, 41(1): 70 (1986): Polypodium induratum Baker

CTENOPTERIS
davalliacea *(F. Mueller & Baker) B.S. Parris in Kew Bull.*, 41(3): 501 (1986): Polypodium davalliaceum F. Mueller & Baker
flavovirens *(Alston) B.S. Parris in Kew Bull.*, 41(1): 69 (1986): Polypodium flavovirens Alston
malayana *B.S. Parris in Kew Bull.*, 41(3): 494 (1986)—Malaya.

CYATHEA
exilis *R.E. Holttum in Kew Bull.*, 41(3): 532 (1986)—Australia (Queensland).

CYCLODIUM
akawaiorum *A.R. Smith in Amer. Fern J.*, 76(2): 71 (1986)—Guyana, Venezuela.
calophyllum *(Morton) A.R. Smith in Amer. Fern J.*, 76(2): 73 (1986): Dryopteris calophylla Morton

CYCLODIUM :-
guianense *(Kl.) L.D. Gómez P. in Phytologia,* 60(5): 371 (1986): Aspidium guianense Kl.
guianense *(Klotzsch) A.R. Smith in Amer. Fern J.*, 76(2): 75 (1986): Aspidium guianense Klotzsch
heterodon
var. abbreviatum *(Presl) A.R. Smith in Amer. Fern J.*, 76(2): 80 (1986): Cyclodium abbreviatum Presl
inerme *(Fée) A.R. Smith in Amer. Fern J.*, 76(2): 82 (1986): Polystichum inerme Fée
meniscioides
var. paludosum *(Morton) A.R. Smith in Amer. Fern J.*, 76(2): 87 (1986): Dryopteris paludosa Morton
var. rigidissimum *(C. Chr.) A.R. Smith in Amer. Fern J.*, 76(2): 88 (1986): Cyclodium rigidissimum C. Chr.
rheophilum *A.R. Smith in Amer. Fern J.*, 76(2): 88 (1986)—French Guiana.
seemannii *(Hook.) A.R. Smith in Amer. Fern J.*, 76(2): 89 (1986): Aspidium seemannii Hook.
trianae *(Mett.) A.R. Smith in Amer. Fern J.*, 76(2): 92 (1986): Aspidium trianae Mett.
var. chocoense *A.R. Smith in Amer. Fern J.*, 76(2): 93 (1986)—Colombia, Panama.
varians *(Fée) A.R. Smith in Amer. Fern J.*, 76(2): 94 (1986): Nephrodium varians Fée

CYCLOSORUS
sinoacuminata *R.C. Ching & Z.Y. Liu in Bull. Bot. Res. North-East. Forest. Inst.*, 6(1): 180 (1986), nom. nov., as 'sino-acuminata': Cyclosorus nanchuanensis *Ching & Liu (1984).*

CYSTODIUM
sorbifolium
subsp. solomonense *J.R. Croft in Kew Bull.*, 41(4): 802 (1986)—British Solomon Is., Bougainville Is.

D

DIPHASIASTRUM
montellii *(Kukk.) N.A. Minyaev & Yu.A. Ivanenko in Bot. Zhurn.*, 71(8): 1126 (1986): Diphasium complanatum subsp. montellii Kukk.

DIPLAZIUM
decompositum *(Copel.) B.S. Parris in Kew Bull.*, 41(1): 69 (1986): Athyrium decompositum Copel.
pedicellatum *(Copel.) B.S. Parris in Kew Bull.*, 41(1): 69 (1986): Athyrium pedicellatum Copel.

DIPLAZIUM :–
squamuligerum *(Rosenst.) B.S. Parris in Kew Bull.*, 41(1): 69 (1986): Asplenium varians var. squamuligerum Rosenst.

DRYOPSIS R.E. Holttum & P.J. Edwards in Kew Bull., 41(1): 179 (1986).
adnata *(Bl.) R.E. Holttum & P.J. Edwards in Kew Bull.*, 41(1): 190 (1986): Aspidium adnatum Bl.
apiciflora *(Wall. ex Mett.) R.E. Holttum & P.J. Edwards in Kew Bull.*, 41(1): 189 (1986): Aspidium apiciflorum Wall. ex Mett.
clarkei *(Bak.) R.E. Holttum & P.J. Edwards in Kew Bull.*, 41(1): 181 (1986): Nephrodium clarkei Bak.
contigua *(Ching) R.E. Holttum & P.J. Edwards in Kew Bull.*, 41(1): 193 (1986): Ctenitis contigua Ching
crassirachis *(Ching) R.E. Holttum & P.J. Edwards in Kew Bull.*, 41(1): 195 (1986): Ctenitis crassirachis Ching
crenata *(Ching) R.E. Holttum & P.J. Edwards in Kew Bull.*, 41(1): 193 (1986): Ctenitis crenata Ching
dentisora *(Ching) R.E. Holttum & P.J. Edwards in Kew Bull.*, 41(1): 195 (1986): Ctenitis dentisora Ching
×fauriei *R.E. Holttum & P.J. Edwards in Kew Bull.*, 41(1): 198 (1986)—Taiwan. (D. apiciflora × maximowicziana).
fengiana *(Ching) R.E. Holttum & P.J. Edwards in Kew Bull.*, 41(1): 183 (1986): Ctenitis fengiana Ching
ferruginea *(Bak.) R.E. Holttum & P.J. Edwards in Kew Bull.*, 41(1): 201 (1986): Nephrodium ferrugineum Bak.
heterolaena *(C. Chr.) R.E. Holttum & P.J. Edwards in Kew Bull.*, 41(1): 185 (1986): Dryopteris heterolaena C. Chr.
kawakamii *(Hayata) R.E. Holttum & P.J. Edwards in Kew Bull.*, 41(1): 186 (1986): Dryopteris kawakamii Hayata
kwangsiensis *(Ching & Wang) R.E. Holttum & P.J. Edwards in Kew Bull.*, 41(1): 186 (1986): Ctenitis kwangsiensis Ching & Wang
manipurensis *(Bedd.) R.E. Holttum & P.J. Edwards in Kew Bull.*, 41(1): 200 (1986): Polypodium manipurense Bedd.
mariformis *(Rosenst.) R.E. Holttum & P.J. Edwards in Kew Bull.*, 41(1): 188 (1986): Dryopteris mariformis Rosenst.
maximowicziana *(Miquel) R.E. Holttum & P.J. Edwards in Kew Bull.*, 41(1): 197 (1986): Aspidium maximowiczianum Miquel
nidus *(Bak.) R.E. Holttum & P.J. Edwards in Kew Bull.*, 41(1): 192 (1986): Nephrodium filix-mas var. nidus Bak.
obtusiloba *(Bak.) R.E. Holttum & P.J. Edwards in Kew Bull.*, 41(1): 202 (1986): Nephrodium obtusilobum Bak.
paucisora *(Copel.) R.E. Holttum & P.J. Edwards in Kew Bull.*, 41(1): 183 (1986): Dryopteris paucisora Copel.

DRYOPSIS :–
scabrosa *(Kunze) R.E. Holttum & P.J. Edwards in Kew Bull.*, 41(1): 199 (1986): Aspidium scabrosum Kunze
silaensis *(Ching) R.E. Holttum & P.J. Edwards in Kew Bull.*, 41(1): 197 (1986): Ctenitis silaensis Ching
sphaeropteroides *(Bak.) R.E. Holttum & P.J. Edwards in Kew Bull.*, 41(1): 199 (1986): Polypodium sphaeropteroides Bak.
submariformis *(Ching & Wang) R.E. Holttum & P.J. Edwards in Kew Bull.*, 41(1): 194 (1986): Ctenitis submariformis Ching & Wang
transmorrisonensis *(Hayata) R.E. Holttum & P.J. Edwards in Kew Bull.*, 41(1): 191 (1986): Polystichum transmorrisonense Hayata
truncata *(Ching & H.S. Kung) R.E. Holttum & P.J. Edwards in Kew Bull.*, 41(1): 188 (1986): Ctenitis truncata Ching & H.S. Kung
wantsingshanica *(Ching & Hsing) R.E. Holttum & P.J. Edwards in Kew Bull.*, 41(1): 194 (1986): Ctenitis wantsingshanica Ching & Hsing

DRYOPTERIS
sect. Aemulae *C.R. Fraser-Jenkins in Bull. Brit. Mus. (Nat. Hist.), Bot.*, 14(3): 194 (1986).
sect. Cinnamomeae *C.R. Fraser-Jenkins in Bull. Brit. Mus. (Nat. Hist.), Bot.*, 14(3): 193 (1986).
subgen. Erythrovariae *(H. Itô) C.R. Fraser-Jenkins in Bull. Brit. Mus. (Nat. Hist.), Bot.*, 14(3): 195 (1986): Dryopteris sect. Erythrovariae H. Itô
sect. Hirtipedes *C.R. Fraser-Jenkins in Bull. Brit. Mus. (Nat. Hist.), Bot.*, 14(3): 190 (1986).
sect. Marginatae *C.R. Fraser-Jenkins in Bull. Brit. Mus. (Nat. Hist.), Bot.*, 14(3): 194 (1986).
subgen. Nephrocystis *(H. Itô) C.R. Fraser-Jenkins in Bull. Brit. Mus. (Nat. Hist.), Bot.*, 14(3): 197 (1986): Dryopteris sect. Nephrocystis H. Itô
sect. Pallidae *C.R. Fraser-Jenkins in Bull. Brit. Mus. (Nat. Hist.), Bot.*, 14(3): 192 (1986).
sect. Pandae *C.R. Fraser-Jenkins in Bull. Brit. Mus. (Nat. Hist.), Bot.*, 14(3): 191 (1986).
sect. Politae *C.R. Fraser-Jenkins in Bull. Brit. Mus. (Nat. Hist.), Bot.*, 14(3): 196 (1986).
sect. Purpurascentes *C.R. Fraser-Jenkins in Bull. Brit. Mus. (Nat. Hist.), Bot.*, 14(3): 197 (1986).
sect. Remotae *C.R. Fraser-Jenkins in Bull. Brit. Mus. (Nat. Hist.), Bot.*, 14(3): 192 (1986).
sect. Splendentes *C.R. Fraser-Jenkins in Bull. Brit. Mus. (Nat. Hist.), Bot.*, 14(3): 193 (1986).

DRYOPTERIS :–
sect. Variae *C.R. Fraser-Jenkins in Bull. Brit. Mus. (Nat. Hist.), Bot.,* 14(3): 196 (1986).
×asturiensis *C.R. Fraser-Jenkins & M. Gibby in Fern Gaz. (U.K.),* 13(2): 113 (1986)—Spain. (D. affinis subsp. affinis × corleyi).
austriaca
subsp. assimilis *(Walker) O. de Bolòs & J. Vigo, Fl. Països Catalans:* 170 (1984): Dryopteris assimilis Walker
crassirhizoma
subsp. whangshangensis *(Ching) C.R. Fraser-Jenkins in Bull. Brit. Mus. (Nat. Hist.), Bot.,* 14(3): 191 (1986): Dryopteris whangshangensis Ching
expansa
var. alpina *(Moore) R.L.L. Viane in Pl. Syst. Evol.,* 153(1–2): 90 (1986): Lastrea dilatata var. alpina Moore
filix-mas
subsp. oreades *(Fomin) O. de Bolòs & J. Vigo, Fl. Països Catalans:* 171 (1984): Dryopteris oreades Fomin
goldiana
subsp. monticola *(Makino) C.R. Fraser-Jenkins in Bull. Brit. Mus. (Nat. Hist.), Bot.,* 14(3): 192 (1986): Nephrodium monticola Makino
jinfoshanensis *R.C. Ching & Z.Y. Liu in Bull. Bot. Res. North-East. Forest. Inst.,* 6(1): 179 (1986), nom. nov.: Dryopteris nanchuanensis *Ching & Liu in Bull. Bot. Res. N.E. Forest. Inst., 4(3): 19 (1984).*
mediterranea
var. robusta *(Fraser-Jenkins) J. Holub in Preslia,* 58(4): 301 (1986): Dryopteris affinis subsp. robusta Fraser-Jenkins
parrisiae *C.R. Fraser-Jenkins in Bull. Brit. Mus. (Nat. Hist.), Bot.,* 14(3): 199 (1986)—New Guinea (Papua New Guinea, W. Irian).
reichsteinii *C.R. Fraser-Jenkins in Bull. Brit. Mus. (Nat. Hist.), Bot.,* 14(3): 191 (1986), nom. nov.: Dryopteris paleacea var. madagascariensis *C. Chr.*
ruwenzoriensis *C. Chr. ex C.R. Fraser-Jenkins in Bull. Brit. Mus. (Nat. Hist.), Bot.,* 14(3): 204 (1986)—Zaïre, Uganda.
sinovaria *R.C. Ching & Z.Y. Liu in Bull. Bot. Res. North-East. Forest. Inst.,* 6(1): 179 (1986), nom. nov., as 'sino-varia': Dryopteris nanchuanensis *Ching & Liu in Bull. Bot. Res. N.E. Forest. Inst., 4(4): 9 (1984).*
stilluppensis *(Sabranski) J. Holub in Preslia,* 58(4): 302 (1986): Aspidium filix-mas var. stilluppense Sabranski
subpycnopteroides *Ching ex C.R. Fraser-Jenkins in Bull. Brit. Mus. (Nat. Hist.), Bot.,* 14(3): 199 (1986)—China.
tingiensis *Ching & S.K. Wu ex C.R. Fraser-Jenkins in Bull. Brit. Mus. (Nat. Hist.), Bot.,* 14(3): 202 (1986)—Xizang.
villarii
subsp. balearica *(Litard.) O. de Bolòs & J. Vigo, Fl. Països Catalans:* 172 (1984): Dryopteris rigida var. balearica Litard.

E

ELAPHOGLOSSUM
gerardianum *L.D. Gómez P. in Phytologia,* 60(1): 74 (1986)—Costa Rica.
lalitae *L.D. Gómez P. in Phytologia,* 60(1): 74 (1986)—Costa Rica.
ometepense *L.D. Gómez P. in Phytologia,* 60(5): 369 (1986)—Nicaragua, Costa Rica.

EQUISETUM
ramosissimum
var. altissimum *(A. Br.) S.S. Bir in Indian Fern J.,* 2(1–2): 42 (1985): Equisetum altissimum A. Br.

G

GRAMMITIS
rivularis *B.S. Parris in Kew Bull.,* 41(3): 514 (1986)—Malaya, Borneo (Sarawak).

GYMNOGRAMMA
lherminieri *Bory ex Kunze, Hort. Univ. Lit. Lips.,* 1839; *Linnaea 16, Lit.:* 107 (1842); *cf. D.J. Mabberley in Taxon,* 32(1): 85 (1983)—Locality not stated.
linkiana *(Presl) Kunze, Ind. Fil. Hort. Bot. Lips.,* 1837-1843: 1 (1843); *cf. D.J. Mabberley in Taxon,* 32(1): 85 (1983): Basionym not stated.

H

HEMIONITIS
japonica *C.P. Thunberg ex A. Murray in Linnaeus, Syst. Veg., ed. 14:* 932 (1784)—Japan. †

HYPODEMATIUM
glabrius *(Copel.) R.E. Holttum in Gard. Bull. Singapore,* 38(2): 148 (1985): Dryopteris glabrior Copel.
hirsutum *(Don) R.C. Ching in Indian Fern J.,* 1(1–2): 49 (1984 publ. 1985): Nephrodium hirsutum Don
laxum *R.C. Ching in Fl. Beijing,* 1: 669, 26 (1984)—China.
pilosum *R.C. Ching in Fl. Beijing,* 1: 669, 25 (1984)—China.

HYPOLEPIS
coerulescens *A. Biswas in J. Econ. Taxon. Bot.,* 7(1): 121 (1985)—Sikkim.
gamblei *A. Biswas in J. Econ. Taxon. Bot.,* 7(1): 124 (1985)—India, Sikkim.

HYPOLEPIS :-
indica *A. Biswas in J. Econ. Taxon. Bot.*, 7(1): 112 (1985)—Sikkim.
longa *A. Biswas in J. Econ. Taxon. Bot.*, 7(1): 112 (1985)—India.
sikkimensis *A. Biswas in J. Econ. Taxon. Bot.*, 7(1): 122 (1985)—Sikkim, India.
viridula *A. Biswas in J. Econ. Taxon. Bot.*, 7(1): 123 (1985)—Sikkim.

I

ISOETES
baculata *R.J. Hickey & H.P. Fuchs-Eckert in Syst. Bot.*, 11(2): 310 (1986)—Brazil.
capensis
var. stephansenii *(Duthie) E.A. Schelpe & N.C. Anthony in Bothalia,* 15(3–4): 555 (1985): Isoetes stephansenii Duthie
eshbaughii *R.J. Hickey in Syst. Bot.*, 11(2): 310 (1986)—Bolivia.

L

LECANOPTERIS
balgooyi *E. Hennipman in Kew Bull.*, 41(4): 783 (1986)—Celebes.
darnaedii *E. Hennipman in Kew Bull.*, 41(4): 785 (1986)—Celebes.

LEPISORUS
tenuipes *Ching & S.P. Khullar in Indian Fern J.*, 1(1–2): 91 (1984 publ. 1985)—India.

LUNATHYRIUM
macdonellii *(Beddome) S.P. Khullar in Indian Fern J.*, 1(1–2): 93 (1984 publ. 1985): Asplenium macdonelli Beddome

LYCOPODIUM
acifolium *C. Rolleri in Rev. Mus. La Plata, Bot.*, 14(87): 2 (1985)—Peru, Bolivia, Ecuador.
cordatum *Cameron in Loud., Hort. Brit., ed. 3:* 647 (1839); *cf. D.J. Mabberley in Taxon,* 30(1): 16 (1981)— Locality not stated.
japonicum *C.P. Thunberg ex A. Murray in Linnaeus, Syst. Veg., ed. 14:* 944 (1784)—Japan.
serratum *C.P. Thunberg ex A. Murray in Linnaeus, Syst. Veg., ed. 14:* 944 (1784)—Japan.
spongiosum *C. Rolleri in Rev. Mus. La Plata, Bot.*, 14(87): 6 (1985)—Colombia.

M

MARSILEA
sect. Clemys *D.M. Johnson, Syst. New World spp. Marsilea (Marsileac.) (Syst. Bot. Monog., 11):* 40 (1986).
sect. Nodorhizae *D.M. Johnson, Syst. New World spp. Marsilea (Marsileac.) (Syst. Bot. Monog., 11):* 48 (1986).
crotophora *D.M. Johnson, Syst. New World spp. Marsilea (Marsileac.) (Syst. Bot. Monog., 11):* 46 (1986)—Brazil, Mexico, Nicaragua, Venezuela, Bolivia, Paraguay.
vestita
subsp. tenuifolia *(Engelmann ex A. Braun) D.M. Johnson, Syst. New World spp. Marsilea (Marsileac.) (Syst. Bot. Monog., 11):* 69 (1986): Marsilea tenuifolia Engelmann ex A. Braun

MICROPHLEBODIUM L.D. Gómez P. in Phytologia, 59(1): 58 (1985).
muenchii *(Christ) L.D. Gómez P. in Phytologia,* 59(1): 58 (1985): Polypodium muenchii Christ

MICROSORUM
papuanum *(Baker) B.S. Parris in Kew Bull.*, 41(1): 70 (1986): Polypodium papuanum Baker
rampans *(Baker) B.S. Parris in Kew Bull.*, 41(1): 70 (1986): Polypodium rampans Baker

N

NEPHRODIUM
aculeatum *(L.) Endl., Cat. Horti Vindob.*, 1: 11 (1842); *cf. D.J. Mabberley in Taxon,* 33(3): 435 (1984): Basionym not stated.
chrysolobum *(Link) Endl., Cat. Horti Vindob.*, 1: 11 (1842); *cf. D.J. Mabberley in Taxon,* 33(3): 435 (1984): Basionym not stated.
kaulfussii *(Link) Endl., Cat. Horti Vindob.*, 1: 11 (1842); *cf. D.J. Mabberley in Taxon,* 33(3): 435 (1984): Basionym not stated.
patens *(Willd.) Endl., Cat. Horti Vindob.*, 1: 11 (1842); *cf. D.J. Mabberley in Taxon,* 33(3): 435 (1984): Basionym not stated.
pectinatum *Sweet, Hort. Brit.*: 462 (1827); *cf. D.J. Mabberley in Taxon,* 30(1): 16 (1981)—Locality not stated.
serra *(Sw.) Endl., Cat. Horti Vindob.*, 1: 11 (1842); *cf. D.J. Mabberley in Taxon,* 33(3): 435 (1984): Basionym not stated.

NIPHOBOLUS
linearis *(L.) G. Don f. in Loud., Hort. Brit.*: 418 (1830); *cf. D.J. Mabberley in Taxon,* 32(1): 87 (1983): Basionym not stated.

O

OPHIOGLOSSUM
japonicum *C.P. Thunberg ex A. Murray in Linnaeus, Syst. Veg., ed. 14:* 926 (1784)—Japan. †

OSMUNDA
japonica *C.P. Thunberg ex A. Murray in Linnaeus, Syst. Veg., ed. 14:* 928 (1784)—Japan. †
lancea *C.P. Thunberg ex A. Murray in Linnaeus, Syst. Veg., ed. 14:* 928 (1784)—Japan. †
ternata *C.P. Thunberg ex A. Murray in Linnaeus, Syst. Veg., ed. 14:* 927 (1784)—Japan. †

P

PELLAEA
seticaulis *(Hook.) S.R. Ghosh in J. Econ. Taxon. Bot.,* 7(3): 681 (1985 publ. 1986): Pteris seticaulis Hook.

PHYMATOPTERIS
ebenipes
var. oakesii *(Clarke) C.K. Satija & S.S. Bir, Polypod. Ferns India (Aspects Pl. Sci. 8):* 61 (1985): Polypodium ebenipes var. oakesii Clarke
integerrima *(Ching) S.S. Bir in C.K. Satija & S.S. Bir, Polypod. Ferns India (Aspects Pl. Sci. 8):* 98 (1985): Phymatopsis integerrima Ching
nigrovenia *(Christ) C.K. Satija & S.S. Bir, Polypod. Ferns India (Aspects Pl. Sci. 8):* 64 (1985): Polypodium shensiense var. nigrovenium Christ
quasidivaricata *(Hayata) C.K. Satija & S.S. Bir, Polypod. Ferns India (Aspects Pl. Sci. 8):* 63 (1985): Polypodium quasidivaricatum Hayata

PHYSEMATIUM
pubescens *(Spring) Bevis in Loud., Hort. Brit., ed. 3:* 656 (1839); *cf. D.J. Mabberley in Taxon,* 30(1): 17 (1981): Basionym not stated.

PLEOPELTIS
gomeziana *W.H. Wagner in Bull. Torrey Bot. Club,* 113(2): 163 (1986)—Costa Rica.

×PLEOPODIUM E.A. Schelpe & N.C. Anthony in Bothalia, 15(3–4): 557 (1985). (Pleopeltis × Polypodium).
simianum *E.A. Schelpe & N.C. Anthony in Bothalia,* 15(3–4): 557 (1985)—South Africa (Natal, Transvaal, Cape Province), Zimbabwe.

POLYPODIUM
cambricum
subsp. serrulatum *(Sch. ex Arcang.) R.E.G. Pichi Sermolli in Webbia,* 40(1): 49 (1986): Polypodium vulgare subsp. serrulatum Sch. ex Arcang.
dichotomum *C.P. Thunberg ex A. Murray in Linnaeus, Syst. Veg., ed. 14:* 938 (1784)—Japan. †
ellipticum *C.P. Thunberg ex A. Murray in Linnaeus, Syst. Veg., ed. 14:* 935 (1784)—Japan. †
glaucum *C.P. Thunberg ex A. Murray in Linnaeus, Syst. Veg., ed. 14:* 938 (1784)—Japan. †
hastatum *C.P. Thunberg ex A. Murray in Linnaeus, Syst. Veg., ed. 14:* 935 (1784)—Japan. †
lacerum *C.P. Thunberg ex A. Murray in Linnaeus, Syst. Veg., ed. 14:* 938 (1784)—Japan. †
lineare *C.P. Thunberg ex A. Murray in Linnaeus, Syst. Veg., ed. 14:* 934 (1784)—Japan. †
marginale *C.P. Thunberg ex A. Murray in Linnaeus, Syst. Veg., ed. 14:* 937 (1784)—Japan. †
punctatum *C.P. Thunberg ex A. Murray in Linnaeus, Syst. Veg., ed. 14:* 939 (1784)—Japan. †
setosum *C.P. Thunberg ex A. Murray in Linnaeus, Syst. Veg., ed. 14:* 938 (1784)—Japan. †

POLYSTICHUM
mehrae *C.R. Fraser-Jenkins & S.P. Khullar in Indian Fern J.,* 2(1–2): 10 (1985)—Pakistan, India, ?Burma, Xizang, China.
monticola *N.C. Anthony & E.A. Schelpe in Bothalia,* 15(3–4): 554 (1985)—South Africa (Cape Province, Natal, Orange Free State), Lesotho.
×potteri *D.S. Barrington in Rhodora,* 88(855): 298 (1986)—U.S.A. (Vermont, New Hampshire, New York, Pennsylvania), Canada (Nova Scotia, New Brunswick, Quebec). (P. acrostichoides × braunii).

PSEUDOCYCLOSORUS
falcilobus
var. latior *(Ching) S. Kaur & S. Chandra in New Bot.,* 12(2–4): 102 (1985 publ. 1986): Thelypteris falciloba var. latior Ching

PTERIS
nervosa *C.P. Thunberg ex A. Murray in Linnaeus, Syst. Veg., ed. 14:* 930 (1784)—Japan. †
rufescens *(Link) Endl., Cat. Horti Vindob.,* 1: 7 (1842); *cf. D.J. Mabberley in Taxon,* 33(3): 435 (1984): Basionym not stated.
sinuata *C.P. Thunberg ex A. Murray in Linnaeus, Syst. Veg., ed. 14:* 931 (1784)—Japan. †

S

SELAGINELLA

subgen. Ericetorum *A.C. Jermy in Fern Gaz. (U.K.)*, 13(2): 117 (1986).

subgen. Tetragonostachys *A.C. Jermy in Fern Gaz. (U.K.)*, 13(2): 118 (1986).

chaii *A.C. Jermy in Kew Bull.*, 41(3): 551 (1986)—Borneo (Sarawak).

coonooriana *R.D. Dixit in Bull. Bot. Surv. India*, 25(1–4): 223 (1983 publ. 1985)—India.

ganguliana *R.D. Dixit in Bull. Bot. Surv. India*, 26(1–2): 104 (1984 publ. 1985)—India.

gigantea *J.A. Steyermark & A.R. Smith in Ann. Missouri Bot. Gard.*, 73(1): 209 (1986)—Venezuela.

jainii *R.D. Dixit in Bull. Bot. Surv. India*, 25(1–4): 225 (1983 publ. 1985)—India.

liboensis *H.S. Kung & P.S. Wang in Acta Phytotax. Sin.*, 24(4): 325 (1986)—China.

lineariformis *A.C. Jermy in Kew Bull.*, 41(3): 549 (1986)—Borneo (Sarawak).

nairii *R.D. Dixit in Bull. Bot. Surv. India*, 26(1–2): 106 (1984 publ. 1985)—India.

panchghaniana *R.D. Dixit in Bull. Bot. Surv. India*, 25(1–4): 226 (1983 publ. 1985)—India.

panigrahii *R.D. Dixit in Bull. Bot. Surv. India*, 25(1–4): 226 (1983 publ. 1985)—India.

parviarticulata *W.R. Buck in Brittonia*, 38(1): 45 (1986)—Paraguay.

SPHAEROPTERIS

aciculosa *(Copel.) R. Tryon in Contrib. Gray Herb.*, 200: 21 (1970): Cyathea aciculosa Copel.

aeneifolia *(vAvR.) R. Tryon in Contrib. Gray Herb.*, 200: 21 (1970): Alsophila aeneifolia vAvR.

agatheti *(Holtt.) R. Tryon in Contrib. Gray Herb.*, 200: 21 (1970): Cyathea agatheti Holtt.

albidosquamata *(Rosenst.) R. Tryon in Contrib. Gray Herb.*, 200: 21 (1970): Cyathea albidosquamata Rosenst.

albifrons *(Fourn.) R. Tryon in Contrib. Gray Herb.*, 200: 24 (1970): Cyathea albifrons Fourn.

albosetacea *(Bedd.) R. Tryon in Contrib. Gray Herb.*, 200: 21 (1970): Alsophila albosetacea Bedd.

alternans *(Hook.) R. Tryon in Contrib. Gray Herb.*, 200: 21 (1970): Hemitelia alternans Hook.

angiensis *(Gepp) R. Tryon in Contrib. Gray Herb.*, 200: 21 (1970): Alsophila angiensis Gepp

angustipinna *(Holtt.) R. Tryon in Contrib. Gray Herb.*, 200: 21 (1970): Cyathea angustipinna Holtt.

aramaganensis *(Kanehira) R. Tryon in Contrib. Gray Herb.*, 200: 24 (1970): Cyathea aramaganensis Kanehira

SPHAEROPTERIS :–

arthropoda *(Copel.) R. Tryon in Contrib. Gray Herb.*, 200: 21 (1970): Cyathea arthropoda Copel.

assimilis *(Hook.) R. Tryon in Contrib. Gray Herb.*, 200: 21 (1970): Cyathea assimilis Hook.

atrospinosa *(Holtt.) R. Tryon in Contrib. Gray Herb.*, 200: 21 (1970): Cyathea atrospinosa Holtt.

atrox *(C. Chr.) R. Tryon in Contrib. Gray Herb.*, 200: 21 (1970): Cyathea atrox C. Chr.

auriculifera *(Copel.) R. Tryon in Contrib. Gray Herb.*, 200: 21 (1970): Cyathea auriculifera Copel.

australis *(Presl) R. Tryon in Contrib. Gray Herb.*, 200: 24 (1970): Hemitelia australis Presl

binuangensis *(vAvR.) R. Tryon in Contrib. Gray Herb.*, 200: 21 (1970): Cyathea binuangensis vAvR.

brackenridgei *(Mett.) R. Tryon in Contrib. Gray Herb.*, 200: 24 (1970): Cyathea brackenridgei Mett.

brunoniana *(Hook.) R. Tryon in Contrib. Gray Herb.*, 200: 21 (1970): Alsophila brunoniana Hook.

capitata *(Copel.) R. Tryon in Contrib. Gray Herb.*, 200: 21 (1970): Cyathea capitata Copel.

carrii *(Holtt.) R. Tryon in Contrib. Gray Herb.*, 200: 21 (1970): Cyathea carrii Holtt.

celebica *(Bl.) R. Tryon in Contrib. Gray Herb.*, 200: 22 (1970): Cyathea celebica Bl.

concinna *(Baker) R. Tryon in Contrib. Gray Herb.*, 200: 22 (1970): Alsophila concinna Baker

cooperi *(F.v.Muell.) R. Tryon in Contrib. Gray Herb.*, 200: 24 (1970): Alsophila cooperi F.v.Muell.

crinita *(Hook.) R. Tryon in Contrib. Gray Herb.*, 200: 21 (1970): Alsophila crinita Hook.

curranii *(Copel.) R. Tryon in Contrib. Gray Herb.*, 200: 22 (1970): Cyathea curranii Copel.

discophora *(Holtt.) R. Tryon in Contrib. Gray Herb.*, 200: 22 (1970): Cyathea discophora Holtt.

elliptica *(Copel.) R. Tryon in Contrib. Gray Herb.*, 200: 22 (1970): Cyathea elliptica Copel.

elmeri *(Copel.) R. Tryon in Contrib. Gray Herb.*, 200: 22 (1970): Alsophila elmeri Copel.

excelsa *(Endl.) R. Tryon in Contrib. Gray Herb.*, 200: 24 (1970): Alsophila excelsa Endl.

feani *(E. Brown) R. Tryon in Contrib. Gray Herb.*, 200: 24 (1970): Cyathea feani E. Brown

fugax *(vAvR.) R. Tryon in Contrib. Gray Herb.*, 200: 22 (1970): Cyathea fugax vAvR.

fusca *(Baker) R. Tryon in Contrib. Gray Herb.*, 200: 22 (1970): Cyathea fusca Baker

SPHAEROPTERIS :–

glauca *(Bl.) R. Tryon in Contrib. Gray Herb.*, 200: 21 (1970): Chnoophora glauca Bl.

hainanensis *(Ching) R. Tryon in Contrib. Gray Herb.*, 200: 21 (1970): Cyathea hainanensis Ching

inaequalis *(Holtt.) R. Tryon in Contrib. Gray Herb.*, 200: 22 (1970): Cyathea inaequalis Holtt.

insularum *(Holtt.) R. Tryon in Contrib. Gray Herb.*, 200: 22 (1970): Cyathea insularum Holtt.

integra *(J. Sm.) R. Tryon in Contrib. Gray Herb.*, 200: 22 (1970): Cyathea integra J. Sm.

intermedia *(Mett.) R. Tryon in Contrib. Gray Herb.*, 200: 24 (1970): Alsophila intermedia Mett.

ledermannii *(Brause) R. Tryon in Contrib. Gray Herb.*, 200: 22 (1970): Hemitelia ledermannii Brause

lepifera *(Hook.) R. Tryon in Contrib. Gray Herb.*, 200: 21 (1970): Alsophila lepifera Hook.

leucolepis *(Mett.) R. Tryon in Contrib. Gray Herb.*, 200: 24 (1970): Cyathea leucolepis Mett.

leucotricha *(Christ) R. Tryon in Contrib. Gray Herb.*, 200: 22 (1970): Cyathea leucotricha Christ

lunulata *(Forst.) R. Tryon in Contrib. Gray Herb.*, 200: 22 (1970): Polypodium lunulatum Forst.

magna *(Copel.) R. Tryon in Contrib. Gray Herb.*, 200: 22 (1970): Cyathea magna Copel.

marginata *(Brause) R. Tryon in Contrib. Gray Herb.*, 200: 22 (1970): Alsophila marginata Brause

megalosora *(Copel.) R. Tryon in Contrib. Gray Herb.*, 200: 22 (1970): Cyathea megalosora Copel

mertensiana *(Kze.) R. Tryon in Contrib. Gray Herb.*, 200: 21 (1970). Alsophila mertensiana Kze.

microlepidota *(Copel.) R. Tryon in Contrib. Gray Herb.*, 200: 24 (1970): Cyathea microlepidota Copel.

moluccana *(Desv.) R. Tryon in Contrib. Gray Herb.*, 200: 22 (1970): Cyathea moluccana Desv.

moseleyi *(Baker) R. Tryon in Contrib. Gray Herb.*, 200: 22 (1970): Cyathea moseleyi Baker

nigricans *(Mett.) R. Tryon in Contrib. Gray Herb.*, 200: 24 (1970): Cyathea nigricans Mett.

novae-caledoniae *(Mett.) R. Tryon in Contrib. Gray Herb.*, 200: 24 (1970): Alsophila novae-caledoniae Mett.

obliqua *(Copel.) R. Tryon in Contrib. Gray Herb.*, 200: 22 (1970): Cyathea obliqua Copel.

obscura *(Bedd.) R. Tryon in Contrib. Gray Herb.*, 200: 22 (1970): Alsophila obscura Bedd.

SPHAEROPTERIS :–

papuana *(Ridley) R. Tryon in Contrib. Gray Herb.*, 200: 22 (1970): Alsophila papuana Ridley

parksiae *(Copel.) R. Tryon in Contrib. Gray Herb.*, 200: 24 (1970): Cyathea parksiae Copel.

parvifolia *(Holtt.) R. Tryon in Contrib. Gray Herb.*, 200: 22 (1970): Alsophila parvifolia Holtt.

parvipinna *(Holtt.) R. Tryon in Contrib. Gray Herb.*, 200: 22 (1970): Cyathea parvipinna Holtt.

persquamulifera *(vAvR.) R. Tryon in Contrib. Gray Herb.*, 200: 23 (1970): Cyathea contaminans var. persquamulifera vAvR.

philippinensis *(Baker) R. Tryon in Contrib. Gray Herb.*, 200: 23 (1970): Cyathea philippinensis Baker

pilulifera *(Copel.) R. Tryon in Contrib. Gray Herb.*, 200: 23 (1970): Cyathea pilulifera Copel.

polypoda *(Baker) R. Tryon in Contrib. Gray Herb.*, 200: 23 (1970): Cyathea polypoda Baker

procera *(Brause) R. Tryon in Contrib. Gray Herb.*, 200: 23 (1970): Cyathea procera Brause

propinqua *(Mett.) R. Tryon in Contrib. Gray Herb.*, 200: 24 (1970): Cyathea propinqua Mett.

pulcherrima *(Copel.) R. Tryon in Contrib. Gray Herb.*, 200: 23 (1970): Cyathea pulcherrima Copel.

robinsonii *(Copel.) R. Tryon in Contrib. Gray Herb.*, 200: 23 (1970): Cyathea robinsonii Copel.

robusta *(Watts) R. Tryon in Contrib. Gray Herb.*, 200: 24 (1970): Alsophila robusta Watts

rosenstockii *(Brause) R. Tryon in Contrib. Gray Herb.*, 200: 23 (1970): Cyathea rosenstockii Brause

runensis *(vAvR.) R. Tryon in Contrib. Gray Herb.*, 200: 23 (1970): Cyathea runensis vAvR.

samoensis *(Brack.) R. Tryon in Contrib. Gray Herb.*, 200: 24 (1970): Alsophila samoensis Brack.

sarasinorum *(Holtt.) R. Tryon in Contrib. Gray Herb.*, 200: 23 (1970): Cyathea sarasinorum Holtt.

senex *(vAvR.) R. Tryon in Contrib. Gray Herb.*, 200: 23 (1970): Cyathea senex vAvR.

setifera *(Holtt.) R. Tryon in Contrib. Gray Herb.*, 200: 23 (1970): Cyathea setifera Holtt.

sibuyanensis *(Copel.) R. Tryon in Contrib. Gray Herb.*, 200: 23 (1970): Cyathea sibuyanensis Copel.

squamulata *(Bl.) R. Tryon in Contrib. Gray Herb.*, 200: 23 (1970): Gymnosphaera squamulata Bl.

stipitipinnula *(Holtt.) R. Tryon in Contrib. Gray Herb.*, 200: 23 (1970): Cyathea stipitipinnula Holtt.

SPHAEROPTERIS :-

strigosa *(Christ) R. Tryon in Contrib. Gray Herb.,* 200: 23 (1970): Cyathea strigosa Christ

subsessilis *(Copel.) R. Tryon in Contrib. Gray Herb.,* 200: 25 (1970): Cyathea subsessilis Copel.

suluensis *(Baker) R. Tryon in Contrib. Gray Herb.,* 200: 23 (1970): Cyathea suluensis Baker

tenggerensis *(Rosenst.) R. Tryon in Contrib. Gray Herb.,* 200: 23 (1970): Alsophila tenggerensis Rosenst.

teysmannii *(Copel.) R. Tryon in Contrib. Gray Herb.,* 200: 23 (1970): Cyathea teysmannii Copel.

tomentosa *(Bl.) R. Tryon in Contrib. Gray Herb.,* 200: 23 (1970): Chnoophora tomentosa Bl.

tomentosissima *(Copel.) R. Tryon in Contrib. Gray Herb.,* 200: 23 (1970): Cyathea tomentosissima Copel.

trichodesma *(Bedd.) R. Tryon in Contrib. Gray Herb.,* 200: 23 (1970): Alsophila trichodesma Bedd.

trichophora *(Copel.) R. Tryon in Contrib. Gray Herb.,* 200: 23 (1970): Cyathea trichophora Copel.

tripinnata *(Copel.) R. Tryon in Contrib. Gray Herb.,* 200: 23 (1970): Cyathea tripinnata Copel.

tripinnatifida *(Roxb.) R. Tryon in Contrib. Gray Herb.,* 200: 23 (1970): Cyathea tripinnatifida Roxb.

truncata *(Brack.) R. Tryon in Contrib. Gray Herb.,* 200: 25 (1970): Alsophila truncata Brack.

vaupelii *(Copel.) R. Tryon in Contrib. Gray Herb.,* 200: 25 (1970): Cyathea vaupelii Copel.

verrucosa *(Holtt.) R. Tryon in Contrib. Gray Herb.,* 200: 24 (1970): Cyathea verrucosa Holtt.

vittata *(Copel.) R. Tryon in Contrib. Gray Herb.,* 200: 25 (1970): Cyathea vittata Copel.

wallacei *(Kuhn) R. Tryon in Contrib. Gray Herb.,* 200: 24 (1970): Alsophila wallacei Kuhn

werneri *(Rosenst.) R. Tryon in Contrib. Gray Herb.,* 200: 24 (1970): Cyathea werneri Rosenst.

womersleyi *(Holtt.) R. Tryon in Contrib. Gray Herb.,* 200: 24 (1970): Cyathea womersleyi Holtt.

zamboangana *(Copel.) R. Tryon in Contrib. Gray Herb.,* 200: 24 (1970): Cyathea zamboangana Copel.

SPHAEROSTEPHANOS

cucullatus *(Blume) S.M. Almeida & M.R. Almeida in J. Bombay Nat. Hist. Soc.,* 82(3): 699 (1985 publ. 1986): Aspidium cucullatum Blume

T

TECTARIA

curtisii *R.E. Holttum in Gard. Bull. Singapore,* 38(2): 145 (1985)—Malaya.

hymenophylla *(Bedd.) R.E. Holttum in Indian Fern J.,* 1(1–2): 35 (1984 publ. 1985): Acrophorus hymenophyllus Bedd.

manilensis *(Presl) R.E. Holttum in Indian Fern J.,* 1(1–2): 36 (1984 publ. 1985): Lastrea manilensis Presl

polymorpha

var. macrocarpa *(Bedd.) S. Chandra & S. Kaur in Indian Fern J.,* 1(1–2): 86 (1984 publ. 1985): Aspidium polymorphum var. macrocarpum Bedd.

translucens *R.E. Holttum in Gard. Bull. Singapore,* 38(2): 145 (1985)—Malaya.

THELYPTERIS

beddomei

f. purpurascenstipes *N. Yoshikawa in J. Phytogeogr. Taxon.,* 33(2): 80 (1985)—Philippine Is.

darwinii *L.D. Gómez P. in Phytologia,* 60(5): 369 (1986)—Mexico.

peripaeoides *L.D. Gómez P. in Phytologia,* 60(5): 370 (1986)—Costa Rica.

TMESIPTERIS

norfolkensis *P.S. Green in J. Arnold Arbor.,* 67(1): 109 (1986)—Norfolk Is.

oblongifolia *A.F. Braithwaite in Fern Gaz. (U.K.),* 13(2): 93 (1986)—Vanuatu, Mquesas Is., Philippine Is.

vanuatensis *A.F. Braithwaite in Fern Gaz. (U.K.),* 13(2): 94 (1986)—Vanuatu.

TRICHOMANES

japonicum *C.P. Thunberg ex A. Murray in Linnaeus, Syst. Veg., ed. 14:* 941 (1784)—Japan. †

strigosum *C.P. Thunberg ex A. Murray in Linnaeus, Syst. Veg., ed. 14:* 941 (1784)—Japan. †

TRIPLOPHYLLUM R.E. Holttum in Kew Bull., 41(2): 239 (1986).

acutilobum *R.E. Holttum in Kew Bull.,* 41(2): 258 (1986)—Brazil, Guyana.

angustifolium *R.E. Holttum in Kew Bull.,* 41(2): 259 (1986)—French Guiana, Guyana, Surinam.

batesii *R.E. Holttum in Kew Bull.,* 41(2): 245 (1986)—Cameroun.

buchholzii *(Kuhn) R.E. Holttum in Kew Bull.,* 41(2): 251 (1986): Aspidium buchholzii Kuhn

crassifolium *R.E. Holttum in Kew Bull.,* 41(2): 257 (1986)—Surinam.

var. caudatum *R.E. Holttum in Kew Bull.,* 41(2): 258 (1986)—French Guiana.

dicksonioides *(Fée) R.E. Holttum in Kew Bull.,* 41(2): 257 (1986): Aspidium dicksonioides Fée

TRIPLOPHYLLUM :–

dimidiatum *(Mett.) R.E. Holttum in Kew Bull.*, 41(2): 251 (1986): Aspidium dimidiatum Mett.

fraternum *(Mett.) R.E. Holttum in Kew Bull.*, 41(2): 253 (1986): Aspidium fraternum Mett.

var. elongatum *(Hook.) R.E. Holttum in Kew Bull.*, 41(2): 254 (1986): Nephrodium subquinquefidum var. elongatum Hook.

funestum *(Kunze) R.E. Holttum in Kew Bull.*, 41(2): 255 (1986): Aspidium funestum Kunze

var. hirsutum *R.E. Holttum in Kew Bull.*, 41(2): 256 (1986)—Guyana.

var. perpilosum *R.E. Holttum in Kew Bull.*, 41(2): 256 (1986)—Colombia.

gabonense *R.E. Holttum in Kew Bull.*, 41(2): 245 (1986)—Gabon, Angola, Cameroun.

jenseniae *(C. Chr.) R.E. Holttum in Kew Bull.*, 41(2): 253 (1986): Dryopteris jenseniae C. Chr.

pentagonum *(Bonap.) R.E. Holttum in Kew Bull.*, 41(2): 243 (1986): Dryopteris pentagona Bonap.

pilosissimum *(Moore) R.E. Holttum in Kew Bull.*, 41(2): 246 (1986): Lastrea pilosissima Moore

principis *R.E. Holttum in Kew Bull.*, 41(2): 246 (1986)—Principé Is.

protensum *(Sw.) R.E. Holttum in Kew Bull.*, 41(2): 247 (1986): Aspidium protensum Sw.

securidiforme *(Hook.) R.E. Holttum in Kew Bull.*, 41(2): 242 (1986): Nephrodium subquinquefidum var. securidiforme Hook.

securidiforme

var. nanum *(Bonap.) R.E. Holttum in Kew Bull.*, 41(2): 243 (1986): Dryopteris securidiformis var. nana Bonap.

TRIPLOPHYLLUM :–

speciosum *(Mett.) R.E. Holttum in Kew Bull.*, 41(2): 247 (1986): Aspidium speciosum Mett.

troupinii *(Pic. Ser.) R.E. Holttum in Kew Bull.*, 41(2): 243 (1986): Ctenitis troupinii Pic. Ser.

varians *(Moore) R.E. Holttum in Kew Bull.*, 41(2): 249 (1986): Dictyopteris varians Moore

vogelii *(Hook.) R.E. Holttum in Kew Bull.*, 41(2): 249 (1986): Aspidium vogelii Hook.

V

VANDENBOSCHIA

anceps *(Clarke) S. Chandra & S. Kaur in Indian Fern J.*, 1(1–2): 87 (1984 publ. 1985), error (Wall.): Trichomanes radicans var. anceps Clarke

X

XIPHOPTERIS

exilis *B.S. Parris in Kew Bull.*, 41(3): 503 (1986)—Malaya.

musgraviana *(Baker) B.S. Parris in Kew Bull.*, 41(1): 69 (1986): Polypodium musgravianum Baker

setulifera *(Alderw.) B.S. Parris in Kew Bull.*, 41(1): 69 (1986): Polypodium setuliferum Alderw.